Additive Manufacturing Technologies

Ian Gibson • David Rosen • Brent Stucker
Mahyar Khorasani

Additive Manufacturing Technologies

Third Edition

 Springer

Ian Gibson
Department of Design, Production
and Management
University of Twente
Enschede, The Netherlands

David Rosen
George W. Woodruff School
of Mechanical Engineering
Georgia Institute of Technology
Atlanta, GA, USA

Brent Stucker
ANSYS
Park City, UT, USA

Mahyar Khorasani
Deakin University
Armstrong Creek, VIC, Australia

ISBN 978-3-030-56129-1   ISBN 978-3-030-56127-7   (eBook)
https://doi.org/10.1007/978-3-030-56127-7

This Springer imprint is published by the registered company Springer Nature Switzerland AG
The registered company address is: Gewerbestrasse 11, 6330 Cham, Switzerland

Preface

Thank you for taking the time to read this textbook on Additive Manufacturing (AM). We hope you benefit from the time and effort it has taken putting it together and that you think it was a worthwhile undertaking. It all started as a discussion at a conference in Portugal when we realized we were putting together books with similar aims and objectives. Since we are friends as well as colleagues, it seemed sensible we join forces rather than compete; sharing the load and playing to each other's strengths undoubtedly means a better all-round effort and result.

We wrote this textbook because we have all been working in the field of AM for many years. Although none of us like to be called "old," we do seem to have around 85 years of experience, collectively, and have each established reputations as educators and researchers in this field. We have each seen the technologies described in this textbook take shape and develop into serious commercial tools, with hundreds of thousands of regular users and many millions of parts being made by AM machines each year. AM is now being incorporated into curricula in many schools, polytechnics, and universities. More and more students are becoming aware of these technologies and yet, as we saw it, there was no single text adequate for such curricula. We believe that the first edition provided such a text, and based upon updated information, the 2nd and 3rd editions were developed.

Additive Manufacturing includes a range of technologies that are capable of translating virtual solid model data into physical models in an automated process. The data are broken down into a series of 2D cross-sections of finite thickness. These cross-sections are fed into AM machines so that they can be created, combined, and added together in a layer-by-layer sequence to form the physical part. The geometry of the digital part is therefore reproduced physically in the AM machine without having to adjust for manufacturing processes, like attention to tooling, undercuts, draft angles, or other features. We can therefore say that the AM machine is a What You See Is What You Build (WYSIWYB) process that is particularly valuable the more complex the geometry. This basic principle drives nearly all AM machines, with variations in each technology in terms of the techniques used for creating layers and bonding them together. Further variations include speed, layer thickness, range of materials, accuracy, and, of course, cost. With so many variables, it is clear to see why this textbook must be so long and detailed. Having said that, we still feel there is much more we could have written.

The first three chapters of this textbook provide a basic overview of AM processes. Without fully describing each technology, we provide an appreciation for why AM is so important to many branches of industry. We outline the rapid development of this technology from humble beginnings, which showed promise but still required much development, to one that is now maturing and showing a real benefit to product development organizations. By reading these chapters, we hope you can learn the basics of how AM works.

The next eight chapters (Chaps. 4, 5, 6, 7, 8, 9, 10 and 11) take each group of technologies in turn and describe them in detail. The fundamentals of each technology are dealt with in terms of the basic process, whether it involves photopolymer curing, inkjet printing, sintering, melting, etc., so that the reader can appreciate what is needed in order to understand, develop, optimize, and use each technology. Most technologies discussed in this textbook have been commercialized by at least one company, and many of these commercial variants are described along with discussion on how to get the best out of them.

The next two chapters are about hybrid AM (the mix of AM and other manufacturing processes in a single machine) and low-cost AM techniques. The low-cost AM chapter focuses on inexpensive machines, which overlaps some of the material in earlier chapters. However, we felt that the increasing interest in these low-cost machines justified the special treatment.

The final nine chapters deal with how to apply AM technology in different settings. We look at the materials available for AM and the post-processing you need to perform after removing a part from the machine, before it can be used. We provide selection methods for sorting through the many machine options. Since all AM machines depend on input from 3D CAD software, we go on to discuss how software is used. We follow this with a discussion of novel applications which utilize AM machine output for end-product use, called Direct Digital Manufacturing. Many of these products were impossible, infeasible, or uneconomic in the past, including mass customization, where a product can be produced according to the tastes of an individual consumer but at a cost-effective price. Next we look at how AM affects the design process, considering how we might improve designs to take advantage of unique benefits of AM. This moves us on nicely to the subjects of applications of AM, including tooling and products in the medical, aerospace, and automotive sectors. We complete the textbook with a chapter on the business and societal implications of AM, investigating how AM enables creative businesses and entrepreneurs to invent new products.

This textbook is primarily aimed at students and educators studying AM, either as a self-contained course or as a module within a larger course on manufacturing technology. There is sufficient depth for an undergraduate or graduate-level course, with many references to point the student further along the path. Each chapter also has a number of exercise questions designed to test the reader's knowledge and to expand their thinking. A companion instructor's guide is being developed as part of the 3rd edition. Researchers into AM may also find this text useful in helping them understand the state of the art and the opportunities for further research. AM industry practitioners will also find this textbook useful as they guide their companies in adopting AM.

We made a wide range of changes in moving from the first edition, completed in 2009, to the 2nd edition published in 2015. We added a number of new sections and chapters, including expanding the chapter on medical applications to include discussion on automotive and aerospace applications. A new chapter on rapid tooling was also added in the 2nd edition.

Between the 1st and 2nd editions, the authors helped establish the ASTM F42 and ISO TC261 committees on Additive Manufacturing. These committees developed standard terminology for AM, and the 2nd edition used this updated terminology. In the 2nd edition, we also edited the text to reduce references to company-specific technologies and instead focused more on technological principles and general understanding. We split the original chapter on printing processes into two chapters on material jetting and on binder jetting to reflect the standard terminology and the evolution of these processes in different directions. We inserted a range of recent technological innovations, including discussion on the Additive Manufacturing File format. As a result of these many additions and changes, we believe the 2nd edition was significantly more comprehensive than the first. The response to this 2nd edition was positive, as sales increased by almost 40% 2 years after publishing, and by 2019 this book was firmly established as the leading AM textbook worldwide.

To keep up with the extensive growth of AM since 2015, we decided to publish a 3rd edition. This took nearly 2 years of effort with a new co-author added to the list. Older examples of vat photopolymerization, material jetting, and binder jetting machines were replaced by newer technologies and applications. New variants of existing AM techniques, such as high-speed sintering, multi-material jetting, cold spray, friction stir AM, and more, were added to relevant chapters. We continue to update our materials to conform to international standards, including ASTM 52900 terminology. To help our readers understand how new terminology relates to older company-specific and technology-specific names, we have added information to Chaps. 5 and 10 cross-referencing nomenclature for AM processes to help address any confusion.

Approximately 200 figures have been redrawn to improve understanding of the contents. We have significantly expanded our discussion of post-processing methods to improve surface quality, dimensional accuracy, and mechanical properties. We provide a better classification for different AM-related software. New topics like the influence of AM within Industry 4.0, design for 4D printing, and a discussion of design of experiments have also been added.

Hybrid AM techniques and materials for AM have expanded dramatically since the 2nd edition, so we significantly expanded our treatment of these topics, resulting in two new chapters. Hybrid AM is a combination of additive and conventional manufacturing into a single process, resulting in significantly improved capabilities. In the new chapter on materials, we classify the feedstock based on the physical state of the material when added to the machine, including liquid, powder, and solid states. This new chapter should help students, researchers, and professionals select a suitable machine/material combination for their research and businesses.

We are currently working on online supplementary information for this textbook. In particular, we hope to publish a solutions manual soon. Please check online for its availability.

Although we have worked hard to make this textbook as comprehensive as possible, we recognize that a textbook about such rapidly changing technology will not remain up-to-date for very long. If you have comments, questions, or suggestions for improvement, they are welcome. We anticipate updating this textbook in the future, and we look forward to hearing how you have used these materials and how we might improve this textbook.

As mentioned earlier, each author is an established expert in AM with many years of research experience. In addition, in many ways, this textbook is only possible due to the many students and colleagues with whom we have collaborated over the years. To introduce you to the authors and some of the others who have made this textbook possible, we will end this preface with brief author acknowledgments.

Enschede, The Netherlands   Ian Gibson
Atlanta, GA, USA   David Rosen
Park City, UT, USA   Brent Stucker
Armstrong Creek, VIC, Australia   Mahyar Khorasani

Acknowledgment

Dr. Mahyar Khorasani, the newest member of our team, acknowledges Deakin University for providing time to edit this textbook. Without this cooperation as well as the support of his mentor, Professor Bernard Rolfe, the time-consuming editing process would not have been possible. Dr. Khorasani thanks his family, Giti, Massoud, and lovely Elmira, for their patience, encouragement, and love during the hours of writing and editing the contents and chapters. Dr. Khorasani would like to thank Amir Hossein Ghasemi for his great support and help in this edition.

Prof. Brent Stucker thanks Utah State University, VTT Technical Research Center, and the University of Louisville for providing the academic freedom, sabbaticals, and environment needed to complete the 1st and 2nd editions and Ansys for providing support to complete the 3rd edition. This textbook would not have been possible without the many graduate students and postdoctoral researchers who have worked with Prof. Stucker over the years. In particular, he would like to thank Prof. G.D. Janaki Ram of the Indian Institute of Technology Hyderabad, whose co-authoring of the "Layer-Based Additive Manufacturing Technologies" chapter in the *CRC Materials Processing Handbook* helped lead to the organization of this textbook. Additionally, many students' work directly led to content in this textbook and solution manual, including: Muni Malhotra, Xiuzhi Qu, Carson Esplin, Adam Smith, Joshua George, Christopher Robinson, Yanzhe Yang, Matthew Swank, John Obielodan, Kai Zeng, Haijun Gong, Xiaodong Xing, Hengfeng Gu, Md Anam, Nachiket Patil, and Deepankar Pal. Special thanks are due to Prof. Stucker's wife, Gail, and their children, Tristie, Andrew, Megan, and Emma, who patiently supported many days and evenings on this textbook. Prof. Stucker has sought to merge his Christian faith with his work, and he hopes that this textbook inspires its readers to create amazing things that bring goodness to people around the world and, as a result, bring joy to God.

Prof. David W. Rosen acknowledges support from Georgia Tech, the Singapore University of Technology and Design, and the many graduate students and postdocs who contributed technically to the content in this book. In particular, he thanks Drs. Fei Ding, Mahdi Emami, Amit Jariwala, Scott Johnston, Samyeon Kim, Ameya Limaye, J. Mark Meacham, Sang-In Park, Benay Sager, L. Angela Tse, Hongqing

Wang, Chris Williams, Yong Yang, Xiong Yi, Xiayun Zhao, and Wenchao Zhou as well as Lauren Margolin. He also appreciates the many students who took his graduate-level course at Georgia Tech for their feedback on the textbook and their project ideas. Special thanks goes out to his wife Joan and children Erik and Krista for their patience while he worked on this book.

Prof. Ian Gibson would like to acknowledge the support of the Fraunhofer Project Center at the University of Twente and Deakin University in providing sufficient time for him to work on this book. He sincerely wishes to thank his lovely wife, Lina, for her patience, love, and understanding during the long hours preparing the material and writing the chapters. He also dedicates this textbook to his parents, June Gibson and Robert Ervin Gibson, as well as his children, Louise, Claire, and James. He hopes they would be proud of this wonderful achievement.

Contents

Chapter 1
Introduction and Basic Principles

Abstract The technology described in this book was originally referred to as Rapid Prototyping. The term Rapid Prototyping (or RP) is used to describe a process for rapidly creating a system or part representation before final release or commercialization. The emphasis is on creating something quickly for use as a prototype or basis model from which further models and eventually the final product will be derived. In product development, the term Rapid Prototyping describes technologies which create physical prototypes. This text is about technologies which can directly produce objects from digital data. These technologies were first developed for prototyping but are now used for many more purposes.

1.1 What Is Additive Manufacturing?

Additive Manufacturing is the formalized term for what used to be called Rapid Prototyping and what is popularly called 3D Printing. The term Rapid Prototyping (or RP) is used in a variety of industries to describe a process for rapidly creating a system or part representation before final release or commercialization. In other words, the emphasis is on creating something quickly, and that the output is a prototype or basis model from which further models and eventually the final product will be derived. Management consultants and software engineers both also use the term Rapid Prototyping to describe a process of developing business and software solutions in a piecewise fashion that allows clients and other stakeholders to test ideas and provide feedback during the development process. In a product development context, the term Rapid Prototyping was used widely to describe technologies which created physical prototypes directly from digital model data. This text is about these latter technologies, first developed for prototyping but now used for many more purposes.

Users of RP technology have come to realize that this term is inadequate and in particular does not effectively describe more recent applications of the technology.

© Springer Nature Switzerland AG 2021
I. Gibson et al., *Additive Manufacturing Technologies*,
https://doi.org/10.1007/978-3-030-56127-7_1

Improvements in the quality of the output from these machines have meant that there is often a much closer link to the final product. Many parts are in fact now directly manufactured in these machines, so it is not possible for us to label them as "prototypes." The term Rapid Prototyping also overlooks the basic principle of these technologies in that they all fabricate parts using an additive approach. A Technical Committee within ASTM International agreed that new terminology should be adopted. As a result, ASTM consensus standards now use the term Additive Manufacturing [1] as do most standards bodies worldwide.

Referred to in short as AM, the basic principle of this technology is that a model, initially generated using a three-dimensional Computer Aided Design (3D CAD) system, can be fabricated directly without the need for process planning. Although this is not in reality as simple as it first sounds, AM technology significantly simplifies the process of producing complex 3D objects directly from CAD data. Other manufacturing processes require a careful and detailed analysis of the part geometry to determine things like the order in which different features can be fabricated, what tools and processes must be used, and what additional fixtures may be required to complete the part. In contrast, AM needs only some basic dimensional details and a small amount of understanding as to how the AM machine works and the materials that are used to build the part.

The key to how AM works is that parts are made by adding material in layers; each layer is a thin cross-section of the part derived from the original CAD data. Obviously in the physical world, each layer must have a finite thickness to it, and so the resulting part will be an approximation of the original data, as illustrated by Fig. 1.1. The thinner each layer is, the closer the final part will be to the original. All commercialized AM machines to date use a layer-based approach, and the major ways that they differ are in the materials that can be used, how the layers are created,

Fig. 1.1 CAD image of a teacup with further images showing the effects of building using different layer thicknesses

and how the layers are bonded to each other. Such differences will determine factors like the accuracy of the final part plus its material properties and mechanical properties. They will also determine factors like how quickly the part can be made, how much post-processing is required, the size of the AM machine used, and the overall cost of the machine and process.

This chapter will introduce the basic concepts of Additive Manufacturing and describe a generic AM process from design to application. It will go on to discuss the implications of AM on design and manufacturing and attempt to help in understanding how it has changed the entire product development process. Since AM is an increasingly important tool for product development, the chapter ends with a discussion of some related tools in the product development process.

1.2   What Are AM Parts Used For?

Throughout this book you will find a wide variety of applications for AM. You will also realize that the number of applications is increasing as the processes develop and improve. Initially, AM was used specifically to create visualization models for products as they were being developed. It is widely known that models can be much more helpful than drawings or renderings in fully understanding the intent of the designer when presenting the conceptual design. While drawings are quicker and easier to create, models are nearly always required in the end to fully validate the design.

Following this initial purpose of simple model making, AM technology has developed over time as materials, accuracy, and the overall quality of the output improved. Models were quickly employed to supply information about what is known as the "3 Fs" of Form, Fit, and Function. The initial models were used to help fully appreciate the shape and general purpose of a design (Form). Improved accuracy in the process meant that components were capable of being built to the tolerances required for assembly purposes (Fit). Improved material properties meant that parts could be properly handled so that they could be assessed according to how they would eventually work (Function).

To say that AM technology is only useful for making models, though, would be inaccurate and undervaluing the technology. AM, when used in conjunction with other technologies to form process chains, can be used to significantly shorten product development times and costs. More recently, some of these technologies have been developed to the extent that the output is suitable for end use. This explains why the terminology has essentially evolved from Rapid Prototyping to Additive Manufacturing.

1.3   The Generic AM Process

AM involves a number of steps that move from the virtual CAD description to the physical resultant part. Different products will involve AM in different ways and to different degrees. Small, relatively simple products may only make use of AM for visualization models, while larger, more complex products with greater engineering content may involve AM during numerous stages and iterations throughout the development process. Furthermore, early stages of the product development process may only require rough parts, with AM being used because of the speed at which they can be fabricated. At later stages of the process, parts may require careful cleaning and post-processing (including sanding, surface preparation, and painting) before they are used, with AM being useful here because of the complexity of form that can be created without having to consider tooling. Later on, we will investigate thoroughly the different stages of the AM process, but to summarize, most AM processes involve, to some degree at least, the following eight steps (as illustrated in Fig. 1.2).

1.3.1 Step 1: CAD

All AM parts must start from a software model that fully describes the external geometry. This can involve the use of almost any professional CAD solid modeling software, but the output must be a 3D solid or surface representation. Reverse engineering equipment (e.g., laser and optical scanning) can also be used to create this representation.

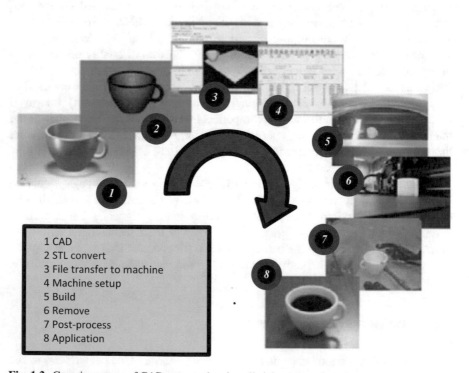

1 CAD
2 STL convert
3 File transfer to machine
4 Machine setup
5 Build
6 Remove
7 Post-process
8 Application

Fig. 1.2 Generic process of CAD to part, showing all eight stages

1.3.2 Step 2: Conversion to STL

Nearly every AM machine accepts the STL file format, which has become a de facto standard, and nowadays nearly every CAD system can output such a file format. This file describes the external closed surfaces of the original CAD model and forms the basis for calculation of the slices.

1.3.3 Step 3: Transfer to AM Machine and STL File Manipulation

The STL file describing the part must be transferred to the AM machine. Here, there may be some general manipulation of the file so that it is the correct size, position, and orientation for building.

1.3.4 Step 4: Machine Setup

The AM machine must be properly set up prior to the build process. Such settings would relate to the build parameters like the material constraints, energy source, layer thickness, timings, etc.

1.3.5 Step 5: Build

Building the part is mainly an automated process, and the machine can largely carry on without supervision. Only superficial monitoring of the machine needs to take place at this time to ensure no errors have taken place like running out of material, power or software glitches, etc.

1.3.6 Step 6: Removal

Once the AM machine has completed the build, the parts must be removed. This may require interaction with the machine, which may have safety interlocks to ensure, for example, that the operating temperatures are sufficiently low or that there are no actively moving parts.

1.3.7   Step 7: Post-Processing

Once removed from the machine, parts may require an amount of additional cleaning up before they are ready for use. Parts may be weak at this stage, or they may have supporting features that must be removed. They may require priming and painting to give an acceptable surface texture and finish. This step may also involve heat treatment. Post-processing may be costly, laborious, and lengthy if the finishing requirements are very demanding.

1.3.8   Step 8: Application

Parts are now ready for use. This may require them to be assembled together with other mechanical or electronic components to form a final model or product.

While the numerous stages in the AM process have now been discussed, it is important to realize that other steps may be needed for certain machines and processes. For example, certain machine setup software and AM machines may be compatible with native CAD information, and thus the STL file may be skipped. In addition, many AM machines require careful maintenance. Many AM machines use fragile laser or printer technology that must be carefully monitored and that should preferably not be used in a dirty or noisy environment. While machines are generally designed to operate unattended, it is important to include regular checks in the maintenance schedule, and that different technologies and/or applications require different levels of maintenance. There is an increasing use of materials and process standards for AM, and the ASTM F42 Technical Committee on Additive Manufacturing Technologies is working to add more [1]. Maintenance standards may be more strict for certain applications than for others. In addition, many machine vendors recommend and provide test patterns and maintenance guidance that can be used periodically to confirm that machines are operating within acceptable limits.

In addition to the machinery, materials may also require careful handling. The raw materials used in some AM processes have limited shelf life and may also be required to be kept in conditions that prevent them from unwanted chemical reactions. Exposure to moisture, excess light, and other contaminants should also be avoided. Most processes use materials that can be reused for more than one build. However, it may be that reuse could degrade the properties if performed many times over, and therefore a procedure for maintaining consistent material quality through recycling should also be observed.

1.4   Why Use the Term Additive Manufacturing?

By now, you should realize that the technology we are referring to is primarily the use of additive processes, combining materials layer-by-layer. The term Additive Manufacturing, or AM, seems to describe this quite well, but there are many other terms which are in use. This section discusses other terms that have been used to describe this technology as a way of explaining the overall purpose and benefits of the technology for product development.

1.4.1   Automated Fabrication (Autofab)

This term was popularized by Marshall Burns in his book of the same name, which was one of the first texts to cover AM technology in the early 1990s [2]. The emphasis here is on the use of automation to manufacture products, thus implying the simplification or removal of manual tasks from the process. Computers and microcontrollers are used to control the actuators and to monitor the system variables. This term can also be used to describe other forms of Computer Numerical Controlled (CNC) machining centers since there is no direct reference as to how parts are built or the number of stages it would take to build them, although Burns does primarily focus on the technologies also covered by this book. Some key technologies are however omitted since they arose after the book was written.

1.4.2   Freeform Fabrication or Solid Freeform Fabrication

The emphasis here is in the capability of the processes to fabricate complex geometric shapes. Sometimes the advantage of these technologies is described in terms of providing "complexity for free," implying that it doesn't particularly matter what the shape of the input object actually is. A simple cube or cylinder would take almost as much time and effort to fabricate within the machine as a complex anatomical structure with the same enclosing volume. The reference to "freeform" relates to the independence of form from the manufacturing process. This is very different from most conventional manufacturing processes that become much more involved as the geometric complexity increases.

1.4.3   Additive Manufacturing or Layer-Based Manufacturing

These descriptions relate to the way the processes fabricate parts by adding material in layers. This is in contrast to machining technology that removes or subtracts material from a block of raw material. It should be noted that some of the processes are not purely additive, in that they may add material at one point but also use subtractive processes at some stage as well. Currently, every commercial process works in a layer-wise fashion. However, there is nothing to suggest that this is an essential approach to use and that future systems may add material in other ways and yet still come under a broad classification that is appropriate to this text. A slight variation on this, Additive Fabrication, is a term that was popularized by Terry Wohlers, a well-known industry consultant in this field and who compiles a widely regarded annual industry report on the state of this industry [3]. However, the term "manufacturing" tends to be the more general term, and, in fact, "Additive Manufacturing" is now the international standard term used to refer to this class of manufacturing processes [4]. Originally approved in 2009 by the ASTM F42 Committee, Additive Manufacturing was adopted as the official name for the technology, along with the names of the seven identified classes of AM processes.

1.4.4   Rapid Prototyping

Rapid Prototyping was termed because of the process this technology was designed to enhance or replace. Manufacturers and product developers used to find prototyping a complex, tedious, and expensive process that often impeded the developmental and creative phases during the introduction of a new product. RP was found to significantly speed up this process, and thus the term was adopted. However, users and developers of this technology now realize that AM technology can be used for much more than just prototyping. Significant improvements in accuracy and material properties have seen this technology catapulted into testing, tooling, manufacturing, and other realms that are outside the "prototyping" definition.

1.4.5   Stereolithography or 3D Printing

These two terms were initially used to describe specific machines. Stereolithography (SL) was termed by the US company 3D Systems [5, 6], and 3D Printing (3DP) was widely used by researchers at MIT [7] who invented an inkjet printing-based technology. Both terms allude to the use of 2D processes (lithography and printing) and extending them into the third dimension. Since most people are very familiar with printing technology, the idea of printing a physical three-dimensional object should make sense. Many consider that eventually the term 3D Printing will become the most commonly used wording to describe AM technologies. Recent media interest in the technology has proven this to be true, and the general public is much more likely to know the term 3D Printing than any other term mentioned in this book.

   We use Additive Manufacturing or its abbreviation AM throughout this book as the generic term for the suite of technologies covered by this book as this is the terminology used by experts and standards bodies. It should be noted that, in the literature, most of the terms introduced above are interchangeable; but different terminology may emphasize the approach used in a particular instance. Thus, both in this book and while searching for or reading other literature, the reader must consider the context to best understand what each of these terms means.

1.5   The Benefits of AM

Many people have described AM as revolutionizing product development and manufacturing. Some have even gone on to say that manufacturing, as we know it today, may not exist if we follow AM to its ultimate conclusion and that we are experiencing a new industrial revolution. AM is now frequently referred to as one of a series of disruptive technologies that are changing the way we design products and set up new businesses. We might, therefore, like to ask "why is this the case?" What is it about AM that enthuses and inspires some to make these kinds of statements?

First, let's consider the "rapid" character of this technology. The speed advantage is not just in terms of the time it takes to build parts. The speeding up of the whole product development process relies much on the fact that we are using computers throughout. Since 3D CAD is being used as the starting point and the transfer to AM is relatively seamless, there is much less concern over data conversion or interpretation of the design intent. Just as 3D CAD is becoming What You See Is What You Get (WYSIWYG), so it is the same with AM, and we might just as easily say that What you See Is What You Build (WYSIWYB).

The seamlessness can also be seen in terms of the reduction in process steps. Regardless of the complexity of parts to be built, building within an AM machine is generally performed in a single step. Most other manufacturing processes would require multiple and iterative stages to be carried out. As you include more features in a design, the number of these stages may increase dramatically. Even a relatively simple change in the design may result in a significant increase in the time required to build using conventional methods. The amount of time to fabricate models using AM, however, is relatively insensitive to simple design changes that may be implemented during this formative stage of product development.

Similarly, the number of processes and resources required can be significantly reduced when using AM. If a skilled craftsman was requested to build a prototype according to a set of CAD drawings, he may find that he must manufacture the part in a number of stages. This may be because he must employ a variety of construction methods, ranging from hand carving, through molding and forming techniques, to CNC machining. Hand carving and similar operations are tedious, difficult, and prone to error. Molding technology can be messy and obviously requires the building of one or more molds. CNC machining requires careful planning and a sequential approach that may also require construction of fixtures before the part itself can be made. All this of course presupposes that these technologies are within the repertoire of the craftsman and readily available.

AM can be used to remove or at least simplify many of these multistage processes. With the addition of some supporting technologies like silicone–rubber molding, drills, polishers, grinders, etc., it can be possible to manufacture a vast range of different parts with different characteristics. Workshops which adopt AM technology can be much cleaner, more streamlined, and more versatile than before.

1.6   Distinction Between AM and Conventional Manufacturing Processes

As mentioned in the discussion on Automated Fabrication, AM shares some of its DNA with conventional manufacturing technologies like Computer Numerical Controlled (CNC) machining. CNC technologies in general are computer-based manufacturing technologies. Conventional technologies are often divided into sub-tractive, casting, and forming technologies. All can use CNC controllers to increase the accuracy and better control the process. In subtractive manufacturing, such as CNC machining, materials are removed by contact with cutting tools, and therefore chips are produced. In forming the forming tool changes the shape of sheet or blocks of material by exerting high forces, and chips are not generated. In casting, molten material is directed into a mold. CNC machining is capable of making complex parts directly from CAD data but in a subtractive rather than additive way. CNC machining requires a block of material that must be at least as big as the part that is to be made. In AM powders, filaments, liquids, or other feedstocks are added together to create a part that is larger than the feedstock in one or more dimensions.

This section discusses a range of topics where comparisons between CNC machining and AM are made. The purpose is not to influence choice of one technology over another but rather to establish how they may be implemented for different stages in the product development process or for different types of product.

1.6.1   Material

AM technology was originally developed around polymeric materials, waxes, and paper laminates. Subsequently, there has been introduction of composites, metals, and ceramics. CNC machining can be used for soft materials, like medium-density fiber-board (MDF), machineable foams, machineable waxes, and even some polymers. However, use of CNC to shape softer materials is typically focused on preparing these parts for use in a multistage process like casting. When using CNC machining to make final products, it works particularly well for hard, relatively brittle materials like steels and other metal alloys to produce high accuracy parts with well-defined properties. Some AM parts, in contrast, may have voids or anisotropy that are a function of part orientation, process parameters, or how the design was input to the machine, whereas CNC parts will normally be more homogeneous and predictable in quality.

1.6.2   Speed

High-speed CNC machining can generally remove material much faster than AM machines can add a similar volume of material. However, this is only part of the picture, as AM technology can be used to produce a part in a single stage. CNC machines require considerable setup and process planning, particularly as parts become more complex in their geometry. Speed must therefore be considered in terms of the whole process rather than just the physical interaction with the part material. CNC is likely to be a multistage manufacturing process, requiring repositioning or relocation of parts within one machine or use of more than one machine. To make a part in an AM machine, it may only take a few hours; and in fact multiple parts are often batched together inside a single AM build. Finishing may take a few days if the requirement is for high quality. Using CNC machining, even 5-axis high-speed machining, this same process may take weeks for high complexity parts, with considerably more uncertainty over the completion time. This is due to the complex planning involved and custom jigs and fixtures needed to make the part the first time. For the second and subsequent parts, however, CNC machining may be faster than AM.

1.6.3   Complexity

As mentioned above, the higher the geometric complexity, the greater the advantage AM has over CNC. If CNC is being used to create a part directly in a single piece, then there may be some geometric features that cannot be fabricated. Since a machining tool must be carried in a spindle, there may be certain accessibility constraints or clashes preventing the tool from being located on the machining surface of a part. AM processes are not constrained in the same way, and undercuts and internal features can be easily built without specific process planning. Certain parts cannot be fabricated by CNC unless they are broken up into components and reassembled at a later stage. Consider, for example, the possibility of machining a ship inside a bottle. How would you machine the ship while it is still inside the bottle? Most likely you would machine both elements separately and work out a way to combine them together as an assembly and/or joining process. With AM you can build the ship and the bottle all at once. An expert in machining must therefore analyze each part prior to it being built to ensure that it indeed can be built and to determine what methods need to be used. While it is still possible that some parts cannot be built with AM, the likelihood is much lower, and there are generally ways in which this may be overcome without too much difficulty.

1.6.4   Accuracy

AM machines generally operate with a resolution of a few tens of microns. It is common for AM machines to also have different resolution along different orthogonal axes. Typically, the vertical build axis corresponds to layer thickness, and this would be of a lower resolution compared with the resolution of the two axes in the build plane. Accuracy in the build plane is determined by the positioning of the build mechanism, which will normally involve gearboxes and motors of some kind. This mechanism may also determine the minimum feature size as well. For example, stereolithography uses a laser as part of the build mechanism that will normally be positioned using galvanometric mirror drives. The resolution of the galvanometers would determine the overall dimensions of parts built, while the diameter of the laser beam would determine the minimum wall thickness. The accuracy of CNC machines on the other hand is mainly determined by a similar positioning resolution along all three orthogonal axes and by the diameter of the rotary cutting tools. There are factors that are defined by the tool geometry, like the radius of internal corners, but wall thickness can be thinner than the tool diameter since it is a subtractive process. In both cases very fine detail will also be a function of the desired geometry and properties of the build material.

1.6.5 Geometry

AM machines essentially break up a complex, 3D problem into a series of simple 2D cross-sections with a nominal thickness. In this way, the connection of surfaces in 3D is removed, and continuity is determined by how close the proximity of one cross-section is with an adjacent one. Since this cannot be easily done in CNC, machining of surfaces must normally be generated in 3D space. With simple geometries, like cylinders, cuboids, cones, etc., this is a relatively easy process defined by joining points along a path, these points being quite far apart and the tool orientation being fixed. In cases of freeform surfaces, these points can become very close together with many changes in orientation. Such geometry can become extremely difficult to produce with CNC, even with 5-axis interpolated control or greater. Undercuts, enclosures, sharp internal corners, and other features can all fail if these features are beyond a certain limit. Consider, for example, the features represented in the part in Fig. 1.3. Many of them would be very difficult to machine without manipulation of the part at various stages.

1.6.6 Programming

Determining the program sequence for a CNC machine can be very involved, including tool selection, machine speed settings, approach position and angle, etc. Many AM machines also have options that must be selected, but the range, complexity, and implications surrounding their choice are minimized when using pre-configured process parameter combinations. The worst that is likely to happen in most AM machines is that the part will not be built very well if the programming and process parameter selection is not done properly. Incorrect programming of a CNC machine could result in severe damage to the machine and may even be a human safety risk.

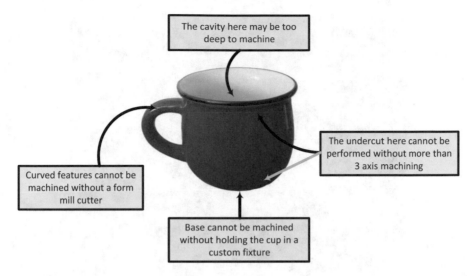

Fig. 1.3 Features that represent problems using CNC machining

1.7   Example AM Parts

Figure 1.4 shows a montage of parts fabricated using some of the common AM processes. Part a. was fabricated using a stereolithography machine and depicts a simplified fuselage for an unmanned aerial vehicle where the skin is reinforced with a conformal lattice structure (see Chap. 4 for more information about the process). A more complete description of this part is included in the Design for Additive Manufacturing chapter. Parts b. and c. were fabricated using material jetting (Chap. 7). Part b. demonstrates the capability of depositing multiple materials simultaneously, where one set of nozzles deposited the clear material, while another set deposited the black material for the lines and the Objet name. Part c. is a section of chain. Both parts b. and c. have working revolute joints that were fabricated using clearances for the joints and dissolvable support structures. Part d. is a metal part that was fabricated in a metal Powder Bed Fusion (PBF) machine using an electron beam as its energy source (Chap. 5). The part is a model of a facial implant. Part e. was fabricated in an Mcor Technologies Sheet Lamination machine that has inkjet printing capability for the multiple colors (Chap. 9). Parts f. and g. were fabricated using Material Extrusion (MEX) (Chap. 6). Part f. is a ratchet mechanism that was fabricated in a single build in an industrial machine. Again, the working mechanism is achieved through proper joint designs and dissolvable support structures. Part g. was fabricated in a low-cost, MEX machine (that one of the authors has at home). Parts h. and i. were fabricated using polymer PBF. Part h. is the well-known "brain gear" model of a three-dimensional gear train. When one gear is rotated, all other gears rotate as well. Since parts fabricated in polymer PBF do not need supports, working revolute and gear joints can be created by managing clearances and remov-

Fig. 1.4 Montage of AM parts

ing the loose powder from the joint regions. Part i. is another conformal lattice structure showing the shape complexity capability of AM technologies.

1.8 Other Related Technologies

The most common input method for AM technology is to accept a file converted into the STL file format originally built within a conventional 3D CAD system. There are, however, other ways in which the STL files can be generated and other technologies that can be used in conjunction with AM technology. This section will describe a few of these.

1.8.1 Reverse Engineering Technology

More and more models are being built from data generated using reverse engineering (RE) 3D imaging equipment and software. In this context, RE is the process of capturing geometric data from another object. These data are usually initially available in what is termed "point cloud" form, meaning an unconnected set of points representing the object surfaces. These points need to be connected together using RE software like Geomagic [8], which may also be used to combine point clouds from different scans and to perform other functions like hole-filling and smoothing.

In many cases, the data will not be entirely complete. Samples may, for example, need to be placed in a holding fixture, and thus the surfaces adjacent to this fixture may not be scanned. In addition, some surfaces may obscure others, like with deep crevices and internal features, so that the representation may not turn out exactly how the object is in reality. Recently there have been huge improvements in scanning technology. An adapted smartphone using its in-built camera can now produce a high-quality 3D scan for just a few hundred dollars that even just a few years ago would have required an expensive laser scanning or stereoscopic camera system costing $100,000 or more.

Engineered objects would normally be scanned using laser scanning or touch probe technology. Objects that have complex internal features or anatomical models may make use of Computerized Tomography (CT), which was initially developed for medical imaging but is also available for scanning industrially produced objects. This technique essentially works in a similar way to AM, by scanning layer-by-layer and using software to join these layers and identify the surface boundaries. Boundaries from adjacent layers are then connected together to form surfaces. The advantage of CT technology is that internal features can also be generated. High-energy X-rays are used in industrial technology to create high-resolution images of around 1 μm. Another approach that can help digitize objects is the Capture Geometry Inside [9] technology that also works very much like a reverse of AM technology, where 2D imaging is used to capture cross-sections of a part as it is machined away layer-by-layer. Obviously this is a destructive approach to geometry capture so it cannot be used for every type of product.

AM can be used to reproduce the articles that were scanned, which essentially would form a kind of 3D facsimile (3D Fax) process. More likely, however, the data will be modified and/or combined with other data to form complex, freeform artifacts that are taking advantage of the "complexity for free" feature of the technology. An example may be where individual patient data are combined with an engineering design to form a customized medical implant. This is something that will be discussed in more detail later in this book.

1.8.2   Computer-Aided Engineering/Technologies (CAX)

CAX is the use of computer technologies to aid in the design, analysis, and manufacture of products. Computer-Aided Engineering (CAE) is the wide usage of computer software to help engineers analyze their tasks using simulation. Other major CAX sub-categories include Computer-Aided Design (CAD) and Computer-Aided Manufacturing (CAM).

1.8.2.1   Computer-Aided Design (CAD)

CAD is the use of a computer to aid in the creation, design, modification, analysis, and optimization of a component. CAD software helps increase the efficiency and quality of the design process and enhances communication through documentation by generating organized databases. One of the features of most CAD software is an ability to create electronic output files for printing, machining, or other manufacturing operations. CT images can be converted to CAD files using various software packages, and subsequent modifications can be performed. For instance, to fill the worn part of a bone in the spinal column, CT scan images can be converted to 3D images, and the worn part can be modified and filled. Then in the next step, the file can be saved as an STL file and sent directly to an AM machine for printing.

3D CAD is an extremely valuable resource for product design and development. One major benefit to using software-based design is the ability to implement changes easily and cheaply. If we are able to keep the design primarily in a software format for a larger proportion of the product development cycle, we can ensure that any design changes are performed virtually on the software description rather than physically on the product itself. The more we know about how the product is going to perform before it is built, the more effective that product is going to be. This is also the most cost-effective way to deal with product development. If problems are only noticed after parts are physically manufactured, this can be very costly. 3D CAD can make use of AM to help visualize and perform basic tests on candidate designs prior to full-scale commitment to manufacturing. However, the more complex and performance-related the design, the less likely we are to gain sufficient insight using these methods.

3D CAD is commonly linked to other software packages, often using techniques like the Finite Element Method (FEM) to calculate the mechanical response of a design to certain stimuli, collectively known as Computer-Aided Engineering (CAE). Forces, dynamics, stresses, flow, heat, and other responses can be predicted to determine how well a design will perform under certain conditions. While such software cannot easily predict the exact behavior of a part, for analysis of critical parts, a combination of CAE, backed up with AM-based experimental analysis, may be a useful solution. Further, with the advent of Direct Digital Manufacture, where AM can be used to directly produce final products, there is an increasing need for CAE tools to evaluate how these parts would perform prior to AM so that we can build these products right the first time.

1.8.3   Haptic-Based CAD

3D CAD systems are generally built on the principle that models are constructed from basic geometric shapes that are then combined in different ways to make more complex forms. This works very well for the engineered products we are familiar with but may not be so effective for more unusual designs. Many consumer products are developed from ideas generated by artists and designers rather than engineers. We also note that AM has provided a mechanism for greater freedom of expression. AM is in fact now becoming a popular tool for artists and sculptors, like, for example, Bathsheba Grossman [10] who takes advantage of the geometric freedom to create visually exciting sculptures (including the one shown on the cover of this book). One problem we face today is that some computer-based design tools constrain or restrict creative processes and there is scope for a CAD system that provides greater freedom. Haptic-based CAD modeling systems like the experimental system shown in Fig. 1.5 [11] work in a similar way to commercially available freeform [12] modeling systems to provide a design environment that is more intuitive than standard CAD systems. In this image a robotic haptic feedback device called the Phantom provides force feedback relating to the virtual modeling environment. An object can be seen on screen but also felt in 3D space using the Phantom. The modeling environment includes what is known as Virtual Clay that deforms under force applied using the haptic cursor. This provides a mechanism for direct interaction with the modeling material, much like how a sculptor interacts with actual clay. The results using this system are generally much more organic and freeform than the surfaces that can be incorporated into product designs by using traditional engineering CAD tools. As consumers become more demanding and discerning, we can see that CAD tools for non-engineers like designers, sculptors, and even members of the general public are likely to become much more commonplace.

Fig. 1.5   Freeform modeling system

1.8.3.1   Computer-Aided Manufacturing (CAM)

CAM is the use of a computer to control machine tools in order to fabricate a component. The initial purpose of CAM was increasing the speed of manufacturing processes and simultaneously increasing the accuracy, consistency, and precision of manufacturing combined with a reduction of lead time and energy.

The first CAM systems used an individual programmer to create and develop a solution by code writing and manually changing controller options. After significant trial and error, a successful "recipe" would be developed that could be used repeatedly to create identical components. Initial systems were numerically controlled (NC), and the codes were fed into a machine using punch cards. The programs in the NC machine could not be stored and had less flexibility, accuracy, and productivity and also required an expert programmer. The next generation was Computer Numerical Controlled (CNC), where a computer directly controlled the motion of the tool. The programs were fed into the machine directly from computers. In CNC, modification of the programs were greatly simplified. Today, the connection between CAD and CAM is simple and efficient in many software tools.

1.8.3.2   Computer-Aided Engineering (CAE)

CAE is the use of a computer to simulate the effects of various physics applied to a component or system. Today's multi-physics simulation tools are powerful complements to AM. Topology optimization and generative design are commonly used by designers to guide them to a part shape which is the lowest weight component capable of achieving a certain task. AM process simulation tools are a relatively new set of CAE tools, introduced into the market over the past few years. These AM process simulation tools can predict distortion, microstructure, porosity, and other characteristics of an AM part prior to its creation. By simulating the effects of an AM process on a specific geometry, build failures can be reduced, and shape changes that occur in the part can be compensated for prior to production, so that the part can be produced to a higher tolerance and with a higher probability of success [13].

1.9   About This Book

There have been a number of texts describing additive manufacturing processes, either as dedicated books or as sections in other books. Prior to the first edition of this book, however, there were no texts dedicated to presenting this technology in a comprehensive way within a university setting. Universities are incorporating additive manufacturing into various curricula. This has varied from segments of single modules to complete postgraduate courses. This text is aimed at supporting these curricula with a comprehensive coverage of as many aspects of this technology as possible. The authors of this text have all been involved in setting up programs in their home universities and have written this book because they feel that there were no books that covered the required material in sufficient breadth and depth. Furthermore, with the increasing interest in Additive Manufacturing and 3D Printing, we believe that this text can also provide a comprehensive understanding of the technologies involved. Despite increased popularity, it is clear that there is a significant lack of basic understanding by many of the breadth that AM has to offer.

Early chapters in this book discuss general aspects of AM, followed by chapters which focus on specific AM technologies. The final chapters focus more on generic processes and applications. It is anticipated that the reader will be familiar with 3D solid modeling CAD technology and have at least a small amount of knowledge about product design, development, and manufacturing. The majority of readers would be expected to have an engineering or design background, more specifically product design, or mechanical, materials, or manufacturing engineering. Since AM technology also involves significant electronic and information technology components, readers with a background in computer applications and mechatronics may also find this text beneficial.

This third edition has been comprehensively overhauled in an attempt to bring it up to date with the extremely fast-moving landscape of AM. All references have been checked and updated, with many more web-based sources. All images have been checked, with many diagrams redrawn for greater clarity. Replacement photographs show more recent technologies where appropriate. There is much more detail regarding processes and, in particular, on materials. A new chapter on AM materials helps readers understand how new materials are being developed. In particular, there is increasing focus on high performance materials, including ceramics and multi-phase, multi-material structures. We have included as many of these new concepts as we could to show how this exciting domain continues to develop. Finally, we recognize that many educators are including this book as a core text. To help, we have revamped the end of chapter questions as part of a larger process aimed at helping with curriculum development.

1.10   Questions

1. Find three other definitions for Rapid Prototyping other than that of Additive Manufacturing as covered by this book.
2. From the web, find different examples of applications of AM that illustrate their use for "Form," "Fit," and "Function."
3. What functions can be carried out on point cloud data using reverse engineering software? How do these tools differ from conventional 3D CAD software?
4. What is your favorite term (AM, Freeform Fabrication, RP, etc.) for describing this technology and why?
5. Create a web link list of videos showing operation of different AM technologies and representative process chains.

6. Make a list of different characteristics of AM technologies as a means to compare with CNC machining. Under what circumstances does AM have the advantage and under what would CNC?
7. How does the Phantom desktop haptic device work and why might it be more useful for creating freeform models than conventional 3D CAD?
8. With a basic understanding of Additive Manufacturing, what is an application you can think of where additive manufacturing could be used in your daily life? Explain how.
9. What are the differences between end-use parts and prototypes?
10. What was Additive Manufacturing initially used for, and how is it utilized today?
11. Why is the term "Rapid Prototyping" not suitable for additive manufacturing anymore?
12. What is "concurrent engineering"? How can Additive Manufacturing help with concurrent engineering?
13. What type of file is typically required to use as input to process a part in an AM machine?
14. List one academic institution with research activities in AM and some of their research projects that interest you.
15. List one company which produces Additive Manufacturing machines and associated products such as materials and also list one of their products.
16. A company manufactures missile casings with internal channels. Why might Additive Manufacturing be a better manufacturing process than CNC machining?
17. Why have 3D printers not been widely applied for household use today? What benefits do you think AM will bring to your daily life in the future?
18. Visualize a process in which small robots dispersed materials in various places to construct a component. Is this an additive manufacturing process?
19. Recently historical museums have started using additive manufacturing and reverse engineering techniques. What are museums using them for?
20. Why does NASA wish to create an AM lab in outer space?
21. What changes are additive manufacturing making to the mechanical engineering field?
22. Design a product that is very difficult to make using traditional manufacturing methods but that can be printed using AM.
23. Find one Additive Manufacturing application example not covered in this book. List the benefits and disadvantages of this application compared with its traditional manufacturing method.
24. Find the newest definition for Additive Manufacturing by the ASTM F42 Committee. Make a table of all the categories of AM technologies covered in the latest standard and list one typical technique for each category.
25. Discuss how AM can improve manufacturing environmental impact and sustainability.

References

1. ASTM International (2012). Standard terminology for additive manufacturing technologies (F42 on Additive Manufacturing Technologies. Subcommittee F42. 91 on Terminology). ASTM International.
2. Burns, M. (1993). *Automated fabrication: improving productivity in manufacturing*.
3. Wohlers, T. (2009). Wohlers report. State of The Industry: Annual worldwide progress report. Wohlers Associates.
4. ASTM International (2015). *ISO/ASTM52900–15, standard terminology for additive manufacturing – general principles – terminology*. West Conshohocken: ASTM International.
5. Jacobs, P. F. (1995). *Stereolithography and other RP&M technologies: from rapid prototyping to rapid tooling*. Dearborn: Society of Manufacturing Engineers.
6. 3D Systems. (2020). http://www.3dsystems.com
7. Sachs, E., et al. (1992). Three dimensional printing: Rapid tooling and prototypes directly from a CAD model. *Journal of Engineering for Industry, 114*(4), 481–488.
8. Geomagic Reverse Engineering software. (2020). www.geomagic.com/
9. Capture geometry inside. (2020). https://www.moldmakingtechnology.com/suppliers/cgicap
10. Bathesheba Grossman. (2020). http://www.bathsheba.com
11. Gao, Z., & Gibson, I. (2007). A 6-DOF haptic interface and its applications in CAD. *International Journal of Computer Applications in Technology, 30*(3), 163–170.
12. Sensable, 3D Systems. (2020). http://www.sensable.com
13. Goehrke, S. (2020). *Additive Manufacturing Simulation Smartens Up In Latest ANSYS Software Release,* https://www.forbes.com/sites/sarahgoehrke/2019/01/29/additive-manufacturing-simulation-smartens-up-in-latest-ansys-software-release/#1f32adad74eb

Chapter 2
Development of Additive Manufacturing Technology

Abstract It is important to understand that AM was not developed in isolation from other technologies. It would not be possible for AM to exist were it not for innovations in areas like 3D graphics and Computer-Aided Design software. This chapter highlights some of the key moments that catalogue the development of Additive Manufacturing. It describes how the different technologies converged to a state where they could be integrated into AM machines. It will also discuss milestone AM technologies and how they have contributed to increase the range of AM applications. Furthermore, we will discuss how the application of Additive Manufacturing has evolved to include greater functionality and embrace a wider range of applications beyond the initial intent of prototyping.

2.1 Introduction

Additive Manufacturing (AM) came about as a result of developments in a variety of technology sectors. Like with many manufacturing technologies, improvements in computing power and reduction in mass storage costs paved the way for processing the large amounts of data typical of modern 3D Computer-Aided Design (CAD) models within reasonable time frames. Nowadays, we have become quite accustomed to having powerful computers and other complex automated machines around us, and sometimes it may be difficult for us to imagine how the pioneers struggled to develop the first AM machines.

This chapter highlights some of the key moments that catalogue the development of Additive Manufacturing technology. It will describe how the different technologies converged to a state where they could be integrated into AM machines. It will also discuss milestone AM technologies. Furthermore, we will discuss how the application of Additive Manufacturing has evolved to include greater functionality and embrace a wider range of applications beyond the initial intention of just proto-

© Springer Nature Switzerland AG 2021
I. Gibson et al., *Additive Manufacturing Technologies*,
https://doi.org/10.1007/978-3-030-56127-7_2

typing. We close the chapter with a summary of standards that have been developed to support the adoption of AM.

2.2 Computers

Like many other technologies, AM came about as a result of the invention of the computer. However, there was little indication that the first computers built in the 1940s (like the Zuse Z3 [1], ENIAC [2], and EDSAC [3] computers) would change lives in the way that they so obviously have. Inventions like the thermionic valve, transistor, and microchip made it possible for computers to become faster, smaller, and cheaper with greater functionality. This development has been so quick that even Bill Gates of Microsoft was caught off guard when he thought in 1981 that 640 kb of memory would be sufficient for any Windows-based computer. In 1989, he admitted his error when addressing the University of Waterloo Computer Science Club [4]. Similarly in 1977, Ken Olsen of Digital Electronics Corp. (DEC) stated that there would never be any reason for people to have computers in their homes when he addressed the World Future Society in Boston [5]. That remarkable misjudgment may have caused Olsen to lose his job not long afterward.

One key to the development of computers as serviceable tools lies in their ability to perform tasks in real time. In the early days, serious computational tasks took many hours or even days to prepare, run, and complete. This served as a limitation to everyday computer use, and it is only since it was shown that tasks can complete in real time that computers have been accepted as everyday items rather than just for academics or big business. This has included the ability to display results not just numerically but graphically as well. For this we owe a debt of thanks at least in part to the gaming industry, which has pioneered many developments in graphics technology with the aim to display more detailed and more "realistic" images to enhance the gaming experience.

AM takes full advantage of many of the important features of computer technology, both directly (in the AM machines themselves) and indirectly (within the supporting technology), including:

- *Processing power*: Part data files can be very large and require a reasonable amount of processing power to manipulate while setting up the machine and when slicing the data before building. Earlier machines would have had difficulty handling large CAD data files.
- *Graphics capability*: AM machine operation does not require a big graphics engine except to see the file while positioning within the virtual machine space. However, all machines benefit from a good graphical user interface (GUI) that can make the machine easier to set up, operate, and maintain.
- *Machine control*: AM technology requires precise positioning of equipment in a similar way to a Computer Numerical Controlled (CNC) machining center or even a high-end photocopy machine or laser printer. Such equipment requires controllers that take information from sensors for determining status and actuators for positioning and other output functions. Computation is generally required

in order to determine the control requirements. Conducting these control tasks even in real time does not normally require significant amounts of processing power by today's standards. Dedicated functions like positioning of motors, lenses, etc. would normally require individual controller modules. A computer would be used to oversee the communication to and from these controllers and pass data related to the part build function.

- *Networking*: Nearly every computer these days has a method for communicating with other computers around the world. Files for building would normally be designed on another computer to that running the AM machine. Earlier systems would have required the files to be loaded from disk or tape. Nowadays almost all files will be sent using an Ethernet connection, often via the Internet.
- *Integration*: As is indicated by the variety of functions, the computer forms a central component that ties different processes together. The purpose of the computer would be to communicate with other parts of the system, to process data, and to send that data from one part of the system to the other. Figure 2.1 shows how the abovementioned technologies are integrated to form an AM machine.

Earlier computer-based design environments required physically large mainframe and mini computers. Workstations that generally ran the graphics and input/output functions were connected to these computers. The computer then ran the complex calculations for manipulating the models. This was a costly solution based around the fact that the processor and memory components were very expensive elements. With the reduction in the cost of these components, personal computers

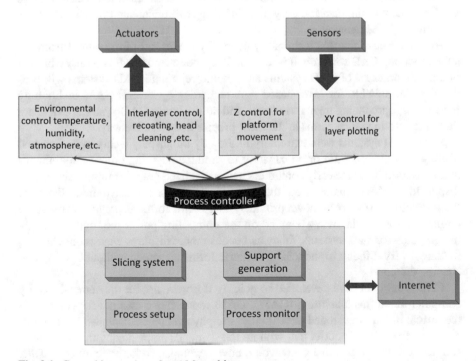

Fig. 2.1 General integration of an AM machine

(PCs) became viable solutions. Earlier PCs were not powerful enough to replace the complex functions that workstation-based computers could perform, but the speedy development of PCs soon overcame all but the most computationally expensive requirements.

Without computers there would be no capability to display 3D graphic images. Without 3D graphics, there would be no 3D Computer-Aided Design. Without this ability to represent objects digitally in 3D, we would have a limited desire to use machines to fabricate anything but the simplest shapes. It is safe to say, therefore, that without the computers we have today, we would not have seen Additive Manufacturing develop.

2.3   Computer-Aided Design Technology

Today, every engineering student must learn how to use computers for many of their tasks, including the development of new designs. CAD technologies are available for assisting in the design of large buildings and of nano-scale microprocessors. CAD technology holds within it the knowledge associated with a particular type of product, including geometric, electrical, thermal, dynamic, and static behavior. CAD systems may contain rules associated with such behaviors that allow the user to focus on design and functionality without worrying too much whether a product can or cannot work. CAD also allows the user to focus on small features of a large product, maintaining data integrity and ordering it to understand how subsystems integrate with the remainder.

Additive Manufacturing technology primarily makes use of the output from 3D solid modeling CAD software. It is important to understand that this is only a branch of a much larger set of CAD systems and, therefore, not all CAD systems will produce output suitable for layer-based AM technology. Currently, AM technology focuses on reproducing geometric form, and so the better CAD systems to use are those that produce such forms in the most precise and effective way.

Early CAD systems were extremely limited by the display technology. The first display systems had little or no capacity to produce anything other than alphanumeric text output. Some early computers had specialized graphic output devices that displayed graphics separate from the text commands used to drive them. Even so, the geometric forms were shown primarily in a vector form, displaying wire-frame output. As well as the heavy demand on the computing power required to display the graphics for such systems, this was because most displays were monochrome, making it very difficult to show 3D geometric forms on screen without lighting and shading effects.

CAD would not have developed so quickly if it were not for the demands set by computer-aided manufacturing (CAM). CAM represents a channel for converting the virtual models developed in CAD into the physical products that we use in our everyday lives. It is doubtful that without the demands associated with this conversion from virtual to real that CAD would have developed so far or so quickly. This, in turn, was fueled and driven by the developments in associated technologies, like processor, memory, and display technologies. CAM systems produce the code for numerically controlled (NC) machinery, essentially combining coordinate data with commands to select and actuate the cutting tools. Early NC technologies would take CAM data relating to the location of machined features, like holes, slots, pockets, etc. These features would then be fabricated by machining from a stock material. As NC machines proved their value in their precise, automated functionality, so the sophistication of the features increased. This has now extended to the ability to machine highly complex, freeform surfaces. However, there are two key limitations to all NC machining:

- Almost every part must be made in stages, often requiring multiple passes for material removal and setups.
- All machining is performed from an approach direction (sometimes referred to as 2.5D rather than fully 3D manufacture). This requires that the stock material be held in a particular orientation and that not all the material can be accessible at any one stage in the process.

NC machining, therefore, only requires surface modeling software. All early CAM systems were based on surface modeling CAD. AM technology was the first automated computer-aided manufacturing process that truly required 3D solid modeling CAD. It was necessary to have a fully enclosed surface to generate the driving coordinates for AM. This can be achieved using surface modeling systems, but because surfaces are described by boundary curves, it is often difficult to precisely and seamlessly connect these together. Even if the gaps are imperceptible, the resulting models may be difficult to build using AM. At the very least, any inaccuracies in the 3D model would be passed on to the AM part that was constructed. Early AM applications often displayed difficulties because of associated problems with surface modeling software.

Since it is important for AM systems to have accurate models that are fully enclosed, the preference is for solid modeling CAD. Solid modeling CAD ensures that all models made have a volume and, therefore, by definition are fully enclosed surfaces. While surface modeling can be used in part construction, we cannot always be sure that the final model is faithfully represented as a solid. Such models are generally necessary for computer-aided engineering (CAE) tools like finite element analysis (FEA) but are also very important for other CAM processes.

Most CAD systems can now quite readily run on PCs. This is generally a result of the improvements in computer technology mentioned earlier but is also a result in improvements in the way CAD data is presented, manipulated, and stored. Most CAD systems utilize non-uniform rational basis splines or NURBS [6]. NURBS are an excellent way of precisely defining the curves and surfaces that correspond to the outer shell of a CAD model. Since model definitions can include freeform surfaces as well as simple geometric shapes, the representation must accommodate this, and splines are complex enough to represent such shapes without making the files too large and unwieldy. They are also easy to manipulate to modify the resulting shape.

CAD technology has rapidly improved along the following lines:

- *Realism*: With lighting, shading effects, ray tracing, and other photorealistic imaging techniques, it is becoming possible to generate images of the CAD models that are difficult to distinguish from actual photographs. In some ways, this reduces the requirements on AM models for visualization purposes.
- *Usability and user interface*: Early CAD software required the input of text-based instructions through a dialog box. Development of Windows-based graphical user interfaces (GUIs) has led to graphics-based dialogs and even direct manipulation of models within virtual 3D environments. Instructions are issued through the use of drop-down menu systems and context-related commands. To suit different user preferences and styles, it is often possible to execute the same instruction in different ways. Keyboards are still necessary for input of specific measurements, but the usability of CAD systems has improved dramatically. There is still some way to go to make CAD systems available to those without engineering knowledge or without training, however.
- *Engineering content*: Since CAD is almost an essential part of a modern engineer's training, it is vital that the software includes as much engineering content as possible. With solid modeling CAD, it is possible to calculate the volumes and masses of models, to investigate fits and clearances according to tolerance variations, and to export files with mesh data for finite element analysis. FEA is often even possible without having to leave the CAD system.
- *Speed*: As mentioned previously, the use of NURBS assists in optimizing CAD data manipulation. CAD systems are constantly being optimized in various ways, mainly by exploiting the hardware developments of computers.
- *Accuracy*: If high tolerances are expected for a design, then it is important that calculations are precise. High precision can make heavy demands on processing time and memory.
- *Complexity*: All of the above characteristics can lead to extremely complex systems. It is a challenge to software vendors to incorporate these features without making them unwieldy and unworkable.
- *Usability*: Recent developments in CAD technology have focused on making the systems available to a wider range of users. In particular the aim has been to allow untrained users to be able to design complex geometry parts for themselves. There are now 3D solid modeling CAD systems that run entirely within a web browser with similar capabilities to workstation systems of only 10 years ago.

Many CAD software vendors are focusing on producing highly integrated design environments that allow designers to work as teams and to share designs across different platforms and for different departments. Industrial designers must work with sales and marketing, engineering designers, analysts, manufacturing engineers, and many other branches of an organization to bring a design to fruition as a product. Such branches may even be in different regions of the world and may be part of the same organization or acting as subcontractors. The Internet must therefore also be integrated with these software systems, with appropriate measures for fast and accurate transmission and protection of intellectual property.

Fig. 2.2 A CAD model on the left converted into STL format on the right

It is quite possible to directly manipulate the CAD file to generate the slice data that will drive an AM machine, and this is commonly referred to as direct slicing [7]. However, this would mean every CAD system must have a direct slicing algorithm that would have to be compatible with all the different types of AM technology. Alternatively, each AM system vendor would have to write a routine for every CAD system. Both of these approaches are impractical. The solution is to use a generic format that is specific to the technology. This generic format was developed by 3D Systems, USA, who was the first company to commercialize AM technology and called the file format "STL" after their stereolithography technology (an example of which is shown in Fig. 2.2).

The STL file format was made public domain to allow all CAD vendors to access it easily and hopefully integrate it into their systems. This strategy has been successful, and STL is now a standard output for nearly all solid modeling CAD systems and has also been adopted by AM system vendors [8]. STL uses triangles to describe the surfaces to be built. Each triangle is described as three points and a facet normal vector indicating the outward side of the triangle, in a manner similar to the following:

facet normal − 4.470293E−02 7.003503E−01 − 7.123981E−01,
outer loop,
vertex − 2.812284E+00 2.298693E+01 0.000000E+00,
vertex − 2.812284E+00 2.296699E+01−1.960784E−02,
vertex − 3.124760E+00 2.296699E+01 0.000000E+00,
endloop,
endfacet.

The demands on CAD technology in the future are set to change with respect to AM. As we move toward more and more functionality in the parts produced by AM, we must understand that the CAD system must include rules associated with AM. To date, the focus has been on the external geometry. In the future, we may need to know rules associated with how the AM systems function so that the output can be optimized.

2.4   Other Associated Technologies

Aside from computer technology, there are a number of other technologies that have developed along with AM that are worthy of note here since they have served to contribute to further improvement of AM systems.

2.4.1   Printing Technologies

Inkjet and droplet printing technologies have rapidly developed in recent years. Improvements in resolution and reduction in costs has meant that high-resolution printing, typically with multiple colors, is available as part of our everyday lives. Such improvement in resolution has also been supported by improvement in material handling capacity and reliability. Initially, colored inks were low viscosity and fed into the print heads at ambient temperatures. Now it is possible to generate much higher pressures within the droplet formation chamber so that materials with much higher viscosity and even molten materials can be printed. This means that droplet deposition can now be used to print photocurable and molten resins as well as binders for powder systems. Since print heads are relatively compact devices with all the droplet control technology highly integrated into these heads (like the one shown in Fig. 2.3), it is possible to produce low-cost, high-resolution, high-throughput AM technology.

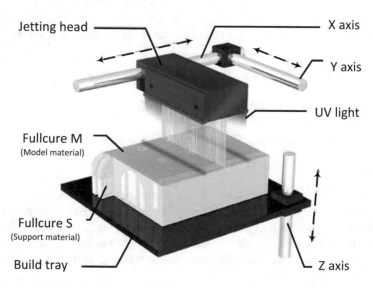

Fig. 2.3 Printer technology used on an AM machine. (Photo courtesy of Objet)

2.4.2   Programmable Logic Controllers

The input CAD models for AM are large data files generated using standard computer technology. Once they are on the AM machine, however, these files are reduced to a series of process stages that require sensor input and signaling of actuators. This process and machine control is often best carried out using microcontroller systems rather than microprocessor systems. Industrial microcontroller systems form the basis of programmable logic controllers (PLCs), which are used to reliably control industrial processes. Designing and building industrial machinery, like AM machines, are much easier using building blocks based around modern PLCs for coordinating and controlling the various steps in the machine process.

2.4.3   Materials

Early AM technologies were built around materials that were already available and that had been developed to suit other processes. However, AM processes are somewhat unique, and these original materials were far from ideal for these new applications. For example, the early photocurable resins resulted in models that were brittle and that warped easily. Powders used in Laser-Based Powder Bed Fusion (LB-PBF) processes degraded quickly within the machine, and many of the materials resulted in parts that were quite weak. As we came to understand the technology better, materials were developed specifically to suit AM processes. Materials have been tuned to suit more closely the operating parameters of the different processes and to provide better output parts. As a result, parts are now much more accurate, stronger, and longer lasting. In turn, these new materials have resulted in the processes being tuned to produce higher-temperature materials (including metals), smaller feature sizes, and faster throughput.

2.4.4   Computer Numerically Controlled Machining

One of the reasons AM technology was originally developed was because CNC technology was not able to produce satisfactory output within the required time frames. CNC machining was slow, cumbersome, and difficult to operate. AM technology on the other hand was quite easy to set up with quick results but had poor accuracy and limited material capability. As improvements in AM technologies came about, vendors of CNC machining technology realized that there was now growing competition. CNC machining has dramatically improved, just as AM technologies have matured. It could be argued that high-speed CNC would have developed anyway, but some have argued that the perceived threat from AM technology caused CNC machining vendors to rethink how their machines were made. The combination of high-speed machining and AM for certain applications, such as for

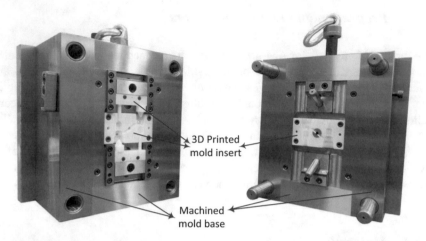

Fig. 2.4 Standardized mold system and 3D printed mold inserts permitting rapid, low-cost prototypes. (Photo courtesy of HASCO) [10]

making large, complex, and durable molds for standardized mold system (plastic injection molding), as shown in Fig. 2.4 [9], illustrates how the two can be used interchangeably to take advantage of the benefits of both technologies. For geometries that can be machined using a single setup orientation, CNC machining is often the fastest, most cost-effective method. For parts with complex geometries or parts which require a large proportion of the overall material volume to be machined away as scrap, AM can be used to more quickly and economically produce the part than when using CNC.

2.5   The Use of Layers

A key enabling principle of AM part manufacture is the use of layers as finite 2D cross-sections of the 3D model. Almost every AM technology builds parts using layers of material added together, primarily due to the simplification of building 3D objects. Using 2D representations to represent cross-sections of a more complex 3D feature has been common in many applications outside AM. The most obvious example of this is how cartographers use a line of constant height to represent hills and other geographical reliefs. These contour lines, or iso-heights, can be used as plates that can be stacked to form representations of geographical regions. The gaps between these 2D cross-sections cannot be precisely represented and are therefore approximated, or interpolated, in the form of continuity curves connecting these layers. Such techniques can also be used to provide a 3D representation of other physical properties, like isobars or isotherms on weather maps.

Architects have also used such methods to represent landscapes of actual or planned areas, like that used by an architect firm in Fig. 2.5 [11]. The concept is

Fig. 2.5 An architectural landscape model, illustrating the use of layers. (Photo courtesy of LiD)

particularly logical to manufacturers of buildings who also use an additive approach, albeit not using layers. Consider how the pyramids in Egypt and in South America were created. Notwithstanding how they were fabricated, it's clear that they were created using a layered approach, adding material as they went.

2.6 Classification of AM Processes

There are numerous ways to classify AM technologies. A popular approach is to classify according to baseline technology, like whether the process uses lasers, printer technology, extrusion technology, etc. [12, 13]. Another approach is to collect processes together according to the type of raw material input [14, 15]. The problem with these classification methods is that some processes get lumped together in what seems to be odd combinations (like Selective Laser Sintering being grouped together with 3D Printing) or that some processes that may appear to produce similar results end up being separated (like stereolithography and material jetting (MJT) with photopolymers). It is probably inappropriate, therefore, to use a single classification approach.

An early classification method was described by Pham [16], which uses a two-dimensional classification method as shown in Fig. 2.6. The first dimension relates to the method by which the layers are constructed. Earlier technologies used a single point source to draw across the surface of the base material. Later systems increased the number of sources to increase the throughput, which was made possible with the use of droplet deposition technology, for example, which can be constructed into a one-dimensional array of deposition heads. Further throughput improvements are possible with the use of 2D array technology using the likes of digital micromirror devices (DMDs) and high-resolution display technology, capable of exposing an entire surface in a single pass. However, just using this classifica-

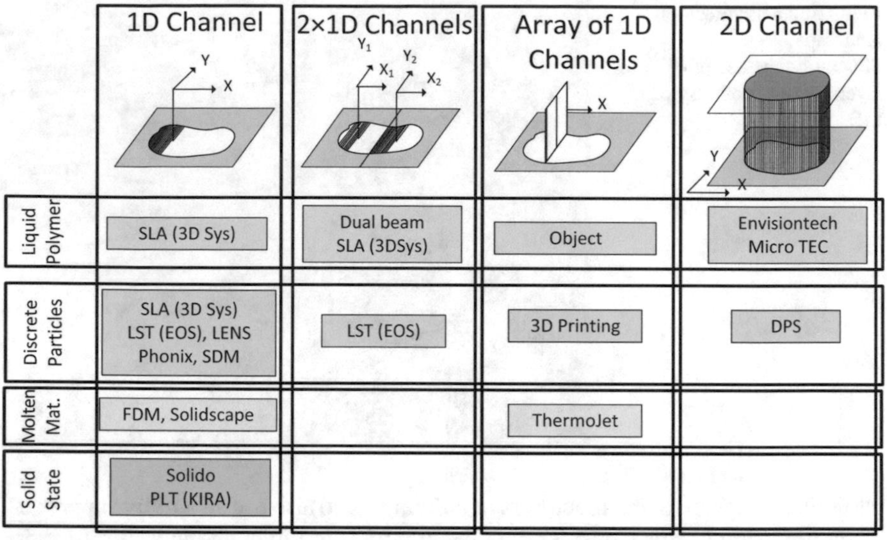

1D Channel	2×1D Channels	Array of 1D Channels	2D Channel
SLA (3D Sys)	Dual beam SLA (3DSys)	Object	Envisiontech Micro TEC
SLA (3D Sys) LST (EOS), LENS Phonix, SDM	LST (EOS)	3D Printing	DPS
FDM, Solidscape		ThermoJet	
Solido PLT (KIRA)			

(Row labels, top to bottom: Liquid Polymer; Discrete Particles; Molten Mat.; Solid State)

Fig. 2.6 Layered manufacturing (LM) processes as classified by Pham (note that this diagram has been amended to include some recent AM technologies)

tion results in the previously mentioned anomalies where numerous dissimilar processes are grouped together. This is solved by introducing a second dimension of raw material to the classification. Pham uses four separate material classifications: liquid polymer, discrete particles, molten material, and laminated sheets. Some more exotic systems mentioned in this book may not fit directly into this classification. An example is the possible deposition of composite material, like concrete, using an extrusion-based technology. This fits well as a 1D channel, but the material is not explicitly listed as it is not really a molten material. Furthermore, future systems may be developed that use 3D holography to project and fabricate complete objects in a single pass. As with many classifications, there can sometimes be processes or systems that lie outside them. If there are sufficient systems to warrant an extension to this classification, then it should not be a problem.

It should be noted that, in particular, 1D and 2 × 1D channel systems combine both vector- and raster-based scanning methods. Often, the outline of a layer is traced first before being filled in with regular or irregular scanning patterns. The outline is generally referred to as vector scanned, while the fill pattern can often be generalized as a raster pattern. The array methods tend not to separate the outline and the fill.

Most AM technology started using a 1D channel approach, although one of the earliest and now obsolete technologies, Solid Ground Curing from Cubital, used liquid photopolymers and essentially (although perhaps arguably) a 2D channel method. As technology developed, so more of the boxes in the classification array began to be filled. The empty boxes in this array may serve as a guide to researchers and developers for further technological advances. Most of the 1D array methods use at least 2 × 1D lines. This is similar to conventional 2D printing where each line deposits a different material in different locations within a layer. The Connex process using the PolyJet technology from Stratasys is a good example of this where it is possible to create parts with different material properties in a single step using this approach. Color 3D Printing is possible using multiple 1D arrays with ink or separately colored material in each. Note however that the part coloration in the sheet laminating Mcor process [17] is separated from the layer formation process, which is why it is defined as a 1D channel approach.

2.6.1   Liquid Polymer Systems

As can be seen from Fig. 2.6, liquid polymers appear to be a popular material. The first commercial system was the 3D Systems Stereolithography process based on liquid photopolymers. A large portion of systems in use today are, in fact, not just liquid polymer systems but more specifically liquid photopolymer systems. However, this classification should not be restricted to just photopolymers, since a number of experimental systems are using hydrogels that would also fit into this category. Furthermore, the Fab@Home system developed at Cornell University in the USA and the open-source RepRap systems originating from Bath University in the UK also use liquid polymers with curing techniques other than UV or other wavelength optical curing methods [18, 19].

Using this material and a 1D channel or 2 × 1D channel scanning method, a great option is to use a laser like in the stereolithography process. Droplet deposition of polymers using an array of 1D channels can simplify the curing process to a floodlight (for photopolymers) or similar method. This approach was commercialized by the Israeli company Objet (now part of Stratasys) who use printer technology to print fine droplets of photopolymer "ink" [20]. One unique feature of the Objet system is the ability to vary the material properties within a single part. Parts can, for example, have soft-feel, rubber-like features combined with more solid resins to achieve a result similar to an overmolding effect.

Controlling the area to be exposed using digital micromirror devices or other high-resolution display technology obviates the need for any scanning at all, thus increasing throughput and reducing the number of moving parts. DMDs are generally applied to micron-scale additive approaches, like those used by Microtec in Germany [21]. For normal-scale systems, EnvisionTEC uses high-resolution DMD displays to cure photopolymer resin in their low-cost AM machines. The 3D Systems V-Flash process is also a variation on this approach, exposing thin sheets of polymer spread onto a build surface.

2.6.2   Discrete Particle Systems

Discrete particles are normally powders that are generally graded into a relatively uniform particle size and shape and narrow size distribution. The finer the particles, the better, but there will be problems if the dimensions get too small in terms of controlling the distribution and dispersion. Again, the conventional 1D channel approach is to use a laser, this time to produce thermal energy in a controlled manner and, therefore, raise the temperature sufficiently to melt the powder. Polymer powders must therefore exhibit thermoplastic behavior so that they can be melted and remelted to permit bonding of one layer to another. There are a wide variety of such systems that generally differ in terms of the material that can be processed. The two main polymer-based systems commercially available are the Selective Laser Sintering (SLS) technology marketed by 3D Systems [22] and the EOSINT processes developed by the German company EOS [23].

Application of printer technology to powder beds resulted in the (original) 3D Printing (3DP) process. This technique was developed by researchers at MIT in the USA [24]. Droplet printing technology is used to print a binder, or glue, onto a powder bed. The glue sticks the powder particles together to form a 3D structure. This basic technique has been developed for different applications dependent on the type of powder and binder combination. The most successful approaches use low-cost, starch-, and plaster-based powders with inexpensive glues, as commercialized by Z Corp, USA [25], which is now part of 3D Systems. Ceramic powders and appropriate binders were similarly used in the Direct Shell Production Casting (DSPC) process by Soligen [26] as part of a service to create shells for casting of metal parts. Alternatively, if the binder were to contain an amount of drug, 3DP can be used to create controlled delivery-rate drugs like in the process developed by the US company Therics. Neither of these last two processes has proven to be as successful as that licensed by Z Corp/3D Systems. One particular advantage of the Z Corp technology is that the binders can be jetted from multi-nozzle print heads. Binders coming from different nozzles can be different, and, therefore, subtle variations can be incorporated into the resulting part. The most obvious of these is the color that can be incorporated into parts.

2.6.3 Molten Material Systems

Molten material systems are characterized by a preheating chamber that raises the material temperature to the melting point so that it can flow through a delivery system. The most well-known method for doing this is the Fused Deposition Modeling (FDM) Material Extrusion (MEX) technology developed by the US company Stratasys [27]. This approach extrudes the material through a nozzle in a controlled manner. Two extrusion heads are often used so that support structures can be fabricated from a different material to facilitate part cleanup and removal. It should be noted that there are now a huge number of variants of this technology due to the lapse of key FDM patents. This competition has driven the price of these machines down to such a level that individual buyers can afford to have their own machines at home.

Printer technology has also been adapted to suit this material delivery approach. One technique, developed initially as the Sanders prototyping machine, that later became Solidscape, USA [28], and which is now part of Stratasys, is a 1D channel system. A single jet piezoelectric deposition head lays down wax material. Another head lays down a second wax material with a lower melting temperature that is used for support structures. The droplets from these print heads are very small so the resulting parts are fine in detail. To further maintain the part precision, a planar cutting process is used to level each layer once the printing has been completed. Supports are removed by inserting the complete part into a temperature-controlled bath that melts the support material away, leaving the part material intact. The use of wax along with the precision of Solidscape machines makes this approach ideal for precision casting applications like jewelry, medical devices, and dental castings. Few machines are sold outside of these niche areas.

The 1D channel approach, however, is very slow in comparison with other methods, and applying a parallel element does significantly improve throughput. The Thermojet technology from 3D Systems also deposits a wax material through droplet-based printing heads. The use of parallel print heads as an array of 1D channels effectively multiplies the deposition rate. The Thermojet approach, however, is not widely used because wax materials are difficult and fragile when handled. Thermojet machines are no longer being made, although existing machines are commonly used for investment casting patterns.

2.6.4 Solid Sheet Systems

One of the earliest AM technologies was the Laminated Object Manufacturing (LOM) system from Helisys, USA. This technology used a laser to cut out profiles from sheet paper, supplied from a continuous roll, which formed the layers of the final part. Layers were bonded together using a heat-activated resin that was coated on one surface of the paper. Once all the layers were bonded together, the result was very like a wooden block. A hatch pattern cut into the excess material allowed the user to separate away waste material and reveal the part.

A similar approach was used by the Japanese company Kira, in their Solid Center machine [29], and by the Israeli company Solidimension with their Solido machine. The major difference is that both these machines cut out the part profile using a blade similar to those found in vinyl sign-making machines, driven using a 2D plotter drive. The Kira machine used a heat-activated adhesive applied using laser printing technology to bond the paper layers together. Both the Solido and Kira machines have been discontinued for reasons including poor reliability, material wastage, and the need for excessive amounts of manual post-processing. Mcor Technologies expanded upon this approach and produces a modern version of this technology, using low-cost color printing to make it possible to make laminated color parts [17].

2.6.5 New AM Classification Schemes

In this book, we use a version of Pham's classification introduced in Fig. 2.6. Instead of using the 1D and 2 × 1D channel terminology, we will typically use the terminology "point" or "point-wise" systems. For arrays of 1D channels, such as when using inkjet print heads, we refer to this as "line" processing. 2D channel technologies will be referred to as "layer" processing. Last, although no current commercialized processes are capable of this, holographic-like techniques are considered "volume" processing.

The technology-specific descriptions starting in Chap. 4 are based, in part, upon a separation of technologies into groups where processes which use a common type of machine architecture and similar materials transformation physics are grouped together. This grouping is a refinement of an approach introduced by Stucker and Janaki Ram in the CRC Materials Processing Handbook [30]. In this grouping scheme, for example, processes which use a common machine architecture developed for stacking layers of powdered material and a materials transformation mechanism using heat to fuse those powders together are all discussed in the PBF chapter. These are grouped together even though these processes encompass polymer, metal, ceramic, and composite materials, multiple types of energy sources (e.g., lasers and infrared heaters), and point-wise and layer processing approaches. Using this classification scheme, all AM processes fall into one of seven categories. An understanding of these seven categories should enable a person familiar with the concepts introduced in this book to quickly grasp and understand an unfamiliar AM process by comparing its similarities, benefits, drawbacks, and processing characteristics to the other processes in the grouping into which it falls.

This classification scheme from the first edition of this textbook had an important impact on the development and adoption of ASTM/ISO standard terminology. The authors were involved in these consensus standards activities, and we have agreed to adopt the modified terminology from ASTM F42 and ISO TC 261 in the second and subsequent editions [31]. Of course, in the future, we will continue to support the ASTM/ISO standardization efforts and keep the textbook up to date.

The seven process categories are presented here. Chapters 4, 5, 6, 7, 8, 9, and 10 cover each one in detail:

- Vat Photopolymerization (VPP): processes that utilize a liquid photopolymer that is contained in a vat and processed by selectively delivering energy to cure specific regions of a part cross-section.
- Powder Bed Fusion (PBF): processes that utilize a container filled with powder that is processed selectively using an energy source, most commonly a scanning laser or electron beam.
- Material Extrusion (MEX): processes that deposit a material by extruding it through a nozzle, typically while scanning the nozzle in a pattern that produces a part cross-section.
- Material Jetting (MJT): processes that selectively deposit droplets of feedstock material.

- Binder Jetting (BJT): processes where a liquid bonding agent is printed into a powder bed in order to form part cross-sections.
- Sheet Lamination (SHL): processes that bond sheets of material to form a part.
- Directed Energy Deposition (DED): processes that simultaneously deposit a material (usually powder or wire) and provide energy to process that material through a single deposition device.

2.7 Heat Sources

2.7.1 Lasers

Many of the earliest AM systems were based on laser technology. The reasons are that lasers provide a high intensity and highly collimated beam of energy that can be moved very quickly in a controlled manner with the use of directional mirrors. Since AM requires the material in each layer to be solidified or joined in a selective manner, lasers are ideal candidates for use, provided the laser energy is compatible with the material transformation mechanisms. There are two kinds of laser processing used in AM: curing and heating. With photopolymer resins the requirement is for laser energy of a specific frequency that will cause the liquid resin to solidify or "cure." Usually this laser is in the ultraviolet range, but other frequencies can be used. For heating, the requirement is for the laser to carry sufficient thermal energy to cut through a layer of solid material, to cause powder to melt, or to cause sheets of material to fuse. For powder processes, for example, the key is to melt the material in a controlled fashion without creating too great a buildup of heat, so that when the laser energy is removed, the molten material rapidly solidifies again. For cutting, the intention is to separate a region of material from another in the form of laser cutting. Earlier AM machines used tube lasers to provide the required energy, but most manufacturers have switched to solid-state technology, which provides greater efficiency, lifetime, and reliability.

A laser generally has the following components: (I) gain medium, (II) a pumping energy source, and (III) optical resonator. The gain medium is located in the optical resonator and amplifies the light beam by stimulating emission. Lasers are classified according to the type of gain medium. Figure 2.7 shows the classification of lasers.

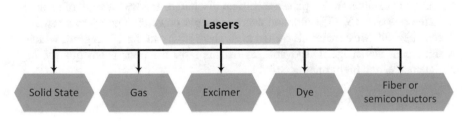

Fig. 2.7 Different lasers

In Additive Manufacturing gas, solid-state and fiber lasers are the most commonly used.

Early AM machines used gas lasers, primarily CO_2. The most common gas lasers produce infrared output wavelengths ranging from 9000 to 11,000 nm with an efficiency of 5–20% and power ranging from 100 to 20,000 W. The simplicity of a gas laser system brings low cost, highly reliability, and system compactness, which are the main reasons that CO_2 lasers are the workhorse of precision manufacturing. However, their output power is relatively non-stable because of thermal expansion and contraction of the laser structure due to the heat generated in the process of energy pumping to a large volume of CO_2 gas. Low light absorption in the infrared region is another problem of CO_2 lasers, with high reflectivity among metals in particular.

Nd:YAG is a solid-state laser, and the gain medium is made of rod-shaped Nd:YAG crystals. The gain medium is pumped by a flash lamp or 808 nm laser diode to produce a near-infrared output wavelength of 1064 nm. The significant advantages of this laser over CO_2 are system compactness and efficiency in delivering the beam through optical fibers. This laser can be operated both in continuous mode using crystals doped with low concentrations and in pulsed mode using highly doped crystals. The maximum output power for continuous and pulse lasers is a few kW and 20 kW, respectively. This laser suffers from relatively low electrical-to-optical power conversion efficiency which leads to lower beam quality. The reason is most of the unabsorbed energy is dissipated as heat. This heat leads to thermal lensing and poor beam quality. Diode-pumped solid-state (DPSS) lasers are a solution for these problems.

The gain medium for fiber lasers is made from rare earth-doped optical fibers such as Yb-fibers. This fiber has high quantum efficiency (−95%). They can produce a near-infrared laser beam in the range 1030–1070 nm output wavelength and have high electrical-to-optical efficiency (−25%), excellent beam quality, robustness against environmental disturbances, and system compactness. Light propagation inside the fiber, unexpected polarization change from fiber bending, vibration, and temperature variation are typical problems of Yb-fiber lasers.

Excited dimer (Excimers) are gas mixtures containing a noble gas (e.g., argon, krypton, or xenon), a halogen (e.g., fluorine or chlorine), and a buffer gas (typically neon or helium). If the gain medium is made by excimers, the system is called an excimer laser. In this laser nanosecond pulses in the ultraviolet region are produced by pumping excimers using pulsed electrical discharge. The wavelength of excimers is in the range of 157–351 nm, and average output power can go up to a few hundreds of Watts. Most materials have a high absorptivity in the UV region, which is a significant advantage. Disadvantages of this laser are poor beam quality, high maintenance, and high running cost.

2.7.2 Electron Beam

An electron beam is another energy source used to melt materials in Additive Manufacturing. In this method free electrons in a vacuum can be controlled and positioned by electric and magnetic fields to shape a beam. When a high-velocity beam of electrons contacts solid matter, the kinetic energy is transformed to heat. This high heat, when concentrated, is an efficient means for melting materials that are electrically conductive. Electron beams require a vacuum chamber, since electrons scatter and diffuse if they interact with gas atoms between the electron gun and the material being heated. Electron beam melting (EBM) has a higher production rate compared to laser beam melting due to a higher energy density. A greater energy density leads to Marangoni convection in the melting area, which results in lower surface quality (roughness) and thus higher dimensional deviations. More Additive Manufacturing machines are equipped with lasers than electron beams.

2.7.3 Electric Arc/Plasma Arc

The electrical breakdown of a gas in a nonconductive medium like air produces plasma and an electric arc, plasma arc, or arc discharge. A normal gas molecule at room temperature consists of two or more atoms. When the temperature of the molecule increases to 2000 °C, atoms are separated. If the gas temperature goes up to 3000 °C, some of the atoms are separated, and gas is ionized and decomposition of atoms happens. This gas is called plasma and is a conductor of electricity, and so plasma is the fourth state of matter.

The plasma is accompanied by electrical discharge. The source of heat production in plasma is the recombination of electrons and ions to form atoms and the recombining of atoms to form molecules. This released bonding energy increases the kinetic energy of atoms or molecules formed by recombination and can be used in different material processing. This temperature can go up to 30,000 °C and be used for melting and evaporation of almost all materials.

Electrical and plasma arcs are used in Additive Manufacturing mainly in Directed Energy Deposition (DED) using wire. This is also referred to as Wire Arc Additive Manufacturing (WAAM). In WAAM motion can be provided either by robotic systems or Computer Numerical Controlled gantries. In this method wire plays the role of a consumable electrode [32].

Electrical or plasma arcs are used particularly for building large components due to their high deposition rate, low material and equipment costs, and good structural integrity. These specifications characterize it as a suitable candidate for replacing the conventional methods of manufacturing from solid billets or large forgings, notably with regard to low and medium complexity parts.

2.8 Metal Systems

One of the most important historical developments in AM was the proliferation of
direct metal processes. Machines like the EOSINT-M [21] and Laser Engineered
Net Shaping (LENS) have been around for a number of years [33, 34]. Recent additions from other companies and improvements in laser technology, machine accuracy, speed, and cost have opened up this market, and today, metal AM is a major
driver of AM adoption.

Most direct metal systems work using a point-wise method, and nearly all of
them utilize metal powders as input. The main exception to this approach is the
Sheet Lamination (SHL) processes, particularly the Ultrasonic Consolidation process commercialized by Solidica, USA (now Fabrisonic), which uses sheet metal
laminates that are ultrasonically welded together [35]. Of the powder systems,
almost every newer machine uses a powder spreading approach similar to the
Selective Laser Sintering process, followed by melting using an energy beam. This
energy is normally a high-power laser, except in the case of the EBM process by the
Swedish company Arcam [36] (now part of GE). Another approach is the LENS
powder delivery system used by Optomec [31]. This machine employs powder
delivery through a nozzle placed above the part. The powder is melted where the
material converges with the laser and the substrate. This approach allows the process to be used to add material to an existing part, which means it can be used for
repair of expensive metal components that may have been damaged, like chipped
turbine blades and injection mold tool inserts.

2.9 Hybrid Systems

Some of the machines described above are, in fact, hybrid additive/subtractive processes rather than purely additive. Including a subtractive component can assist in
making the process more precise. An example is the use of planar milling at the end
of each additive layer in the Sanders and Objet machines. This stage makes for a
smooth planar surface onto which the next layer can be added, negating cumulative
effects from errors in droplet deposition height.

It should be noted that when subtractive methods are used, waste will be generated. Machining processes require removal of material that in general cannot easily
be recycled. Similarly, many additive processes require the use of support structures, and these too must be removed or "subtracted."

It can be said that with the Objet process, for instance, the additive element is
dominant and that the subtractive component is important but relatively insignificant. There have been a number of attempts to merge subtractive and additive technologies together where the subtractive component is the dominant element. An
excellent example of this is the Stratoconception approach [37], where the original
CAD models are divided into thick machinable layers. Once these layers are

machined, they are bonded together to form the complete solid part. This approach works very well for very large parts that may have features that would be difficult to machine using a multi-axis machining center due to the accessibility of the tool. This approach can be applied to foam- and wood-based materials or to metals. For structural components it is important to consider the bonding methods. For high-strength metal parts, diffusion bonding may be an alternative.

A lower-cost solution that works in a similar way is Subtractive RP (SRP) from Roland [38], who is also famous for plotter technology. SRP makes use of Roland desktop milling machines to machine sheets of material that can be sandwiched together, similar to Stratoconception. The key is to use the exterior material as a frame that can be used to register each slice to others and to hold the part in place. With this method not all the material is machined away, and a web of connecting spars are used to maintain this registration.

Another variation of this method that was never commercialized was Shape Deposition Manufacturing (SDM), developed mainly at Stanford and Carnegie-Mellon Universities in the USA [39]. With SDM, the geometry of the part is devolved into a sequence of easier to manufacture parts that can in some way be combined together. A decision is made concerning whether each subpart should be manufactured using additive or subtractive technology dependent on such factors as the accuracy, material, geometrical features, functional requirements, etc. Furthermore, parts can be made from multiple materials, combined together using a variety of processes, including the use of plastics, metals, and even ceramics. Some of the materials can also be used in a sacrificial way to create cavities and clearances. Additionally, the "layers" are not necessarily planar, nor constant in thickness. Such a system would be unwieldy and difficult to realize commercially, but the ideas generated during this research have influenced many studies and systems thereafter.

In this book, for technologies where both additive and subtractive approaches are used, these technologies are discussed in the chapter where their additive approach best fits.

2.10   Milestones in AM Development

We can look at the historical development of AM in a variety ways. The origins may be difficult to properly define, and there was certainly a lot of activity in the 1950s and 1960s related to joining materials together to form objects. But development of the associated technologies (computers, lasers, controllers, etc.) caught up with the concept in the early 1980s. Interestingly, parallel patents were filed in 1984 in Japan (Murutani), France (Andre et al.) and in the USA (Masters in July and Hull in August). All of these patents described a similar concept of fabricating a 3D object by selectively adding material layer-by-layer. While earlier work in Japan is quite well-documented, proving that this concept could be realized, it was the patent by Charles Hull that is generally recognized as the most influential since it gave rise to

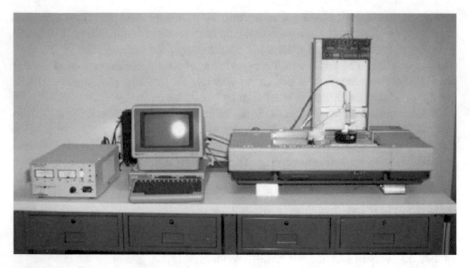

Fig. 2.8 The first AM technology from Hull, who founded 3D Systems (photo courtesy of 3D Systems)

3D Systems. This was the first company to commercialize AM technology with the Stereolithography apparatus (Fig. 2.8).

Further patents came along in 1986, resulting in three more companies, Helisys (Laminated Object Manufacturing or LOM), Cubital (with Solid Ground Curing, SGC), and DTM with their Selective Laser Sintering (SLS) process. It is interesting to note neither Helisys nor Cubital exist anymore, and only SLS remains as a commercial process with DTM merging with 3D Systems in 2001. In 1989, Scott Crump patented the Fused Deposition Modeling (FDM) process, forming the Stratasys Company. Also in 1989 a group from MIT patented the 3D Printing (3DP) process. These processes from 1989 are heavily used today, with FDM variants currently being the most successful. Rather than forming a company, the MIT group licensed the 3DP technology to a number of different companies, who applied it in different ways to form the basis for different applications of their AM technology. The most successful of these were Z Corp, which focused mainly on low-cost technology, and ExOne, which focused on powdered metal and sandcasting applications.

Inkjet technology has become employed to deposit droplets of material directly onto a substrate, where that material hardens and becomes the part itself rather than just as a binder. Sanders developed this process in 1994, and Objet also used this technique to print photocurable resins in droplet form in 2001.

There have been numerous failures and successes in AM history, with the previous paragraphs mentioning only a small number. However, it is hard to know whether a specific technology may have failed because of poor business models or by poor timing rather than having a poor process. Helisys appears to have failed with their LOM machine, but there have been at least five variants from Singapore, China, Japan, Israel, and Ireland. The most recent Mcor process laminates colored sheets together rather than the monochrome paper sheets used in the original LOM machine. However, Mcor went into receivership, and their IP was acquired by CleanGreen3D in late 2019. Perhaps this is a better application, and perhaps the technology is in a better position to become successful in the future; however the business track record for LOM-based businesses has been consistently poor. Another example is the defunct Ballistic Particle Manufacturing process, which used a 5-axis mechanism to direct wax droplets onto a substrate. Although no company currently uses such an approach for polymers, similar 5-axis deposition schemes are being used for depositing metal and composites.

Another important trend that is impacting the development of AM technology is the expiration of many of the foundational patents for key AM processes. We saw an explosion of MEX vendors systems and systems after the first FDM patents expired in the early 2010s. Initial patents in the stereolithography, laser sintering, and LOM have also expired, which is leading to a proliferation of technologies, processes, machines, companies, and competition worldwide.

2.11 AM around the World

As was mentioned, early patents were filed in Europe (France), the USA, and Asia (Japan) almost simultaneously. In early years, most pioneering and commercially successful systems came out of the USA. Companies like Stratasys, 3D Systems, and Z Corp have spearheaded the way forward. These companies have generally strengthened over the years, but most new companies have come from outside the USA.

In Europe, the first company with a worldwide impact in AM was EOS, Germany. EOS stopped making SL machines following settlement of disputes with 3D Systems but continues to make PBF systems which use lasers to melt polymers, binder-coated sand, and metals. Companies from elsewhere in Europe are smaller but are competitive in their respective marketplaces. Examples of these companies include Phenix [40] (now part of 3D Systems), Arcam, Strataconception, SLM Solutions, Trumpf, Additive Industries, and Materialise. The last of these, Materialise from Belgium [41], has seen considerable success in developing software tools to support AM technology.

In the early 1980s and 1990s, a number of Japanese companies focused on AM technology. Large companies like Sony and Kira, who established subsidiaries to build AM technology, also became involved. Much of the Japanese technology was based around the photopolymer curing processes. With 3D Systems dominant in much of the rest of the world, these Japanese companies struggled to find market, and many of them failed to become commercially viable, even though their technology showed some initial promise. Some of this demise may have resulted in the unusually slow uptake of CAD technology within Japanese industry in general. Although the Japanese company CMET [42] still seems to be doing well, you will likely find more non-Japanese made machines in Japan than home-grown ones.

AM technology has also been developed in other parts of Asia, most notably in Korea and China. Korean AM companies have had little international impact; however, quite a few Chinese manufacturers have been active for a number of years and are beginning to make inroads into markets in Europe and the Americas. Patent conflicts with the earlier US, Japanese, and European designs have meant that many of these machines were not able to be sold outside of China. Earlier Chinese machines were also thought to be of questionable quality, but more recent machines have markedly improved performance (like the machine shown in Fig. 2.9). Chinese machines primarily reflect the SL, FDM, and SLS technologies found elsewhere in the world, and the expiration of initial patents for almost every type of AM technology has created an opening for Chinese companies to compete worldwide.

A particular country of interest in terms of AM technology development is Israel. One of the earliest AM machines was developed by the Israeli company Cubital. Although this technology was not a commercial success, in spite of early installations around the world, they demonstrated a number of innovations not found in other machines, including layer processing through a mask, removable secondary support materials and milling after each layer to maintain a constant layer thickness. Some of the concepts used in Cubital can be found in Sanders machines as well as machines from another Israeli company, Objet. Objet joined with Stratasys and moved Stratasy's headquarters to Israel. The success of Objet's droplet deposition technology as well as other software and hardware startups in the AM arena makes Israel an outsized player in AM relative to its size.

Fig. 2.9 AM technology from Beijing Yinhua Co. Ltd., China

2.12   AM Standards

Standards in conventional manufacturing play significant roles in helping organizations develop, manufacture, and supply goods and services in a more efficient, effective, and safer manner. Standards in AM can fulfill the same roles. Companies use AM standards to learn to adopt AM technologies, gain confidence in AM usage, communicate with suppliers and customers using a common language, and specify materials and their properties, among others. More specifically, current AM standards assist the specification of types of technical requirements expected when ordering parts, specification of what is needed to qualify a material or process, understanding of how to ensure safe machine operations and environments, and characterization of properties of fabricated parts.

In 2009, the authors of this book worked with ASTM International to create the F42 Committee to develop AM standards. Several other standards organizations followed in the next years, including ISO, SAE, and ASME. In 2013, ASTM and ISO formed a partnership to develop and promote AM standards in a landmark agreement between the organizations. The Society of Automotive Engineers (SAE) is developing standards for the aerospace industry for metal AM technologies. The American Society of Mechanical Engineers (ASME) is investigating modifications to their dimensioning and tolerancing standards to accommodate unique characteristics of AM. In 2017, a partnership between the American National Standards Institute (ANSI) and America Makes [43] released a standardization roadmap that identified 89 gaps, i.e., topics for which standards are needed, and corresponding recommendations [44]; their second roadmap in 2018 continues to guide standardization efforts worldwide.

Within ASTM F42, standards development occurs within six subcommittees on test methods, design, materials and processes, environment health and safety, applications (which has nine sub-subcommittees for various industries), and data. The test methods subcommittee develops new test standards for AM-fabricated specimens and provides guides on how to adopt existing standards to the unique characteristics of AM parts. The design subcommittee develops data exchange standards, including the Additive Manufacturing Format (AMF) [45]; guides on how to use data exchange formats for multiple materials, medical imaging, etc.; and design for AM guides. In the materials and processes subcommittee, standards are developed for AM-fabricated materials, technical guides for specific processes, finished part properties, post-processing methods, and requirements for purchased AM parts. The last topic is represented by a joint ISO/ASTM standard [46] that specifies the types of information that suppliers and customers of AM parts should consider to meet acceptance requirements. The environment, health, and safety subcommittee develops standards for operator safety, allowable indoor air quality requirements for AM printers, and related topics. In the applications subcommittee, many standards under development focus on qualification issues that arise in a wide range of industries. Additionally, they have participation from other US federal agencies that oversee specific industries, including the US Food and Drug Administration (FDA) for

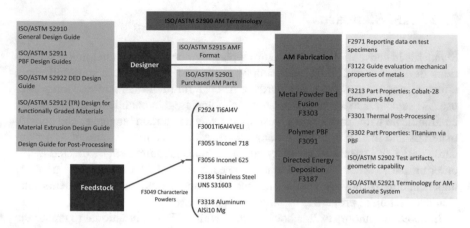

Fig. 2.10 Current ASTM and ISO standards and their relationships

medical devices and the Federal Aviation Administration (FAA) for aerospace applications.

A snapshot of ASTM and ISO standards available in 2020 is shown in Fig. 2.10, arranged according to the product development stage of relevance. The AM terminology standard provides the foundation for the others. The designer receives guidance from the various design guides during the design process and then can rely on the AMF file format to exchange their parts with the manufacturer. For contracting, the purchased AM parts standard provides guidance. Many standards for materials have been developed, with a focus on metals so far, as well as a standard on powder characterization. For AM fabrication, standards support test specimen reporting data, guidance on evaluating mechanical properties, finished part properties, post-processing methods, etc.

As mentioned, several organizations are actively developing standards that should impact AM technology and practices. The specifics presented about ASTM and ISO are meant to illustrate some of the current accomplishments and organization of the activity. In the future, a much more comprehensive set of standards will be available to guide the AM field; these standards will be the product of concerted efforts among many professionals in many standards organizations, companies, universities, and government agencies.

2.13   The Future? Rapid Prototyping Develops into Direct Digital Manufacturing

How might the future of AM look? The ability to "grow" parts may form the core to the answer to that question. The true benefit behind AM is the fact that we do not really need to design the part according to how it is to be manufactured. We would prefer to design the part to perform a particular function. Avoiding the need to consider how the part can be manufactured certainly simplifies the process of design and allows the designer to focus more on the intended application. The design flexibility of AM is making this more and more possible.

An example of geometric flexibility is customization of a product. If a product is specifically designed to suit the needs of a unique individual, then it can truly be said to be customized. Imagine being able to modify your mobile phone so that it fits snugly into your hand based on the dimensions gathered directly from your hand. Imagine a hearing aid that can fit precisely inside your ear because it was made from an impression of your ear canal (like those shown in Fig. 2.11). Such things are possible using AM because it has the capacity to make one-off parts, directly from digital models that may not only include geometric features but may also include biometric data gathered from a specific individual.

With improvements in AM technology, the speed, quality, accuracy, and material properties have all developed to the extent that parts can be made for final use and not just for prototyping. The terms Rapid Manufacturing and Direct Digital Manufacturing (RM and DDM) gained popularity prior to the adoption of the term Additive Manufacturing to represent the use of AM to produce parts which will be used as an end product. Certainly we will continue to use this technology for prototyping for years to come, but we are already entering a time when it is commonplace to manufacture products in low volumes or unique products using AM. We even see these machines being used as home fabrication devices for hobbyists.

2.14   Questions

1. (a) Based upon an Internet search, describe the Solid Ground Curing process developed by Cubital. (b) Solid Ground Curing has been described as a 2D channel (layer) technique. Could it also be described in another category? Why?
2. Make a list of five different metal AM technologies that are currently available on the market today. How can you distinguish between the different systems? What different materials can be processed in these machines?
3. NC machining is often referred to as a 2.5D process. What does this mean? Why might it not be regarded as fully 3D?

Fig. 2.11 RM of custom hearing aids, from a wax ear impression, on to the machine to the final product. (Photo courtesy of EnvisionTEC and Sonova) [47, 48]

4. Provide three instances where a layer-based approach has been used in fabrication, other than AM.
5. Find five countries not specifically mentioned in this chapter where AM technology has been developed commercially and describe the machines.
6. Home fabrication using AM was an area of extensive market speculation in the late 2010s. But adoption of AM as a home fabrication method remains very low as of 2020. Why do you think this is? Do you see a large future opportunity for home fabrication, or do you think it will stay a niche, hobbyist market? Why or why not?
7. How might the future of Additive Manufacturing (AM) look? Give one specific example of geometric flexibility as a customization of a product.
8. Which of the Additive Manufacturing processes do you believe to be most suitable for home/domestic use and why?
9. What are four ways AM relies on computers that are different from traditional manufacturing?
10. Briefly describe some of the technologies that have contributed to the advancement of AM techniques.
11. MIT performed a study on using silkworm patterns for distributing silk in 2013 (available http://www.dezeen.com/2013/03/13/mit-researchers-to-3d-print-a-pavilion-by-imitating-silkworms/). How does this apply to Additive Manufacturing?
12. Find three parts that were created via three differing Additive Manufacturing Technologies. Present the picture of the part, the process used for production, the material used, and what is it about the part that gives an idea of the process used to create it.
13. Describe three Additive Manufacturing process types and attributes. Include example companies, materials utilized in machines, and typical markets.
14. What is the AMF format? What are the differences between the AMF format and an STL file? Compare the AMF format to STL files by describing at least 5 advantages AMF over STL file.

References

1. Zuse Z3 Computer. (2020). http://www.zib.de/zuse
2. Goldstine, H. H., & Goldstine, A. (1946). The electronic numerical integrator and computer (eniac). *Mathematical Tables and Other Aids to Computation, 2*(15), 97–110.
3. Wilkes, M. V., & Renwick, W. (1949). The EDSAC-an electronic calculating machine. *Journal of Scientific Instruments, 26*(12), 385.
4. Waterloo Computer Science Club. (2020). Talk by Bill Gates. http://csclub.uwaterloo.ca/media
5. Gatlin, J. (1999). *Bill Gates: The Path to the Future*. Avon Books.
6. Piegl, L., & Tiller, W. (1997). *The NURBS Book* (2nd ed.). Berlin: Springer.
7. Jamieson, R., & Hacker, H. (1995). Direct slicing of CAD models for rapid prototyping. *Rapid Prototyping Journal, 1*(2), 4–12.
8. Roscoe, L. (1988). Stereolithography interface specification. *America-3D Systems Inc, 27*, 10.
9. Protoform. (2020). *Space Puzzle Moulding*. http://www.protoform.de
10. HASCO. (2020) *Precision for Mouldmaking*. https://www.hasco.com/en/
11. LiD. (2020). *Architects*. http://www.lid-architecture.net

12. Burns, M. (1993). *Automated fabrication: improving productivity in manufacturing*. Englewood Cliffs: Prentice Hall.
13. Kruth, J.-P., Leu, M.-C., & Nakagawa, T. (1998). Progress in additive manufacturing and rapid prototyping. *CIRP Annals, 47*(2), 525–540.
14. Chua, C. K., & Leong, K. F. (1998). *Rapid prototyping: principles and applications in manufacturing*. New York: Wiley.
15. Chua, C. K., Leong, K. F., & Lim, C. S. (2003). *Rapid prototyping: principles and applications. Vol. 1*. Singapore: World Scientific.
16. Pham, D. T., & Gault, R. S. (1998). A comparison of rapid prototyping technologies. *International Journal of Machine Tools and Manufacture, 38*(10–11), 1257–1287.
17. MCor Technologies. (2020). http://www.mcortechnologies.com
18. Fab@Home. (2020). http://www.fabathome.org
19. RepRap. (2020). http://www.reprap.org
20. Stratasys. (2020), *3D Printing & Additive Manufacturing, Stratasys*. https://www.stratasys.com/
21. Microtec. (2020). http://www.microtec-d.com
22. 3D Systems, Stereolithography and selective laser sintering machines. (2020). http://www.3dsystems.com
23. EOS. (2020). http://www.eos.info
24. Sachs, E., et al. (1992). Three dimensional printing: Rapid tooling and prototypes directly from a CAD model. *Journal of Engineering for Industry, 114*(4), 481–488.
25. ZCrop (3D Systems). (2020). http://www.zcorp.com
26. Soligen. (2020). http://www.soligen.com
27. Stratasys. (2020). http://www.stratasys.com
28. Solidscape. (2020). http://www.solid-scape.com
29. Kira. (2020). *Solid Center machine*. www.kiracorp.co.jp/EG/pro/rp/top.html
30. Groza, J. R., & Shackelford, J. F. (2007). *Materials processing handbook*. Boca Raton: CRC Press.
31. ASTM International. (2015). *ISO/ASTM52900–15, standard terminology for additive manufacturing – general principles – terminology*. West Conshohocken: ASTM International.
32. Wu, B., et al. (2018). A review of the wire arc additive manufacturing of metals: Properties, defects and quality improvement. *Journal of Manufacturing Processes, 35*, 127–139.
33. Atwood, C., et al. *Laser engineered net shaping (LENS™): A tool for direct fabrication of metal parts*. In *International Congress on Applications of Lasers & Electro-Optics*. 1998. LIA
34. Optomec, LENS process. (2020). http://www.optomec.com
35. White, D. (2003). *Ultrasonic object consolidation*. US Patents.
36. Arcam. (2020). *Electron Beam Melting*. http://www.arcam.com
37. *Stratoconception, Thick layer hybrid AM*. http://www.stratoconception.com
38. *Roland, SRP technology*. (2020). http://www.rolanddga.com/solutions/rapidprototyping/
39. Prinz, F. B., & Weiss, L. E. (1998). Novel applications and implementations of shape deposition manufacturing. *Naval Research Reviews, 50*, 19–26.
40. *Phenix, Metal RP technology (Owned By 3D System)*. (2020). http://www.phenix-systems.com.
41. *Materialise*. (2020). *AM software systems and service provider*. http://www.materialise.com.
42. CMET. (2020). *Stereolithography technology*. http://www.cmet.co.jp
43. *America Makes – National Additive Manufacturing Innovation Institute*. (2020). www.americamakes.us
44. *America Makes & ANSI Additive Manufacturing Standardization Collaborative, Standardization Roadmap for Additive Manufacturing, Version 2.0*. (2018).
45. ASTM International. (2015). *ISO/ASTM52915–16 Standard Specification for Additive Manufacturing File Format (AMF) Version 1.2*. West Conshohocken: ASTM International.
46. ASTM International. (2016). *ISO/ASTM52901–16 standard guide for additive manufacturing – general principles – requirements for purchased AM parts*. West Conshohocken: ASTM International.
47. Hearing Aid, (2020), *EnvisionTEC*. https://envisiontec.com/
48. *3D printing technology for improved hearing, Sonova*. (2020). https://www.sonova.biz/en/features/3d-printing-technology-improved-hearing

Chapter 3
Generalized Additive Manufacturing Process Chain

Abstract Although there are an increasing number of AM technologies and variants, they nearly all use a similar process chain. This eight-step process is introduced in this chapter, which aims to provide a framework for the rest of the book. Later chapters address specific process stages in much more detail, leading to ways of overcoming inherent problems within the AM process chain. Furthermore, we look at some application sectors and discuss how the demands of different industries are driving the development of AM.

3.1 Introduction

Every product development process involving an Additive Manufacturing machine requires the operator to go through a set sequence of tasks. Easy-to-use "personal" 3D printing machines emphasize the simplicity of this task sequence. These desktop-sized machines are characterized by their low cost, simplicity of use, and ability to be placed in a home or office environment. The larger and more "industrial" AM machines are more capable of being tuned to suit different user requirements and therefore require more expertise to operate, but with a wider variety of possible results and effects that may be put to good use by an experienced operator. Such machines usually require more careful installation in industrial environments.

This chapter will take the reader through the different stages of the process that were described in Chap. 1. Different steps in the process will be described with reference to different processes and machines. The objective is to allow the reader to understand how these machines may differ and also to see how each task works and how it may be exploited to the benefit of higher-quality results. As mentioned before, we will refer to eight key steps in the process sequence:

- Conceptualization and CAD
- Conversion to STL/AMF

© Springer Nature Switzerland AG 2021
I. Gibson et al., *Additive Manufacturing Technologies*,
https://doi.org/10.1007/978-3-030-56127-7_3

- Transfer and manipulation of STL/AMF file on AM machine
- Machine setup
- Build
- Part removal and cleanup
- Post-processing of part
- Application

There are other ways to break down this process flow, depending on your perspective and equipment familiarity. For example, if you are a designer, you may see more stages in the early product design aspects. Model makers may see more steps in the post-build part of the process. Different AM technologies handle this process sequence differently, so this chapter will also discuss how choice of machine affects the generic process.

The use of AM in place of conventional manufacturing processes, such as machining and injection molding, enables designers to ignore some of the constraints of conventional manufacturing. However, conventional manufacturing will remain core to how many products are manufactured. Thus, we must also understand how conventional technologies, such as machining, integrate with AM. This may be particularly relevant to the increasingly popular metal AM processes. Thus, we will discuss how to deal with metal systems in detail.

3.2   The Eight Steps in Additive Manufacture

The abovementioned sequence of steps is generally appropriate to all AM technologies. There will be some variations dependent on which technology is being used and also on the design of the particular part. Some steps can be quite involved for some machines but may be trivial for others.

3.2.1   Step 1: Conceptualization and CAD

The first step in any product development process is to come up with an idea for how the product will look and function. Conceptualization, sometimes referred to as ideation, can take many forms, from textual and narrative descriptions to sketches and representative models. If AM is to be used, the product description must be in a digital form that allows a physical model to be made. It may be that AM technology will be used to prototype and not build the final product, but in either case, there are many stages in a product development process where digital models are required.

AM technology would not exist if it were not for 3D CAD. Only after we gained the ability to represent solid objects in computers were we able to develop technology to physically reproduce such objects. Initially, this was the principle surrounding CNC machining technology in general. AM can thus be described as a direct or streamlined Computer-Aided Design to Computer-Aided Manufacturing (CAD/

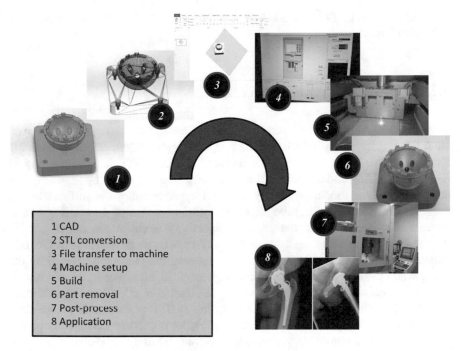

1 CAD
2 STL conversion
3 File transfer to machine
4 Machine setup
5 Build
6 Part removal
7 Post-process
8 Application

Fig. 3.1 The eight stages of the AM process [1]

CAM) process. Unlike most other CAD/CAM technologies, there is little or no intervention between the design and manufacturing stages for AM.

The generic AM process must therefore start with 3D CAD information, as shown in Fig. 3.1. There may be a variety of ways for how the 3D source data can be created. This model description could be generated by a design expert via a user interface, by software as part of an automated optimization algorithm, by 3D scanning of an existing physical part, or by some combination of all of these. Most 3D CAD systems are solid modeling systems with surface modeling components; solid models are often constructed by combining surfaces together or by adding thickness to a surface. In the past, 3D CAD modeling software had difficulty creating fully enclosed solid models, and often models would appear to the casual observer to be enclosed but in fact were not mathematically closed. Such models could result in unpredictable output from AM machines, with different AM technologies treating gaps in different ways.

Most modern solid modeling CAD tools can now create files without gaps (e.g., "water tight"), resulting in geometrically unambiguous representations of a part. Most CAD packages treat surfaces as construction tools that are used to act on solid models, and this has the effect of maintaining the integrity of the solid data. Provided it can fit inside the machine, typically any CAD model can be made using AM technology without too many difficulties. However, there still remain some older or poorly developed 3D CAD software that may result in solids that are not fully enclosed and produce unreliable AM output. Problems of this manner are normally detected once the CAD model has been converted into the STL format for building using AM technology.

3.2.2 Step 2: Conversion to STL/AMF

Nearly every AM technology uses the STL file format. The term STL was derived from STereoLithograhy, which was the first commercial AM technology from 3D Systems in the 1990s. Considered a de facto standard, STL is a simple way of describing a CAD model in terms of its geometry alone. It works by removing any construction data, modeling history, etc. and approximating the surfaces of the model with a series of triangular facets. The minimum size of these triangles can be set within most CAD software, and the objective is to ensure the models created do not show any obvious triangles on the surface. The triangle size is in fact calculated in terms of the minimum distance between the plane represented by the triangle and the surface it is supposed to represent. In other words, a basic rule of thumb is to ensure that the minimum triangle offset is smaller than the resolution of the AM machine. The process of converting to STL is automatic within most CAD systems, but there is a possibility of errors occurring during this phase. There have been a number of software tools developed to detect such errors and to rectify them if possible.

STL files are an unordered collection of triangle vertices and surface normal vectors. As such, an STL file has no units, color, material, or other feature information. These limitations of an STL file have led to the adoption of a new "AMF" file format. This format is an international ASTM/ISO standard format which extends the STL format to include dimensions, color, material, and many other useful features [2]. Many CAD companies and AM hardware vendors support the AMF standard, offering the capability to write or read AMF files as an alternative to STL or other exchange formats. Thus, although the term STL is used throughout the remainder of this textbook, the AMF file could be simply substituted wherever STL appears, as the AMF format has all of the benefits of the STL file format with fewer limitations.

STL file repair software, like the MAGICS software from the Belgian company Materialise [3], is used when there are problems with the STL file that may prevent the part from being built correctly. With complex geometries, it may be difficult for a human to detect such problems when inspecting the CAD or the subsequently generated STL data. If the errors are small, then they may even go unnoticed until after the part has been built. Such software may therefore be applied as a checking stage to ensure that there are no problems with the STL file data before the build is performed.

Since STL is essentially a surface description, the corresponding triangles in the files must be pointing in the correct direction; in other words, the surface normal vector associated with the triangle must indicate which side of the triangle is outside vs. inside the part. The cross-section that corresponds to the part layers of a region near an inverted normal vector may therefore be the inversion of what is desired.

Additionally, complex and highly discontinuous geometry may result in triangle vertices that do not align correctly. This may result in gaps in the surface. Various AM technologies may react to these problems in different ways. Some machines may process the STL data in such a way that the gaps are bridged. This bridge may not represent the desired surface; however, and it may be possible that additional, unwanted material may be included in the part.

While most errors can be detected and rectified automatically, there may also be a requirement for manual intervention. Software should therefore highlight the problem, indicating what is thought to be inverted triangles, for instance. Since geometries can become very complex, it may be difficult for the software to establish whether the result is in fact an error or something that was part of the original design intent.

3.2.3 Step 3: Transfer to AM Machine and STL File Manipulation

Once the STL file has been created and repaired, it can be sent directly to the target AM machine. Ideally, it should be possible to press a "print" button, and the machine should build the part straight away. This is not usually the case however, and there may be a number of actions required prior to building the part.

The first task would be to verify that the part is correct. AM system software normally has a visualization tool that allows the user to view and manipulate the part. The user may wish to reposition the part or even change the orientation to allow it to be built at a specific location within the machine. It is quite common to build more than one part in an AM machine at a time. This may be multiples of the same part (thus requiring a copy function) or completely different STL files. STL files can be linearly scaled quite easily. Some applications may require the AM part to be slightly larger or slightly smaller than the original to account for process shrinkage or coatings; and so scaling may be required prior to building. Applications may also require that the part be identified in some way and some software tools have been developed to add text and simple features to STL formatted data for this purpose. This would be done in the form of adding 3D embossed characters. More unusual cases may even require segmentation of STL files (e.g., for parts that may be too large) or even merging of multiple STL files. For some AM technologies, supports are required to hold the part, while it is being built. This can be done automatically for certain technologies or may require substantial expert decision making for other technologies and applications. For technologies which require supports, part orientation becomes an important decision at this point as well. It should be noted that not all AM machines will have all the functions mentioned here, but numerous STL file manipulation software tools are available for purchase or, in some cases, for free download to perform these functions prior to sending the file to a machine.

3.2.4   Step 4: Machine Setup

All AM machines will have at least some setup parameters that are specific to that machine or process. Some machines are only designed to run a few specific materials and give the user few options to vary layer thickness or other build parameters. These types of machines will have very few setup changes to make from build to build. Other machines are designed to run with a variety of materials and may also have some parameters that require optimization to suit the type of part that is to be built or permit parts to be built quicker but with poorer resolution. Such machines can have numerous setup options available. It is common in the more complex cases to have default settings or save files from previously defined setups to help speed up the machine setup process and to prevent mistakes being made. Normally, an incorrect setup procedure will still result in a part being built. The final quality of that part may, however, be unacceptable.

In addition to setting up machine process parameters, most machines must be physically prepared for a build. The operator must check to make sure sufficient build material is loaded into the machine to complete the build. For machines which use powder, the powder is often sifted and subsequently loaded and leveled in the machine as part of the setup operation. For processes which utilize a build plate, the plate must be inserted and leveled with respect to the machine axes. Some of these machine setup operations are automated as part of the start-up of a build, but for most machines these operations are done manually by a trained operator.

Other aspects such as the direction of printing and selection of proper process parameters have to be considered by operators. Different printing directions can affect the direction of formation of material microstructure which may change the mechanical properties such as breaking elongation and yield stress. Process parameters are important, and unsuitable selection of process parameters may result in formation of pores, unprocessed regions, material deformation, etc. This in turn may lead to dimensional deviation, low surface quality, delamination, cracks, etc. Scan pattern may also change cooling or curing rate and also affect the mechanical properties of the printed parts. Many manufacturers produce industrial machines with the capability of using multiple print-heads, lasers, and other modules which should be considered in order to provide an optimum process strategy for the part geometry and material to be used.

3.2.5   Step 5: Build

Although benefitting from the assistance of computers, the first few stages of the AM process are semiautomated tasks that may require considerable manual control, interaction, and decision-making. Once these steps are completed, the process switches to the computer-controlled building phase. This is where the previously mentioned layer-based manufacturing takes place. All AM machines will have a similar sequence of layering, including a height-adjustable platform or deposition head, material deposition/spreading mechanisms, and layer cross-section formation. Some machines will combine the material deposition and layer formation simultaneously, while others will separate them. As long as no errors are detected during the build, AM machines will repeat the layering process until the build is complete.

3.2.6   Step 6: Removal and Cleanup

Ideally, the output from the AM machine should be ready for use with minimal manual intervention. While sometimes this may be the case, more often than not parts will require a significant amount of post-processing before they are ready for use. In all cases, the part must be either separated from a build platform on which the part was produced or removed from excess build material surrounding the part. Some AM processes use additional material other than that used to make the part itself (secondary support materials). Later chapters describe how various AM processes need these support structures to help keep the part from collapsing or warping during the build process. At this stage, it is not necessary to understand exactly how support structures work, but it is necessary to know that they need to be dealt with. While some processes have been developed to produce easy-to-remove supports, there is often a significant amount of manual work required at this stage. For metal supports, a wire EDM machine, band saw, and/or milling equipment may be required to remove the part from the base plate and the supports from the part. There is a degree of operator skill required in part removal, since mishandling of parts and poor technique can result in damage to the part. Different AM parts have different cleanup requirements, but suffice it to say that all processes have some requirement at this stage. The cleanup stage may also be considered as the initial part of the post-processing stage.

3.2.7   Step 7: Post-processing

Post-processing refers to the (usually manual) stages of finishing the parts for application purposes. This may involve abrasive finishing, like polishing and sandpapering, or application of coatings. This stage in the process is very application-specific. Some applications may only require a minimum of post-processing. Other applications may require very careful handling of the parts to maintain good precision and finish. Some post-processing may involve chemical or thermal treatment of the part to achieve final part properties. Different AM processes have different results in terms of accuracy, and thus machining to final dimensions may be required. Some processes produce relatively fragile components that may require the use of infiltration and/or surface coatings to strengthen the final part. As already stated, this is often a manually intensive task due to the complexity of most AM parts. However, some of the tasks can benefit from the use of power tools, CNC milling, and additional equipment, like polishing tubs or drying and baking ovens.

3.2.8 Step 8: Application

Following post-processing, parts are ready for use. It should be noted that, although parts may be made from similar materials to those available from other manufacturing processes (like molding and casting), parts may not behave according to standard material specifications. Some AM processes inherently create parts with small voids trapped inside them, which could be the source for part failure under mechanical stress. In addition, some processes may cause the material to degrade during build or for materials not to bond, link, or crystallize in an optimum way. In almost every case, the properties are anisotropic (different properties in different directions). For most metal AM processes, rapid cooling results in different microstructures than those from conventional manufacturing. As a result, AM-produced parts behave differently than parts made using a more conventional manufacturing approach. This behavior may be better or worse for a particular application, and thus a designer should be aware of these differences and take them into account during the design stage. AM materials and processes are improving rapidly, and thus designers must be aware of recent advancements in materials and processes to best determine how to use AM for their needs.

3.3 Variations from One AM Machine to Another

The above generic process steps can be applied to every commercial AM technology. As has been noted, different technologies may require more or less attention for a number of these stages. Here we discuss the implications of these variations, not only from process to process but also in some cases within a specific technology.

The nominal layer thickness for most machines is around 0.1 mm. However, it should be noted that this is just a rule of thumb. For example, the layer thickness for some Material Extrusion (MEX) machines is 0.254 mm, whereas layer thicknesses between 0.01 and 0.1 mm are commonly used for Vat Photopolymerization (VPP) processes, and small intricate parts made for investment casting and precision engineering applications may have layer thicknesses of less than 0.01 mm. Many technologies have the capacity to vary the layer thickness. The reasoning is that thicker layer parts are quicker to build but are less precise. This may not be a problem for some applications where it may be more important to make the parts as quickly as possible.

Fine detail in a design may cause problems with some AM technologies, such as wall thickness, particularly if there is no choice but to build the part vertically. This is because even though positioning within the machine may be very precise, there is a finite dimension to the droplet size, laser diameter, or extrusion head that essentially defines the finest detail or thinnest wall that can be fabricated.

There are other factors that may not only affect the choice of process but also influence some of the steps in the process chain. In particular, the use of different materials even within the same process may affect the time, resources, and skill required to carry out a stage. For example, the use of water-soluble supports in MEX processes may require specialist equipment but will also provide better finish to parts with less hand finishing required than when using conventional supports. Alternatively, some polymers require special attention, like the use (or avoidance) of particular solvents or infiltration compounds. A number of processes benefit from application of sealants or even infiltration of liquid polymers. These materials must be compatible with the part material both chemically and mechanically. Post-processing that involves heat must include awareness of the heat resistance or melting temperature of the materials involved. Abrasive or machining-based processing must also require knowledge of the mechanical properties of the materials involved. If considerable finishing is required, it may also be necessary to include an allowance in the part geometry, perhaps by using scaling of the STL file or offsetting of the part's surfaces, so that the part does not become worn away too much.

Variations between AM technologies will become clarified further in the following chapters, but a general understanding can be achieved by considering whether the build material is processed as a powder, molten material, solid sheet, vat of liquid photopolymer, or ink-jet deposited photopolymer.

3.3.1 Photopolymer-Based Systems

It is quite easy to set up systems which utilize photopolymers as the build material. Photopolymer-based systems, however, require files to be created which represent the support structures. All liquid vat systems must use supports from essentially the same material as that used for the part. For Material Jetting (MJT) systems, it is possible to use a secondary support material from parallel ink-jet print heads so that the supports will come off easier. An advantage of photopolymer systems is that accuracy is generally very good, with thin layers and fine precision where required compared with other systems. Photopolymers have historically had poor material properties when compared with other AM materials; however newer resins have been developed that offer improved temperature resistance, strength, and ductility. The main drawback of photopolymer materials is that degradation can occur quite rapidly if UV protective coatings are not applied.

3.3.2 Powder-Based Systems

There is no need to use supports for powder systems which deposit a bed of powder layer-by-layer (with the exception of supports for metal systems, as addressed below). Thus, powder bed-based systems are among the easiest to set up for a simple build. Parts made using Binder Jetting (BJT) into a powder bed can be colored by using colored binder material. If color is used then coding the file may take a longer time, as standard STL data does not include color. There are, however, other file formats based around VRML that allow colored geometries to be built, in addition to AMF. Powder Bed Fusion (PBF) processes have a significant amount of unused powder in every build that has been subjected to some level of thermal history. This thermal history may cause changes in the powder. Thus, a well-designed recycling strategy based upon one of several proven methods can help ensure that the material being used is within appropriate limits to guarantee good builds [4].

It is also important to understand the way powders behave inside a machine. For example, some machines use powder feed chambers at either side of the build platform. The powder at the top of these chambers is likely to be less dense than the powder at the bottom, which will have been compressed under the weight of the powder on top. This in turn may affect the amount of material deposited at each layer and density of the final part built in the machine. For very tall builds, this may be a particular problem that can be solved by carefully compacting the powder in the feed chambers before starting the machine and also by adjusting temperatures and powder feed settings during the build.

3.3.3 Molten Material Systems

Systems which melt and deposit material in a molten state require support structures. For droplet-based systems like with the Thermojet process, these supports are automatically generated; but with MEX processes or Directed Energy Deposition (DED) systems, supports can either be generated automatically, or the user can use some flexibility to change how supports are made. With water-soluble supports, it is not too important where the supports go, but with breakaway support systems made from the same material as the build material, it is worthwhile to check where the supports go, as surface damage to the part will occur to some extent where these supports were attached before breaking them away. Also, fill patterns for MEX may require some attention, based upon the design intent. Parts can be easily made using default settings, but there may be some benefit in changing aspects of the build sequence if a part or region of a part requires specific characteristics. For example, there are typically small voids in FDM parts that can be minimized by increasing the amount of material extruded in a particular region. This will minimize voids, but at the expenses of part accuracy. Although wax parts made using MJT are good for reproducing fine features, they are difficult to handle because of their low strength and brittleness. ABS parts made using MEX, on the other hand, are among the strongest AM polymer parts available, but when they are desired as a functional end-use part, this may mean they need substantial finishing compared with other processes as they exhibit lower accuracy than some other AM technologies.

3.3.4   Solid Sheets

With Sheet Lamination (SHL) methods where the sheets are first placed and then cut, there is no need for supports. Instead, there is a need to process the waste material in such a way that it can be removed from the part. This is generally a straight-forward automated process, but there may be a need for close attention to fine detail within a part. Cleaning up the parts can be the most laborious process, and there is a general need to know exactly what the final part is supposed to look like so that damage is not caused to the part during the waste removal stage. Paper-based systems experience problems with handling if they are not carefully and comprehensively finished using sealants and coatings. For polymer SHL, the parts are typically not as sensitive to damage. For metal SHL processes, typically the sheets are cut first and then stacked to form the 3D shape, and thus support removal becomes unnecessary.

3.4   Metal Systems

As previously mentioned, operation of metal-based AM systems is conceptually similar to polymer systems. However, the following points are worth considering.

3.4.1   The Use of Substrates

Most metal systems make use of a base platform or substrate onto which parts are built and from which they must be removed using machining, wire cutting, or a similar method. The need to attach the parts to a base platform is mainly because of the high-temperature gradients between the temporarily molten material and its sur-roundings, resulting in large residual stress upon cooling. If the material is not rigidly attached to a solid platform, then the part warps as it cools, which means further layers of powder or deposited material could not be spread evenly over top. Therefore, even though these processes may build within a powder bed, there is still a need for supports.

Heat transfer from the substrate affects solidification from the molten state and therefore the microstructure and mechanical properties of the built part. As shown in Fig. 3.2, the cross-section of this part is smaller where it is attached to the substrate compared to the points which are far from the substrate. In the top region, the heat conduction to the substrate is slower, and thus the cooling rate decreases. This results in grain and crystal lattice size differences in different regions of the part [5, 6].

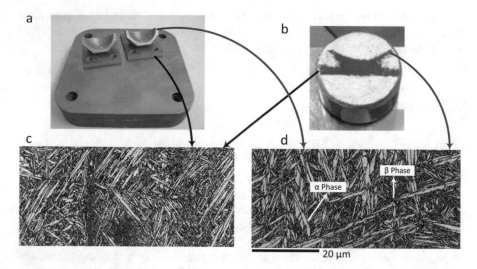

Fig. 3.2 (**a**) LPBF- SLM Ti-6Al-4V samples. (**b**) Cross-section. (**c**) Microstructure for close to substrate. (**d**) Microstructure far from substrate (Elsevier license number 4523990304932) [7]

Mechanical properties are highly dependent on microstructure, where finer microstructure results in higher tensile strengths, but lower ductility [8, 9]. In addition, due to high cooling rates, some intermetallic phases may appear in AM-produced parts that are not in conventionally produced components, which may significantly affect mechanical properties like elongation before failure [10].

3.4.2   Energy Density

Energy density can be defined as the amount of energy per unit of distance, area, or volume that is input into a part as it is being built. Energy density is key factor which determines localized temperature and the size of the melt pool being formed [11]. The energy density required to melt metals is much higher than for melting polymers. The high temperatures achieved during metal melting may require more stringent heat shielding, insulation, temperature control, and atmospheric control than for polymer systems.

3.4.3   Weight

Metal powder systems may process lightweight titanium powders, but they also process high-density tool steels and even gold powders. The powder handling technology must be capable of withstanding the mass of these materials. This means that power requirements for positioning and handling equipment must be quite substantial or gear ratios must be high (and corresponding travel speeds lower) to deal with these tasks. Some companies like Trumpf, 3D Systems, Renishaw, and Additive Industries have included automatic powder handling system for individual or multiple machines. These systems aim to enable a homogeneous build process, reduction of operator time, and increased system security and reliability. Furthermore, some of these systems include automatic sifting and recycling of powder resulting in further reduction of manual handling of materials [12–14].

3.4.4   Accuracy

Accuracy of produced parts must consider both surface quality and dimensional accuracy. Metal powder systems are generally at least as accurate as polymer powder systems. Surface finish is characteristically grainy, but part density and part accuracy are very good. Surface roughness is on the order of a few tens to a few hundreds of microns depending on the process and can be likened in general appearance to precision casting technology. Some companies like 3D MicroPrint [15] provide a superior surface finish of <15 µm. For metal parts, this may still not be satisfactory, and at least some shot-peening or other surface finishing may be required to smooth the surface. Key mating features on metal parts often require surface machining or grinding. The part density will be high (generally over 99%), although some voids may still be seen. To improve the surface quality process, parameters should be selected if possible to avoid defects such as pores, voids, cracks, particle spatter from the melt pool, etc. Both surface and dimensional accuracy can be modified after the build process by using material removal systems such as hand polishing, grinding, milling, or turning. If the structure is complex, containing internal lattices, for example, then the best method may be to use abrasive particle slurry methods [16, 17].

3.4.5   Speed

Since there are heavy demands on the amount of energy to melt the powder particles and to handle those powders within the machine, the build speed of metal systems is generally slower than a comparable polymer system. Laser powers are usually just a few hundred watts (polymer systems start at around 50 W of laser power). This means that the laser scanning speed is lower than for polymer systems, to ensure enough energy is delivered to the powder. The maximum power of heavy industrial machines has increased to the range of 1–1.5 kW for pulsed lasers and over 10 kW for CW lasers with additional features like multi-laser processing and up to 1 m^3 chamber size [18]. Also, some new machines are equipped with high scanning speeds up to a few thousand mm/s [19, 20].

3.4.6 Build Rate

Build rate is the rate at which material is added together in an AM process, measured in volume per unit time. PBF systems have a typical build rate of 150 cm³/ hour. Nonpowder-bed metal DED systems generally have higher build rates, in the order of 1000 cm³/h. For VPP, MJT, and BJT processes, maximum deposition rates of 100–150 L/h are obtained.

3.5 Maintenance of Equipment

Many machines require careful maintenance to ensure repeatability and reliability over time. Some machines use sensitive laser or printer technology that must be carefully monitored and that should preferably not be used in a dirty or noisy (both electrical noise and mechanical vibration) environment. Similarly, many of the feed materials require careful handling and should be used in low humidity conditions. While machines are designed to operate unattended, it is important to include regular checks in the maintenance schedule. Many machine vendors recommend and provide test patterns that should be used periodically to confirm that the machines are operating within acceptable limits.

Laser-based systems are generally expensive because of the cost of the laser and scanner system. Furthermore, maintenance of a laser can be very expensive, particularly for lasers with limited lifetimes. Print heads are also components that have finite lifetimes for MJT and BJT systems. The fine nozzle dimensions and the use of relatively high viscosity fluids mean they are prone to clogging and contamination effects. Replacement costs are, however, generally quite low.

3.6 Materials Handling Issues

In addition to the machinery, AM materials often require careful handling. The raw materials used in some AM processes have limited shelf life and must also be kept in conditions that prevent them from chemical reaction or degradation. Exposure to moisture and to excess light should be avoided. Most processes use materials that can be used for more than one build. However, it may be that this could degrade the material if used many times over and therefore a procedure for maintaining consistent material quality through recycling should also be observed.

While there are some health concerns with extended exposure to some photopolymer resins, most AM polymer raw materials are safe to handle. Powder materials may in general be medically inert, but excess amounts of powder can make the workplace slippery, contaminate mechanisms, and create a breathing hazard. In addition, reactive powders can be a fire hazard. These issues may cause problems if machines are to be used in a design center environment rather than in a workshop. AM system vendors have spent considerable effort to simplify and facilitate material handling. Loading new materials is often a procedure that can be done offline or with minimal changeover time so that machines can run continuously. Software systems are often tuned to the materials so that they can recognize different materials and adjust build parameters accordingly.

Many materials are carefully tuned to work with a specific AM technology. There are often warranty issues surrounding the use of third-party materials that users should be aware of. For example, some polymer laser sintering powders may have additives that prevent degradation due to oxidation since they are kept at elevated temperatures for long periods of time. Also, MEX filaments need a very tight diametric tolerance not normally available from conventional extruders. Since a MEX drive pushes the filament through the machine, variations in diameter may cause slippage. Furthermore, build parameters are designed around the standard materials used. Since there are huge numbers of material formulations, changing one material for another, even though they appear to be the same, may require careful build setup and process parameter optimization.

Some machines allow the user to recycle some or all of the material used in a machine but not consumed during the build of a prior part. This is particularly true with the powder-based systems. Also photopolymer resins can be reused. However, there may be artifacts and other contaminants in the recycled materials, and it is important to carefully inspect, sift, or sieve the material before returning it to the machine. Many laser sintering builds have been spoiled, for example, by hairs that have come off a paintbrush used to clean the parts from a previous build.

Using a machine that is capable of processing more than one material can result in cross-contamination from one material to the next. In metal systems precipitates may occur which can combine poorly with another material, producing inferior mechanical properties. Figure 3.3 shows the contamination of Fe within the microstructure of Ti-6Al-4V. This is related to using a single machine for different materials, and the high percentage of iron (2%) is related to using stainless steel 316L

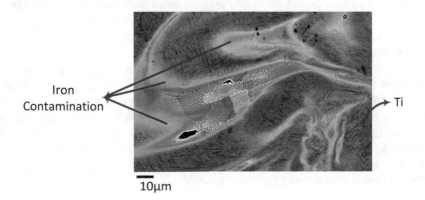

Fig. 3.3 Contamination of Fe on LB-PBF Ti-6Al-4V

before Ti-6Al-4V. The deposited iron doesn't disperse well and produces inferior mechanical properties.

3.7 Design for AM

Designers and operators should consider a number of build-related factors when considering the setup of an AM machine, including the following sections. This is a brief introduction, but more information can be found in Chap. 19.

3.7.1 Part Orientation

If a cylinder was built on its end, then it would consist of a series of circular layers built on top of each other. Although layer edges may not be precisely vertical for all AM processes, the result would normally be a very well-defined cylinder with a relatively smooth edge. The same cylinder built on its side will have distinct layer stair-step patterning on the sides. This will result in less accurate reproduction of the original CAD data with a poorer aesthetic appearance. Additionally, as the layering process for most AM machines takes additional time, a long cylinder built vertically will take more time to build than if it is laid horizontally. For MEX processes, however, the time to build a part is solely a factor of the total build volume (including supports), and thus a cylinder should always be built vertically if possible.

Orientation of the part within the machine can affect part accuracy. Since many parts will have complex features along multiple axes, there may not be an ideal orientation for a particular part. Furthermore, it may be more important to maintain the geometry of some features when compared with others, so correct orientation may be a judgment call. This judgment may also be in contrast with other factors like the time it takes to build a part (e.g., taller builds take longer than shorter ones so high aspect ratio parts may be better built lying down), whether a certain orientation will generate more supports, or whether certain surfaces should be built faceup to ensure good surface finish in areas that are not in contact with support structures.

In general upward-facing features in AM have the best quality. The reason for this depends upon the process. For instance, upward-facing features are not in contact with the supports required for many processes. For powder beds, the upward-facing features are smooth since they solidify against air, whereas downward-facing and sideways-facing features solidify against powder and thus have a powdery texture. For MEX, upward-facing surfaces are smoothed by the extrusion tip. Thus, this upward-facing feature quality rule is one of the few rules of thumb that are generically applicable to every AM process.

3.7.2   Removal of Supports

For those technologies that require supports, it is a good idea to try and minimize the amount. Wherever the supports meet the part, there will be small marks, and reducing the amount of supports would reduce the amount of part cleanup and post-process finishing. However, as mentioned above, some surfaces may not be as important as others, and so positioning of the part must be weighed against the relative importance of an affected surface. In addition, too few supports may mean that the part becomes detached from the base plate and will move around during subsequent layering. If distortion causes a part to extend in the z direction enough that it hits the layering mechanism (such as a powder-spreading blade), then the build will fail.

Parts that require supports may also require planning for their removal. Supports may be located in difficult to reach regions within the part. For example, a hollow cylinder with end caps built vertically will require supports for the top surface. However, if there is no access hole, then these supports cannot be removed. Inclusion of access holes (which could be plugged later) is a possible solution to this, as may be breaking up the part so the supports can be removed before reassembly. Similarly, parts made using VPP processes may require drain holes for any trapped liquid resin.

3.7.3   Hollowing Out Parts

Parts that have thick walls may be designed to include hollow features if this does not reduce the part's functionality. The main benefits of doing this are the reduced build time, the reduced cost from the use of less material, and the reduced mass in the final component. As mentioned previously, some liquid-based resin systems would require drain holes to remove excess resin from inside the part, and the same is true for powder. A honeycomb- or truss-like internal structure can assist in providing support and strength within a part while reducing its overall mass and volume. All of these approaches must be balanced against the additional time that it would take to design such a part. However, there are software systems that would allow this to be done automatically for most types of parts.

3.7.4   Inclusion of Undercuts and Other Manufacturing Constraining Features

AM models can be used at various stages of the product development process. When evaluating initial designs, focus may be on the aesthetics or ultimate functionality of the part. Consideration of how to include manufacturing-related features would have lower priority at this stage. Conventional manufacturing would require considerable planning to ensure that a part is fabricated correctly. Undercuts, draft angles, holes, pockets, etc. must be created in a specific order when using multiple-stage conventional processes. While this can be ignored when designing the part for AM, it is important not to forget them if AM is being used just as a prototype process. AM can be used in the design process to help determine where and what type of rib, boss, and other strengthening approaches should be used on the final part. If the final part is to be injection molded, the AM part can be used to determine the best location for the parting lines in the mold.

3.7.5   Interlocking Features

AM machines have a finite build volume, and large parts may not be capable of being built inside them. A solution may be to break the design up into segments that can fit into the machine and manually assemble them together later. The designer must therefore consider the best way to break up the parts. The regions where the breaks are made can be designed in such a way as to facilitate reassembly. Techniques can include incorporation of interlocking features and maximizing surface area so that adhesives can be most effective. Such regions should also be in easy to reach but difficult to observe locations.

This approach of breaking parts up may be helpful even when they can still fit inside the machine. Consider the design shown in Fig. 3.4. If it was built as a single part, it would take a long time and may require a significant amount of supports (as shown in the left-hand figure). If the part were built as two separate pieces, the resulting height would be significantly reduced, and there would be few supports. The part could be glued together later. This glued region may be slightly weakened, but the individual segments may be stronger. Since the example has a thin wall section, the top of the upright band shown in the left side of the figure will exhibit stair-stepping and may also be weaker than the rest of the part, whereas the part built

Fig. 3.4 The build on the left (shown with support materials within the arch) can be broken into the two parts on the right, which may be stronger and can be glued together later. Note the reduction in the amount of supports and the reduced build height

lying down would typically be stronger. For the bonded region, it is possible to include large overlapping regions that will enable more effective bonding.

3.7.6   Reduction of Part Count in an Assembly

There are numerous sections in this book that discuss the use of AM for direct manufacture of parts for end-use applications. The AM process is therefore toward the end of the product development process, and the design does not need to consider alternative manufacturing processes. This in turn means that if part assembly can be simplified using AM, then this should be done. For example, it is possible to build fully assembled hinge structures by providing clearance around the moving features. In addition, complex assemblies made up of multiple injection molded parts, for instance, could be built as a single component. Thus, when producing components with AM, designers should always look for ways to consolidate multiple parts into a single part and to include additional part complexity where it can improve system performance. Several of the parts in Fig. 1.4 provide good examples of these concepts.

3.7.7   Identification Markings/Numbers

Although AM parts are often unique, it may be difficult for a company to keep track of them when they are possibly building hundreds of parts per week. It is a straightforward process to include identifying features on the parts. This can be done when designing the CAD model, but that may not be possible since the models may come from a third party. There are a number of software systems that provide tools for labeling parts by embossing alphanumeric characters onto them as 3D models. In addition, some service providers build all the parts ordered by a particular customer (or small parts which might otherwise get lost) within a mesh box so that they are easy to find and identify during part cleanup.

3.8   Application Areas for AM-Enabled Product Development

Additive Manufacturing technology opens up opportunities for many applications that do not take the standard product development route. The capability of integrating AM with customizing data or data from unusual sources makes for rapid response and an economical solution. The following sections are examples where different AM based design approaches are applicable.

3.8.1 Medical Modeling

AM is increasingly used to make parts based on an individual person's medical data. Such data are based on 3D scanning obtained from systems like computerized tomography (CT), magnetic resonance imaging (MRI), 3D ultrasound, etc. These datasets often need considerable processing to extract the relevant sections before it can be built as a model or further incorporated into a product design. There are a few software systems that can process medical data in a suitable way, and a range of applications have emerged. For example, Materialise [1] developed software used in the production of hearing aids. AM technology helps in customizing these hearing aids from data that are collected from the ear canals of individual patients.

3.8.2 Reverse Engineering Data

Medical data from patients is just one application that benefits from being able to collect and process complex surface information. For nonmedical data collection, the more common approach is to use laser scanning technology. Such technology has the ability to faithfully collect surface data from many types of surfaces that are difficult to model because they cannot be easily defined geometrically. Similar to medical data, although the models can just be reproduced within the AM machine (like a kind of 3D copy machine), the typical intent is to merge this data into product design. Interestingly, laser scanners for reverse engineering and inspection run the gamut from very expensive, very high-quality systems (e.g., from Leica and Steinbichler) to mid-range systems (from Faro and Creaform) to multi-image systems based around mobile phone cameras.

3.8.3 Architectural Modeling

Architectural models are usually created to emphasize certain features within a building design, and so designs are modified to show textures, colors, and shapes that may not be exact reproductions of the final design. Therefore, architectural packages may require features that are tuned to the AM technology.

3.8.4 Automotive

Using traditional production processes, complex geometries can be expensive to do, or impossible to produce with a specific technology. Using AM technologies, hollow structures, which are less expensive and lighter than solid ones, can be easily made. In the automotive industry, a major goal is to make the lightest practical car while maintaining safety. AM technologies provide a chance for manufacturers to create complex parts such as honeycomb or other shapes with cavities and cutouts that can improve the strength to weight ratio.

3.8.5   Aerospace

AM components for the commercial aerospace market have unique opportunities with a positive long-term outlook. Rising global gross domestic products, record air traffic, and resultant aircraft orders have created an environment with 8–10-year forward production for original equipment manufacturers and supply chains. This presents challenges: the industry is executing an unprecedented ramp-up of production over the next 5 years when it is already operating at capacity. Additionally, there is a transition from older aircraft to new variants with more fuel-efficient engines. AM now allows aerospace components to have complex geometry and the potential of the component to be hollow, thus providing significant weight reductions. Almost 60% of the components (at least 160,000 in a commercial airliner) in an aircraft may be suitable for AM production. However, an interesting dilemma facing AM in production is the surface quality of the final part, particularly when the component has internal structures [21].

3.9   Further Discussion

AM technologies are beginning to move beyond a common set of basic process steps. In the future we will likely see more processes using variations of the conventional AM approach and combinations of AM with conventional manufacturing operations. Some technologies are being developed to process regions rather than layers of a part. As a result, more intelligent and complex software systems will be required to effectively deal with segmentation.

We can expect processes to become more complex within a single machine. We already see numerous additive processes combined with subtractive elements. As technology develops further, we will see increased utilization of hybrid technologies that include additive, subtractive, and robotic handling phases in a complex coordinated and controlled fashion. This will require much more attention to software descriptions, but may also lead to highly optimized parts with multiple functionality and vastly improved quality with very little manual intervention during the build and post-process steps.

Another trend we are likely to see is the development of customized AM systems. Presently, AM machines are designed to produce as wide a variety of possible part geometries with as wide a range of materials as possible. Reduction of these variables may result in machines that are designed only to build a subset of parts or

Fig. 3.5 FigurePrints model, post-processed for output to an AM machine [23]

materials very efficiently or inexpensively. This has already started with the proliferation of "personal" versus "industrial" MEX systems. In addition, many machines
are being targeted for the dental or hearing aid markets, and system manufacturers
have redesigned their basic machine architectures and/or software tools to enable
rapid setup, building, and post-processing of patient-specific small parts.

Software is increasingly being optimized specifically for AM processing. Special
software has been designed to increase the efficiency of hearing aid design and manufacture. There is also special software designed to convert the designs of World of
Warcraft models into "FigurePrints" (see Fig. 3.5) as well as specially designed post-
processing techniques [22]. As Direct Digital Manufacturing becomes more common,
we will see the need to develop standardized software processes based around AM, so
that we can better control, track, regulate, and predict the manufacturing process.

3.10   Questions

1. Investigate some of the websites associated with different AM technologies.
   Find out information on how to handle the processes and resulting parts according to the eight stages mentioned in this chapter. What are four different tasks
   that you would need to carry out using a VPP process that you wouldn't have to
   do using a BJT technology and vice versa?
2. Explain why surface modeling software is not ideal for describing models that
   are to be made using AM, even though the STL file format is itself a surface
   approximation. What kind of problems may occur when using surface modeling
   only?

3. What is the VRML file format like? How is it more suitable for specifying color models to be built using color AM machines than the STL standard? How does it compare with the AMF format?
4. What extra considerations might you need to give when producing medical models using AM instead of conventionally engineered products?
5. Consider the FigurePrints part shown in Fig. 3.5, which is made using a color BJT process. What finishing methods would you use for this application?
6. A vehicle manufacturer desires to reverse engineering a complex part with internal structures. What scanning technology method should be chosen and why?
7. What are some of the effects of part building orientation in AM processes? Name some situations where the hollowing out of parts may be required and/or beneficial in AM processes.
8. Name and briefly describe the four types of systems discussed in Sect. 3.3. Name one benefit and one drawback of each system type.
9. Summarize the general steps for an Additive Manufacturing process, and give a good example of a part that will be better fabricated by AM than traditional manufacturing. Describe which specific characteristics of that part are better produced due to the general AM manufacturing steps than traditional steps.

References

1. Vincent, J., et al. (2018). Preservation of the acetabular cup during revision Total hip arthroplasty using a novel mini-navigation tool: A case report. *Journal of Orthopaedic Case Reports, 8*(1), 53.
2. ASTM International. (2015). *ISO/ASTM52915-16 standard specification for additive manufacturing file format (AMF) version 1.2*. West Conshohocken: ASTM International.
3. Materialise. (2020). *AM software systems and service provider*. https://www.materialise.com/en
4. Choren, J., et al. (2001). SLS powder life study. In *Twelfth annual international solid freeform fabrication symposium*, University of Texas at Austin.
5. Dilip, J. J. S., et al. (2017). Selective laser melting of HY100 steel: Process parameters, microstructure and mechanical properties. *Additive Manufacturing, 13*, 49–60.
6. Khorasani, A. M., Gibson, I., & Ghaderi, A. R. (2018). Rheological characterization of process parameters influence on surface quality of Ti-6Al-4V parts manufactured by selective laser melting. *The International Journal of Advanced Manufacturing Technology, 97*, 3761–3775.
7. Khorasani, A. M., et al. (2016). A survey on mechanisms and critical parameters on solidification of selective laser melting during fabrication of Ti-6Al-4V prosthetic acetabular cup. *Materials and Design, 103*, 348–355.
8. Akram, J., et al. (2018). Understanding grain evolution in additive manufacturing through modeling. *Additive Manufacturing, 21*, 255–268.
9. Khorasani, A. M., et al. (2018). Mass transfer and flow in additive manufacturing of a spherical component. *The International Journal of Advanced Manufacturing Technology, 96*, 3711–3718.
10. Khorasani, A. M., et al. (2017). On the role of different annealing heat treatments on mechanical properties and microstructure of selective laser melted and conventional wrought Ti-6Al-4V. *Rapid Prototyping Journal, 23*(2), 295–304.

11. Teng, C., et al. (2017). Simulating melt pool shape and lack of fusion porosity for selective laser melting of cobalt chromium components. *Journal of Manufacturing Science and Engineering, 139*(1), 011009.
12. Renishaw. (2020). *Additive manufacturing products.* https://www.renishaw.com/en/additive-manufacturing-products%2D%2D17475
13. 3D Systems. (2020). *3D printers.* https://www.3dsystems.com/3d-printers#metal-3d-printers
14. Trumpf. (2020). https://www.trumpf.com/en_SE/products/machines-systems/additive-production-systems/
15. 3D Micro Print. (2020). https://www.3dmicroprint.com/products/machines/.
16. Khorasani, A. M., et al. (2018). A comprehensive study on surface quality in 5-axis milling of SLM Ti-6Al-4V spherical components. *The International Journal of Advanced Manufacturing Technology, 94*(9–12), 3765–3784.
17. Khorasani, A. M., et al. (2018). Characterizing the effect of cutting condition, tool path, and heat treatment on cutting forces of selective laser melting spherical component in five-Axis milling. *Journal of Manufacturing Science and Engineering, 140*(5), 051011.
18. GE Additive. (2020). *Additive manufacturing machines.* https://www.ge.com/additive/additive-manufacturing/machines.
19. SLM Solutions. (2020). https://slm-solutions.com/products/machines
20. Coherent, OR Laser. (2020). http://creator.or-laser.com/en/
21. Young, E. (2016). *Top 10 risks in aerospace and defense (A&D).* London, UK: Ernst & Young.
22. Figureprints. (2020). *3DP models from World of Warcraft figures.* http://www.figureprints.com/wow/
23. World of Warcraft. (2020). https://worldofwarcraft.com/en-us/

Chapter 4
Vat Photopolymerization

Abstract Photopolymerization processes make use of liquid polymers that react to radiation to become solid. This reaction is called photopolymerization, and liquids which photopolymerize are known as photopolymers. Photopolymers are widely applied in coating and printing industries, as well as for other purposes. Stereolithography was the first type of Vat Photopolymerization (VPP) process and was developed by Charles Hull as an extension to work he had done previously with photopolymers. In VPP, a container (vat) containing photopolymers is exposed to repeated patterns of radiation corresponding to cross-sections of the part being built. In the case of stereolithography, the radiation was provided by a UV laser. Subsequent variants of VPP processes have involved different radiation sources and several types of layering mechanisms. VPP processes were the first commercialized AM technologies, and they continue to be broadly used across many industries and applications.

4.1 Introduction

Photopolymerization processes make use of liquid, radiation curable resins, or photopolymers, as their primary materials. Most photopolymers react to radiation in the ultraviolet (UV) range of wavelengths, but some visible light systems are used as well. Upon irradiation, these materials undergo a chemical reaction to become solid. This reaction is called photopolymerization and is typically complex, involving many chemical participants.

Photopolymers were developed in the late 1960s and soon became widely applied in several commercial areas, most notably for coating and printing. Many of the glossy coatings on paper and cardboard, for example, are photopolymers. Additionally, photocurable resins are used in dentistry, such as for sealing the top surfaces of teeth to fill in deep grooves and prevent cavities. In these applications, coatings are cured by

© Springer Nature Switzerland AG 2021
I. Gibson et al., *Additive Manufacturing Technologies*,
https://doi.org/10.1007/978-3-030-56127-7_4

radiation that blankets the resin without the need for patterning either the material or the radiation. This changed with the introduction of stereolithography.

In the mid-1980s, Charles (Chuck) Hull was experimenting with UV curable materials by exposing them to a scanning laser, similar to the system found in laser printers. He discovered that solid polymer patterns could be produced. By curing one layer over a previous layer, he could fabricate a solid 3D part. This was the beginning of stereolithography (SL) technology. The company 3D Systems was created shortly thereafter to market SL machines as "rapid prototyping" machines to the product development industry. Since then, a wide variety of SL-related processes and technologies has been developed. The term "Vat Photopolymerization" is a general term that encompasses SL and these related processes. SL will be used to refer specifically to macro-scale, laser scan VPP; otherwise, the term Vat Photopolymerization will be used and will be abbreviated as VPP.

Various types of radiation may be used to cure commercial photopolymers, including gamma rays, X-rays, electron beams, UV, and in some cases visible light. In VPP systems, UV and visible light radiation are used most commonly. In the microelectronics industry, photomask materials are often photopolymers and are typically irradiated using far UV and electron beams. In contrast, the field of dentistry uses visible light predominantly.

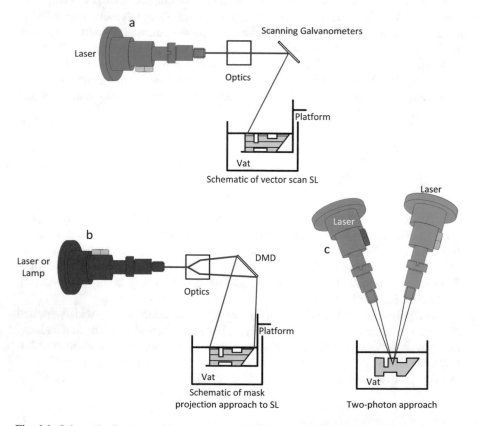

Fig. 4.1 Schematic diagrams of three approaches to photopolymerization processes

Two primary configurations were developed for photopolymerization processes in a vat, plus one additional configuration that has seen some research interest. Although photopolymers are also used in some inkjet printing processes, this method of line-wise processing is not covered in this chapter, as the basic processing steps are more similar to the printing processes covered in Chap. 7. The configurations discussed in this chapter include:

- Vector scan, or point-wise, approaches typical of commercial SL machines
- Mask projection, or layer-wise, approaches that irradiate entire layers at one time
- Two-photon approaches that are essentially high-resolution point-by-point approaches

These three configurations are shown schematically in Fig. 4.1. Note that in the vector scan and two-photon approaches, scanning laser beams are needed, while the mask projection approach utilizes a large radiation beam that is patterned by another device, in this case a Digital Micromirror Device™ (DMD). In the two-photon case, photopolymerization occurs at the intersection of two scanning laser beams, although other configurations use a single laser and different photoinitiator chemistries. Another distinction is the need to recoat or apply a new layer of resin in the vector scan and mask projection approaches, while in the two-photon approach, the part is fabricated below the resin surface, making recoating unnecessary. Approaches that avoid recoating could theoretically be faster and less complicated from a machine design perspective, but the chemical reactions involved with these approaches typically are much slower, and thus these potential speed improvements are yet to be realized.

In this chapter, we first introduce photopolymer materials and then present the vector scan SL machines, technologies, and processes. Mask projection approaches are presented and contrasted with the vector scan approach. Additional configurations, along with their applications, are presented at the end of the chapter. Advantages, disadvantages, and uniquenesses of each approach and technology are highlighted.

4.2 Vat Photopolymerization Materials

Some background of UV photopolymers will be presented in this section that is common to all configurations of photopolymerization processes. Two subsections on reaction rates and characterization methods conclude this section. Much of this material is from the Jacobs book [1] and from a Masters thesis from the early 2000s [2].

4.2.1 UV Curable Photopolymers

As mentioned, photopolymers were developed in the late 1960s. In addition to the applications mentioned in Sect. 4.1, they are used as photoresists in the microelectronics industry. This application has had a major impact on the development of epoxy-based photopolymers. Photoresists are essentially one-layer VPP but with critical requirements on accuracy and feature resolution.

Various types of radiation may be used to cure commercial photopolymers, including gamma rays, X-rays, electron beams, UV, and in some cases visible light, although UV and electron beam are the most prevalent. In AM, many of these radiation sources have been utilized in research; however only UV and visible light systems are utilized in commercial systems. In SL systems, for example, UV radiation is used exclusively although, in principle, other types could be used. In the SLA-250 from 3D Systems, a helium-cadmium (HeCd) laser is used with a wavelength of 325 nm. In contrast, the solid-state lasers used in the other SL models are Nd-YVO$_4$. In mask projection DMD-based systems, UV and visible light radiation are used.

Thermoplastic polymers that are typically injection molded have a linear or branched molecular structure that allows them to melt and solidify repeatedly. In contrast, VPP photopolymers are cross-linked and, as a result, do not melt and exhibit much less creep and stress relaxation. Figure 4.2 shows the three polymer structures mentioned [3].

The first US patents describing SL resins were published in 1989 and 1990 [4, 5]. These resins were prepared from acrylates, which had high reactivity but typically produced weak parts due to the inaccuracy caused by shrinkage and curling. The acrylate-based resins typically could only be cured to 46% completion when the image was transferred through the laser [6]. When a fresh coating was put on the exposed layer, some radiation went through the new coating and initiated new pho-

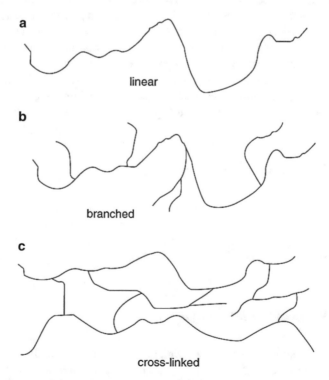

a

linear

b

branched

c

cross-linked

Fig. 4.2 Schematics of polymer types

tochemical reactions in the layer that was already partially cured. This layer was less susceptible to oxygen inhibition after it had been coated. The additional cross-linking on this layer caused extra shrinkage, which increased stresses in the layer and caused curling that was observed either during or after the part fabrication process [7].

The first patents that prepared an epoxide composition for SL resins appeared in 1988 [8, 9] (Japanese). The epoxy resins produced more accurate, harder, and stronger parts than the acrylate resins. While the polymerization of acrylate compositions leads to 5–20% shrinkage, the ring-opening polymerization of epoxy compositions only leads to a shrinkage of 1–2% [10]. This low level of shrinkage associated with epoxy chemistry contributes to excellent adhesion and reduced tendency for flexible substrates to curl during cure. Furthermore, the polymerization of the epoxy-based resins is not inhibited by atmospheric oxygen. This enables low photoinitiator concentrations, giving lower residual odor than acrylic formulations [11].

Epoxy resins have disadvantages of slow photospeed and brittleness of the cured parts. The addition of some acrylate to epoxy resins is required to rapidly build part strength so that they will have enough integrity to be handled without distortion during fabrication. The acrylates are also useful to reduce the brittleness of the epoxy parts [7]. Another disadvantage of epoxy resins is their sensitivity to humidity, which can inhibit polymerization [11].

As a result, most SL and VPP resins commercially available today are epoxides with some acrylate content. It is necessary to have both materials present in the same formulation to combine the advantages of both curing types. The improvement in accuracy resulting from the use of hybrid resins has given VPP a tremendous boost.

4.2.2   Overview of Photopolymer Chemistry

VPP photopolymers are composed of several types of ingredients: photoinitiators, reactive diluents, flexibilizers, stabilizers, and liquid monomers. Broadly speaking, when UV radiation impinges on VPP resin, the photoinitiators undergo a chemical transformation and become "reactive" with the liquid monomers. A "reactive" photoinitiator reacts with a monomer molecule to start a polymer chain. Subsequent reactions occur to build polymer chains and then to cross-link by creation of strong covalent bonds between polymer chains. Polymerization is the term used to describe the process of linking small molecules (monomers) into larger molecules (polymers) composed of many monomer units [1, 12, 13]. Two main types of photopolymer chemistry are commercially evident:

- Free-radical photopolymerization – acrylate
- Cationic photopolymerization – epoxy and vinyl ether

The molecular structures of these types of photopolymers are shown in Fig. 4.3. Symbols C and H denote carbon and hydrogen atoms, respectively, while R denotes

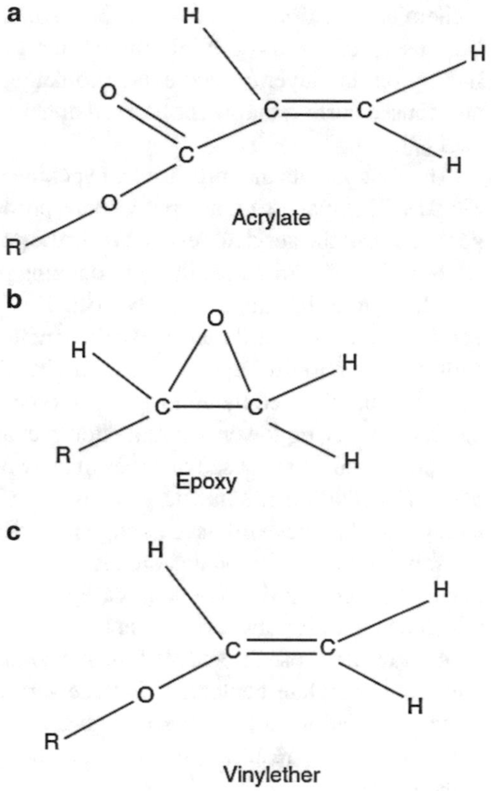

Fig. 4.3 Molecular structure of VPP monomers

a

Acrylate

b

Epoxy

c

Vinylether

a molecular group which typically consists of one or more vinyl groups. A vinyl group is a molecular structure with a carbon–carbon double bond. It is these vinyl groups in the R structures that enable photopolymers to become cross-linked.

Free-radical photopolymerization was the first type that was commercially developed. Such SL resins were acrylates. Acrylates form long polymer chains once the photoinitiator becomes "reactive," building the molecule linearly by adding monomer segments. Cross-linking typically happens after the polymer chains grow enough so that they become close to one another. Acrylate photopolymers exhibit high photospeed (react quickly when exposed to UV radiation) but have a number of disadvantages including significant shrinkage and a tendency to warp and curl. As a result, they are rarely used now without epoxy or other photopolymer elements.

The most common cationic photopolymers are epoxies, although vinyl ethers are also commercially available. Epoxy monomers have rings, as shown in Fig. 4.3. When reacted, these rings open, resulting in sites for other chemical bonds. Ring opening is known to impart minimal volume change on reaction, because the number and types of chemical bonds are essentially identical before and after reaction [14]. As a result, epoxy VPP resins typically have much smaller shrinkages and much less tendency to warp and curl. Almost all commercially available VPP resins have significant amounts of epoxies.

> $P\text{-}I \rightarrow$ -$I\bullet$ (free radical formation)
>
> $I\bullet$ +M $\rightarrow$ $I\text{-}M\bullet$ (initiation)
>
> $I\text{-}M\bullet$ $\rightarrow$ $\rightarrow$ $I\text{-}M\text{-}M\text{-}M\text{-}M...\text{-}M\bullet$ (propagation)
>
> $\rightarrow$ $I\text{-}M\text{-}M\text{-}M\text{-}M...\text{-}M\text{-}I$ (termination)

Fig. 4.4 Free-radical polymerization process

Polymerization of VPP monomers is an exothermic reaction, with heats of reaction around 85 kJ/mol for an example acrylate monomer. Despite high heats of reaction, a catalyst is necessary to initiate the reaction. As described earlier, a photoinitiator acts as the catalyst.

Schematically, the free radical-initiated polymerization process can be illustrated as shown in Fig. 4.4 [1]. On average, for every two photons (from the laser), one radical will be produced. That radical can lead to the polymerization of over 1000 monomers, as shown in the intermediate steps of the process, called propagation. In general, longer polymer molecules are preferred, yielding higher molecular weights. This indicates a more complete reaction. In Fig. 4.4, the P–I term indicates a photoinitiator, the $-I\bullet$ symbol is a free radical, and M is a monomer.

Polymerization terminates from one of three causes, recombination, disproportionation, or occlusion. Recombination occurs when two polymer chains merge by joining two radicals. Disproportionation involves essentially the cancelation of one radical by another, without joining. Occlusion occurs when free radicals become "trapped" within a solidified polymer, meaning that reaction sites remain available but are prevented from reacting with other monomers or polymers by the limited mobility within the polymer network. These occluded sites will most certainly react eventually, but not with another polymer chain or monomer. Instead, they will react with oxygen or another reactive species that diffuses into the occluded region. This may be a cause of aging or other changes in mechanical properties of cured parts, which is a topic of research.

Cationic photopolymerization shares the same broad structure as free-radical polymerization, where a photoinitiator generates a cation as a result of laser energy, the cation reacts with a monomer, propagation occurs to generate a polymer, and a termination process completes the reaction. A typical catalyst for a cationic polymerization is a Lewis acid, such as BF_3 [15]. Initially, cationic photopolymerization received little attention, but that changed during the 1990s due to advances in the microelectronics industry, as well as interest in VPP technology. We will not investigate the specifics of cationic reactions here but will note that the ring-opening reaction mechanism of epoxy monomers is similar to radical propagation in acrylates.

4.2.3 Resin Formulations and Reaction Mechanisms

Basic raw materials such as polyols, epoxides, (meth) acrylic acids and their esters, diisocyanates, etc. are used to produce the monomers and oligomers used for radiation curing. Most of the monomers are multifunctional monomers (MFM) or polyol polyacrylates which give a cross-linking polymerization. The main chemical families of oligomers are polyester acrylate (PEA), epoxy acrylates (EA), urethane acrylates (UA), amino acrylates (used as a photoaccelerator in the photoinitiator system), and cycloaliphatic epoxies [11].

Resin suppliers create ready-to-use formulations by mixing the oligomers and monomers with a photoinitiator, as well as other materials to affect reaction rates and part properties. In practice, photosensitizers are often used in combination with the photoinitiator to shift the absorption toward longer wavelengths. In addition, supporting materials may be mixed with the initiator to achieve improved solubility in the formulation. Furthermore, mixtures of different types of photoinitiators may also be employed for a given application. Thus, photoinitiating systems are, in practice, often highly elaborate mixtures of various compounds which provide optimum performance for specific applications [10].

Other additives facilitate the application process and achieve products of good properties. A reactive diluent, for example, is usually added to adjust the viscosity of the mixtures to an acceptable level for application [16]; it also participates in the polymerization reaction.

4.2.3.1   Photoinitiator System

The role of the photoinitiator is to convert the physical energy of the incident light into chemical energy in the form of reactive intermediates. The photoinitiator must exhibit a strong absorption at the laser emission wavelength and undergo a fast photolysis to generate the initiating species with a great quantum yield [17]. The reactive intermediates are either radicals capable of adding to vinylic or acrylic double bonds, thereby initiating radical polymerization, or reactive cationic species which can initiate polymerization reactions among epoxy molecules [10].

The free-radical polymerization process was outlined in Fig. 4.4, with the formation of free radicals as the first step. In the typical case in VPP, radical photoinitiator systems include compounds that undergo unimolecular bond cleavage upon irradiation. This class includes aromatic carbonyl compounds that are known to undergo a homolytic C–C bond scission upon UV exposure [18]. The benzoyl radical is the major initiating species, while the other fragment may, in some cases, also contribute to the initiation. The most efficient photoinitiators include benzoin ether derivatives, benzyl ketals, hydroxyalkylphenones, α-amino ketones, and acylphosphine oxides [18, 19]. The Irgacure family of radical photoinitiators from Ciba Specialty Chemicals is commonly used in VPP.

While photoinitiated free-radical polymerizations have been investigated for more than 60 years, the corresponding photoinduced cationic polymerizations have received much less attention. The main reason for the slow development in this area was the lack of suitable photoinitiators capable of efficiently inducing cationic polymerization [20]. Beginning in 1965, with the earliest work on diazonium salt initiators, this situation has markedly changed. The discovery in the

1970s of onium salts or organometallic compounds with excellent photoresponse and high efficiency has initiated the very rapid and promising development of cationic photopolymerization and made possible the concurrent radical and cationic reaction in hybrid systems [21]. Excellent reviews have been published in this field [10, 20, 22–25]. The most important cationic photoinitiators are the onium salts, particularly the triarylsulfonium and diaryliodonium salts. Examples of the cationic photoinitiator are triarylsulfonium hexafluorophosphate solutions in propylene carbonate such as Degacure KI 85 (Degussa), SP-55 (Asahi Denka), Sarcat KI-85 (Sartomer), and 53,113–8 (Aldrich) or mixtures of sulfonium salts such as SR-1010 (Sartomer, currently unavailable), UVI 6976 (B-V), and UVI 6992 (B-VI) (Dow).

Initiation of cationic polymerization takes place not only from the primary products of the photolysis of triarylsulfonium salts but also from secondary products of the reaction of those reactive species with solvents, monomers, or even other photolysis species. Probably the most ubiquitous species present is the protonic acid derived from the anion of the original salt. Undoubtedly, the largest portion of the initiating activity in cationic polymerization by photolysis of triarylsulfonium salts is due to protonic acids [20].

4.2.3.2   Monomer Formulations

The monomer formulations presented here are from a set of patents from the mid to late 1990s. Both di-functional and higher functionality monomers are used typically in VPP resins. Poly(meth)acrylates may be tri-, pentafunctional monomeric or oligomeric aliphatic, cycloaliphatic or aromatic (meth)acrylates, or polyfunctional urethane (meth)acrylates [26–29]. One specific compound in the Huntsman SL-7510 resin includes the dipentaerythritol monohydroxy penta(meth)acrylates [29], such as Dipentaerythritol Pentaacrylate (SR-399, Sartomer).

The cationically curable epoxy resins may have an aliphatic, aromatic, cycloaliphatic, araliphatic, or heterocyclic structure; they on average possess more than one epoxide group (oxirane ring) in the molecule and comprise epoxide groups as side groups, or those groups form part of an alicyclic or heterocyclic ring system. Examples of epoxy resins of this type are also given by these patents such as polyglycidyl esters or ethers, poly(N or S-glycidyl) compounds, and epoxide compounds in which the epoxide groups form part of an alicyclic or heterocyclic ring system. One specific composition includes at least 50% by weight of a cycloaliphatic diepoxide [29] such as bis(2,3-epoxycyclopentyl) ether (formula A-I), 3,4-epoxycyclohexyl-methyl 3,4-epoxycyclohexanecarboxylate (A-II), dicyclopentadiene diepoxide (A-III), and bis-(3,4-epoxycyclohexylmethyl) adipate (A-IV).

Additional insight into compositions can be gained by investigating the patent literature further.

4.2.3.3   Interpenetrating Polymer Network Formation

As described earlier, acrylates polymerize radically, while epoxides cationically polymerize to form their respective polymer networks. In the presence of each other during the curing process, an interpenetrating polymer network (IPN) is finally obtained [30, 31]. An IPN can be defined as a combination of two polymers in network form, at least one of which is synthesized and/or cross-linked in the immediate presence of the other [32]. It is therefore a special class of polymer blends in which both polymers generally are in network form [32–34] and which is originally generated by the concurrent reactions instead of by a simple mechanical mixing process. In addition, it is a polymer blend rather than a copolymer that is generated from the hybrid curing [35], which indicates that acrylate and epoxy monomers undergo independent polymerization instead of copolymerization. However, in special cases, copolymerization can occur, thus leading to a chemical bonding of the two networks [36].

It is likely that in typical SL resins, the acrylate and epoxide react independently. Interestingly, however, these two monomers definitely affect each other physically during the curing process. The reaction of acrylate will enhance the photospeed and reduce the energy requirement of the epoxy reaction. Also, the presence of acrylate monomer may decrease the inhibitory effect of humidity on the epoxy polymerization. On the other hand, the epoxy monomer acts as a plasticizer during the early polymerization of the acrylate monomer where the acrylate forms a network while the epoxy is still at liquid stage [31]. This plasticizing effect, by increasing molecular mobility, favors the chain propagation reaction [37]. As a result, the acrylate polymerizes more extensively in the presence of epoxy than in the neat acrylate monomer. Furthermore, the reduced sensitivity of acrylate to oxygen in the hybrid system than in the neat composition may be due to the simultaneous polymerization of the epoxide which makes the viscosity rise, thus slowing down the diffusion of atmospheric oxygen into the coating [33].

In addition, it has been shown [33] that the acrylate/epoxide hybrid system requires a shorter exposure to be cured than either of the two monomers taken separately. It might be due to the plasticizing effect of epoxy monomer and the contribution of acrylate monomer to the photospeed of the epoxy polymerization. The two monomers benefit from each other by a synergistic effect.

It should be noted that if the concentration of the radical photoinitiator was decreased so that the two polymer networks were generated simultaneously, the plasticizing effect of the epoxy monomer would become less pronounced. As a result, it would be more difficult to achieve complete polymerization of the acrylate monomer and thus require longer exposure time.

Although the acrylate/epoxy hybrid system proceeds via a heterogeneous mechanism, the resultant product (IPN) seems to be a uniphase component [38]. The properties appear to be extended rather than compromised [33, 36]. The optimal properties of IPNs for specific applications can be obtained by selecting two appropriate components and adjusting their proportions [36]. For example, increasing the acrylate content increases the cure speed but decreases the adhesion characteristics, while increasing the epoxy content reduces the shrinkage of curing and improves the adhesion but decreases the cure speed [38].

4.3   Reaction Rates

As is evident, the photopolymerization reaction in VPP resins is very complex. To date, no one has published an analytical photopolymerization model that describes reaction results and reaction rates. However, qualitative understanding of reaction rates is straightforward for simple formulations. Broadly speaking, reaction rates for photopolymers are controlled by concentrations of photoinitiators [I] and monomers [M]. The rate of polymerization is the rate of monomer consumption, which can be shown as [3]:

$$R_p = -d[M] / dt \, \alpha [M] \left(k[I] \right)^{1/2} \tag{4.1}$$

where k = constant that is a function of radical generation efficiency, rate of radical initiation, and rate of radical termination. Hence, the polymerization rate is proportional to the concentration of monomer but is only proportional to the square root of initiator concentration.

Using similar reasoning, it can be shown that the average molecular weight of polymers is the ratio of the rate of propagation and the rate of initiation. This average weight is called the kinetic average chain length, v_o, and is given in (4.2):

$$v_o = R_p / R_i \; \alpha [M] / [I]^{1/2} \tag{4.2}$$

where R_i is the rate of initiation of macromonomers.

Equations (4.1) and (4.2) have important consequences for the VPP process. The higher the rate of polymerization, the faster parts can be built. Since VPP resins are predominantly composed of monomers, the monomer concentration cannot be changed much. Hence, the only other direct method for controlling the polymerization rate and the kinetic average chain length is through the concentration of initiator. However, (4.1 and 4.2) indicate a trade-off between these characteristics. Doubling the initiator concentration only increases the polymerization rate by a factor of 1.4 but reduces the molecular weight of resulting polymers by the same amount. Strictly speaking, this analysis is more appropriate for acrylate resins, since epoxies continue to react after laser exposure, so (4.2) does not apply well for epoxies. However, reaction of epoxies is still limited, so it can be concluded that a trade-off does exist between polymerization rate and molecular weight for epoxy resins.

4.4   Laser Scan Vat Photopolymerization

Laser scan VPP creates solid parts by selectively solidifying a liquid photopolymer resin using an UV laser. As with many other AM processes, the physical parts are manufactured by fabricating cross-sectional contours, or slices, one on top of another. These slices are created by tracing 2D contours of a CAD model in a vat of photopolymer resin with a laser. The part being built rests on a platform that is dipped into the vat of resin, as shown schematically in Fig. 4.1a. After each slice is created, the platform is lowered, the surface of the vat is recoated, and then the laser starts to trace the next slice of the CAD model, building the prototype from the bottom up. A more complete description of the VPP process may be found in [14]. In VPP, supports are needed to attach the part to the platform and to keep the part from moving and warping during subsequent layering processes.

After building the part, the part must be cleaned, post-cured, and finished. During the cleaning and finishing phase, the VPP machine operator may remove support structures. During finishing, the operator may spend considerable time sanding and filing the part to provide the desired surface finishes.

4.5   Photopolymerization Process Modeling

The background on VPP materials and energy sources enables us to investigate the curing process of photopolymers in VPP machines. We will begin with an investigation into the fundamental interactions of laser energy with photopolymer resins. Through the application of the Beer–Lambert law, the theoretical relationship between resin characteristics and exposure can be developed, which can be used to specify laser scan speeds. This understanding can then be applied to investigate mechanical properties of cured resins. From here, we will briefly investigate the ranges of size scales and time scales of relevance to the VPP process. Much of this section is adapted from [1].

Nomenclature

C_d	cure depth = depth of resin cure as a result of laser irradiation [mm]
D_p	depth of penetration of laser into a resin until a reduction in irradiance of $1/e$ is reached = key resin characteristic [mm]
E	exposure, possibly as a function of spatial coordinates [energy/unit area] [mJ/mm^2]
E_c	critical exposure = exposure at which resin solidification starts to occur [mJ/mm^2]
E_{max}	peak exposure of laser shining on the resin surface (center of laser spot) [mJ/mm^2]
$H(x,y,z)$	irradiance (radiant power per unit area) at an arbitrary point in the resin = time derivative of $E(x,y,z)$.[W/mm^2]
P_L	output power of laser [W]
V_s	scan speed of laser [mm/s]
W_0	radius of laser beam focused on the resin surface [mm]

4.5.1 Irradiance and Exposure

As a laser beam is scanned across the resin surface, it cures a line of resin to a depth that depends on many factors. However, it is also important to consider the width of the cured line as well as its profile. The shape of the cured line depends on resin characteristics, laser energy characteristics, and the scan speed. We will investigate the relationships among all of these factors in this subsection.

The first concept of interest here is *irradiance*, the radiant power of the laser per unit area, $H(x,y,z)$. As the laser scans a line, the radiant power is distributed over a finite area (beam spots are not infinitesimal). Figure 4.5 shows a laser scanning a line along the x-axis at a speed V_s [1]. Consider the z-axis oriented perpendicular to the resin surface and into the resin and consider the origin such that the point of interest, p', has an x coordinate of 0. The irradiance at any point x,y,z in the resin is related to the irradiance at the surface, assuming that the resin absorbs radiation according to the Beer–Lambert law. The general form of the irradiance equation for a Gaussian laser beam is given here as (4.3).

$$H(x,y,z) = H(x,y,0)e^{-z/D_p} \qquad (4.3)$$

From this relationship, we can understand the meaning of the penetration depth, D_p. Setting $z = D_p$, we get that the irradiance at a depth D_p is about 37% ($e^{-1} = 0.36788$) of the irradiance at the resin surface. Thus, D_p is the depth into the resin at which the irradiance is 37% of the irradiance at the surface. Furthermore, since we are assuming the Beer–Lambert law holds, D_p is only a function of the resin.

Without loss of generality, we will assume that the laser scans along the x-axis from the origin to point b. Then, the irradiance at coordinate x along the scan line is given by:

$$H(x,y,0) = H(x,y) = H_0 e^{-2x^2/W_0^2} e^{-2y^2/W_0^2} \qquad (4.4)$$

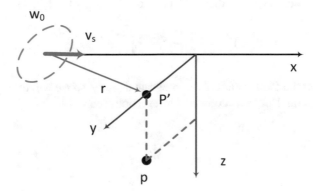

Fig. 4.5 Scan line of Gaussian laser

where $H_0 = H(0,0)$ when $x = 0$, and W_0 is the $1/e^2$ Gaussian half-width of the beam spot. Note that when $x = W_0$, $H(x,0) = H_0 e^{-2} = 0.13534 H_0$.

The maximum irradiance, H_0, occurs at the center of the beam spot ($x = 0$). H_0 can be determined by integrating the irradiance function over the area covered by the beam at any particular point in time. Changing from Cartesian to polar coordinates, the integral can be set equal to the laser power, P_L, as shown in (4.5):

$$P_L = \int_{r=0}^{r=\infty} H(r,0)\,dA \tag{4.5}$$

When solved, H_o turns out to be a simple function of laser power and beam half-width, as in (4.6):

$$H_0 = \frac{2P_L}{\pi W_0^2} \tag{4.6}$$

As a result, the irradiance at any point x,y between $x = 0$ and $x = b$ is given by:

$$H(x,y) = \frac{2P_L}{\pi W_0^2} e^{-2x^2/W_0^2} e^{-2y^2/W_0^2} \tag{4.7}$$

However, we are interested in *exposure* at an arbitrary point, p, not irradiance, since exposure controls the extent of resin cure. Exposure is the energy per unit area; when exposure at a point in the resin vat exceeds a critical value, called E_c, we assume that resin cures. Exposure can be determined at point p by appropriately integrating (4.7) along the scan line, from time 0 to time t_b, when the laser reaches point b:

$$E(y,0) = \int_{t=0}^{t=t_b} H[x(t),0]\,dt \tag{4.8}$$

It is far more convenient to integrate over distance than over time. If we assume a constant laser scan velocity, then it is easy to substitute t for x, as in (4.9):

$$E(y,0) = \frac{2P_L}{\pi V_s W_0^2} e^{-2y^2/W_0^2} \int_{x=0}^{x=b} e^{-2x^2/W_0^2}\,dx \tag{4.9}$$

The exponential term is difficult to integrate directly, so we will change the variable of integration. Define a variable of integration, v, as:

$$v^2 \equiv \frac{2x^2}{W_0^2}$$

Then, take the square root of both sides, take the derivative, and rearrange to give:

$$dx = \frac{W_0}{\sqrt{2}} dv$$

Due to the change of variables, it is also necessary to convert the integration limit to: $b = \sqrt{2} / W_0 x_e$.

Several steps in the derivation will be skipped. After integration, the exposure received at a point x,y between $x = (0,b)$ can be computed as:

$$E(y,0) = \frac{P_L}{\sqrt{2\pi} W_0 V_s} e^{-\frac{2y^2}{W_0^2}} \left[erf(b) \right] \tag{4.10}$$

where $erf(x)$ is the error function evaluated at x. $erf(x)$, is close to -1 for negative values of x, is close to 1 for positive values of x, and rapidly transitions from -1 to 1 for values of x close to 0. This behavior localizes the exposure within a narrow range around the scan vector. This makes sense since the laser beam is small, and we expect that the energy received from the laser drops off quickly outside of its radius.

Equation (4.10) is not quite as easy to apply as a form of the exposure equation that results from assuming an infinitely long scan vector. If we make this assumption, then (4.10) becomes:

$$E(y,0) = \frac{2P_L}{\pi V_s W_0^2} e^{-2y^2/W_0^2} \int_{x=\infty}^{x=\infty} e^{-2x^2/W_0^2} dx$$

and after integration, exposure is given by:

$$E(y,0) = \sqrt{\frac{2}{\pi}} \frac{P_L}{W_0 V_s} e^{-2y^2/W_0^2} \tag{4.11}$$

Combining this with (4.3) yields the fundamental general exposure equation:

$$E(x,y,z) = \sqrt{\frac{2}{\pi}} \frac{P_L}{W_0 V_s} e^{-2y^2/W_0^2} e^{-z/D_p} \tag{4.12}$$

4.5.2   Laser–Resin Interaction

In this subsection, we will utilize the irradiance and exposure relationships to determine the shape of a scanned vector line and its width. As we will see, the cross-sectional shape of a cured line becomes a parabola.

Starting with (4.12), the locus of points in the resin that is just at its gel point, where $E = E_c$, is denoted by y^* and z^*. Equation (4.12) can be rearranged, with y^*, z^*, and E_c substituted to give (4.13):

$$e^{2y^{*2}/W_0^2 + z*/D_P} = \sqrt{\frac{2}{\pi}} \frac{P_L}{W_0 V_s E_c} \tag{4.13}$$

Taking natural logarithms of both sides yields:

$$2\frac{y^{*2}}{W_0^2} + \frac{z^*}{D_p} = \ln\left[\sqrt{\frac{2}{\pi}} \frac{P_L}{W_0 V_s E_c}\right] \tag{4.14}$$

This is the equation of a parabolic cylinder in y^* and z^*, which can be seen more clearly in the following form:

$$ay^{*2} + bz^* = c \tag{4.15}$$

where a, b, and c are constants, immediately derivable from (4.14). Figure 4.6 illustrates the parabolic shape of a cured scan line.

To determine the maximum depth of cure, we can solve (4.14) for z^* and set $y^* = 0$, since the maximum cure depth will occur along the center of the scan vector. Cure depth, C_d, is given by:

$$C_d = D_p \ln\left[\sqrt{\frac{2}{\pi}} \frac{P_L}{W_0 V_s E_c}\right] \tag{4.16}$$

As is probably intuitive, the width of a cured line of resin is maximum at the resin surface, i.e., y_{max} occurs at $z = 0$. To determine line width, we start with the line

Fig. 4.6 Cured line showing parabolic shape, cure depth, and line width

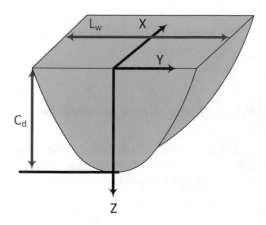

shape function, (4.14). Setting $z = 0$ and letting line width, L_w, equal $2y_{max}$, the line width can be found:

$$L_W = W_0 \sqrt{2C_d / D_p} \tag{4.17}$$

As a result, three important aspects become clear. First, line width is proportional to the beam spot size. Second, for typical resins, line width of a single scan is less than the laser beam diameter. Third, if a greater cure depth is desired, line width must increase, all else remaining the same. This becomes very important when performing line width compensation during process planning.

The final concept to be presented in this subsection is fundamental to commercial VPP. It is the *working curve*, which relates exposure to cure depth, and includes the two key resin constants, D_p and E_c. At the resin surface and in the center of the scan line:

$$E(0,0) \equiv E_{max} = \sqrt{\frac{2}{\pi}} \frac{P_L}{W_0 V_s} \tag{4.18}$$

which is most of the expression within the logarithm term in (4.16). Substituting (4.18) into (4.16) yields the working curve equation:

$$C_d = D_p \ln\left(\frac{E_{max}}{E_c}\right) \tag{4.19}$$

In summary, a laser of power P_L scans across the resin surface at some speed V_s solidifying resin to a depth C_d, the cure depth, assuming that the total energy incident along the scan vector exceeds a critical value called the critical exposure, E_c. If the laser scans too quickly, no polymerization reaction takes place, i.e., exposure E is less than E_c. E_c is assumed to be a characteristic quantity of a particular resin.

An example working curve is shown in Fig. 4.7, where measured cure depths at a given exposure are indicated by "*." The working curve equation, (4.19), has several major properties [1]:

1. The cure depth is proportional to the natural logarithm of the maximum exposure on the centerline of a scanned laser beam.
2. A semilog plot of C_d vs. E_{max} should be a straight line. This plot is known as the *working curve* for a given resin.
3. The slope of the working curve is precisely D_p at the laser wavelength being used to generate the working curve.
4. The x-axis intercept of the working curve is E_c, the critical exposure of the resin at that wavelength. Theoretically, the cure depth is 0 at E_c, but this does indicate the gel point of the resin.
5. Since D_p and E_c are purely resin parameters, the slope and intercept of the working curve are independent of laser power.

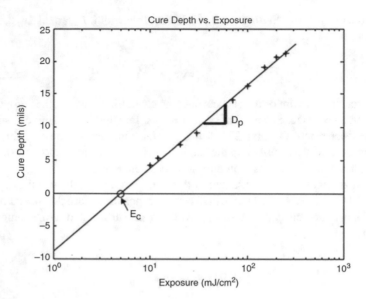

Fig. 4.7 Resin "working curve" of cure depth vs. exposure

In practice, various E_{max} values can be generated easily by varying the laser scan speed, as indicated by (4.19).

4.5.3 Photospeed

Photospeed is typically used as an intuitive approximation of VPP photosensitivity. But it is useful in that it relates to the speed at which the laser can be scanned across the polymer surface to give a specified cure depth. The faster the laser can be scanned to give a desired cure depth, the higher the photospeed. Photospeed is a characteristic of the resin and does not depend upon the specifics of the laser or optics subsystems. In particular, photospeed is indicated by the resin constants E_c and D_p, where higher levels of D_p and lower values of E_c indicate higher photospeed.

To determine scan velocity for a desired cure depth, it is straightforward to solve (4.16) for V_s. Recall that at the maximum cure depth, the exposure received equals the cure threshold, E_c. Scan velocity is given by (4.20):

$$V_s = \sqrt{\frac{2}{\pi}} \frac{P_L}{W_0 E_c} e^{-C_d/D_p} \tag{4.20}$$

This discussion can be related back to the working curve. Both E_c and D_p must be determined experimentally. 3D Systems has developed a procedure called the WINDOWPANE procedure for finding E_c and D_p values [39]. The cure depth, C_d, can be measured directly from specimens built on a VPP machine that are one layer thickness in depth. The WINDOWPANE procedure uses a specific part shape, but the principle is simply to build a part with different amounts of laser exposure in different places in the part. By measuring the part thickness, C_d, and correlating that with the exposure values, a "working curve" can easily be plotted. Note that (4.19) is log-linear. Hence, C_d is plotted linearly vs. the logarithm of exposure to generate a working curve.

So how is exposure varied? Exposure is varied by simply using different scan velocities in different regions of the WINDOWPANE part. The different scan velocities will result in different cure depths. In practice, (4.20) is very useful since we want to directly control cure depth and want to determine how fast to scan the laser to give that cure depth. Of course, for the WINDOWPANE experiment, it is more useful to use (4.16) or (4.19).

4.5.4   Time Scales

It is interesting to investigate the time scales at which VPP operates. On the short end of the time scale, the time it takes for a photon of laser light to traverse a photopolymer layer is about a picosecond (10^{-12} s). Photon absorption by the photoinitiator and the generation of free radicals or cations occur at about the same time frame. A measure of photopolymer reaction speed is the kinetic reaction rates, t_k, which are typically several microseconds.

The time it takes for the laser to scan past a particular point on the resin surface is related to the size of the laser beam. We will call this time the characteristic exposure time, t_e. Values of t_e are typically 50–2000 µs, depending on the scan speed (500–5000 mm/s). Laser exposure continues long after the onset of polymerization. Continued exposure generates more free radicals or cations and, presumably, generates these at points deeper in the photopolymer. During and after the laser beam traverses the point of interest, cross-linking occurs in the photopolymer.

The onset of measurable shrinkage, $t_{s,o}$, lags exposure by several orders of magnitude. This appears to be due to the rate of cross-linking but, for the epoxy-based resins, may have more complicated characteristics. Time at corresponding completion of shrinkage is denoted $t_{s,c}$. For the acrylate-based resins of the early 1990s, times for the onset and completion of shrinkage were typically 0.4–1 and 4–10 s, respectively. Recall that epoxies can take hours or days to polymerize. Since shrinkage lags exposure, this is clearly a phenomenon that complicates the VPP process. Shrinkage leads directly to accuracy problems, including deviation from nominal dimensions, warpage, and curl.

The final time dimension is that of scan time for a layer, denoted t_d, which typically spans 10–300 s. The time scales can be summarized as:

$$t_t \ll t_k \ll t_e \ll t_{s,o} < t_{s,c} \ll t_d \tag{4.21}$$

As a result, characteristic times for the VPP process span about 14 orders of magnitude.

4.6   Vector Scan VPP Machines

At present (2020), 3D Systems is the predominant manufacturer of laser scanning VPP machines in the world, although several other companies in Japan and elsewhere in Asia also market VPP machines. Several Japanese companies produce or produced machines, including Denken Engineering, CMET (Mitsubishi), Sony, Meiko Corp., Mitsui Zosen, and Teijin Seiki (licensed from DuPont). Formlabs is a relatively new company that markets several small, high-resolution VPP machines.

A schematic of a typical VPP machine was illustrated in Fig. 4.1a, which shows the main subsystems, including the laser and optics, the platform and elevator, the vat and resin-handling subsystem, and the recoater. The machine subsystem hierarchy is given in Fig. 4.8. Note that the five main subsystems are recoating system, platform system, vat system, laser and optics system, and control system.

Typically, recoating is done using a shallow dip and recoater blade sweeping. Recoating issues are discussed in [40]. The process can be described as follows:

- After a layer has been cured, the platform dips down by a layer thickness.

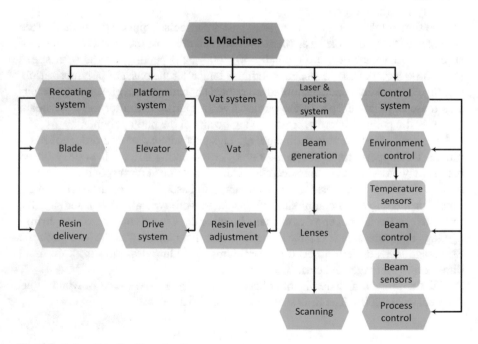

Fig. 4.8 Subsystems for SL technology

- The recoater blade slides over the whole build, spreading a new layer of resin and smoothing the surface of the vat.

A common recoater blade type is the zephyr blade, which is a hollow blade that is filled with resin. A vacuum system pulls resin into the blade from the vat. As the blade translates over the vat to perform recoating, resin is deposited in regions where the previous part cross-section was built. When the blade encounters a region in the vat without resin, the resin falls into this region since its weight is stronger than the vacuum force. Blade alignment is critical to avoid "blade crashes," when the blade hits the part being built and often delaminates the previous layer. The blade gap (distance between the bottom of the blade and the resin surface) and speed are important variables under user control.

The platform system consists of a build platform that supports the part being built and an elevator that lowers and raises the platform. The elevator is driven by a leadscrew. The vat system is simply the vat that holds the resin, combined with a level adjustment device and usually an automated refill capability.

The optics system includes a laser, focusing and adjustment optics, and two galvanometers that scan the laser beam across the surface of the vat. Modern VPP machines have solid-state lasers that have more stable characteristics than their predecessors, various gas lasers. SL machines from 3D Systems have Nd-YVO$_4$ lasers that output radiation at about 1062 nm wavelength (near infrared). Additional optical devices triple the frequency to 354 nm, in the UV range. These lasers have relatively low power, in the range of 0.1–1 W, compared with lasers used in other AM and material processing applications.

The control system consists of three main subsystems. First, a process controller controls the sequence of machine operations. Typically, this involves executing the sequence of operations that are described in the build file that was prepared for a specific part or set of parts. Commands are sent to the various subsystems to actuate the recoating blade, to adjust resin level or change the vat height, or to activate the beam controller. Sensors are used to detect resin height and to detect forces on the recoater blade to detect blade crashes. Second, the beam controller converts operation descriptions into actions that adjust beam spot size, focus depth, and scan speed, with some sensors providing feedback. Third, the environment controller adjusts resin vat temperature and, depending on machine model, adjusts environment temperature and humidity.

Two of the main advantages of VPP technology over other AM technologies are part accuracy and surface finish, in combination with moderate mechanical properties. These characteristics led to the widespread usage of VPP parts as form, fit, and, to a lesser extent, functional prototypes. Typical dimensional accuracies for VPP machines are often quoted as a ratio of an error per unit length. For example, accuracy of a SLA-250 is typically quoted as 0.002 in./in [41]. Modern VPP machines are somewhat more accurate. Surface finish of VPP parts ranges from submicron Ra for up-facing surfaces to over 100 μm Ra for surfaces at slanted angles [42].

The laser scan commercial VPP product line from 3D Systems consists of four models: ProJet 6000 HD, ProJet 7000HD, ProX 800, and ProX 950. These machines are summarized in Table 4.1 [43]. In these machines, the optics system automatically switches between the "normal" beam of approximately 0.13 mm diameter for borders and fills and the "wide" beam of 0.76 mm diameter for hatch vectors (filling in large areas). The wide beam enables much faster builds. Note that the ProX 950 has two lasers, which enables accurate printing across the entire 1.5 m long vat. The ProJet and ProX lines replace other machines, including the popular SLA-3500, SLA-5000, SLA-7000, and Viper Si2 models, the SLA Viper Pro, and most recently the iPro models. Additionally, the SLA-250 was a very popular model that was discontinued in 2001 with the introduction of the Viper Si2 model. (Fig. 4.9 shows the latest SLA 3D Systems machines.)

4.7   Scan Patterns

4.7.1   Layer-Based Build Phenomena and Errors

Several phenomena are common to all radiation and layer-based AM processes. The most obvious phenomenon is discretization, e.g., a stack of layers causes "stair steps" on slanted or curved surfaces. So, the layer-wise nature of most AM processes causes edges of layers to be visible. Conventionally, commercial AM processes build parts in a "material safe" mode, meaning that the stair steps are on the outside of the CAD part surfaces. Technicians can sand or finish parts; the material they remove is outside of the desired part geometry. Other discretization examples

Table 4.1 The latest 3D Systems SLA machine models and specifications [44]

Specifications	ProJet 6000 HD	ProJet 7000 HD	ProX 800	ProX 950
Max Build Size (xyz)	250 × 250 × 250 mm	380 × 380 × 250 mm	650 × 750 × 550 mm	1500 × 750 × 550 mm
Max. Build volume	40 liters	84 liters	414 liters	935 liters
Max part mass	9.6 kg	21.6 kg	75 kg	150 kg
Max resolution	4000 DPI	4000 DPI	4000 DPI	4000 DPI
Min. Layer thickness	0.05 mm	0.05 mm	0.03 mm	0.03 mm
Laser power (at vat)	1 Watt			
Border spot diameter	0.075 mm	0.075 mm	0.125 mm	0.125 mm
Large spot diameter	0.76 mm	0.76 mm	0.76 mm	0.76 mm
Max. Scan speed (border)	3.5 m/sec			
Max. Scan speed (large)	25 m/sec			
Accuracy	0.001–0.002 inch per inch (0.001–0.002 mm per mm) of part dimension			
Input data file formats supported	STL, CTL, OBJ, PLY, ZPR, ZBD, AMF, WRL, 3DS, FBX, MJPDDD, 3DPRINT, BFF, IGES, IGS, STEP, and STP			

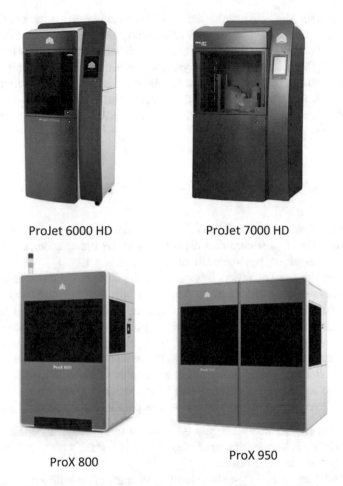

ProJet 6000 HD   ProJet 7000 HD

ProX 800   ProX 950

Fig. 4.9 The latest SL 3D Systems machines (photo courtesy of 3D Systems) [44]

are the set of laser scans or the pixels of a DMD. In most processes, individual laser scans or pixels are not visible on part surfaces, but in other processes such as Material Extrusion (MEX), the individual filaments can be noticeable.

As a laser scans a cross-section, or a lamp illuminates a layer, the material solidifies and, as a result, shrinks. When resins photopolymerize, they shrink since the volume occupied by monomer molecules is larger than that of reacted polymer. Similarly, after powder melts, it cools and freezes, which reduces the volume of the material. When the current layer is processed, its shrinkage pulls on the previous layers, causing stresses to build up in the part. Typically, those stresses remain and are called residual stresses. Also, those stresses can cause part edges to curl upward. Other warpage or part deformations can occur due to these residual stresses, as well.

The last phenomenon to be discussed is that of print-through errors. In photopolymerization processes, it is necessary to have the current layer cure into the previous layer. In Powder Bed Fusion (PBF) processes, the current layer needs to melt into the previous layer so that one solid part results, instead of a stack of disconnected solid layers. The extra energy that extends below the current layer results in thicker part sections. This extra thickness is called print-through error in VPP and "bonus Z" in PBF. Most process planning systems compensate for print-through by giving users the option of skipping the first few layers of a part, which works well unless important features are contained within those layers.

These phenomena will be illustrated in this section through an investigation of scan patterns in SL.

4.7.2   Weave

Prior to the development of WEAVE, scan patterns were largely an ad hoc development. As a result, post-cure curl distortion was the major accuracy problem. The WEAVE scan pattern became available for use in late 1990 [1].

The development of WEAVE began with the observation that distortion in post-cured parts was proportional to the percent of uncured resin after removal from the vat. Another motivating factor was the observation that shrinkage lags exposure and that this time lag must be considered when planning the pattern of laser scans. The key idea in WEAVE development was to separate the curing of the majority of a layer from the adherence of that layer to the previous layer. Additionally, to prevent laser scan lines from interfering with one another while each is shrinking, parallel scans were separated from one another by more than a line width.

The WEAVE style consists of two sets of parallel laser scans:

- First, parallel to the x-axis, spaced 1 mil (1 mil = 0.001 in. = 0.25 mm, which historically is a standard unit of measure in SL) apart, with a cure depth of 1 mil less than the layer thickness.
- Second, parallel to the y-axis, spaced 1 mil apart, again with a cure depth of 1 mil less than the layer thickness.

However, it is important to understand the relationships between cure depth and exposure. On the first pass, a certain cure depth is achieved, C_{d1}, based on an amount of exposure, E_{max1}. On the second pass, the same amount of exposure is provided, and the cure depth increases to C_{d2}. A simple relationship can be derived among these quantities, as shown in (4.21):

$$C_{d2} = D_p \ln\left(2E_{max1} / E_c\right) = D_p \ln(2) + D_p \ln\left(E_{max1} / E_c\right) \tag{4.21}$$

$$C_{d2} = C_{d1} + D_p \ln(2) \tag{4.22}$$

It is the second pass that provides enough exposure to adhere the current layer to the previous one. The incremental cure depth caused by the second pass is just $\ln(2)D_p$ or about $0.6931D_p$. This distance is always greater than 1 mil.

The WEAVE build style cures about 99% of the resin at the vat surface and about 96% of the resin volume through the layer thickness. Compared with previous build styles, WEAVE provided far superior results in terms of eliminating curl and warpage. Figure 4.10 shows a typical WEAVE pattern, illustrating how WEAVE gets its name.

Even though WEAVE was a tremendous improvement, several flaws were observed with its usage. Corners were distorted on large flat surfaces, one of these corners always exhibited larger distortion, and it was always the same corner. Some microfissures occurred; on a flat plate with a hole, a macrofissure tangent to the hole would appear.

It was concluded that significant internal stresses developed within parts during part building, not only post-cure. As a result, improvements to WEAVE were investigated, leading to the development of STAR-WEAVE.

4.7.3 Star-Weave

STAR-WEAVE was released in October 1991 [1]. STAR-WEAVE addressed all of the known deficiencies of WEAVE and worked very well with the resins available at the time. WEAVE's deficiencies were traced to the consequences of two related phenomena: the presence of shrinkage and the lag of shrinkage relative to exposure. These phenomena led directly to the presence of large internal stresses in parts. STAR-WEAVE gets its name from the three main improvements from WEAVE:

1. STaggered hatch
2. Alternating sequence
3. Retracted hatch

Staggered hatch directly addresses the observed microfissures. Consider Fig. 4.11 which shows a cross-sectional view of the hatch vectors from two layers. In Fig. 4.11

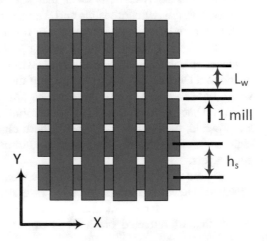

Fig. 4.10 WEAVE scan pattern

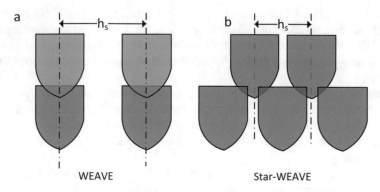

Fig. 4.11 Cross-sectional view of WEAVE and STAR-WEAVE patterns

(a), the hatch vectors in WEAVE form vertical "walls" that do not directly touch. In STAR-WEAVE, Fig. 4.11 (b), the hatch vectors are staggered such that they directly adhere to the layer below. This resulting overlap from one layer to the next eliminated microfissures and eliminated stress concentrations in the regions between vectors.

Upon close inspection, it became clear why the WEAVE scan pattern tended to cause internal stresses, particularly if a part had a large cross-section. Consider a thin cross-section which is short in the x direction but long in the y direction. The WEAVE pattern was set up to always proceed in a certain manner. First, the x-axis vectors were drawn left to right and front to back. Then, the y-axis vectors were drawn front to back and left to right. Consider what happens as the y-axis vectors are drawn and the fact that shrinkage lags exposure. As successive vectors are drawn, previous vectors are shrinking, but these vectors have adhered to the x-axis vectors and to the previous layer. In effect, the successive shrinkage of y-axis vectors causes a "wave" of shrinkage from left to right, effectively setting up significant internal stresses. These stresses cause curl.

Given this behavior, it is clear that square cross-sections will have internal stresses, possibly without visible curl. However, if the part cannot curl, the stresses will remain and may result in warpage or other form errors.

With a better understanding of curing and shrinking behavior, the alternating sequence enhancement to building styles was introduced. This behavior can be alleviated to a large extent simply by varying the x and y scan patterns. There are two vector types: x and y. These types can be drawn left to right, right to left, front to back, and back to front. Looking at all combinations, eight different scan sequences are possible. As a part is being built, these eight scan sequences alternate, so that eight consecutive layers have different patterns, and this pattern is repeated every eight layers.

The good news is that internal stresses were reduced and the macrofissures disappeared. However, internal stresses were still evident. To alleviate the internal stresses to a greater extent, the final improvement in STAR-WEAVE was introduced, that of retracted hatch. It is important to realize that the border of a cross-section is scanned first and then the hatch is scanned. As a result, the x-axis vectors adhere to both the left and right border vectors. When they shrink, they pull on the borders, bending them toward one another, causing internal stresses. To alleviate this, alternating hatch vectors are retracted from the border, as shown in Fig. 4.12. This retracted hatch is performed for both the x and y vectors.

4.7.4   ACES Scan Pattern

With the development of epoxy-based photopolymers in 1992–1993, new scan patterns were needed to best adopt to their curing characteristics. ACES (Accurate, Clear, Epoxy, Solid) was the answer to these needs. ACES is not just a scan pattern but is a family of build styles. The operative word in the ACES acronym is Accurate. ACES was mainly developed to provide yet another leap in part accuracy by overcoming deficiencies in STAR-WEAVE, most particularly, in percent of resin cured in the vat. Rather than achieving 96% solidification, ACES is typically capable of 98%, further reducing post-cure shrinkage and the associated internal stresses, curl, and warpage [14].

Machine operators have a lot of control over the particular scan pattern used, along with several other process variables. For example, while WEAVE and STAR-WEAVE utilized 0.001 in. spacings between solidified lines, ACES allows the user to specify hatch spacing. Table 4.2 shows many of the process variables for the SLA-250, the first machine on which ACES was implemented, along with typical ranges of variable settings.

In Table 4.2, the first four variables are called scan variables since they control the scan pattern, while the remaining variables are recoat variables since they control how the vat and part are recoated. With this set of variables, the machine operator has a tremendous amount of control over the process; however, the number of variables can cause a lot of confusion since it is difficult to predict exactly how the part will behave as a result of changing a variable's value. To address this issue, 3D

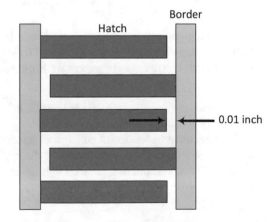

Fig. 4.12 Retracted hatch of the STAR-WEAVE pattern

Table 4.2 ACES process
variables for the SLA-250

Variable	Range
Layer thickness	0.002–0.008 in
Hatch spacing	0.006–0.012 in
Hatch overcure	(−0.003)–(+0.001) in
Fill overcure	0.006–0.012 in
Blade gap %	100–200
Sweep period	5–15 s
Z-Wait	0–20 s

Systems provides nominal values for many of the variables as a function of layer thickness.

The fundamental premise behind ACES is that of curing more resin in a layer before bonding that layer to the previous one. This is accomplished by overlapping hatch vectors, rather than providing 0.001 in. spacing between hatch vectors. As a result, each point in a layer is exposed to laser radiation from multiple scans. Hence, it is necessary to consider these multiple scans when determining cure depth for a layer. ACES also makes use of two passes of scan vectors, one parallel to the x-axis and one parallel to the y-axis. In the first pass, the resin is cured to a depth 1 mil less than the desired layer thickness. Then on the second pass, the remaining resin is cured, and the layer is bonded to the previous one.

As might be imagined, more scan vectors are necessary using the ACES scan pattern, compared with WEAVE and STAR-WEAVE.

The remaining presentation in this section is on the mathematical model of cure depth as a function of hatch spacing to provide insight into the cure behavior of ACES.

Consider Fig. 4.13 that shows multiple, overlapping scan lines with hatch spacing h_s. Also shown is the cure depth of each line, C_{d0}, and the cure depth, C_{d1}, of the entire scan pass. As we know from earlier, the relationship between exposure and cure depth is given by (4.23):

$$C_{d_0} = D_p \ln\left(E_{max} / E_c\right) \tag{4.23}$$

The challenge is to find an expression for cure depth of a scan pass when the scan vectors overlap. This can be accomplished by starting from the relationship describing the spatial distribution of exposure. From earlier, we know that:

$$E(y,0) = E_{max} e^{-\left(2y^2/W_0^2\right)} \tag{4.24}$$

Consider the progression of curing that results from many more scans in Fig. 4.13. If we consider a point P in the region of the central scan, we need to determine the number of scan vectors that provide significant exposure to P. Since the region of influence is proportional to beam spot size, the number of scans depends upon the

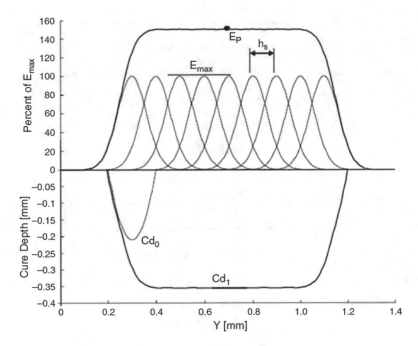

Fig. 4.13 Cure depth and exposure for the ACES scan pattern

beam size and the hatch spacing. Considering that the ratio of hatch spacing to beam half-width, W_0, is rarely less than 0.5 (i.e., $h_s / W_0 \geq 0.5$), then we can determine that point P receives about 99% of its exposure from a distance of 4 h_s or less. In other words, if we start at the center of a scan vector, at most, we need to consider four scans to the left and four scans to the right when determining cure depth.

In this case, we are only concerned with the variation of exposure with y, the dimension perpendicular to the scan direction.

Given that it is necessary to consider nine scans, we know the various values of y in (4.24). We can consider that $y = nh_s$, and let n range from -4 to $+4$. Then, the total exposure received at a point P is the sum of the exposures received over those nine scans, as shown in (4.25) and (4.26):

$$E_P = E_0 + 2E_1 + 2E_2 + 2E_3 + 2E_4 \tag{4.25}$$

$$\text{where } E_n \equiv E(nh_s, 0) = E_{max} e^{-2(nh_s/W_0)^2}$$

$$E_P = E_{max}\left[1 + 2e^{-2(h_s/W_0)^2} + 2e^{-8(h_s/W_0)^2} + 2e^{-18(h_s/W_0)^2} + 2e^{-32(h_s/W_0)^2} \right] \tag{4.26}$$

It is convenient to parameterize exposure vs. E_{max} against the ratio of hatch spacing vs. beam half-width. A simple rearrangement of (4.26) yields (4.27). A plot of (4.27) over the typical range of size ratios (h_s / W_0) is shown in Fig. 4.14:

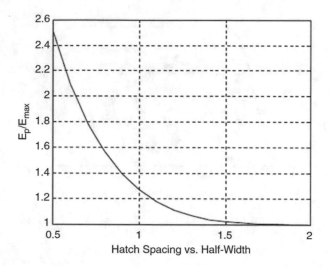

Fig. 4.14 Plot of (4.27): exposure ratios vs. size ratios

$$\frac{E_{\mathrm{p}}}{E_{\max}} = 1 + 2\sum_{n=1}^{4} e^{-2(nh_s/W_0)^2} \tag{4.27}$$

We can now return to our initial objective of determining the cure depth for a single pass of overlapping scan vectors. Further, we can determine the increase in cure depth from a single scan to the entire layer. A cure depth for a single pass, C_{d1}, with overlapping scans is a function of the total exposure given in (4.26). C_{d1} is determined using (4.28):

$$C_{d_1} = D_{\mathrm{P}} \ln\left(E_{\mathrm{p}} / E_{\mathrm{c}}\right) \tag{4.28}$$

The cure depth increase is given by $C_{d1}-C_{d0}$ and can be determined using (4.29):

$$C_{d_1} - C_{d_0} = D_{\mathrm{P}} \ln\left(E_{\mathrm{p}} / E_{\max}\right) \tag{4.29}$$

As an example, consider that we desire a layer thickness to be 4 mils using a resin with a D_{p} of 5.8 mils. Assume further that the desired hatch spacing is 6 mils and the beam half-width is 5 mils, giving a size ratio of $h_s / W_0 = 1.2$. On the first pass, the cure depth, C_{d1}, should be 4–1 = 3 mils. From (4.27), the exposure ratio can be determined to be 1.1123 (or see Fig. 4.14). The cure depth for a single scan vector can be determined by rearranging (4.29) to solve for C_{d0}:

$$\begin{aligned}
C_{d_0} &= C_{d_1} - D_{\mathrm{P}} \ln(E_{\mathrm{P}} / E_{\max} \\
&= 3 - 5.8 * \ln(1.1123) \\
&= 2.383 \text{ mils}
\end{aligned}$$

From this calculation, it is evident that the cure depth of a single scan vector is 1.6 mils less than the desired layer thickness. Rounding up from 1.6 mils, we say that the hatch overcure of this situation is −2 mils. Recall that the hatch overcure is one of the variables that can be adjusted by the SL machine operator.

This concludes the presentation of traditional vector scan VPP. We now proceed to discuss micro-VPP and mask projection-based systems, where areas of the vat surface are illuminated simultaneously to define a part cross-section.

4.8   Vector Scan Micro Vat Photopolymerization

Several processes were developed exclusively for microfabrication applications based on photopolymerization principles using both lasers and X-rays as the energy source. These processes build complex-shaped parts that are typically less than 1 mm in size. They are referred to as Microstereolithography (MSL), Integrated Hardened Stereolithography (IH), LIGA [45], Deep X-ray Lithography (DXRL), and other names. In this section, we will focus on those processes that utilize UV radiation to directly process photopolymer materials.

In contrast to conventional VPP, vector scan technologies for the microscale typically have moved the vat in x, y, and z directions, rather than scanning the laser beam. To focus a typical laser to spot sizes less than 20 μm requires the laser's focal length to be very short, causing difficulties for scanning the laser. For an SLA-250 with a 325 nm wavelength HeCd laser, the beam has a diameter of 0.33 mm and a divergence of 1.25 mrad as it exits the laser. It propagates 280 mm and then encounters a diverging lens (focal length − 25 mm) and a converging lens (focal length 100 mm) which is 85 mm away. Using simple thin-lens approximations, the distance from the converging lens to the focal point, where the laser reaches a spot size of 0.2 mm, is 940 mm, and its Rayleigh range is 72 mm. Hence, the focused laser spot is a long distance from the focusing optics, and the Rayleigh range is long enough to enable a wide scanning region and a large build area.

In contrast, a typical calculation is presented here for a high-resolution micro-VPP system with a laser spot size of 10 μm. A 325 nm wavelength HeCd laser used in SL is included here to give the reader an idea of the challenge. The beam, as it exits the laser, has a diameter of 0.33 mm and a divergence of 1.25 mrad. It propagates 280 mm and then encounters a diverging lens (focal length − 25 mm) and a converging lens (focal length 36.55 mm). The distance from the converging lens to the focal point is 54.3 mm, and its Rayleigh range is only 0.24 mm. It would be very difficult to scan this laser beam across a vat without severe spot distortions.

Scanning micro-VPP systems have been presented in literature since 1993 with the introduction of the Integrated Hardening method of Ikuta and Hirowatari [46]. They used a laser spot focused to a 5-μm diameter, and the resin vat is scanned underneath it to cure a layer. Examples of devices built with this method include tubes, manifolds, and springs and flexible microactuators [47] and fluid channels on silicon [48]. Takagi and Nakajima [49] have demonstrated the use of this technology for connecting MEMS gears together on a substrate. The artifact fabricated using micro-VPP can be used as a mold for subsequent electroplating followed by removal of the resin [50]. This method has been able to achieve sub-1 μm minimum feature size.

The following specifications of a typical point-wise Microstereolithography process have been presented in [51]:

- 5-μm spot size of the UV beam
- Positional accuracy is 0.25 μm (in the x–y directions) and 1.0 μm in the z direction
- Minimum size of the unit of hardened polymer is 5 μm × 5 μm × 3 μm (in x, y, z)
- Maximum size of fabrication structure is 10 mm × 10 mm × 10 mm

The capability of building around inserted components has also been proposed for components such as ultrafiltration membranes and electrical conductors. Applications include fluid chips for protein synthesis [49, 52] and bioanalysis [53]. The bioanalysis system was constructed with integrated valves and pumps that include a stacked modular design, 13×13 mm^2 and 3 mm thick, each of which has different fluid function. However, the full extent of integrated processing on silicon has not yet been demonstrated. The benefits of greater design flexibility and lower cost of fabrication may be realized in the future.

4.9 Mask Projection VPP Technologies and Processes

Technologies to project bitmaps onto a resin surface to cure a layer at a time were first developed in the early 1990s by researchers who wanted to develop special VPP machines to fabricate microscale parts. Several groups in Japan and Europe pursued what was called mask projection stereolithography technology at that time. The main advantage of mask projection methods is speed: since an entire part cross-section can be cured at one time, it can be faster than scanning a laser beam. Dynamic masks can be realized by LCD screens, by spatial light modulators, or by DMDs, such as the Digital Light Processing (DLP™) chips manufactured by Texas Instruments [54].

4.9.1 Mask Projection VPP Technology

Mask projection VPP (MPVPP) systems have been realized by several groups around the world. Some of the earlier systems utilized LCD displays as their dynamic mask [55, 56], while another early system used a spatial light modulator [57]. The remaining systems all used DMDs as their dynamic masks [58–61]. These latest systems all use UV lamps as their radiation source, while others have used lamps in the visible range [58] or lasers in the UV. A good overview of micro-VPP technology, systems, and applications is the book by Varadan et al. [62].

Microscale VPP has been commercialized by Microtec GmbH, Germany. Although machines are not for sale, the company offers customer-specific services. The company has developed machines based on point-wise as well as layer-wise photopolymerization principles. Their Rapid Micro Product Development (RMPD) machines using a HeCd laser enable construction of small parts layer-by-layer (as thin as 1 μm) with a high surface quality in the subnanometer range and with a feature definition of <10 μm.

A schematic and photograph of a MPVPP system from Georgia Tech is shown in Fig. 4.15. Similar to conventional SL, the MPVPP process starts with the CAD model of the part, which is then sliced at various heights. Each resulting slice cross-section is stored as bitmaps to be displayed on the dynamic mask. UV radiation reflects off of the "on" micromirrors and is imaged onto the resin surface to cure a layer. In the system at Georgia Tech, a broadband UV lamp is the light source, a DMD is the dynamic mask, and an automated XYZ stage is used to translate the vat of resin in three dimensions. Standard VPP resins are typically used, although other research groups formulate their own.

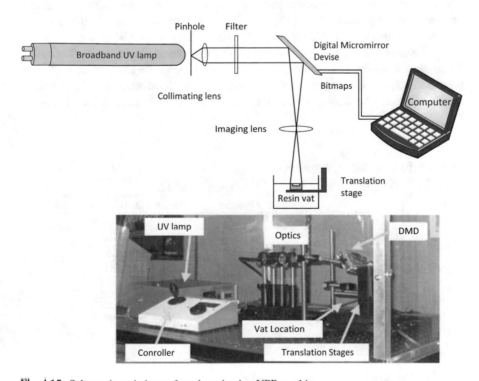

Fig. 4.15 Schematic and photo of mask projection VPP machine

4.9.2   Commercial MPVPP Systems

Several companies market VPP systems based on mask projection technology, including EnvisionTEC and Carbon3D. Newer companies in Europe and Asia also market MPVPP systems.

EnvisionTEC first marketed their MPVPP systems in 2003. They now have several lines of machines with various build envelopes and resolutions based on the MPVPP process that are shown in Table 4.3. Variants of some of these models are available, including specialized Perfactory machines for dental restorations or for hearing aid shells. A photo of the Perfactory Standard P4K machine is shown in Fig. 4.16, and its technical specifications are listed in Table 4.4. Among these models, Micro Plus HD and Advantage have the highest resolutions in X-Y-Z with 30, 60, and 25 µm, respectively.

Schematically, their machines are very similar to the Georgia Tech machine in Fig. 4.15 and utilize a lamp for illuminating the DMD and vat. However, several of their machine models have a very important difference: they build parts upside down and do not use a recoating mechanism. The vat is illuminated vertically upward through a clear window. After the system irradiates a layer, the cured resin sticks to the window and cures into the previous layer. The build platform pulls away from the window at a slight angle to gently separate the part from the window. The advantage of this approach is threefold. First, no separate recoating mechanism is needed since gravity and capillary action forces the resin to fill in the region between the cured part and the window. Second, the top vat surface being irradiated is a flat window, not a free surface, enabling more precise layers to be fabricated. Third, they have devised a build process that eliminates a regular vat. Instead, they have a supply-on-demand material feed system. The disadvantage is that small or fine features may be damaged when the cured layer is separated from the window.

Table 4.3 EnvisionTEC MPVPP machines [63]

Category	Desktop	Perfactory	CDLM
Model	VIDA	P4K 35	Envision One CDLM Mechanical
	VIDA HD	P4K 62	Envision One CDLM Dental
	Micro Plus HD	P4K 75	VIDA UHD CDLM
	Micro Plus XL	P4K 90	Micro Plus CDLM
	VIDA HD Crown & Bridge		VIDA HD CDLM
	Aureus Plus		VIDA CDLM
	Micro Plus Advantage		

Fig. 4.16 EnvisionTEC   (photo courtesy of
Perfactory P4K model   EnvisionTEC) [63]

Table 4.4 Specifications on EnvisionTEC P4K Perfactory machine [63]

Machine properties	P4K 35	P4K 62	P4K 75	P4K 90
Build envelope mm (in.)	90 × 56 × 200 (3.5 × 2.2 × 7.9)	160 × 100 × 200 (6.3 × 3.9 × 7.9)	192 × 120 × 200 (7.6 × 4.7 ×.9)	230 × 143.75 × 200 (9.1 × 5.7 × 7.9)
Native pixel size XY	35 μm	62 μm	75 μm	90 μm
Pixel size with ERM	23 μm	40 μm	49 μm	59 μm
Dynamic voxel resolution in Z (material dependent)	25–150 μm			
Projector resolution	2560 × 1600 (WQXGA)			
Data handling	STL			
Footprint cm (in.)	73 × 48 × 135 (28.7 × 18.9 × 53.1)		73 × 48 × 160 (28.7 × 18.9 × 63.0)	

4.9.3   MPVPP Modeling

Most of the research presented on MPVPP technology is experimental. As in SL, it is possible to develop good predictive models of curing for MPVPP systems. Broadly speaking, models of the MPVPP process can be described by a model that determines the irradiation of the vat surface and its propagation into the resin, followed by a model that determines how the resin reacts to that irradiation. Schematically, the MPVPP model can be given by Fig. 4.17, showing an irradiance model and a cure model.

As a given bitmap pattern is displayed, the resin imaged by the "on" mirrors is irradiated. The exposure received by the resin is simply the product of the irradiance and the time of exposure. The dimensional accuracy of an imaged part cross-section is a function of the radiation uniformity across the DMD, the collimation of the beam, and the capability of the optics system in delivering an undistorted image.

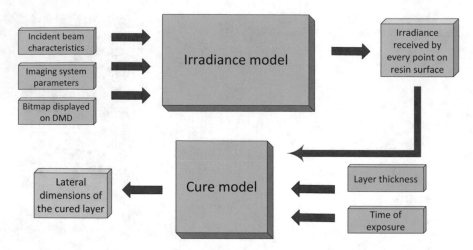

Fig. 4.17 Model of the MPVPP process

If the MPVPP machine's optical system produces a plane wave that is neither converging nor diverging, then it is easy to project rays from the DMD to the resin surface. The irradiance model in this case is very straightforward. However, in most practical cases, it is necessary to model the cone of rays that project from each micromirror on the DMD to the resin. As a result, a point on the resin may receive radiation from several micromirrors. Standard ray-tracing methods can be used to compute the irradiance field that results from a bitmap [64].

After computing the irradiance distribution on the vat surface, the cured shape can be predicted. The depth of cure can be computed in a manner similar to that used in Sect. 4.5. Cure depth is computed as the product of the resin's D_p value and the exponential of the exposure received divided by the resin's E_c value, as in (4.15). The exposure received is simply the product of the irradiance at a point and the time of exposure, T:

$$C_d = D_p e^{-E/E_c} = D_p e^{-H \cdot T/E_c} \qquad (4.30)$$

In the build direction, overcure and print-through errors are evident, as in SL. In principle, however, it is easier to correct for these errors than in point-wise SL systems. A method called the "Compensation Zone" approach was developed to compensate for this unwanted curing [64]. A tailored volume (Compensation Zone) is subtracted from underneath the CAD model to compensate for the increase in the Z dimension that would occur due to print-through. Using this method, more accurate parts and better surface finish can be achieved.

4.9.4   Continuous Liquid Interface Production (CLIP) Technology

CLIP™ is a photochemical process developed by the company Carbon®, Inc. where oxygen is used to control the curing process during VPP. It operates by projecting light through an oxygen-permeable window at the bottom of a reservoir of UV-curable resin. Oxygen inhibits photopolymerization at the window surface, thus keeping the solidifying material from adhering to the window. UV images are projected so that the part solidifies while the build platforms rise in a continuous manner. Parts are built upside down, similar to some of EnvisionTEC's printers.

Using digital light projection, oxygen permeable optics, and programmable liquid resins, parts can be produced with good dimensional accuracy and surface quality and nearly isotropic mechanical properties. A custom LED light engine projects a sequence of UV images. The thin liquid interface of uncured resin between the window and printed part is known as the dead zone. In this dead zone, there is a continuous flow of liquid resin as the platform rises. Just above the dead zone, the UV light causes a partial curing of the part through photopolymerization reactions.

Advances in photopolymer chemistry have led to improved mechanical properties across a variety of material types. At present, materials available include multipurpose, elastomeric, flexible, and rigid polyurethanes, a urethane methacrylate, a cyanate ester, a silicone, and several dental materials [65]. In their fabrication process, part shape is formed, but mechanical properties are not fully defined. A second heat-activated programmable chemistry is used to achieve final properties in a post-processing step in a specially designed forced convection oven [66].

Carbon® CLIP™ technology is used for production applications in medical, aerospace, industrial, consumer, and automotive industries. Some of their materials are specialized for jigs and fixtures and other tooling. Of course, their technology is also used for prototypes in a wide variety of industries. Two interesting applications that Carbon™ has pursued are lattice structures for sport shoe soles and for football helmets.

4.10   Two-Photon Vat Photopolymerization

In the two-photon VPP (2p-VPP) process, the photoinitiator requires two photons to strike it before it decomposes to form a free radical that can initiate polymerization. The effect of this two-photon requirement is to greatly increase the resolution of photopolymerization processes. This is true since only near the center of the laser is the irradiance high enough to provide the photon density necessary to ensure that two photons will strike the same photoinitiator molecule. Feature sizes of 0.2 μm or smaller have been achieved using 2p-VPP.

2p-VPP was first invented in the 1970s for the purpose of fabricating three-dimensional parts [67]. Interestingly, this predates the development of stereolithography by over 10 years. In this approach, two lasers were used to irradiate points in a vat of photopolymer. When the focused laser spots intersected, the photon density was high enough for photopolymerization.

Starting in the late 1990s, 2p-VPP received considerable research attention. A schematic of a typical research setup for this process is shown in Fig. 4.18 [68]. In this system, they used a high-power Ti:Sapphire laser, with wavelength 790 nm, pulse-width 200 fs, and peak power 50 kW. The objective lens had an NA = 0.85. Similarly to other micro-VPP approaches, the vat was moved by a 3D scanning stage, not the laser beam. Parts were built from the bottom-up. The viscosity of the resin was enough to prevent the micropart being cured from floating away. Complicated parts have been produced quickly by various research groups. For example, the micro-bull in Fig. 4.19 was produced in 13 min [69]. The shell of the micro-bull was cured by 2p-VPP, while the interior was cured by flood exposure to UV light.

Typical photopolymer materials can be used in 2p-VPP machines [69–71]. The most commonly used resin was SCR500 from Japan Synthetic Rubber Company, which was a common VPP resin in Japan, where this research originated. SCR500 is a mixture of urethane acrylate oligomers/monomers and common free-radical-generating photoinitiators. The absorption spectrum of the resin shows that it is transparent beyond 550 nm, which is a significant advantage since photons can penetrate the resin to a great depth (D_p is very large). One implication is that parts can be built inside the resin vat, not just at the vat surface, which eliminates the need for recoating.

Photosensitivity of a 2p-VPP resin is measured in terms of the two-photon absorption cross-section (Δ) of the initiator molecule corresponding to the wavelength used to irradiate it. The larger the value of Δ, the more sensitive is the resin to two-photon polymerization, possibly enabling lower-power lasers.

Acrylate photopolymer systems exhibit low photosensitivity as the initiators have small two-photon absorption cross-sections. Consequently, these initiators require high laser power and longer exposure times. Other materials have been

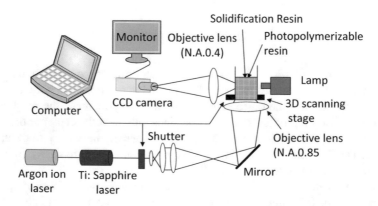

Fig. 4.18 Schematic of typical two-photon equipment

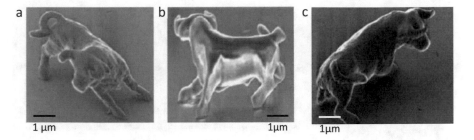

Fig. 4.19 Bull model fabricated by 2p-VPP. The size scale bar is 1 μm

investigated for 2p-VPP, specifically using initiators with larger Δ. New types of photoinitiators tend to be long molecules with certain patterns that make them particularly good candidates for decomposing into free radicals if two photons strike it in rapid succession [72]. By tuning the design of the photoinitiators, large absorption cross-sections and low-polymerization threshold energies can be achieved [73].

The company Nanoscribe has commercialized a 2p-VPP machine called their Photonic Professional GT2 that features a build volume of 100 × 100 × 8 mm and advertised minimum feature size of 160 μm in X and Y directions and 0.5 μm in the Z direction. Researchers have demonstrated features with thickness dimensions of 7–10 nm [74]. The machine uses a near-IR pulsed femtosecond laser which is steered using galvanometers.

Typical applications of 2p-VPP include refractive and diffractive optics, microfluidics, tissue scaffolds, mechanical metamaterials, photonics, and plasmonics. An interesting application is the fabrication of surfaces with structural color capabilities. That is, part features are fabricated as "woodpile structures" (arrays of stacked cylinders) that, when illuminated, projects light of different, designed colors. The examples shown in Fig. 4.20 illustrate the approach [75]. The four different Eifel towers have slightly different cylinder diameters and spacing that produce blue (d), red (e), or multicolored (f, g) versions. An example of the woodpile structure can be seen in the close-up of the Chinese character for "luck" in (j). Note that the scale bars in (d-i) represent 10 μm and scale bar in (j) represents 1 μm.

4.11   Process Benefits and Drawbacks

In addition to the surface finish and part accuracy benefits of VPP mentioned previously, VPP technologies are compatible with many different machine configurations and size scales. Different light sources can be used, including lasers, lamps, or LEDs, as well as different pattern generators, such as scanning galvanometers or DMDs. The size range that has been demonstrated with VPP technologies is vast: from the 1.5 m vat in the ProX 950 machine to the 100 nm features possible with two-photon photopolymerization.

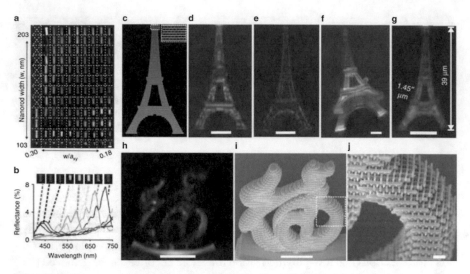

Fig. 4.20 Illustration of structural color capabilities of 2p-VPP [75]

Mask projection VPP technologies have an inherent speed advantage over laser scan SL. By utilizing a mask, an entire part cross-section can be projected, rather than having to sequentially scan the vector pattern for the cross-section. There is a trade-off between resolution and the size of the pattern (and size of the solidified cross-section) due to the mask resolution. For example, typical DMDs have 1280×720 or 1920×1080 resolution, and newer DMDs are being developed with higher resolution. For the 1920×1080 DMD resolution, the maximum part size with 50 μm resolution is 96×54 mm. To fabricate larger parts, a step-and-repeat approach or multiple DMDs are used.

A drawback of VPP processes is their usage of photopolymers, since the chemistries are limited to acrylates and epoxies for most commercial materials. Although quite a few other material systems are photopolymerizable, few have emerged as commercial successes to displace the current chemistries. Generally, the current VPP materials do not have the impact strength and durability of good quality injection molded thermoplastics. Additionally, they are known to age, resulting in degraded mechanical properties over time. The company Carbon™ has developed resins that are more ductile, durable, and with long useful life, which is enabling many production manufacturing applications.

4.12   Summary

Photopolymerization processes make use of liquid, radiation-curable resins called photopolymers to fabricate parts. Upon irradiation, these materials undergo a chemical reaction to become solid. Several methods of illuminating photopolymers for part fabrication were presented, including vector scan point-wise processing, mask projection layer-wise processing, and two-photon approaches. The vector scan approach is used with UV lasers in the VPP process, while DLP micromirror array chips are commonly used for mask projection technologies. Two-photon approaches, which have the highest resolution, remain of research interest only. Advantages, disadvantages, and unique characteristics of these approaches were summarized.

Photopolymerization processes lend themselves to accurate analytical modeling due to the well-defined interactions between radiation and photopolymers. An extensive model for laser scan VPP was presented, while a simpler one for MPVPP was summarized. Discretization errors and scan patterns were covered in this chapter to convey a better understanding of these concepts as they apply to photopolymerization processes, as well as many of the processes still to be presented in this book. The excellent accuracy and surface finish characteristics of VPP processes make them good candidates for components which require tight tolerances and well-defined textures. The degradation of photopolymers over time and their limited mechanical properties make VPP-produced parts well-suitable for form and fit applications, but improved photopolymer formulations are needed to enable their broad usage for functional applications.

4.13   Questions

1. Explain why VPP is a good process to use to fabricate patterns for investment casting of metal parts. (0.5 page+).
2. Explain why two photoinitiators are needed in most commercial VPP resins. Explain what these photoinitiators do.
3. Assume you are building with the STAR-WEAVE build style under the following conditions: layer thickness = 0.006″, D_p = 6.7 mil, E_c = 9.9 mJ/cm^2 (SL-5240), machine = SLA-250/50 (laser power = 30 mW), W_0 = 0.125 mm, vat width = 250 mm. You will also need to know the distance that the laser travels from the scanning galvanometers to the resin vat. Assume this distance is 940 mm.

   (a) Determine the cure depths C_{d1} and C_{d2} needed.
   (b) Compute the laser scan speeds required for C_{d1} and C_{d2}.
   (c) Determine laser scan speeds required C_{d1} and C_{d2} when building along an edge of the vat.

4. Assume you are building with the ACES build style under the following conditions: layer thickness = 0.1 mm, D_p = 4.6 mil, E_c = 8.3 mJ/cm^2 (3D Systems Accura Xtreme White 200 material), machine = ProX 800 with large spot size. Assume laser power is 200 mW at the vat and the hatch ratio (h_s / W_0) is 1.2. Use the distance of 1050 mm from the scanning galvos to the resin vat for the ProX 800.

(a) Determine the cure depths C_{d0} and C_{d2} needed.
(b) Compute the laser scan speeds required for C_{d0} and C_{d2}.
(c) Determine the overcure error.
(d) Determine laser scan speeds required for C_{d0} and C_{d2} when building along an edge of the vat (widest dimension), taking into account the laser beam angle.

5. Repeat problem 4, but with a layer thickness of 0.05 mm.

   (a) Determine the cure depths C_{d0} and C_{d2} needed.
   (b) Compute the laser scan speeds required for C_{d0} and C_{d2}.
   (c) Determine the overcure error.
   (d) Determine laser scan speeds required for C_{d0} and C_{d2} when building along an edge of the vat (widest dimension), taking into account the laser beam angle.
   (e) Do these results make sense? It seems that the ACES build style does not work well for very small layers. Explain.

6. Assume we are interested only in border scans when building in a ProX 800. Further, assume that only one border scan is used and the laser power is reduced to 0.02 W. For the conditions in problem 4, determine the following.

   (a) For a single border scan and layer thickness of 0.05 mm, determine the scan speed required for a cure depth of 0.06 mm.
   (b) For a single border scan at 2000 mm/sec, what would be the cure depth?
   (c) For this 2000 mm/sec scan, what is the line width?

7. Now, assume that a cyclic build style is being used, where borders are scanned using layer thicknesses of 0.05 mm, but interiors of cross-sections are scanned with the large spot size every 4 layers, for an effective layer thickness of 0.2 mm. Repeat problem 4 with a laser power of 1 W and a hatch ratio (h_s / W_0) of 1.4.

   (a) Determine the cure depths C_{d0} and C_{d2} needed.
   (b) Compute the laser scan speeds required for C_{d0} and C_{d2}.
   (c) Determine the overcure error.
   (d) Determine laser scan speeds required for C_{d0} and C_{d2} when building along an edge of the vat (widest dimension), taking into account the laser beam angle.

8. In the derivation of exposure (4.10) for a scan from 0 to $x = b$, several steps were skipped.

   (a) Complete the derivation of (4.10). Note that the integral of e^{-v^2} from 0 to b is
   $$\int_0^b e^{-v^2} dv = \frac{\sqrt{\pi}}{2} erf(v) \Big|_0^b ,$$
   where $erf(v)$ is the error function of variable v (see MATLAB or other math source for explanation of $erf(v)$).

(b) Compute the exposure received from this scan at the origin, at $x = 10$ mm, and at $b = 20$ mm using the conditions in Prob. 3b, where laser power is 60 mW.

(c) Now, let $b = 0.05$ mm and recompute the exposure received at the origin and point b. Compare with results of part (b). Explain the differences observed.

9. Consider a tall thin rib that consists of a stack of 10 vector scans. That is, the rib consists of ten layers, and on each layer, only one vector scan is drawn.

   (a) Derive an expression for the width of the rib at any point z along its height.
   (b) Develop a computer program to solve your rib width equation.
   (c) Using your program, compute the rib widths along the height of the rib and plot a graph of rib width. Use the conditions of Prob. 4 and a scan speed of 1000 mm/s.
   (d) Repeat part (c) using a scan speed of 5000 mm/s. Note the differences between your graphs from (c) and (d).

10. What are some benefits of using photopolymers for Additive Manufacturing? What are the downsides?

11. Explain why the color of VPP-fabricated parts change over time. Also explain what scanning strategy is used in SL and if SL needs support structure or not.

12. Compare the difference between acrylate-based and epoxy-based resins for VPP. Discussed respective benefits and disadvantages with respect to scanning method.

13. What is vector scan Microstereolithography? Compare Microstereolithography technologies and processes with conventional stereolithography.

14. In the process of making molds using indirect soft tooling, it is common to make the "pattern" using VPP. Discuss why this is common. It might be necessary to do a little research to understand what soft tooling is.

15. (a) Why do the parts made by VPP degrade over time? (b) What finishing processes are used for parts fabricated by VPP?

16. Describe the two-photon polymerization and its capability.

17. What attributes does two-photon stereolithography provide over other processes and describe an example that benefits from these attributes?

18. What are the three types of scan patterns used to cure photopolymers, then compare and contrast the three types?

19. Why is it necessary to add some acrylate to epoxy resins?

20. What are the main advantages of VPP technology over other AM technologies?

21. Why are epoxy-type photopolymers less susceptible to curling/shrinkage than other AM materials and processes?

22. One of the main advantages of VPP is that we can use semi-transparent polymers, so that internal details of parts can be discerned readily. List and describe at least two parts where this feature is valuable.

23. The shape of the cure region in the z direction of photopolymer is parabolic, see Fig. 4.6, and explain how this proves an inherent challenge to create layers of even depth and how the use of STAR-WEAVE scan pattern, Fig. 4.10, generally overcomes this challenge.

24. At the end of the chapter, three VPP processes were presented that could create parts with micro or nanometer dimensions: vector scan MSL, mask projection MSL, and two-photon SL. However, not much was mentioned about potential applications. Choose one of these microprocesses and find two research articles that involve that process. Write a one-paragraph summary of each article, stating the purpose of the research and why they chose that particular method, and talk about any findings and/or potential applications. (List the references and limit the age of the research articles to 3 years old at most.)

25. State the working curve equation, describing the cure depth with relation to laser and resin properties (include necessary nomenclature). Describe how all of these laser and resin properties would affect the cure depth when increased (individually increased while all other terms remain constant).

26. In the SL process, what are the irradiance and exposure of the laser energy? Where does the maximum irradiance occur as the laser hits on the resin surface? And calculate the maximum irradiance for an SLA250/50 machine. Similarly, determine the location where the maximum exposure occurs at the same situation, and then calculate this maximum value.

27. ACES is not just a scan pattern but is a family of build styles. It allows the user to specify hatch spacing. Table Q 4.1 shows many of the process variables for the SLA-250 along with typical ranges of variable settings. Consider the layer thickness to be 4 mils using a resin with a Dp of 5.8 mils. Assume further that the desired hatch spacing is 6mils and the beam half-width is 5mils, giving a size ratio of $hs/W_0 = 1.2$. Calculate the cure depth for a single scan vector C_{d0}.

Table Q 4.1 ACES process variables for the SLA-250

Variable	Range
Layer thickness	0.002–0.008 in
Hatch spacing	0.006–0.012 in
Hatch overcure	(−0.003)–(+0.001) in
Fill overcure	0.006–0.012 in
Blade gap %	100–200
Sweep period	5–15 s
Z-Wait	0–20 s
Pre-Dip delay	0–20 s

28. For the following SL machines with the given parameters find:

   (a) Maximum irradiance of the beam spot
   (b) Maximum exposure of the beam spot

1. For SLA 5000 machine, power at vat is 216 mW, beam radius is 0.25 mm and scans speed 5.0 m/sec, calculate the maximal exposure.
2. For SLA 250/50 machine, power at vat is 24 mW, beam radius is 0.24 mm and scans speed 762 mm/sec, calculate the maximal exposure.

29. You are using the WEAVE pattern to lay down the foundation for a new part. Assume that your resin has a D_p = 3.6 mil, cure depth on the first pass was 1.3 mil (0.25 mm), and each resin layer is 5.2 mil thick. What would the cure depth for the second pass (Cd2) be for this process? Will this depth cause any problems with the build? If so, what problems?

30. For ACES scan pattern, we usually will consider four scans to calculate the total exposure and the cure depth. Why we need to consider the nearby scans? Why we choose four scans each side? What will happen if the h_s/W_0 ratio is increasing?

31. Using the method provided above to calculate the building time of the following four cases:

(a) Laser power, 48 mW; volume (part), 5620mm^3; resin used, SL 5170; and maximum z height 45 mm
(b) Laser power, 13 mW; volume (part), 73258.3mm^3; resin used, Somos 9110; and maximum z height 120 mm
(c) Laser power, 33 mW; volume (part), 35869.98mm^3; resin used, Somos 9110; and maximum z height 49 mm
(d) Laser power, 32 mW; volume (part), 16242.622mm^3; resin used, Somos 9110; and maximum z height 44.8 mm

References

1. Jacobs, P. F. (1992). *Rapid prototyping & manufacturing: fundamentals of stereolithography*. Society of Manufacturing Engineers.
2. Tang, Y. (2002). *Stereolithography (SL) cure modeling. Masters Thesis*, in *School of Chemical Engineering*. Georgia Institute of Technology.
3. Beaman, J. J., et al. (1997). *A new direction in manufacturing*. In *Solid freeform fabrication*. Norwell: Kluwer Academic Publishers.
4. Hull, C. W. (1990). *Method for production of three-dimensional objects by stereolithography*. US Patents.
5. Murphy, E. J., Ansel, R. E., & Krajewski, J. J. (1989). *Investment casting utilizing patterns produced by stereolithography*. US Patents.
6. Wohlers, T. (1991). *Wohlers Report. Rapid Prototyping: an Update on RP Applications, Technology Improvements, and Developments in the Industry*. Wohlers Associates.
7. Lu, L., et al. (1995). Origin of shrinkage, distortion and fracture of photopolymerized material. *Materials Research Bulletin, 30*(12), 1561–1569.
8. Denka, A. (1988). *Patent 2,590,216*, Japan's international intellectual property protection association.
9. Denka, A. (1988). *Patent 2,138,471*, Japan's international intellectual property protection association.

10. Crivello, J., & Dietliker, K. (1998). *Photoinitiators for free radical cationic & anionic photopolymerisation.*
11. Dufour, P. (1993). State-of-the-art and trends in the radiation-curing market. *Radiation curing in polymer science and technology, 1,* 2.
12. Jariwala, A. S., Rosen, D. W., & Ding, F. (2016). *Fabricating parts from photopolymer resin.* US Patents.
13. Mahan, K., Rosen, D., & Han, B. (2016). Blister testing for adhesion strength measurement of polymer films subjected to environmental conditions. *Journal of Electronic Packaging, 138*(4), 041003.
14. Jacobs, P. F. (1995). *Stereolithography and other RP&M technologies: from rapid prototyping to rapid tooling.* Society of Manufacturing Engineers.
15. Wilson, J. E., & Burton, M. (1974). Radiation chemistry of monomers, polymers, and plastics. *Physics Today, 27,* 50.
16. Fouassier, J. (1993). An introduction to the basic principles in UV-curing. In *Radiation curing in polymer science and technology.* p. P49.
17. Decker, C., & Elzaouk, B. (1995). Laser curing of photopolymers. In *Current trends in polymer photochemistry.* New York: Ellis Horwood.
18. Andrzejewska, E. (2001). Photopolymerization kinetics of multifunctional monomers. *Progress in Polymer Science, 26*(4), 605–665.
19. Hageman, H. (1989). Photoinitiators and photoinitiation mechanisms of free-radical polymerisation processes. In *Photopolymerisation and photoimaging science and technology* (pp. 1–53). Cham: Springer.
20. Crivello, J. (1993). Latest developments in the chemistry of onium salts. In *Radiation curing in polymer science and technology* (pp. 435–471). London: Elsevier.
21. Bassi, G. L. (1993). Formulation of UV-curable coatings–how to design specific properties. In *Radiation curing in polymer science and technology* (p. P239). London: Elsevier.
22. Crivello, J. (1984). Cationic polymerization—iodonium and sulfonium salt photoinitiators. In *Initiators—poly-reactions—optical activity* (pp. 1–48). Berlin: Springer.
23. Crivello, J., & Lee, J. (1990). The synthesis, characterization, and photoinitiated cationic polymerization of silicon-containing epoxy resins. *Journal of Polymer Science Part A: Polymer Chemistry, 28*(3), 479–503.
24. Crivello, J. V., & Lee, J. L. (1988). *Method for making polymeric photoactive aryl iodonium salts, products obtained therefrom, and use.* US Patents.
25. Crivello, J. V., & Lee, J. (1989). Alkoxy-substituted diaryliodonium salt cationic photoinitiators. *Journal of Polymer Science Part A: Polymer Chemistry, 27*(12), 3951–3968.
26. Melisaris, A. P., Renyi, W., & Pang, T. H. (2000). *Liquid, radiation-curable composition, especially for producing flexible cured articles by stereolithography.* US Patents.
27. Pang, T.H., et al. (2000). *Liquid radiation-curable composition especially for producing cured articles by stereolithography having high heat deflection temperatures.* US Patents.
28. Steinmann, B., & Schulthess, A. (1999). *Liquid, radiation-curable composition, especially for stereolithography.* US Patents.
29. Steinmann, B., et al. (1995). *Photosensitive compositions. Ciba-Geigy Corporation,* US Patents.
30. Decker, C., & Decker, D. (1994). *Kinetic and mechanistic study of the UV-curing of vinyl ether based systems.* In *Proc Rad Tech Conf, Orlando.*
31. Cribello, J., Lee, J., & D. Conlon. (1983). Photoinitiated cationic polymerization with multifunctional vinyl ether monomers. *Journal of Radiation Curing, 10*(1), 6–13.
32. Sperling, L. H. (2012). *Interpenetrating polymer networks and related materials.* Springer Science & Business Media.
33. Decker, C., et al. (2001). UV-radiation curing of acrylate/epoxide systems. *Polymer, 42*(13), 5531–5541.
34. Sperling, L., & Mishra, V. (1996). *Polymeric materials encyclopedia.* CRC Press. p. 3292.

35. Chen, M., et al. (2001). Mechanism and application of hybrid UV curing system. *Journal of Photochemistry and Photobiology A, 19*(3), 208–216.
36. Decker, C. (1996). Photoinitiated crosslinking polymerisation. *Progress in Polymer Science, 21*(4), 593–650.
37. Decker, C., Xuan, H. L., & Viet, T. N. T. (1996). Photocrosslinking of functionalized rubber. III. Polymerization of multifunctional monomers in epoxidized liquid natural rubber. *Journal of Polymer Science Part A: Polymer Chemistry, 34*(9), 1771–1781.
38. Perkins, W. (1981). New developments in photo-induced cationic polymerization. *Journal of Radiation Curing, 8*(1), 16.
39. *3D Systems, Inc. AccuMax Toolkit User Guide, 3D Systems, Inc.* (1996). D. Systems, Editor. Valencia.
40. Renap, K., & Kruth, J.-P. (1995). Recoating issues in stereolithography. *Rapid Prototyping Journal, 1*(3), 4–16.
41. Lynn-Charney, C., & Rosen, D. W. (2000). Usage of accuracy models in stereolithography process planning. *Rapid Prototyping Journal, 6*(2), 77–87.
42. West, A. P. (1999). *A decision support system for fabrication process planning in stereolithography*. Georgia Institute of Technology.
43. *3D Systems*. (2020). http://www.3dsystems.com.
44. *3D Systems, 3D Printers*. (2020). https://www.3dsystems.com/3d-printers#metal-3d-printers.
45. Yi, F., Wu, J., & Xian, D. (1993). LIGA technique for microstructure fabrication. *Microfabrication Technology, 4*(1), 1–7.
46. Ikuta, K. & Hirowatari, K. (1993). Real three dimensional micro fabrication using stereo lithography and metal molding. In *Micro Electro Mechanical Systems, 1993, MEMS'93, Proceedings An Investigation of Micro Structures, Sensors, Actuators, Machines and Systems. IEEE*. IEEE.
47. Suzumori, K., Koga, A., & Riyoko, H. (1994). Microfabrication of integrated FMAs using stereo lithography. In *Micro Electro Mechanical Systems, 1994, MEMS'94, Proceedings, IEEE Workshop on*. IEEE.
48. Ikuta, K., Hirowatari, K., & Ogata, T. (1994). Three dimensional micro integrated fluid systems (MIFS) fabricated by stereo lithography. In *Micro Electro Mechanical Systems, 1994, MEMS'94, Proceedings, IEEE Workshop on*. IEEE.
49. Takagi, T., & Nakajima, N. (1994). Architecture combination by micro photoforming process. In *Micro Electro Mechanical Systems, 1994, MEMS'94, Proceedings, IEEE Workshop on*. IEEE.
50. Ikuta, K., et al. (1999). Micro concentrator with opto-sense micro reactor for biochemical IC chip family. 3D composite structure and experimental verification. In *Micro Electro Mechanical Systems, 1999. MEMS'99. Twelfth IEEE International Conference on*.IEEE.
51. Gardner, J. W., & Varadan, V. K. (2001). *Microsensors, MEMS and smart devices*. Chichester: John Wiley & Sons, Inc.
52. Ikuta, K., et al. (1996). Development of mass productive micro stereo lithography (Mass-IH Process). In *Micro Electro Mechanical Systems, 1996, MEMS'96, Proceedings. An Investigation of Micro Structures, Sensors, Actuators, Machines and Systems. IEEE, The Ninth Annual International Workshop on*. IEEE.
53. Ikuta, K., et al.. (1998). *Chemical IC chip for dynamical control of protein synthesis*. In *Micromechatronics and Human Science, 1998. MHS'98. Proceedings of the 1998 International Symposium on*. IEEE.
54. Dudley, D., Duncan, W. M., & Slaughter, J. Emerging digital micromirror device (DMD) applications. In *MOEMS display and imaging systems*. International Society for Optics and Photonics.
55. Bertsch, A., et al. (1997). Microstereophotolithography using a liquid crystal display as dynamic mask-generator. *Microsystem Technologies, 3*(2), 42–47.

56. Monneret, S., Loubere, V., & Corbel, S. (1999). Microstereolithography using a dynamic mask generator and a noncoherent visible light source. In *Design, Test, and Microfabrication of MEMS and MOEMS*. International Society for Optics and Photonics.
57. Chatwin, C., et al. (1998). UV microstereolithography system that uses spatial light modulator technology. *Applied Optics, 37*(32), 7514–7522.
58. Bertsch, A., et al. (2000). Rapid prototyping of small size objects. *Rapid Prototyping Journal, 6*(4), 259–266.
59. Hadipoespito, G., et al. (2003). Digital Micromirror device based microstereolithography for micro structures of transparent photopolymer and nanocomposites. In *Solid Freeform Fabrication Symposium,* Austin, TX.
60. Limaye, A., & Rosen, D.. (2004). Quantifying dimensional accuracy of a mask projection micro stereolithography system. In *Proceedings of the 15th Solid Freeform Fabrication Symposium,* Austin Texas.
61. Sun, C., et al. (2005). Projection micro-stereolithography using digital micro-mirror dynamic mask. *Sensors and Actuators A: Physical, 121*(1), 113–120.
62. Varadan, V. K., Jiang, X., & Varadan, V. V. (2001). *Microstereolithography and other fabrication techniques for 3D MEMS*. New York: John Wiley & Sons Inc..
63. Envision TEC. 2020. https://envisiontec.com/3d-printers/perfactory-family/
64. Shankar Limaye, A., & Rosen, D. W. (2006). *Compensation zone approach to avoid print-through errors in mask projection stereolithography builds. Rapid Prototyping Journal, 12*(5), 283–291.
65. Carbon's CLIP technology. (2020). https://www.tctmagazine.com/3d-printing-news/carbon-clip-technology-cost-effective-3d-printing-molecular-model-kits/
66. *Carbon*. (2020). https://www.carbon3d.com/our-technology/
67. Swanson, W. K., & Kremer, S. D. (1978). *Three dimensional systems*. US Patents.
68. Maruo, S., Nakamura, O., & Kawata, S. (1997). Three-dimensional microfabrication with two-photon-absorbed photopolymerization. *Optics Letters, 22*(2), 132–134.
69. Kawata, S., et al. (2001). Finer features for functional microdevices. *Nature, 412*(6848), 697.
70. Miwa, M., et al. (2001). Femtosecond two-photon stereo-lithography. *Applied Physics A, 73*(5), 561–566.
71. Sun, H.-B., et al. (2000). Real three-dimensional microstructures fabricated by photopolymerization of resins through two-photon absorption. *Optics Letters, 25*(15), 1110–1112.
72. Albota, M., et al. (1998). Design of organic molecules with large two-photon absorption cross sections. *Science, 281*(5383), 1653–1656.
73. Cumpston, B. H., et al. (1999). Two-photon polymerization initiators for three-dimensional optical data storage and microfabrication. *Nature, 398*(6722), 51.
74. Wang, S., Yu, Y., Liu, H., Lim, K. T. P., Srinivasan, B. M., Zhang, Y. W., & Yang, J. K. W. (2018). Sub-10-nm suspended nano-web formation by direct laser writing. *Nano Futures, 2*(2), 025006.
75. Liu, Y., et al. (2019). Structural color three-dimensional printing by shrinking photonic crystals. *Nature Communications, 10*(1), 1–8.

Chapter 5
Powder Bed Fusion

Abstract Powder Bed Fusion (PBF) was one of the earliest and remains one of the most versatile AM processes, being well-suited for polymers and metals and, to a lesser extent, ceramics and composites. There are an increasing number of machine variants for fusing powders using different energy sources. The most active area of development is for metal PBF processes using lasers. Laser-Based Powder Bed Fusion (LB-PBF) processes are of great interest across many industries as a means of direct manufacturing. This chapter will cover various approaches to PBF, issues surrounding the handling of powders, and the growing types of applications for these technologies.

5.1 Introduction

PBF processes were among the first commercialized AM processes. Developed at the University of Texas at Austin, USA, Selective Laser Sintering (SLS) was the first commercialized PBF process. Its basic method of operation is schematically shown in Fig. 5.1, and all other PBF processes modify this basic approach in one or more ways to enhance machine productivity, to enable different materials to be processed, and/or to avoid specific patented features.

All PBF processes share a basic set of characteristics [1, 2]. These include one or more thermal sources for inducing fusion between powder particles, a method for controlling powder fusion to a prescribed region of each layer, and mechanisms for adding and smoothing powder layers. The most common thermal sources for PBF are lasers. PBF processes which utilize lasers are known as laser sintering (LS) machines. Since polymer laser sintering (pLS) machines and metal laser sintering (mLS) machines are significantly different from each other, we will address each separately. In addition, electron beam and other thermal sources require significantly different machine architectures than laser sintering machines. Non-laser thermal sources will be addressed separately from laser sources at the end of the chapter.

© Springer Nature Switzerland AG 2021
I. Gibson et al., *Additive Manufacturing Technologies*,
https://doi.org/10.1007/978-3-030-56127-7_5

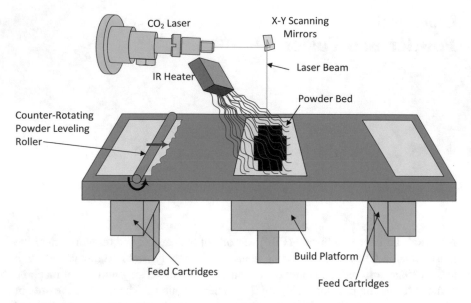

Fig. 5.1 Schematic of the Selective Laser Sintering process

LS processes were originally developed to produce plastic prototypes using a point-wise laser scanning technique. This approach was subsequently extended to metal and ceramic powders; additional thermal sources are now utilized; and variants for layer-wise fusion of powdered materials have been commercially introduced. As a result, PBF processes are widely used worldwide, have a broad range of materials (including polymers, metals, ceramics, and composites) which can be utilized, and are increasingly being used for direct manufacturing of end-use products, as the material properties are comparable to many engineering-grade polymers, metals, and ceramics made using conventional means.

In order to provide a baseline description of PBF processes, polymer laser sintering (pLS) will be described as the paradigm approach to which the other PBF processes will be compared. As shown in Fig. 5.1, pLS fuses thin layers of powder (typically 0.075–0.1 mm thick) which have been spread across the build area using a counter-rotating powder leveling roller. The part building process takes place inside an enclosed chamber filled with nitrogen gas to minimize oxidation and degradation of the powdered material. The powder in the build platform is maintained at an elevated temperature just below the melting point and/or glass transition temperature of the powdered material. Infrared heaters are placed above the build platform to maintain an elevated temperature around the part being formed, as well as above the feed cartridges to preheat the powder prior to spreading over the build area. In some cases, the build platform is also heated using resistive heaters around the build platform. This preheating of powder and maintenance of an elevated, uniform temperature within the build platform are necessary to minimize the laser power requirements of the process (with preheating, less laser energy is required for fusion) and to prevent warping of the part during the build due to non-uniform thermal expansion and contraction (resulting in curling) [3–5].

Once an appropriate powder layer has been formed and preheated, a focused CO_2 laser beam is directed onto the powder bed and is moved using galvanometers in such a way that it thermally fuses the material to form the slice cross-section. Surrounding powder remains loose and serves as support for subsequent layers, thus eliminating the need for the secondary supports which are necessary for Vat Photopolymerization (VPP) processes. After completing a layer, the build platform is lowered by one layer thickness, and a new layer of powder is laid and leveled using the counter-rotating roller. The beam scans the subsequent slice cross-section. This process repeats until the complete part is built. A cool-down period is typically required to allow the parts to uniformly come to a low enough temperature that they can be handled and exposed to ambient temperature and atmosphere. If the parts and powder bed are prematurely exposed to ambient temperature and atmosphere, the powders may degrade in the presence of oxygen, and the parts may warp due to uneven thermal contraction. Finally, the parts are removed from the powder bed, loose powder is cleaned off the parts, and further finishing operations, if necessary, are performed.

5.2 Materials

In principle, all materials that can be melted and resolidified can be used in PBF processes. A brief survey of materials processed using PBF processes will be given here. More details can be found in subsequent sections.

5.2.1 Polymers and Composites

Thermoplastic materials are well-suited for powder bed processing because of their relatively low melting temperatures, low thermal conductivities, and low tendency for balling. Polymers in general can be classified as either a thermoplastic or a thermoset polymer. Thermoset polymers are typically not processed using PBF into parts, since PBF typically operates by melting particles to fabricate part cross-sections, but thermosets degrade, but do not melt, as their temperature is increased. Thermoplastics can be classified further in terms of their crystallinity. Amorphous polymers have a random molecular structure, with polymer chains randomly intertwined. In contrast, crystalline polymers have a regular molecular structure. Much more common are semi-crystalline polymers which have regions of regular structure, called crystallites, and regions of amorphous structure. Amorphous polymers melt over a fairly wide range of temperatures. As the crystallinity of a polymer increases, however, its melting characteristics tend to become more centered around a well-defined melting point.

At present, the most common material used in PBF is polyamide, a thermoplastic polymer, commonly known in the USA as nylon. Most polyamides have fairly high crystallinity and are classified as semi-crystalline materials. They have distinct melting points that enable them to be processed reliably. A given amount of laser energy will melt a certain amount of powder; the melted powder fuses and cools, forming part of a cross-section. In contrast, amorphous polymers tend to melt gradually, cool more slowly, and not form well-defined solidified features. In pLS, amorphous polymers tend to sinter into highly porous shapes, whereas crystalline polymers are typically processed using full melting, which result in higher densities. Polyamide 11 and polyamide 12 are commercially available, where the number designates the number of carbon atoms that are provided by one of the monomers that is reacted to produce polyamide. However, crystalline polymers exhibit greater shrinkage compared to amorphous materials and are more susceptible to curling and distortion and thus require more uniform temperature control. Mechanical properties of pLS parts produced using polyamide powders are similar to those of injection molded thermoplastic parts, but with significantly reduced elongation and unique microstructures.

Polystyrene-based materials with low residual ash content are particularly suitable for making sacrificial patterns for investment casting using pLS. Interestingly, polystyrene is an amorphous polymer but is a successful example material due to its intended application. Porosity in an investment casting pattern aids in melting out the pattern after the ceramic shell is created. Polystyrene parts intended for precision investment casting applications should be sealed to prevent ceramic material seeping in and to achieve a smooth surface finish.

Elastomeric thermoplastic polymers are available for producing highly flexible parts with rubber-like characteristics. These elastomers have good resistance to degradation at elevated temperatures and are resistant to chemicals like gasoline and automotive coolants. Elastomeric materials can be used to produce gaskets, industrial seals, shoe soles, and other components.

Additional polymers that are commercially available include flame-retardant polyamide and polyaryletherketone (known as PAEK or PEEK). Both 3D Systems and EOS GmbH offer most of the materials listed in this section.

Researchers have investigated quite a few polymers for biomedical applications. Several types of biocompatible and biodegradable polymers have been processed using pLS, including polycaprolactone (PCL), polylactide (PLA), and poly-L-lactide (PLLA). Composite materials consisting of PCL and ceramic particles, including hydroxyapatite and calcium silicate, have also been investigated for the fabrication of bone replacement tissue scaffolds.

In addition to neat polymers, polymers in PBF can have fillers that enhance their mechanical properties. For example, the Duraform material from 3D Systems is offered as Duraform PA, which is polyamide 12, as well as Duraform GF, which is polyamide 12 filled with small glass beads. The glass additive enhances the material's stiffness significantly but also causes its ductility to be reduced, compared to polyamide materials without fillers. Additionally, EOS GmbH offers aluminum particle, carbon fiber, and their own glass bead filled polyamide materials.

5.2.2 Metals and Composites

A wide range of metals has been processed using PBF. Generally, any metal that can be welded is considered to be a good candidate for PBF processing. Several types of steels, typically stainless and tool steels, titanium and its alloys, nickel-based alloys, some aluminum alloys, and cobalt-chrome have been processed and are commercially available in some form. Additionally, some companies offer PBF of precious metals, such as silver and gold.

Historically, a number of proprietary metal powders (either thermoplastic binder-coated or binder mixed) were developed before modern mLS machines were available. RapidSteel was one of the first metal/binder systems, developed by DTM Corp. The first version of RapidSteel was available in 1996 and consisted of a thermoplastic binder-coated 1080 carbon steel powder with copper as the infiltrant. RapidSteel achieved fusion using liquid-phase sintering of separate particles, whereas subsequent variants were coated particles, as described in Sect. 5.3.3.1 below. Parts produced using RapidSteel were debinded (350–450 °C), sintered (around 1000 °C), and finally infiltrated with Cu (1120 °C) to produce a final part with approximately 60% low carbon steel and 40% Cu. This is an example of liquid-phase sintering which will be described in the next section.

RapidSteel 2.0 powder was introduced in 1998 for producing functional tooling, parts and mold inserts for injection molding. It was a dry blend of 316 stainless steel powder impact milled with thermoplastic and thermoset organic binders with an average particle size of 33 μm. After green part fabrication, the part was debinded and sintered in a hydrogen-rich atmosphere. The bronze infiltrant was introduced in a separate furnace run to produce a 50% steel and 50% bronze composite. RapidSteel 2.0 was structurally more stable than the original RapidSteel material because the bronze infiltration temperature was less than the sintering temperature of the stainless steel powder. A subsequent material development was LaserForm ST-100, which had a broader particle size range, with fine particles not being screened out. These fine particles allowed ST-100 particles to be furnace sintered at a lower temperature than RapidSteel 2.0, making it possible to carry out sintering and infiltration in a single furnace run. In addition to the above, H13 and A6 tool steel powders with a polymer binder can also be used for tooling applications. The furnace processing operations (sintering and infiltration) must be carefully designed with appropriate choices of temperature, heating and cooling rates, furnace atmosphere pressure, amount of infiltrant, and other factors, to prevent excessive part distortion. After infiltration, the part is finish machined as needed. These issues are further explored in the post-processing chapter.

Several proprietary metal powders were marketed by EOS for their M250 Xtended metal platforms, prior to the introduction of modern mLS machines. These included liquid-phase sintered bronze-based powders and steel-based powders and other proprietary alloys (all without polymer binders). These were suitable for producing tools and inserts for injection molding of plastics. Parts made from these powders were often infiltrated with epoxy to improve the surface finish and seal porosity in the parts. Proprietary Ni-based powders for direct tooling applications and Cu-based powders for parts requiring high thermal and electrical conductivities were also available. All of these materials have been successfully used by many organizations; however, the more recent introduction of mLS and Electron Beam Powder Bed Fusion (EB-PBF) (which is known by electron beam melting (EBM) by Arcam) technology has made these alloys obsolete, as engineering-grade alloys are now able to be processed using a number of manufacturers' machines.

As mentioned, titanium alloys, numerous steel alloys, nickel-based super alloys, CoCrMo, and more are widely available from numerous manufacturers. It should be noted that alloys that crack under high solidification rates are not good candidates for mLS. Due to the high solidification rates in mLS, the crystal structures produced and mechanical properties are different than those for other manufacturing processes. These structures may be metastable, and the heat treatment recipes needed to produce standard microstructures may be different. As mLS and EB-PBF processes advance, the types of metal alloys which are commonly utilized will grow, and new alloys specifically tailored for PBF production have been developed.

5.2.3 Ceramics and Ceramic Composites

Ceramic materials are generally described as compounds that consist of metal oxides, carbides, nitrides and their combinations. Several ceramic materials are available commercially including aluminum oxide and titanium oxide. Commercial machines were developed by a company called Phenix Systems in France, which was acquired by 3D Systems in 2013. At the time of publication, these machines have been discontinued, but we expect other manufacturers to introduce ceramic laser sintering machines in the future.

Ceramics and metal–ceramic composites have been demonstrated in research. Typically, ceramic precipitates form through reactions occurring during the sintering process. One example is the processing of aluminum in a nitrogen atmosphere, which forms an aluminum matrix with small regions of AlN interspersed throughout. This process is called chemically induced sintering and is described further in the next section.

Biocompatible materials have been developed for specific applications. For example, calcium hydroxyapatite, a material very similar to human bone, has been processed using pLS for medical applications.

5.3 Powder Fusion Mechanisms

Since the introduction of LS, each new PBF technology developer has introduced competing terminology to describe the mechanism by which fusion occurs, with variants of "sintering" and "melting" being the most popular. A list of historical terminology for various PBF technologies is shown in Table 5.1 as a reference.

The use of a single word to describe powder fusion is inherently problematic as multiple mechanisms are possible. There are four different fusion mechanisms which are present in PBF processes [6]. These include solid-state sintering, chemically induced binding, liquid-phase sintering, and full melting. Most commercial processes utilize liquid-phase sintering and melting. A brief description of each of these mechanisms and their relevance to AM is as follows.

Table 5.1 Different technologies and commercial names for PBF systems

Terminology	Manufacturer	Acronym
Ceramic laser sintering	NA	CLS
Direct metal laser melting	GE	DMLM
Direct metal laser sintering	EOS GmbH Electro Optical Systems [7]	DMLS
Direct metal laser forming	NA	DMLF
Direct metal printing	NA	DMP
Electron beam melting	GE former name (Arcam)	EBM
LaserCUSISNG	GE former name (Concept Laser)	LaserCUSING
Laser metal fusion	NA	LMF
Laser sintered in solid phase	3D Systems former name (Phenix Systems)	LSSP
Powder Bed Fusion	Trumpf [8]	PBF
Selective laser melting	SLM Solutions, Renishaw, DMG MORI (Former name Realizer)	SLM
Selective laser reaction sintering	3D Systems former name DTM	SLRS
Selective Laser Sintering	3D Systems former name DTM	SLS

5.3.1 Solid-State Sintering

The use of the word sintering to describe powder fusion as a result of thermal processing predates the advent of AM. Sintering, in its classical sense, indicates the fusion of powder particles without melting (i.e., in their "solid-state") at elevated temperatures. This occurs at temperatures between one half of the absolute melting temperature and the melting temperature. The driving force for solid-state sintering is the minimization of total free energy, E_s, of the powder particles. The mechanism for sintering is primarily diffusion between powder particles.

Surface energy E_s is proportional to total particle surface area S_A, through the equation $E_s = \gamma_s \times S_A$ (where γ_s is the surface energy per unit area for a particular material, atmosphere, and temperature). When particles fuse at elevated temperatures (see Fig. 5.2), the total surface area decreases, and thus surface energy decreases.

As the total surface area of the powder bed decreases, the rate of sintering slows. To achieve very low porosity levels, long sintering times or high sintering temperatures are required. The use of external pressure, as is done with hot isostatic pressing, increases the rate of sintering.

As total surface area in a powder bed is a function of particle size, the driving force for sintering is directly related to the surface area to volume ratio for a set of particles. The larger the surface area to volume ratio, the greater the free-energy driving force. Thus, smaller particles experience a greater driving force for necking and consolidation, and hence, smaller particles sinter more rapidly and initiate sintering at lower temperature than larger particles.

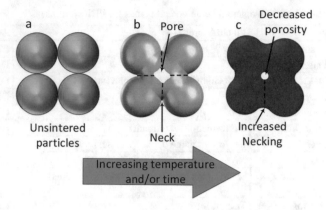

Fig. 5.2 Solid-state sintering. (**a**) Closely packed particles prior to sintering. (**b**) Particles agglomerate due to diffusion at temperatures above one half of the absolute melting temperature, as they seek to minimize free energy by decreasing surface area. (**c**) As sintering progresses, neck size increases and pore size decreases

As diffusion rates exponentially increase with temperature, sintering becomes increasingly rapid as temperatures approach the melting temperature, which can be modeled using a form of the Arrhenius equation. However, even at temperatures approaching the melting temperature, diffusion-induced solid-state sintering is the slowest mechanism for selectively fusing regions of powder within a PBF process.

For AM, the shorter the time it takes to form a layer, the more economically competitive the process becomes. Thus, the heat source which induces fusion should move rapidly and/or induce fusion quickly to increase build rates. Since the time it takes for fusion by sintering is typically much longer than for fusion by melting, few AM processes use sintering as a primary fusion mechanism.

Sintering, however, is still important in most thermal powder processes, even if sintering is not the primary fusion mechanism. There are three secondary ways in which sintering affects a build:

- If the loose powder within the build platform is held at an elevated temperature, the powder bed particles will begin to sinter to one another. This is typically considered a negative effect, as agglomeration of powder particles means that each time the powder is recycled the average particle size increases. This changes the spreading and melting characteristics of the powder each time it is recycled. One positive effect of loose powder sintering, however, is that the powder bed will gain a degree of tensile and compressive strength, thus helping to minimize part curling.

- As a part is being formed in the build platform, thermally induced fusion of the desired cross-sectional geometry causes that region of the powder bed to become much hotter than the surrounding loose powder. If melting is the dominant fusion mechanism (as is typically the case), then the just-formed part cross-section will be quite hot. As a result, the loose powder bed immediately surrounding the fused region heats up considerably, due to conduction from the part being formed.

This region of powder may remain at an elevated temperature for a long time (many hours for polymers) depending upon the size of the part being built, the heater and temperature settings in the process, and the thermal conductivity of the powder bed. Thus, there is sufficient time and energy for the powder immediately next to the part being built to fuse significantly due to solid-state sintering, both to itself and to the part. This results in "part growth," where the originally scanned part grows a "skin" of increasing thickness the longer the powder bed is maintained at an elevated temperature. This phenomenon can be seen in Fig. 5.3 as unmolten particles fused to the edge of a part. For many materials, the skin formed on the part goes from high density, low porosity near the originally scanned region to lower density, and higher porosity further from the part. This part growth can be compensated in the build planning stage by offsetting the laser beam to compensate for part growth or by offsetting the surface of the STL model. In addition, different post-processing methods will remove this skin to a different degree. Thus, the dimensional repeatability of the final part is highly dependent upon effectively compensating for and controlling this part growth. Performing repeatable post-processing to remove the same amount of the skin for every part is thus quite important.

- Rapid fusion of a powder bed using a laser or other heat source makes it difficult to achieve 100% dense, porosity-free parts. Thus, a feature of many parts built using PBF techniques (especially for polymers) is distributed porosity throughout the part. This is typically detrimental to the intended part properties. However, if the part is held at an elevated temperature after scanning, solid-state sintering combined with other high-temperature phenomena (such as grain growth in metals) causes the % porosity in the part to decrease. Since lower layers are maintained at an elevated temperature while additional layers are added, this can result in lower regions of a part being denser than upper regions of a part. This uneven porosity can be controlled, to some extent, by carefully controlling the

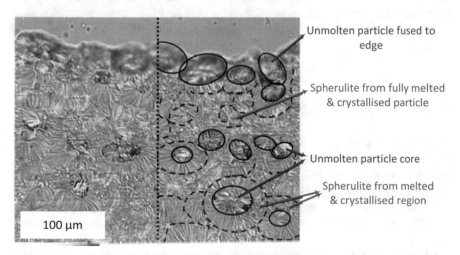

Unmolten particle fused to edge

Spherulite from fully melted & crystallised particle

Unmolten particle core

Spherulite from melted & crystallised region

100 µm

Fig. 5.3 Typical pLS microstructure for nylon polyamide (Elsevier license number 4720630952538) [9]

part bed temperature, cooling rate, and other parameters. Electron beam melting, in particular, often makes use of the positive aspects of elevated-temperature solid-state sintering and grain growth by purposefully maintaining the metal parts that are being built at a high enough temperature that diffusion and grain growth cause the parts being built to reach 100% density.

5.3.2 Chemically Induced Sintering

Chemically induced sintering involves the use of thermally activated chemical reactions between two types of powders or between powders and atmospheric gases to form a by-product which binds the powders together. This fusion mechanism is primarily utilized for ceramic materials. Examples of reactions between powders and atmospheric gases include laser processing of SiC in the presence of oxygen, whereby SiO_2 forms and binds together a composite of SiC and SiO_2; laser processing of ZrB_2 in the presence of oxygen, whereby ZrO_2 forms and binds together a composite of ZrB_2 and ZrO_2; and laser processing of Al in the presence of N_2, whereby AlN forms and binds together the Al and AlN particles.

For chemically induced sintering between powders, various research groups have demonstrated that mixtures of high-temperature structural ceramic and/or intermetallic precursor materials can be made to react using a laser. In this case, raw materials which exothermically react to form the desired by-product are premixed and heated using a laser. By adding chemical reaction energy to the laser energy, high-melting-temperature structures can be created at relatively low-laser energies.

One common characteristic of chemically induced sintering is part porosity. As a result, post-process infiltration or high-temperature furnace sintering to higher densities is often needed to achieve properties that are useful for most applications. This post-process infiltration may involve other reactive elements, forming new chemical compounds after infiltration. The cost and time associated with post-processing have limited the adoption of chemically induced sintering in commercial machines.

5.3.3 Liquid-Phase Sintering and Partial Melting

Liquid-phase sintering (LPS)is arguably the most versatile mechanism for PBF. Liquid-phase sintering is a term used extensively in the powder processing industry to refer to the fusion of powder particles when a portion of constituents within a collection of powder particles becomes molten, while other portions remain solid. In LPS, the molten constituents act as the glue which binds the solid particles together. As a result, high-temperature particles can be bound together without needing to melt or sinter those particles directly. LPS is used in traditional powder metallurgy to form, for instance, cemented carbide cutting tools where Co is used as the lower-melting-point constituent to glue together particles of WC.

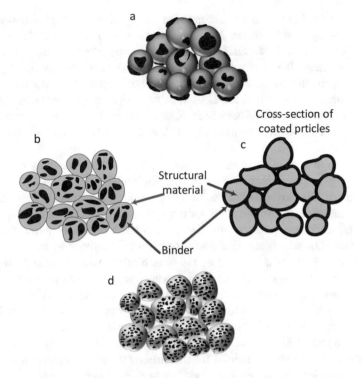

Fig. 5.4 Liquid-phase sintering variations used in PBF processing: (**a**) separate particles, (**b**) composite particles, (**c**) coated particles, and (**d**) indistinct mixtures. Darker regions represent the lower-melting-temperature binder material. Lighter regions represent the high-melting-temperature structural material. For indistinct mixtures, microstructural alloying eliminates distinct binder and structural regions

There are many ways in which LPS can be utilized as a fusion mechanism in AM processes. For purposes of clarity, the classification proposed by Kruth et al. [6] has formed the basis for the distinctions discussed in the following section and shown in Fig. 5.4.

5.3.3.1   Distinct Binder and Structural Materials

In many LPS situations, there is a clear distinction between the binding material and the structural material. The binding and structural material can be combined in three different ways: as separate particles, as composite particles, or as coated particles.

Separate Particles

A simple, well-mixed combination of binder and structural powder particles is sufficient in many cases for LPS. In cases where the structural material has the dominant properties desired in the final structure, it is advantageous for the binder material to be smaller in particle size than the structural material. This enables more efficient packing in the powder bed and less shrinkage and lower porosity after binding. The dispersion of smaller-particle-size binder particles around structural particles also helps the binder flow into the gaps between the structural particles more effectively, thus resulting in better binding of the structural particles. This is often true when, for instance, LS is used to process steel powder with a polymer binder (as discussed more fully in Sect. 5.3.5). This is also true when metal–metal mixtures and metal–ceramic mixtures are directly processed without the use of a polymer binder.

In the case of LPS of separate particles, the heat source passes by quickly, and there is typically insufficient time for the molten binder to flow and surface tension to draw the particles together prior to resolidification of the binder unless the binder has a particularly low viscosity. Thus, composite structures formed from separate particles typically are quite porous. This is often the intent for parts made from separate particles, which are then post-processed in a furnace to achieve the final part properties. Parts held together by polymer binders which require further post-processing (e.g., to lower or fill the porosity) are known as "green" parts.

In some cases, the density of the binder and structural material is quite different. As a result, the binder and structural material may separate during handling. In addition, some powdered materials are most economically manufactured at particle sizes that are too small for effective powder dispensing and leveling (see Sect. 5.5). In either case, it may be beneficial for the structural and/or binder particles to be bound together into larger particle agglomerates. By doing so, composite powder particles made up of both binder and structural material are formed.

Composite Particles

Composite particles contain both the binder and structural material within each powder particle. Mechanical alloying of binder and structural particles or grinding of cast, extruded, or molded mixtures into a powder results in powder particles that are made up of binder and structural materials agglomerated together. The benefits of composite particles are that they typically form higher density green parts and typically have better surface finish after processing than separate particles [6].

Composite particles can consist of mixtures of polymer binders with higher melting point polymer, metal or ceramic structural materials, or metal binders with higher melting point metal or ceramic structural materials. In all cases, the binder and structural portions of each particle, if viewed under a microscope, are distinct from each other and clearly discernable. The most common commercially available composite particle used in PBF processes is glass-filled nylon. In this case, the structural material (glass beads) is used to enhance the properties of the binding material (nylon) rather than the typical use of LPS where the binder is simply a necessary glue to help hold the structural material together in a useful geometric form.

Coated Particles

In some cases, a composite formed by coating structural particles with a binder material is more effective than random agglomerations of binder and structural materials. These coated particles can have several advantages including better absorption of laser energy, more effective binding of the structural particles, and better flow properties.

When composite particles or separate particles are processed, the random distribution of the constituents means that impinging heat energy, such as laser radiation, will be absorbed by whichever constituent has the highest absorptivity and/or most direct "line of sight" to the impinging energy. If the structural materials have a higher absorptivity, a greater amount of energy will be absorbed in the structural particles. If the rate of heating of the structural particles significantly exceeds the rate of conduction to the binder particles, the higher-melting-temperature structural materials may melt prior to the lower-melting-temperature binder materials. As a result, the anticipated microstructure of the processed material will differ significantly from one where the binder had melted, and the structural material had remained solid. This may, in some instances, be desirable but is typically not the intent when formulating a binder/structural material combination. Coated particles can help overcome the structural material heating problem associated with random constituent mixtures and agglomerates. If a structural particle is coated with the binder material, then the impinging energy must first pass through the coating before affecting the structural material. As melting of the binder and not the structural material is the objective of LPS, this helps ensure that the proper constituent melts.

Other benefits of coated particles exist. Since there is a direct correlation between the speed of the impinging energy in AM processing and the build rate, it is desirable for the binder to be molten for only a very short period of time. If the binder is present at the surfaces of the structural material, this is the most effective location for gluing adjacent particles together. If the binder is randomly mixed with the structural materials, and/or the binder's viscosity is too high to flow to the contact points during the short time it is molten, then the binder will not be as effective. As a result, the binder % content required for effective fusion of coated particles is usually less than the binder content required for effective fusion of randomly mixed particles.

Many structural metal powders are spherical. Spherical powders are easier to deposit and smooth using powder spreading techniques. Coated particles retain the spherical nature of the underlying particle shape and thus can be easier to handle and spread.

5.3.3.2   Indistinct Binder and Structural Materials

In polymers, due to their low thermal conductivity, it is possible to melt smaller powder particles and the outer regions of larger powder particles without melting the entire structure (see Fig. 5.3). Whether to more properly label this phenomenon

LPS or just "partial melting" is a matter of debate. Also with polymers, fusion can occur between polymer particles above their glass transition temperature but below their melting temperature. Similarly, amorphous polymers have no distinct melting point, becoming less viscous the higher the temperature goes above the glass transition temperature. As a result, in each of these cases, there can be fusion between polymer powder particles in cases where there is partial but not full melting, which falls within the historical scope of the term "liquid-phase sintering."

In metals, LPS can occur between particles where no distinct binder or structural materials are present. This is possible during partial melting of a single particle type or when an alloyed structure has lower-melting-temperature constituents. For non-eutectic alloy compositions, melting occurs between the liquidus and solidus temperature of the alloy, where only a portion of the alloy will melt when the temperature is maintained in this range. Regions of the alloy with higher concentrations of the lower-melting-temperature constituent(s) will melt first. As a result, it is commonly observed that many metal alloys can be processed in such a way that only a portion of the alloy melts when an appropriate energy level is applied. This type of LPS of metal alloys was the method used in the early EOS M250 direct metal laser sintering machines. Subsequent metal laser sintering commercialized processes are all designed to fully melt the metal alloys they process.

5.3.4 Full Melting

Full melting is the mechanism most commonly associated with PBF processing of engineering metal alloys and semi-crystalline polymers. In these materials, the entire region of material subjected to impinging heat energy is melted to a depth exceeding the layer thickness. Thermal energy of subsequent scans of a laser or electron beam (next to or above the just-scanned area) is typically sufficient to remelt a portion of the previously solidified solid structure; and thus, this type of full melting is very effective at creating well-bonded, high-density structures from engineering metals and polymers.

The most common material used in PBF processing is nylon polyamide [10, 11]. As a semi-crystalline material, it has a distinct melting point. In order to produce parts with the highest possible strength, these materials should be fully melted during processing. However, elevated temperatures associated with full melting result in part growth, and thus, for practical purposes, many accuracy versus strength optimization studies result in parameters which are at the threshold between full melting and LPS, as can be seen from Fig. 5.3.

For metal PBF processes, the engineering alloys that are utilized in these machines (Ti, stainless steel, CoCr, etc.) are typically fully melted. The rapid melting and solidification of these metal alloys result in unique properties that are distinct from and can sometime be more desirable than cast or wrought parts made from identical alloys.

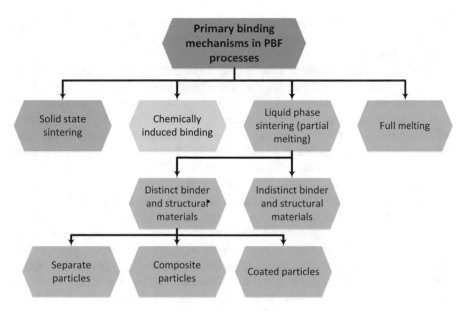

Fig. 5.5 Primary binding mechanisms in PBF processes. (Adapted from [6])

Figure 5.5 summarizes the various binding mechanisms which are utilized in PBF processes. Regardless of whether a technology is known as "Selective Laser Sintering," "selective laser melting," "direct metal laser sintering," "LaserCUSING," "electron beam melting," or some other name, it is possible for any of these mechanisms to be utilized (and, in fact, often more than one is present) depending upon the powder particle combinations and energy input utilized to form a part [12, 13].

5.3.5   High-Speed Sintering

High-speed sintering (HSS) is a PBF technology that utilizes the benefits of inkjet printing to form a composite structure of the spread powder and the printed ink as the cross-section. As opposed to Binder Jetting (BJT) technology, where the ink glues the powder particles together, in HSS, the ink acts as a heat absorption enhancer. After printing, a heater is run across the surface of the powder bed. Wherever ink has been printed, the amount of heat absorbed is sufficient to cause fusion between adjacent particles. Wherever ink is not printed, the reflectivity of the powder remains high enough that absorbed energy is too low to fuse powders.

HSS is being developed and commercialized by a number of companies. Most notable are the HSS processes developed by Xaar and Voxeljet as well as the MultiJet Fusion (MJF) process from Hewlett–Packard. These technologies are opening up new opportunities for functionality and flexibility in AM applications. By combining the advantages of both PBF and DJT, they can achieve the bonding strength of sintering at the speed of BJT. The benefit of combining industrial inkjet technology with a simple heating step is that it offers high precision build rates with consistent layer timing.

Most HSS processes follow a similar build methodology. The process starts by depositing a fine layer of loose powder on the build platform. An inkjet print head then moves over the powder and according to the designed shape prints infrared light-absorbing ink (Fig. 5.6a, b). The entire build area is then irradiated with infrared light causing the printed region to absorb sufficient energy to fuse the underlying powder (Fig. 5.6c). After this process the build platform is lowered one layer thickness. This process is repeated layer-by-layer until the build is complete, and the sintered block is then cooled down before part removal. In contrast to laser-based processes, the entire building process for each layer can be printed in a single pass which provides a constant layer time, regardless of the complexity and size of the components. Commercial print heads have a high resolution, and large arrays of individually controllable nozzles enable printing across the entire powder bed simultaneously. This method is typically performed in air rather than in a controlled build chamber atmosphere, which reduces the cost of the process. Complex elastic and functional models can be created using different polymers such as PA12 or TPU. Figure 5.6 shows the schematic of the HSS process.

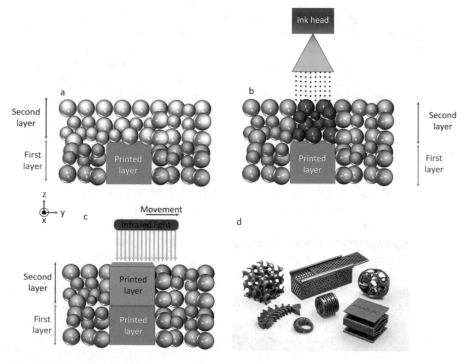

Fig. 5.6 Schematic of HSS process (**a**) loose powder above a previously printed layer, (**b**) adding ink, (**c**) infrared irradiation, and (**d**) 3D printed parts for industrial applications: post-processed and dyed black. Produced by Xaar 3D. (Photo courtesy of Xaar 3D)

5.4   Metal and Ceramic Part Fabrication

5.4.1   Metal Parts

There are four common approaches for using PBF processes in the creation of complex metal components: full melting, liquid-phase sintering, indirect processing, and pattern methods. In the full melting and liquid-phase sintering (when metals are used as both the high-temperature and low-temperature constituents) approaches, a metal part is typically usable in the state in which it comes out of the machine, after separation from a build plate.

In indirect processing, a polymer-coated metallic powder or a mixture of metallic and polymer powders (as described in Sect. 5.3.3.1 above) is used for part construction. Figure 5.7 shows the steps involved in indirect processing of metal powders. During indirect processing, the polymer binder is melted and binds the particles together, and the metal powder remains solid. The metallic powder particles remain

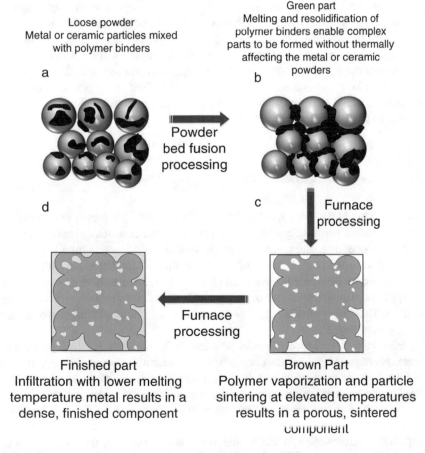

Fig. 5.7   Indirect processing of metal and ceramic powders using PBF

largely unaffected by the heat of the laser, whereas the binder melts and flows to regions between the metal powders due to capillary action. The parts produced are generally porous (sometimes exceeding 50 vol.% porosity). The polymer-bound green parts are subsequently furnace processed. Furnace processing occurs in two stages: (1) debinding and (2) infiltration or consolidation. During debinding, the polymer binder is vaporized to remove it from the green part. Typically, the temperature is also raised to the extent that a small degree of necking (sintering) occurs between the metal particles. Subsequently, the remaining porosity is either filled by infiltration of a lower-melting-point metal to produce a fully dense metallic part or by further sintering and densification to reduce the part porosity. Infiltration is easier to control, dimensionally, as the overall shrinkage is much less than during consolidation. However, infiltrated structures are always composite in nature, whereas consolidated structures can be made up of a single-material type.

The last approach to metal part creation using PBF is the pattern approach. For the previous three approaches, metal powder is utilized in the PBF process; but in this final approach, the part created in the PBF process is a pattern used to create the metal part. The two most common ways PBF-created parts are utilized as patterns for metal part creation are as investment casting patterns or as sand-casting molds. In the case of investment casting, polystyrene or wax-based powders are used in the machine, subsequently invested in ceramic during post-processing, and melted out during casting. In the case of sand-casting molds, mixtures of sand and a thermosetting binder are directly processed in the machine to form a sand-casting core, cavity, or insert. These molds are then assembled, and molten metal is cast into the mold, creating a metal part. Both indirect and pattern-based processes are further discussed in Chap. 20.

5.4.2 Ceramic Parts

Similar to metal parts, there are a number of ways that PBF processes are utilized to create ceramic parts. These include direct sintering, chemically induced sintering, indirect processing, and pattern methods. In direct sintering, a high temperature is maintained in the powder bed, and a laser is utilized to accelerate sintering of the powder bed in the prescribed location of each layer. The resultant ceramic parts will be quite porous and thus are often post-processed in a furnace to achieve higher density. This high porosity is also seen in chemically induced sintering of ceramics, as described earlier.

Indirect processing of ceramic powders is identical to indirect processing of metal powders (Fig. 5.7). After debinding, the ceramic brown part is consolidated to reduce porosity or is infiltrated. In the case of infiltration, when metal powders are used as the infiltrant, a ceramic–metal composite structure can be formed. In some cases, such as when creating SiC structures, a polymer binder can be selected, which leaves behind a significant amount of carbon residue within the brown part. Infiltration with molten Si will result in a reaction between the molten Si and the remaining carbon to produce more SiC, thus increasing the overall SiC content and reducing the fraction of metal Si in the final part. These related approaches have been used to form interesting ceramic–matrix composites and ceramic–metal structures for a number of different applications.

5.5   Process Parameters and Analysis

The use of optimum process parameters is extremely important for producing satisfactory parts using PBF processes. In this section, we will discuss "laser" processing and parameters, but by analogy the parameters and models discussed below could also be applied to other thermal energy sources, such as electron beams or infrared heaters.

5.5.1   Process Parameters

In PBF, process parameters can be lumped into four categories: (1) laser-related parameters (laser power, spot size, pulse duration, pulse frequency, etc.), (2) scan-related parameters (scan speed, scan spacing, and scan pattern), (3) powder-related parameters (particle shape, size and distribution, powder bed density, layer thickness, material properties, etc.), and (4) temperature-related parameters (powder bed temperature, powder feeder temperature, temperature uniformity, etc.). It should be noted that most of these parameters are strongly interdependent and are mutually interacting. The required laser power, for instance, typically increases with melting point of the material and lower powder bed temperature and also varies depending upon the absorptivity characteristics of the powder bed, which is influenced by material type and powder shape, size, and packing density.

A typical PBF machine includes two galvanometers (one for the x-axis and one for the y-axis motion). Similar to stereolithography, scanning often occurs in two modes, contour mode and fill mode, as shown in Fig. 5.8. In contour mode, the outline of the part cross-section for a particular layer is scanned. This is typically done for accuracy and surface finish reasons around the perimeter. The rest of the cross-section is then scanned using a fill pattern. A common fill pattern is a rastering technique whereby one axis is incrementally moved a laser scan width, and the other axis is continuously swept back and forth across the part being formed. In some cases the fill section is subdivided into stripes (where each stripe is scanned sequentially and the stripe angle is rotated every layer) or squares (with each square being processed separately). Randomized scanning is sometimes utilized so that there is no preferential direction for residual stresses induced by the scanning. The use of stripes or a square-based (chessboard) strategy is primarily for metal parts, whereas a simple raster pattern for the entire part (without subdividing into stripes or squares) is typically used for polymers and other low-temperature processing.

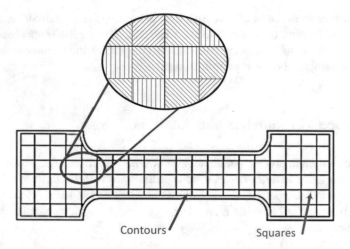

Fig. 5.8 Scan strategies employed in PBF techniques

In addition to melt pool characteristics, scan pattern and scan strategy can have a profound impact on residual stress accumulation within a part. For instance, if a part is moved from one location to another within a machine, the exact laser paths to build the part may change. These laser path changes may cause the part to distort more in one location than another. Thus it is possible for a part to build successfully in one location but not in another location in the same machine due simply to how the scan strategy is applied in different locations.

Powder shape, size, and size distribution strongly influence laser absorption characteristics as well as powder bed density, powder bed thermal conductivity, and powder spreading. Finer particles provide greater surface area and absorb laser energy more efficiently than coarser particles. Powder bed temperature, laser power, scan speed, and scan spacing must be balanced to provide the best trade-off between melt pool size, dimensional accuracy, surface finish, build rate, and mechanical properties. The powder bed temperature should be kept uniform and constant to achieve repeatable results. Generally, high-laser-power/high-bed-temperature combinations produce dense parts but can result in part growth, poor recyclability, and difficulty cleaning parts. On the other hand, low-laser-power/low-bed-temperature combinations produce better dimensional accuracy, but result in lower density parts and a higher tendency for layer delamination. High-laser-power and low-part-bed-temperatures result in an increased tendency for non-uniform shrinkage and the buildup of residual stresses, leading to curling of parts.

Laser power, spot size and scan speed, and bed temperature together determine the energy input needed to fuse the powder into a useable part. The longer the laser dwells in a particular location, the deeper the fusion depth and the larger the melt pool diameter. Typical layer thicknesses range from 0.02 to 0.15 mm. Operating at lower laser powers requires the use of lower scan speeds in order to ensure proper particle fusion. Melt pool size is highly dependent upon settings of laser power, scan speed, spot size, and bed temperature. Scan spacing should be selected to ensure a sufficient degree of melt pool overlap between adjacent lines of fused material to ensure robust mechanical properties.

The powder bed density, as governed by powder shape, size, distribution, and spreading mechanism, can strongly influence the part quality. Powder bed densities typically range between 50% and 60% for most commercially available powders but may be as low as 30% for irregular ceramic powders. Generally the higher the powder packing density, the higher the bed thermal conductivity and the better the part mechanical properties.

Most commercialized PBF processes use continuous wave (CW) lasers. Laser processing research with pulsed lasers, however, has demonstrated a number of potential benefits over CW lasers. In particular, the tendency of molten metal to form disconnected balls of molten metal, rather than a flat molten region on a powder bed surface, can be partially overcome by pulsed energy. CW laser processing is faster than pulsed laser processing, so most machines are configured to use CW lasers to increase machine productivity.

5.5.2 Applied Energy Correlations and Scan Patterns

Many common physics, thermodynamics, and heat transfer models are relevant to PBF techniques. In particular, solutions for stationary and moving point heat sources in infinite media and homogenization equations (to estimate, for instance, powder bed thermophysical properties based upon powder morphology, packing density, etc.) are commonly utilized. The solidification modeling discussed in the Directed Energy Deposition (DED) chapter (Chap. 10) can also be applied to PBF processes. For the purposes of this chapter, a highly simplified model which estimates the energy input characteristics of PBF processes is introduced and discussed with respect to process optimization for PBF processes.

Melt pool formation and characteristics are primarily determined by the total amount of applied energy which is absorbed by the powder bed as the laser beam passes. There is a difference between energy input and absorbed energy, so it is important to make a distinction between the two cases. Input energy is related to heat source power, speed of translation of the heat source, and area of the heat source energy (e.g., diameter of the beam), while absorbed energy is a function of heat source power, speed, diameter of the beam, layer thickness, hatch distance, absorptivity of the material, thermophysical properties of the material, etc. Therefore, it is important to clarify which sort of energy density is discussed.

Both the melt pool size and melt pool depth are a function of absorbed energy density. A simplified energy density equation has been used by numerous investigators as a simple method for correlating input process parameters to the density and strength of produced parts [14]. In their simplified model, applied energy density E_A (also known as the Andrews number) can be found using (5.1):

$$E_A = \frac{P}{SS \times HS} \tag{5.1}$$

where P is laser power, SS is scan speed, and HS is the scan spacing/hatch spacing between parallel scan lines. In this simplified model, applied energy increases with increasing laser power and decreases with increasing velocity and scan spacing. For pLS, typical scan spacing values are ~100 µm, whereas typical laser spot sizes are ~300 µm. Thus, typically every point is scanned by multiple passes of the laser beam.

Although (5.1) does not include powder absorptivity, heat of fusion, laser spot size, bed temperature, or other important characteristics, it provides the simplest analytical approach for optimizing machine performance for a material. For a given material, laser spot size, and machine configuration, a series of experiments can be run to determine the minimum applied energy necessary to achieve adequate material fusion for the desired material properties. Subsequently, build speed can be maximized by utilizing the fastest combination of laser power, scan rate, and scan spacing for a particular machine architecture based upon (5.1).

Optimization of build speed using applied energy is reasonably effective for PBF of polymer materials. However, when a molten pool of metal is present on a powder bed, a phenomenon called balling often occurs. When surface tension forces overcome a combination of dynamic fluid, gravitational, and adhesion forces, the molten metal will form a ball. The surface energy driving force for metal powders to limit their surface area to volume ratio (which is minimized as a sphere) is much greater than the driving force for polymers, and thus this phenomenon is unimportant for polymers but critically important for metals. An example of balling tendency at various power and scan speed combinations is shown in Fig. 5.9 [15]. This figure illustrates five typical types of tracks which are formed at various process parameter combinations.

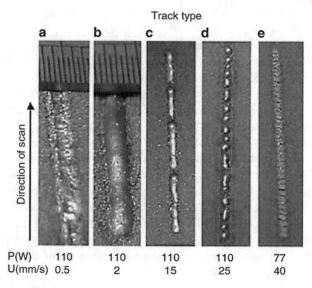

Fig. 5.9 Five examples of test tracks made in −150/+75 µm M2 steel powder in an argon atmosphere with a CO_2 laser beam of 1.1 mm spot size, at similar magnifications. (© Professional Engineering Publishing, reproduced from Childs et al. [15])

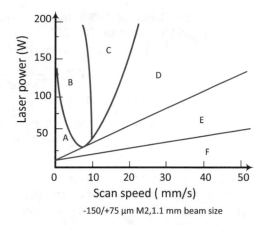

Fig. 5.10 Process map for track types. (© Professional Engineering Publishing, reproduced from Childs et al. [15])

-150/+75 µm M2,1.1 mm beam size

A process map showing regions of power and scan speed combinations which result in each of these track types is shown in Fig. 5.10. As described by Childs et al., tracks of type A were continuous and flat topped or slightly concave. At slightly higher speeds, type B tracks became rounded and sank into the bed. As the speed increased, type C tracks became occasionally broken, although not with the regularity of type D tracks at higher speeds, whose regularly and frequently broken tracks are perfect examples of the balling effect. At even higher speeds, fragile tracks were formed (type E) where the maximum temperatures exceed the solidus temperature but do not reach the liquidus temperature (i.e., partially melted or liquid-phase sintered tracks). In region F, at the highest speed, lowest power combinations, no melting occurred.

Numerous researchers have investigated residual stresses and distortion in laser PBF processes using analytical and finite element methods. These studies have shown that residual stresses and subsequent part deflection increase with increase in track length. Based on these observations, dividing the scan area into small squares (Island scanning strategy) or stripes and then scanning each segment with short tracks are highly beneficial. Thus, there are multiple reasons for subdividing the layer cross-section into small regions for metals.

Randomization of square scanning (rather than scanning contiguous squares one after the other) and changing the primary scan direction between squares help alleviate preferential buildup of residual stresses, as shown in Fig. 5.8. In addition, scanning of stripes whereby the angle of the stripe changes each layer has a positive effect on the buildup of residual stress. As a result, stripes and square scan patterns are extensively utilized in PBF processes for metals.

When considering these results, it is clear that build speed optimization for metals is complex, as a simple maximization of scan speed for a particular power and scan spacing based on Eq. 5.1 are not possible. However, within process map regions A and B, Eq. 5.1 could still be used as a guide for process optimization.

Equation 5.1 solves for the energy density per surface area in ($J/_{m^2}$). However, since the applied energy from a heat source is absorbed by a volume of the material, it is also helpful to calculate volumetric energy density ($J/_{m^3}$). Equations 5.2 and 5.3

show simplified absorbed volumetric energy density (ED$_v$) calculations for continuous and pulsed heat sources [16, 17].

$$ED_v = \frac{P \times \eta_p}{SS \times HS \times LT} \tag{5.2}$$

$$ED_v = \frac{P}{\dfrac{PD}{ET} \times HS \times LT} \tag{5.3}$$

where LT is layer thickness, η_p is powder absorptivity, PD is point distance (the distance traveled between spot centers during pulsed heating), and ET is exposure time per pulse. When we assume a melt pool is in the liquid phase until after a subsequent hatch, Eqs. 5.2 and 5.3 show the energy density. However, if the melt pool is solidified before the fusion of the next laser pass, Eq. 5.4 presents the absorbed volumetric energy density [17, 18].

$$ED_v = \frac{P \times \eta_p}{SS \times BD \times LT} \tag{5.4}$$

where BD is beam diameter. In real conditions, to have a proper bonding between two subsequent layers, the melt pool depth must be more than the layer thickness. When calculating volumetric energy density as a function of melt pool depth (MD), Eq. 5.5 is used.

$$ED_v = \frac{P \times \eta_p}{SS \times BD \times MD} \tag{5.5}$$

Melt pool depth is a function of heat penetration depth (δ_h) and is obtained according to Eq. 5.6. Melt pool depth typically ranges between 0.5 (δ_h) and 0.8 (δ_h) Heat penetration depth is a function of thermophysical properties of the material such as density (ρ), specific heat capacity (C_p), heat conductivity (K), and interaction time (t_i) of the heat source and material.

$$\delta_h = 2\sqrt{\kappa t_i} = 2\sqrt{\frac{K}{\rho C_p} t_i} \tag{5.6}$$

In the case of pulsed heat sources (such as a pulsed laser), t_i is pulse time. For a continuous heat source, $t_i = \dfrac{BD}{SS}$. Generally, the input volumetric energy density is independent of the layer thickness and hatch spacing which is obtained according to Eq. 5.7 [19].

$$ED_v = \frac{P}{SS \times BD \times MD} \tag{5.7}$$

Equations 5.2, 5.3, 5.4, 5.5, 5.6 and 5.7 are simplified by neglecting heat of fusion, bed temperature/preheat condition, etc.

Another important phenomenon in PBF is heat dispersion. Most of the heat sources in PBF are monochromatic electromagnetic radiation beams, and the dispersion of the beam follows the Gaussian beam rules. Equation 5.8 shows the absorbed volumetric energy density by Gaussian dispersion (ED_{vG}).

$$ED_{vG} = \frac{2P \times \eta_p}{\pi \times BD^2 \times MD} \exp\left(\frac{-2\left((x-(SS \times t))^2 + y^2 \right)}{BR^2} \right) \tag{5.8}$$

where BR is beam radius. Equation 5.8 illustrates that the highest energy density is applied in the center of the beam [20, 21].

5.6 Powder Handling

5.6.1 Powder Handling Challenges

Several different systems for powder delivery in PBF processes have been developed. The lack of a single solution for powder delivery goes beyond simply avoiding patented embodiments of the counter-rotating roller. The development of other approaches has resulted in a broader range of powder types and morphologies which can be delivered.

Any powder delivery system for PBF must meet at least four characteristics:

- It must have a powder reservoir of sufficient volume to enable the process to build to the maximum build height without a need to pause the machine to refill the powder reservoir.
- The correct volume of powder must be transported from the powder reservoir to the build platform sufficient to cover the previous layer but without wasteful excess material.
- The powder must be spread to form a smooth, thin, repeatable layer thickness of powder.
- The powder spreading must not create excessive shear forces that disturb the previously processed layers.

In addition, any powder delivery system must be able to deal with these universal characteristics of powder feeding:

- As particle size decreases, interparticle friction and electrostatic forces increase. These result in a situation where powder can lose its flowability. (To illustrate this loss of flowability, compare the flow characteristics of a spoon full of granulated sugar to a spoon full of fine flour. The larger particle size sugar will flow out of the spoon at a relatively shallow angle, whereas the flour will stay in the spoon until the spoon is tipped at a large angle, at which point the flour will fall out as a large clump unless some perturbation (vibration, tapping, etc.) causes it to come out a small amount at a time.) Thus, any effective powder delivery system must make the powder flowable for effective delivery to occur.

- When the surface area to volume ratio of a particle increases, its surface energy increases and becomes more reactive. For certain materials, this means that the powder becomes explosive in the presence of oxygen; or it will burn if there is a spark. As a result, certain powders must be kept in an inert atmosphere while being processed, and powder handling should not result in the generation of sparks.
- When handled, small particles have a tendency to become airborne and float as a cloud of particles. In PBF machines, airborne particles will settle on surrounding surfaces, which may cloud optics, reduce the sensitivity of sensors, deflect laser beams, and damage moving parts. In addition, airborne particles have an effective surface area greater than packed powders, increasing their tendency to explode or burn. As a result, the powder delivery system should be designed in such a way that it minimizes the creation of airborne particles.
- Smaller powder particle sizes enable better surface finish, higher accuracy, and thinner layers. However, smaller powder particle sizes exacerbate all the problems just mentioned. As a result, each design for a powder delivery system is inherently a different approach to effectively feed the smallest possible powder particle sizes while minimizing the negative effects of these small powder particles.

5.6.2   Powder Handling Systems

The earliest commercialized LS powder delivery system, illustrated in Fig. 5.1, is one approach to optimizing these powder handling issues. The two feed cartridges represent the powder reservoir with sufficient material to completely fill the build platform to its greatest build height. The correct amount of powder for each layer is provided by accurately incrementing the feed cartridge up a prescribed amount and the build platform down by the layer thickness. The raised powder is then pushed by the counter-rotating roller over the build platform, depositing the powder. As long as the height of the roller remains constant, layers will be created at the thickness with which the build platform moves. The counter-rotating action of the roller creates a "wave" of powder flowing in front of the cylinder. The counterrotation pushes the powder up, fluidizing the powder being pushed, making it more flowable for a particular particle size and shape. The shear forces on the previously processed layers created by this counter-rotating roller are small, and thus the previously processed layers are relatively undisturbed.

Another commonly utilized solution for powder spreading is a doctor blade. A doctor blade is simply a thin piece of metal that is used to scrape material across the surface of a powder bed. When a doctor blade is used, the powder is not fluidized. Thus, the shear forces applied to the previously deposited layer are greater than for a counter-rotating roller. This increased shear can be reduced if the doctor blade is ultrasonically vibrated, thus partly fluidizing the powder being pushed.

An alternative approach to using a feed cartridge as a powder reservoir is to use a hopper feeding system. A hopper system delivers powder to the powder bed from above rather than beneath. The powder reservoir is typically separate from the build area, and a "feeding system" is used to fill the hopper. The hopper is then used to deposit powder in front of a roller or doctor blade, or a doctor blade or roller can be integrated with a hopper system for combined feeding and spreading. The feeding system (not shown) can have an additional reservoir that is external to the machine, so that powder can be added and then flooded with inert gas prior to being fed into the hopper. This type of powder handling system can also be combined with sieving, filtering, and other systems which enable automated powder recycling (see Sect. 5.5.3). For both feeding and spreading, ultrasonic vibration can be utilized with any of these approaches to help fluidize the powders. Two types of hopper-based powder delivery systems are illustrated in Fig. 5.11.

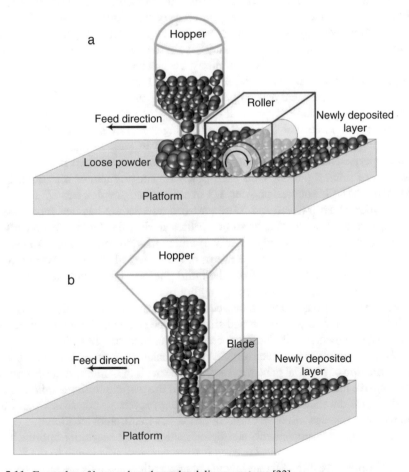

Fig. 5.11 Examples of hopper-based powder delivery systems [22]

In the case of multi-material powder bed processing, the only effective method is to use multiple hoppers with separate materials. In a multi-hopper system, the material type can be changed layer-by-layer. Although this has been demonstrated in a research environment and by some companies for very small parts, to date, all PBF technologies offered for sale commercially utilize a single-material powder feeding system.

5.6.3   Powder Recycling

As mentioned in Sect. 5.3.1, elevated-temperature sintering of the powder surrounding a part being built can cause the powder bed to fuse. In addition, elevated temperatures, particularly in the presence of reacting atmospheric gases, will also change the chemical nature of the powder particles. Similarly, holding polymer materials at elevated temperatures can change the molecular weight of the polymer. These combined effects mean that the properties of many different types of powders (particularly polymers) used in PBF processes change their properties when they are recycled and reused. For some materials these changes are small and thus are considered highly recyclable or infinitely recyclable. In other materials these changes are dramatic, and thus a highly controlled recycling methodology must be used to maintain consistent part properties between builds.

For the most popular PBF polymer material, nylon polyamide, both the effective particle size and molecular weight change during processing. As a result, a number of recycling methodologies have been developed to seek to maintain consistent build properties. The simplest approach to this recycling problem is to mix a specific ratio of unused powder with used powders. An example of a fraction-based mixture might be 1/3 unused powder, 1/3 overflow/feed powder, and 1/3 build platform powder. Overflow/feed and loose part-bed powder are handled separately, as they experience different temperature profiles during the build. The recaptured overflow/feed materials are only slightly modified from the original material as they have been subjected to lower temperatures only in the feed and overflow cartridges, whereas loose part-bed powder from the build platform has been maintained at an elevated temperature, sometimes for many hours.

Part-bed powder is typically processed using a particle sorting method, most commonly either a vibratory screen-based sifting device or an air classifier, before mixing with other powders. Air classifiers can be better than simple sifting, as they mix the powders together more effectively and help break up agglomerates, thus enabling a larger fraction of material to be recycled. However, air classifiers are more complex and expensive than sifting systems. Regardless of the particle sorting method used, it is critical that the material be well-mixed during recycling; otherwise, parts built from recycled powder will have different properties in different locations.

Although easy to implement, a simple fraction-based recycling approach will always result in some amount of mixing inconsistencies. This is due to the fact that different builds have different part layout characteristics, and thus the loose part-bed powder being recycled from one build has a different thermal history than loose part-bed powder being recycled from a different build.

In order to overcome some of the build-to-build inconsistencies inherent in fraction-based mixing, a recycling methodology based upon a powder's melt flow index (MFI) has been developed [23]. MFI is a measure of molten thermoplastic material flow through an extrusion apparatus under prescribed conditions. ASTM and ISO standards, for instance, can be followed to ensure repeatability. When using an MFI-based recycling methodology, a user determines a target MFI, based upon their experience. Used powders (part-bed and overflow/feed materials) are mixed and tested. Unused powder is also tested. The MFI for both is determined, and a well-blended mixture of unused and used powder is created and subsequently tested to achieve the target MFI. This may have to be done iteratively if the target MFI is not reached by the first mixture of unused to used powder. Using this methodology, the closer the target MFI is to the new powder MFI, the higher the new powder fraction and thus the more expensive the part. The MFI method is generally considered more effective for ensuring consistent build-to-build properties than fractional mixing.

Typically, most users find that they need less of the used build platform powder in their mixture than is produced. Thus, this excess build material becomes scrap. In addition, repeated recycling over a long period of time may result in some powder becoming unusable. As a result, the recyclability of a powder and the target MFI or fractional mixing selected by a user can have a significant effect on part properties and cost.

5.7   Powder Bed Fusion Process Variants and Commercial Machines

A large variety of PBF processes has been developed. To understand the practical differences between these processes, it is important to know how the powder delivery method, heating process, energy input type, atmospheric conditions, optics, and other features vary with respect to one another. An overview of commercial processes and a few notable systems under development are discussed in the following section.

5.7.1   Polymer Laser Sintering (pLS)

Prior to 2014 there were only two major producers of pLS machines, EOS and 3D Systems. The expiration of key patents in 2014 opened the door for many new companies to enter the marketplace. Polymer laser sintering machines are designed for directly processing polymers and for indirect processing of metals and ceramics.

Most commercial polymers were developed for processing via injection molding. The thermal and stress conditions for a material processed via pLS, however, are much different than the thermal and stress conditions for a material processed via injection molding. In injection molding the material is slowly heated under pressure, flows under high shear forces into a mold, and is cooled quickly. In pLS the material is heated very quickly as a laser beam passes, it flows via surface tension under gravitational forces, and it cools slowly over a period of hours to days. Since polymer microstructural features depend upon the time the material is held at elevated temperatures, polymer parts made using LS can have very different properties than polymer parts made using injection molding.

Many polymers which are easy to process using injection molding may not be processable using pLS. Figure 5.12 illustrates a schematic of a differential scanning calorimetry (DSC) curve for the types of melting characteristics which are desirable in a polymer for LS. In order to reduce residual stress-induced curling, pLS machines hold the powder bed temperature (T_{bed}) just below the temperature where melting begins ($T_{Melt Onset}$). When the laser melts a region of the powder bed, it should raise the temperature of the material above the melting temperature but below the temperature at which the material begins to deteriorate. If there is a small difference between the melting and deterioration temperatures, then the material will be difficult to successfully process in pLS.

After scanning, the molten cross-section will return over a relatively short period of time to the bed temperature (T_{Bed}). If the bed temperature is above the crystallization temperature of the material, then it will remain in a partially molten state for a very long time. This is advantageous for two reasons. First, by keeping the material partially molten, the part will not experience layer-by-layer accumulation of resid-

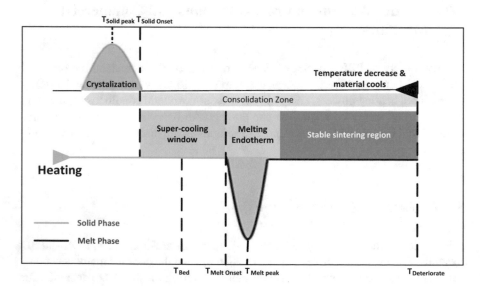

Fig. 5.12 Melting and solidification characteristics for an idealized polymer DSC curve for polymer laser sintering. (Photo courtesy of Neil Hopkinson, Sheffield University)

ual stresses and thus will be more accurate. Secondly, by holding the material in a semi-molten state for a long period of time, the part will achieve higher overall density. As a result, parts at the bottom of a build platform (which were built first and experience a longer time at elevated temperature) are denser than the last parts to be built. Thus, a key characteristic of a good polymer for pLS is that it has a broad "super-cooling window" as illustrated in Fig. 5.12. For most commercially available polymers, the melting curve overlaps the crystallization curve, and there is no super-cooling window. In addition, for amorphous polymers, there is no sharp onset of melting or crystallization. Thus pLS works best for polymers that are crystalline with a large super-cooling window and a high deterioration temperature.

The Selective Laser Sintering Sinterstation 2000 machine was the first commercial PBF system, introduced by the DTM Corporation, USA, in 1992. Subsequently, other variants were commercialized, and these systems are still manufactured and supplied by 3D Systems, USA, which purchased DTM in 2001. Newer machines offer several improvements over previous systems in terms of part accuracy, temperature uniformity, build speed, process repeatability, feature definition, and surface finish, but the basic processing features and system configuration remain unchanged from the description in Sect. 5.1. A typical pLS machine is limited to polymers with a melting temperature below 200 °C, whereas "high-temperature" pLS machines can process polymers with much higher melting temperature. Due to the use of CO_2 lasers and a nitrogen atmosphere with approximately 0.1–3.0% oxygen, pLS machines are incapable of directly processing pure metals or ceramics. Nylon polyamide materials are the most popular pLS materials, but these processes can also be used for many other types of polymer materials as well as indirect processing of metal and ceramic powders with polymer binders.

EOS GmbH, Germany, introduced its first EOSINT P machine in 1994 for producing plastic prototypes. In 1995, the company introduced its EOSINT M 250 machine for direct manufacture of metal casting molds from foundry sand. In 1998, the EOSINT M 250 Xtended machine was launched for direct metal laser sintering (DMLS), which was a liquid-phase sintering approach to processing metallic powders. These early metal machines used a special alloy mixture comprised of bronze and nickel powders developed by Electrolux Rapid Prototyping and licensed exclusively to EOS. The powder could be processed at low temperatures, required no preheating, and exhibited negligible shrinkage during processing; however, the end product was porous and was not representative of any common engineering metal alloys. Subsequently, EOS introduced many other materials and models, including platforms for foundry sand and full melting of metal powders (which will be discussed in the following section). One unique feature of EOS's large-platform systems for polymers and foundry sand is the use of multiple laser beams for faster part construction (as illustrated in the 2 × 1D channels example in Fig. 2.6). This multi-machine approach to PBF has made EOS the market leader in this technology segment. A schematic of an EOS machine illustrating their approach to laser sintering powder delivery and processing for foundry sand is shown in Fig. 5.13.

Working Principle of Laser-Sintering

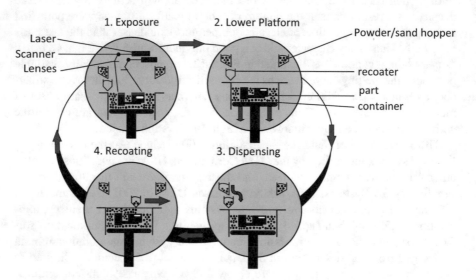

Fig. 5.13 Laser sintering schematic showing the dual-laser system option, hopper powder delivery, and a recoater that combines a movable hopper and doctor blade system

More recently, large-platform pLS systems commonly use a modular design. This modularity can include removable build platforms so that part cool-down and warm-up can occur outside of the chamber, enabling a fresh build platform to be inserted and used with minimal laser downtime; multiple build platform sizes; automated recycling and feeding of powder using a connected powder handling system; and better thermal control options.

In addition to commercial PBF machines, open-source polymer PBF machines are being developed to mimic the success of the RepRap effort for Material Extrusion (MEX) machines. In addition, inventors have applied PBF techniques to nonengineering applications via the CandyFab machine. Sugar is used as the powdered material, and a hot air nozzle is used as the energy source. By scanning the nozzle across the bed in a layer-by-layer fashion, sugar structures are made.

5.7.2   Laser-Based Systems for Metals and Ceramics

There are many companies which make commercially available laser-based systems for direct melting and sintering of metal powders: EOS (Germany), Renishaw (UK), Concept Laser (Germany), SLM Solutions (Germany), Realizer (Germany), 3D Systems (France/USA), Trumpf (Germany), Additive Industries (Netherlands), Velo3D (USA), Farsoon (China), and others are actively competing for market share. Although some of these companies have their own terminology for their machines, they are all laser PBF technologies. For simplicity, we will use metal laser sintering (mLS) to refer to the technologies in general and not to any particular variant.

mLS research in the late 1980s and early 1990s by various research groups was mostly unsuccessful. Compared to polymers, the high thermal conductivity, propensity to oxidize, high surface tension, and high laser reflectivity of metal powders make them significantly more difficult to process than polymers. Most commercially available mLS systems today are variants of the Selective Laser Powder Remelting (SLPR) approach developed by the Fraunhofer Institute for Laser Technology, Germany. Their research developed the basic processing techniques necessary for successful laser-based, point-wise melting of metals. The use of lasers with wavelengths better tuned to the absorptivity of metal powders was one key for enabling mLS. Fraunhofer used an Nd:YAG laser instead of the CO_2 laser used in pLS, which resulted in a much better absorptivity for metal powders (see Fig. 5.14). Subsequently, almost all mLS machines use fiber lasers, which in general are cheaper to purchase and maintain, more compact, and energy efficient and have better beam quality than Nd:YAG lasers. The other key enablers for mLS, compared to pLS, are different laser scan patterns (discussed previously), the use of f-theta lenses to minimize beam distortion during scanning, and low oxygen, inert atmosphere control.

One common practice among mLS manufacturers is the rigid attachment of their parts to a base plate at the bottom of the build platform. This is done to keep the metal part being built from distorting due to residual stresses. This means that the design flexibility for parts made from mLS is not as broad as the design flexibility for parts made using laser sintering of polymers, due to the need to remove these rigid supports using a machining or cutting operation.

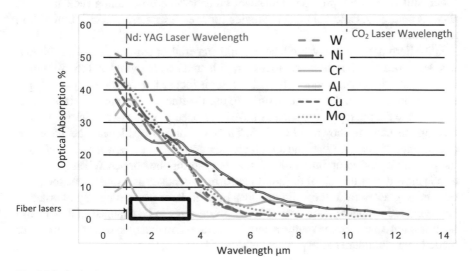

Fig. 5.14 Optical absorption % (absorptivity) of selected metals vs. wavelength (units are micrometers). (Courtesy of Optomec)

Fig. 5.15 3D-Micromac Powder Feed System. In this picture, only one of the powder feeders (located over the build cylinder) is filled with powder. (Photo courtesy of Laserinstitut Mittelsachsen e.V)

Over the years, various mLS machine manufacturers have sought to differentiate themselves from others by the features they offer. This differentiation includes laser power, number of lasers offered, powder handling systems, scanning strategies offered, maximum build volume, and more. Some machine manufacturers give users more control over the process parameters than other manufacturers, enabling more experimentation by the user, whereas other manufacturers only provide "proven" materials and process parameters. For instance, Renishaw machines have safety features to help minimize the risk of powder fires. EOS, as the world's most successful metal PBF provider, has spent considerable time tuning their machine process parameters and scanning strategies for specific materials which they sell to their customers.

3D-Micromac, Germany, produces a multi-material, small-scale mLS machine. It has developed small-scale mLS processes with small build cylinders (25–50 mm in diameter and 40 mm in height). Their fiber laser is focused to a particularly small spot size, for small feature definition. In order to use the fine powder particle sizes necessary for fine feature reproduction, they have developed a unique two-material powder feeding mechanism, shown in Fig. 5.15. The build platform is located between two powder feed cylinders. When the rotating rocker arm is above a powder feed cylinder, the powder is pushed up into the feeder, thus charging the hopper. When the rocker arm is moved over the top of the build platform, it deposits and smoothens the powder, moving away from the build cylinder prior to laser processing. By alternating between feed cylinders, the material being processed can be changed in a layer-by-layer fashion, thus forming multi-material structures. An example of a small impeller made using aluminum oxide powders is shown in Fig. 5.16.

Fig. 5.16 Example
3D-Micromac part made
from aluminum oxide
powders. (Photo courtesy
of Laserinstitut
Mittelsachsen e.V)

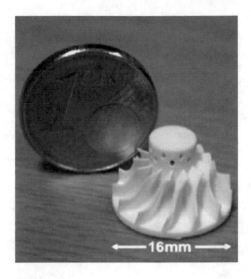

5.7.3   Electron Beam Powder Bed Fusion

Electron Beam Powder Bed Fusion (EB-PBF) which is known as electron beam melting (EBM) by Arcam has become a successful approach to PBF. In contrast to laser-based systems, EB-PBF uses a high-energy electron beam to induce fusion between metal powder particles. This process was developed at Chalmers University of Technology, Sweden, and was commercialized by Arcam AB, Sweden in 2001, and now owned by GE.

Similarly to mLS, in the EB-PBF process, a focused electron beam scans across a thin layer of pre-laid powder, causing localized melting and resolidification per the slice cross-section. There are a number of differences between how mLS and EB-PBF are typically practiced, which are summarized in Table 5.2. Many of these differences are due to EB-PBF having an energy source of electrons, but other differences are due to engineering trade-offs as practiced in EB-PBF and mLS and are not necessarily inherent to the processing. A schematic illustration of an EB-PBF apparatus is shown as Fig. 5.17.

Laser beams heat the powder when photons are absorbed by powder particles. Electron beams, however, heat powder by transfer of kinetic energy from incoming electrons into powder particles. As powder particles absorb electrons, they gain an increasingly negative charge. This has two potentially detrimental effects: (1) if the repulsive force of neighboring negatively charged particles overcomes the gravitational and frictional forces holding them in place, there will be a rapid expulsion of powder particles from the powder bed, creating a powder cloud (which is worse for fine powders than coarser powders); and (2) increasing negative charges in the powder particles will tend to repel the incoming negatively charged electrons, thus creating a more diffuse beam. There are no such complimentary phenomena with photons. As a result, the conductivity of the powder bed in EB-PBF must be high enough that powder particles do not become highly negatively charged, and scan

Table 5.2 Differences between EB-PBF and MLS

Characteristic	Electron beam melting	Metal laser sintering
Thermal source	Electron beam	Laser
Atmosphere	Vacuum	Inert gas
Scanning	Deflection coils	Galvanometers
Energy absorption	Conductivity-limited	Absorptivity-limited
Powder preheating	Use electron beam	Use infrared or resistive heaters
Scan speeds	Very fast, magnetically driven	Limited by galvanometer inertia
Energy costs	Moderate	High
Surface finish	Moderate to poor	Excellent to moderate
Feature resolution	Moderate	Excellent
Materials	Metals (conductors)	Polymers, metals, and ceramics
Powder particle size	Medium	Fine

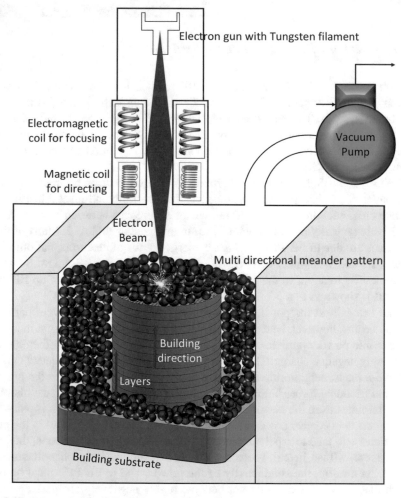

Fig. 5.17 Schematic of an EB-PBF apparatus

strategies must be used to avoid buildup of regions of negatively charged particles. In practice, electron beam energy is more diffuse, in part, so as not to build up too great a negative charge in any one location. As a result, the effective melt pool size increases, creating a larger heat-affected zone. Consequently, the minimum feature size, median powder particle size, layer thickness, resolution, and surface finish of an EB-PBF process are typically larger than for an mLS process.

As mentioned above, in EB-PBF the powder bed must be conductive. Thus, EB-PBF can only be used to process conductive materials (e.g., metals), whereas lasers can be used with any material that absorbs energy at the laser wavelength (e.g., metals, polymers, and ceramics).

Electron beam generation is typically a much more efficient process than laser beam generation. When a voltage difference is applied to the heated filament in an electron beam system, most of the electrical energy is converted into the electron beam; and higher beam energies (above 1 kW) are available at a moderate cost. In contrast, it is common for only 10–20% of the total electrical energy input for laser systems to be converted into beam energy, with the remaining energy lost in the form of heat. In addition, lasers with beam energies above 1 kW are typically much more expensive than comparable electron beams with similar energies. Thus, electron beams are a less costly high-energy source than laser beams. Newer fiber lasers, however, are more simple in their design, more reliable to maintain, and more efficient to use (with conversion efficiencies reported of 70–80% for some fiber lasers). Thus, this energy advantage for electron beams may not be a major advantage in the future.

EB-PBF powder beds are maintained at a higher temperature than mLS powder beds. There are several reasons for this. First, the higher energy input of the beam used in the EB-PBF system naturally heats the surrounding loose powder to a higher temperature than the lower-energy laser beams. In order to maintain a steady-state uniform temperature throughout the build (rather than having the build become hotter as the build height increases), the EB-PBF process uses the electron beam to heat the metal substrate at the bottom of the build platform before laying a powder bed. By defocusing the electron beam and scanning it very rapidly over the entire surface of the substrate (or the powder bed for subsequently layers), the bed can be preheated rapidly and uniformly to any pre-set temperature. As a result, the radiative and resistive heaters present in some mLS systems for substrate and powder bed heating are not used in EB-PBF. By maintaining the powder bed at an elevated temperature, however, the resulting microstructure of a typical EB-PBF part is significantly different from a typical mLS part (see Fig. 5.18). In particular, in mLS the individual laser scan lines are typically easily distinguishable, whereas individual scan lines are often indistinguishable in EB-PBF microstructures. Rapid cooling in mLS creates smaller grain sizes, and subsequent layer scans only partially remelt the previously deposited layer. The powder bed is held at a low enough temperature that elevated-temperature grain growth does not erase the layering effects. In EB-PBF, the higher temperature of the powder bed, and the larger and more diffuse heat input results in a contiguous grain pattern that is more representative of a cast microstructure, with less porosity than an mLS microstructure.

a   b

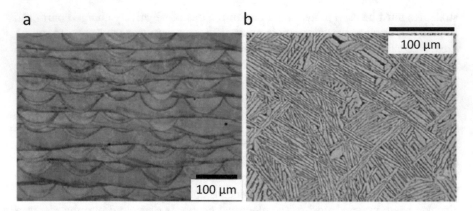

Fig. 5.18 Representative CoCrMo mLS microstructure ((**a**) photo courtesy of EOS) and Ti6Al4V EB-PBF microstructure ((**b**), photo courtesy of Arcam)

Although the microstructures presented in Fig. 5.18 are representative of mLS and EB-PBF, it should be noted that the presence of beam traces in the final microstructure (as seen in the left image of Fig. 5.18) is process parameter and material dependent. For certain alloys, such as titanium, it is not uncommon for contiguous grain growth across layers even for mLS. For other materials, such as those that have a higher melting point, the layering may be more prevalent. In addition, layering is more prevalent for process parameter combinations of lower bed temperature, lower beam energy, faster scan rate, thicker layers, and/or larger scan spacing for both mLS and EB-PBF. The reader is also referred to the presentation of material microstructures and process parameter effects of the DED processes in Sects. 10.6 and 10.7, since the phenomena seen in mLS and EB-PBF are similar to those observed in DED processes.

One of the most promising aspects of EB-PBF is the ability to move the beam nearly instantaneously. The current control system for EB-PBF machines makes use of this capability to keep multiple melt pools moving simultaneously for part contour scanning. Future improvements to scanning strategies may dramatically increase the build speed of EB-PBF over mLS, helping to distinguish it even more for certain applications. For instance, when nonsolid cross-sections are created, in particular when scanning truss-like structures (with designed internal porosity), nearly instantaneous beam motion from one scan location to another can dramatically speed up the production of the overall product.

In EB-PBF, residual stresses are much lower than for mLS due to the elevated bed temperature. Supports are needed to provide electrical conduction through the powder bed to the base plate, to eliminate electron charging, but the mass of these supports is an order of magnitude less than what is needed for mLS of a similar geometry. Future scan strategies for mLS may help reduce the need for supports to a degree where they can be removed easily, but at present EB-PBF has a clear advantage when it comes to minimizing residual stress and supports.

5.7.4   Line-Wise and Layer-Wise PBF Processes for Polymers

PBF processes have proven to be the most flexible general approach to AM. For production of end-use components, PBF processes surpass the applicability of any other approach. However, the use of expensive lasers in most processes, the fact that these lasers can only process one "point" of material at any instant in time, and the overall cost of the systems mean that there is considerable room for improvement. High-speed sintering, described above, and other variants of PBF are being researched and commercialized to fuse lines or layers of polymer material at a time. To date no commercial systems for metal line-wise or layer-wise processes have been introduced, but this is also an area of research interest. Polymer PBF processing in a line-wise or layer-wise manner dramatically increases the build rate of PBF processes, thus making them more cost-competitive. Three general approaches are discussed below. All three utilize infrared energy to induce fusion in powder beds; the key differences lay in their approach to controlling which portions of the powder bed fuse and which remain unfused, as illustrated in Fig. 5.19.

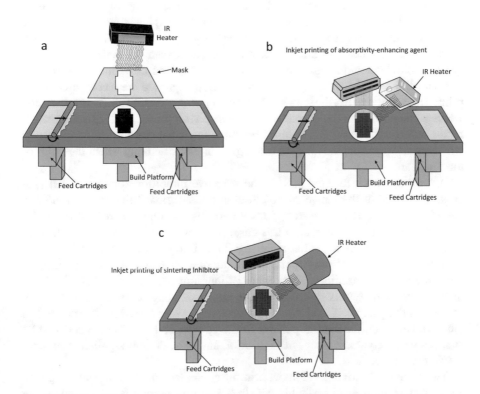

Fig. 5.19 Three different approaches to line- and layer-wise PBF processing (**a**) mask-based sintering, (**b**) printing of an absorptivity enhancing agent in the part region, and (**c**) printing of a sintering inhibitor outside the part region

Sintermask GmbH, Germany, founded in 2009, sold several Selective Mask Sintering (SMS) machines, based upon technology developed at Speedparts AB before going out of business. The key characteristics of their technology were exposure of an entire layer at a time to infrared thermal energy through a mask and rapid layering of powdered material. Their powder delivery system deposited a new layer of powder in 3 s. Heat energy was provided by an infrared heater. A dynamic mask system, similar to those used in a photocopier to transfer ink to paper, was used between the heater and the powder bed. This was a rebirth of an idea commercialized by Cubital for layer-wise photopolymerization in the early days of AM, as mentioned in Chap. 2. The SMS mask allows infrared energy to impinge on the powder bed only in the region prescribed by the layer cross-section, fusing powder in approximately 1 s [24, 25]. From a materials standpoint, the use of an infrared energy source means that the powder must readily absorb and quickly sinter or melt in the presence of infrared energy. Most materials with this characteristic are dark colored (e.g., gray or black), and thus color choice limitations may be a factor for some adopters of the technology. It appears that development of this technology is on hold, as of the writing of this book [26].

High-speed sintering was developed at Loughborough University and Sheffield University. As mentioned above, in HSS an inkjet printer is used to deposit ink onto the powder bed, representing a part's cross-section for that layer. Inks are especially formulated to significantly enhance infrared absorption compared with the surrounding powder bed. An infrared heater is used to scan the entire powder bed quickly, following inkjetting. Thus, this process is an example of line-wise processing. The difference between the absorptivity of the unprinted areas compared to the printed areas means that the unprinted areas do not absorb enough energy to sinter, whereas the powder in the printed areas sinters and/or melts. As the distinguishing factor between the fused and unfused region is the enhanced absorption of energy where printing occurs, the inks are often gray or black and thus affect the color of the final part.

A third approach to rapid PBF is the selective inhibition sintering (SIS) process, developed at the University of Southern California. In contrast to HSS, a sintering inhibitor is printed in regions where fusion is not desired, followed by exposure to infrared radiation. In this case, the inhibitor interferes with diffusion and surface properties to inhibit sintering. In addition, researchers have also utilized movable plates to mask portions of the powder bed where no sintering is desired, in order to minimize the amount of inhibitor required. One benefit of SIS over the previous two is that it does not involve adding an infrared absorption agent into the part itself, and thus the untreated powder becomes the material in the part. However, the unused powder in the powder bed is not easily recyclable, as it has been "contaminated" with inhibitor, and thus, there is significant unrecyclable material created.

Two additional variations of inkjet printing combined with PBF methodology are also practiced in SIS and by fcubic AB. In SIS, if no sintering is performed during the build (i.e., inhibitor is printed but no thermal infrared energy is scanned), the entire part bed can be moved into an oven where the powder is sintered to achieve fusion within the part, but not in areas where inhibitor has been printed.

fcubic AB, Sweden, uses inkjet printing plus sintering in a furnace to compete with traditional powder metallurgy for stainless steel components. A sintering aid is printed in the regions representing the part cross-section, so that this region will fuse more rapidly in a furnace. A sintering aid is an element or alloy which increases the rate at which solid-state sintering occurs between particles by changing surface characteristics and/or by reacting with the particles. Thus, sintering in the part will occur at lower temperatures and times than for the surrounding powder that has not received a sintering aid.

Both SIS and fcubic are similar to the BJT processes described in Chap. 8 (such as practiced by ExOne and Voxeljet) where a binder joins powders in regions of the powder bed where the part is located followed by furnace processing. There is, however, one key aspect of SIS and the fcubic processing which is different than these approaches. In the SIS and fcubic processes, the printed material is a sintering aid or inhibitor rather than a binder, and the part *remains embedded* within the powder bed when sintering in the furnace. Using the ExOne process, for instance, the machine prints a binder to glue powder particles together; and the bound regions are *removed from the powder bed* as a green part before sintering in a furnace (much like the indirect metal processing discussed earlier).

Common to all of the line-wise and layer-wise PBF processes is the need to differentiate between fusion in the part and the remaining powder. Too low total energy input will leave the part weak and only partially sintered. Too high energy levels will result in part growth by sintering of excess surrounding powder to the part and/or degradation of the surrounding powder to the point where it cannot be easily recycled. Most importantly, in all cases, it is the *difference* between fusion induced in the part and fusion induced in the surrounding powder bed that is the key factor to control.

5.8 Process Benefits and Drawbacks

Due to its nature, PBF can process a very wide variety of materials, in contrast to many other AM processes. Although it is easier to control the processing of semi-crystalline polymers, the PBF processing of amorphous polymers has been successful. Many metals can be processed; as mentioned, if a metal can be welded, it is a good candidate for mLS. Few ceramic materials are commercially available, but quite a few others have been demonstrated in research.

During part building, loose powder is a sufficient support material for polymer PBF. This saves significant time during part building and post-processing and enables advanced geometries that are difficult to post-process when supports are necessary. As a result, internal cooling channels and other complex features that would be impossible to machine are possible in polymer PBF.

Supports, however, are required for most metal PBF processes. The high residual stresses experienced when processing metals mean that support structures are typically required to keep the part from excessive warping. This means that post processing of metal parts after AM can be expensive and time-consuming. Small features (including internal cooling channels) can usually be formed without supports; but the part itself is usually constrained to a substrate at the bottom of the build platform to keep it from warping. As a result, the orientation of the part and the location of supports are key factors when setting up a build.

Accuracy and surface finish of powder-based AM processes are typically inferior to liquid-based processes. However, accuracy and surface finish are strongly influenced by the operating conditions and the powder particle size. Finer particle sizes produce smoother, more accurate parts but are difficult to spread and handle. Larger particle sizes facilitate easier powder processing and delivery but hurt surface finish, minimum feature size, and minimum layer thickness. The build materials used in these processes typically exhibit 3–4% shrinkage, which can lead to part distortion. Materials with low thermal conductivity result in better accuracy as melt pool and solidification are more controllable and part growth is minimized when heat conduction is minimized.

With PBF processes, total part construction time can take longer than other Additive Manufacturing processes because of the preheat and cool-down cycles involved. However, as is the case with several newer machine designs, removable build platforms enable preheat and cool-down to occur off-line, thus enabling much greater machine productivity. Additionally, the ability to nest polymer parts in three dimensions, as no support structures are needed, means that many parts can be produced in a single build, thus dramatically improving the productivity of these processes when compared with processes that require supports. Figure 5.20 shows examples of polymer and metal parts made using PBF.

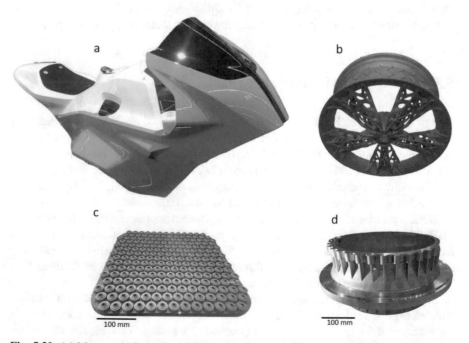

Fig. 5.20 (**a**) Motorcycle frame from FKM Laser Sintering, (**b**) wheel rims from BLT, (**c**) series of prototypes from Additive Industries, and (**d**) Vulcan rocket propellant nozzles from GE

5.9   Summary

PBF processes were one of the earliest AM processes and continue to be one of the most popular. Polymer-based laser sintering is commonly used for prototyping and end-use applications in many industries, competing with injection molding and other polymer manufacturing processes. PBF processes are particularly competitive for low-to-medium volume geometrically complex parts.

Metal-based processes, including laser and electron beam, are one of the fastest growing areas of AM around the world. Metal PBF processes are becoming increasingly common for aerospace and biomedical applications, due to their inherent geometric complexity benefits and their excellent material properties.

As methods for moving from point-wise to line-wise to layer-wise PBF are improved and commercialized, build times and cost will decrease. This will make PBF processing even more competitive. The future for PBF remains bright; and it is likely that PBF processes will remain one of the most common types of AM technologies for the foreseeable future.

5.10   Questions

1. Find a reference which describes an application of the Arrhenius equation to solid-state sintering. If an acceptable level of sintering is achieved within time T_1 at a temperature of 750 K, what temperature would be required to achieve the same level of sintering in half the time?
2. Estimate the energy driving force difference between two different powder beds made up of spherical particles with the same total mass, where the difference in surface area to volume ratio difference between one powder bed and the other is a factor of 2.
3. Explain the pros and cons of the various binder and structural material alternatives in liquid-phase sintering (Sect. 5.3.3.1) for a bone tissue scaffold application, where the binder (matrix material) is PCL and the structural material is hydroxyapatite.
4. Using standard kitchen ingredients, explore the powder characteristics described in Sect. 5.5.1 and powder handling options described in Sect. 5.5.2. Using at least three different ingredients, describe whether or not the issues described are reproducible in your experiments.
5. Using an internet search, find a set of recommended processing parameters for nylon polyamide using laser sintering. Based upon Eq. (5.1), are these parameters limited by machine laser power, scan spacing, or scan speed? Why? What machine characteristics could be changed to increase the build rate for this material and machine combination?
6. Using Fig. 5.10 and the explanatory text, estimate the minimum laser dwell time (how long a spot is under the laser as it passes) needed to maintain a type B scan track at 100 W.

7. What are some of the differences between fiber lasers and Nd:YAG lasers?
8. Why is electron beam generation typically a much more energy efficient process than laser beam generation?
9. What are the inherent differences between Laser-Based Powder Bed Fusion (LB-PBF) and Electron Beam Powder Bed Fusion (EB-PBF)?
10. What are the benefits of using the powder in polymer PBF as supporting material?
11. Consider the use of different materials, spreading techniques, and material thicknesses. What are the primary factors to be considered for a powder delivery system in a Powder Bed Fusion process? What are the benefits associated with using a counter-rotating roller compared to a doctor blade when it comes to distributing new layers of powder?
12. Discuss how Selective Laser Sintering keeps parts from curling, for both polymer and metal parts. How do these processes reduce the amount of curling?
13. How does the orientation of the part inside an SLS machine affect the final part?
14. Which Powder Bed Fusion process can be used on the macro- and microscale and describe their differences?
15. How are different powders handled in the case of multi-material powder bed processing?
16. What is Melt Flow Index (MFI)? How is it used in powder recycling in the Powder Bed Fusion process?
17. Why should the powder bed be conductive in EB-PBF?
18. Why are thermoplastic materials well-suited for Powder Bed Fusion process?
19. What are the most common lasers in Additive Manufacturing? Compare these lasers, and give the specific uses in AM where each type of laser has an advantage.
20. What is the balling effect in PBF? How is it avoided?
21. A set of experiment using a LB-PBF machine with M2 tool steel (1C-4.15Cr-6.4W-5Mo-2V-bal.Fe) tool steel is shown in Table Q5.1. Determine the energy density values for the single layer experiments. What is the optimal energy

Table Q5.1 Question 21

Experiment	Laser power (W)	Scan speed (mm/s)	Scan spacing (% of beam width)	Layer thickness (mm)
Single tracks	32, 58, 77, 110, 143, 170	0.5, 1–12	N/A	N/A
Single layer	58, 77, 110, 143	0.5, 1, 3, 5, 8, 10	25, 50, 100	N/A
Multiple layers	143	5, 8, 10	25, 50	0.4
Two beam widths of 0.55 and 1.1 mm were used				

required to fully melt this material? Why do you suppose they performed these experiments at the prescribed energy densities?

22. The polymers used in Selective Laser Sintering are semi-crystalline polymers which exhibit glass transition temperature (T_g). In this case, part bed temperature should be set close to T_g

The glass transition temperature of polyethylene terephthalate (PET) and poly vinyl alcohol (PVA) are 70° and 85°, respectively. Assume that the T_g of a mixture is proportional to the volume ratio of the two constituent materials.

Note: W_1 and W_2 are the volume fraction of two materials. $\dfrac{1}{T_g} = \dfrac{W1}{T_{g1}} + \dfrac{W2}{T_{g2}}$

(a) Calculate the glass temperature when the volume ration of PET is 10%, 20%, 30%, 40%, and 50%.
(b) Plot T_g as a function of T_{g1} and W_1.

23. Use the formula provided in the text book for the solidification rate, and explain how changing the scanning rate would affect the cooling rate and consequently grain size. Use the figures provided in the text book for justification of your answer.

References

1. Gibson, I., & Khorasani, A. M. (2019). Metallic additive manufacturing: Design, process, and post-processing. *Metals, 9*(2), 137.
2. Dilip, J., et al. (2016). A short study on the fabrication of single track deposits in SLM and characterization. In *Proceedings of the solid freeform fabrication symposium*.
3. Khorasani, A. M., Gibson, I., & Ghaderi, A. R. (2018). Rheological characterization of process parameters influence on surface quality of Ti-6Al-4V parts manufactured by selective laser melting. *The International Journal of Advanced Manufacturing Technology, 97*(9–12), 3761–3775.
4. Khorasani, A. M., et al. (2018). Mass transfer and flow in additive manufacturing of a spherical component. *The International Journal of Advanced Manufacturing Technology, 96*(9), 3711–3718.
5. Thompson, M. K., et al. (2016). Design for additive manufacturing: Trends, opportunities, considerations, and constraints. *CIRP Annals, 65*(2), 737–760.
6. Kruth, J.P, et al. (2005). Binding mechanisms in selective laser sintering and selective laser melting. *Rapid Prototyping Journal, 11*(1), 26–36.
7. Lu, Y., et al. (2018). Experimental sampling of the Z-axis error and laser positioning error of an EOSINT M280 DMLS machine. *Additive Manufacturing, 21*, 501–516.
8. Bandari, Y. K., et al. (2015). Additive manufacture of large structures: Robotic or CNC systems. In *Proceedings of the 26th international solid freeform fabrication symposium*, Austin
9. Zarringhalam, H., et al. (2006). Effects of processing on microstructure and properties of SLS Nylon 12. *Materials Science Engineering: A, 435*, 172–180.
10. Gong, H., et al. (2017). Influence of small particles inclusion on selective laser melting of Ti-6Al-4V powder. *IOP Conference Series: Materials Science and Engineering, 272*, 012024.
11 Dilip, J. J. S., et al. (2017). A novel method to fabricate TiAl intermetallic alloy 3D parts using additive manufacturing. *Defence Technology, 13*(2), 72–76.

12. Gong, H., et al. (2015). Influence of defects on mechanical properties of Ti–6Al–4V compo-
    nents produced by selective laser melting and electron beam melting. *Materials & Design, 86*,
    545–554.
13. Dilip, J. J. S., et al. (2017). Selective laser melting of HY100 steel: Process parameters, micro-
    structure and mechanical properties. *Additive Manufacturing, 13*, 49–60.
14. Williams, J. D., & Deckard, C. R. (1998). Advances in modeling the effects of selected param-
    eters on the SLS process. *Rapid Prototyping Journal, 4*(2), 90–100.
15. Childs, T., Hauser, C., & Badrossamay, M. (2005). Selective laser sintering (melting) of stain-
    less and tool steel powders: Experiments and modelling. *Proceedings of the Institution of
    Mechanical Engineers, Part B: Journal of Engineering Manufacture, 219*(4), 339–357.
16. Karimi, P., et al. (2018). Influence of laser exposure time and point distance on 75-μm-thick
    layer of selective laser melted Alloy 718. *The International Journal of Advanced Manufacturing
    Technology, 94*(5–8), 2199–2207.
17. Liu, Z., et al. (2018). Energy consumption in additive manufacturing of metal parts. *Procedia
    Manufacturing, 26*, 834–845.
18. Kurzynowski, T., et al. (2019). Effect of scanning and support strategies on relative density of
    SLM-ed H13 steel in relation to specimen size. *Materials, 12*(2), 239.
19. Römer, G. R. B. E. (2016). *Laser material processing*. University of Twente. Enschede,
    Netherlands
20. Wolff, S. J., et al. (2019). Experimentally validated predictions of thermal history and micro-
    hardness in laser-deposited Inconel 718 on carbon steel. *Additive Manufacturing, 27*, 540–551.
21. Dai, K., & Shaw, L. (2005). Finite element analysis of the effect of volume shrinkage during
    laser densification. *Acta Materialia, 53*(18), 4743–4754.
22. Yan, Y., et al. (1998). Study on multifunctional rapid prototyping manufacturing system.
    Integrated Manufacturing Systems, 9(4), 236–241.
23. Gornet, T., et al. (2002). Characterization of selective laser sintering materials to determine
    process stability. In *Solid freeform fabrication symposium*.
24. Jariwala, A. S., Rosen, D. W., & Ding, F. (2016). Fabricating parts from photopolymer resin. .
25. Zhao, X., & Rosen, D. W. (2017). Experimental validation and characterization of a real-time
    metrology system for photopolymerization-based stereolithographic additive manufacturing
    process. *The International Journal of Advanced Manufacturing Technology, 91*(1), 1255–1273.
26. Zhao, X., & Rosen, D. (2017). Real-time metrology for photopolymer additive manufactur-
    ing with exposure controlled projection lithography. In *Abstracts of Papers of the American
    Chemical Society*. Amer Chemical SOC 1155 16TH ST, NW, Washington, DC 20036 USA.

Chapter 6
Material Extrusion

Abstract Material Extrusion (MEX) machines have, by far, the largest install base of any AM technology. Inexpensive machines are sold to hobbyists at hardware stores, like Home Depot, and through multiple vendors on Amazon. Higher-end industrial machines are available through dozens of companies. While there are other techniques for creating an extrusion, heat is normally used to melt bulk material just before or during the process of being forced through a nozzle. In most systems, a round material filament is pushed by a set of pinch rollers, which creates the pressure to extrude. In this chapter, we will deal with AM technologies that use extrusion to form parts. We will cover the basic theory and attempt to give the reader a good understanding for why it is a leading AM technology.

6.1 Introduction

Material Extrusion technologies work similarly to how icing is applied to cakes, in that material contained in a reservoir is forced out through a nozzle when pressure is applied. If the pressure remains constant, then the resulting extruded material (commonly referred to as "roads" but more technically known as an "extrudate") will flow at a constant rate and will remain a constant cross-sectional diameter. This diameter will remain constant if the travel of the nozzle across a depositing surface is also kept at a constant speed that corresponds to the flow rate. The material that is being extruded must be in a semisolid state when it comes out of the nozzle. This material must fully solidify while remaining in the deposited shape. Furthermore, the material must bond to the material that has already been extruded so that a solid structure can result.

Since the material is extruded, the AM machine must be capable of scanning in a horizontal plane as well as starting and stopping the flow of material while scanning. Once a layer is completed, the machine must index upward, or move the part downward, so that a further layer can be produced.

© Springer Nature Switzerland AG 2021
I. Gibson et al., *Additive Manufacturing Technologies*,
https://doi.org/10.1007/978-3-030-56127-7_6

There are two primary approaches when using an extrusion process. The most commonly used approach is to use temperature as a way of controlling the material state. Molten material is liquefied inside a reservoir so that it can flow out through the nozzle and bond with adjacent material before solidifying. This approach is similar to conventional polymer extrusion processes, except the extruder is vertically mounted on a plotting system rather than remaining in a fixed horizontal position.

An alternative approach is to use a chemical change to cause solidification. In such cases, a curing agent, residual solvent, reaction with air, or simply drying of a "wet" material permits bonding to occur. Parts may therefore cure or dry out to become fully stable. This approach can be utilized with paste materials. Additionally, it may be more applicable to biochemical applications where materials must have biocompatibility with living cells and so choice of material is very restricted. However, industrial applications may also exist, perhaps using reaction injection molding-related processes rather than relying entirely on thermal effects.

This chapter will start off by describing the basic principles of MEX Additive Manufacturing. Following this will be a description of the most widely used technology, developed and commercialized by the Stratasys company. Bioplotting equipment for tissue engineering and scaffold applications commonly use extrusion technology, and a discussion on how this differs from the Stratasys approach will ensue. Finally, there have been a number of interesting research projects employing, adapting, and developing this technology, and this will be covered at the end of the chapter.

6.2 Basic Principles

There are a number of key features that are common to any MEX system:

- Loading of material
- Liquification of the material
- Application of pressure to move the material through the nozzle
- Extrusion
- Plotting according to a predefined path and in a controlled manner
- Bonding of the material to itself or secondary build materials to form a coherent solid structure
- Inclusion of support structures to enable complex geometrical features

These will be considered in separate sections to fully understand the intricacies of MEX AM.

A mathematical or physics-based understanding of extrusion processes can quickly become complex, since it can involve many nonlinear terms. The basic science involves extrusion of highly viscous materials through a nozzle. It is reasonable to assume that the material flows as a Newtonian fluid in most cases [1]. Most of the discussion in these sections will assume the extrusion is of molten material and may therefore include temperature terms. For solidification, these temperature terms are generally expressed relative to time; and so temperature could be replaced by other time-dependent factors to describe curing or drying processes.

6.2.1 Material Loading

Since extrusion is used, there must be a reservoir from which the material is extruded. This could be preloaded with material, but it would be more useful if there was a continuous supply of material into this reservoir. If the material is in a liquid form, then the ideal approach is to pump this material. Most bulk material is, however, supplied as a solid, and the most suitable methods of supply are in pellet or powder form or where the material is fed in as a continuous filament. The reservoir itself is therefore the main location for the liquification process. Pellets, granules, or powders are fed through the reservoir under gravity or with the aid of a screw or similar propelling process. Materials that are fed through the system under gravity require a plunger or compressed gas to force it through the narrow nozzle. Screw feeding not only pushes the material through to the base of the reservoir but can be sufficient to generate the pressure needed to push it through the nozzle as well. A continuous filament, which is by far the most common feedstock material, can be pushed into the reservoir chamber using pinch rollers, thus providing a mechanism for generating an input pressure for the nozzle.

6.2.2 Liquification

The extrusion method works on the principle that what is held in the reservoir will become a liquid that can eventually be pushed through the die or nozzle. As mentioned earlier, this material could be in the form of a solution that will quickly solidify following the extrusion, but more likely this material will be semiliquid because of heat applied to the reservoir chamber. Such heat would normally be applied by heater coils wrapped around the reservoir, and ideally this heat should be applied to maintain a constant temperature in the melt (see Fig. 6.1). The larger the reservoir, the more difficult this can become for numerous reasons related to heat transfer, thermal currents within the melt, change in physical state of the molten material, location of temperature sensors, etc.

The material inside the reservoir should be kept in a molten state, but care should be taken to maintain it at as low a temperature as possible since some polymers degrade quickly at higher temperatures and could also burn, leaving residue on the inside of the reservoir that would be difficult to remove and that would contaminate further melt. A higher temperature inside the reservoir also requires additional cooling following extrusion.

Fig. 6.1 Schematic of
MEX systems

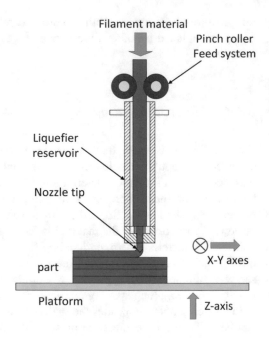

6.2.3 *Extrusion*

The extrusion nozzle determines the shape and size of the extruded filament. A
larger nozzle diameter will enable material to flow more rapidly but would result in
a part with lower precision compared with the original CAD drawing. The diameter
of the nozzle also determines the minimum feature size that can be created. No
feature can be smaller than this diameter, and in practice features should normally
be large relative to the nozzle diameter to faithfully reproduce them with satisfactory strength. MEX processes are therefore more suitable for larger parts that have
features and wall thicknesses that are at least twice the nominal diameter of the
extrusion nozzle used.

Material flow through the nozzle is controlled by the pressure drop between the
reservoir and the surrounding atmosphere. However, the extrusion process used for
AM may not be the same as conventional extrusion. For example, the pressure
developed to push the molten material through the nozzle is typically not generated
by a screw mechanism. Instead, the pinch roller feed system shown in Fig. 6.1
drives a continuous filament into the liquefier reservoir.

A simple volumetric flow rate model can be developed of material flow through
the liquefier. Mass flow through a nozzle is related to pressure drop, nozzle geometry, and material viscosity. The subscript f will be used to denote variables related
to the filament fed into the extruder, and r will denote variables associated with
roads being deposited. The volumetric flow rate into the liquefier is [1, 2]:

$$Q = v_f \pi r_f^2 \tag{6.1}$$

where v_f is the filament feed velocity and r_f is the radius of the filament. At the nozzle, the volumetric flow rate, Q, is:

$$Q = v_r WH \tag{6.2}$$

where v_r is the deposition velocity, W is the width of the deposited road, and H is its height, assuming that the deposited filament cross-section is more rectangular than circular [2]. By equating the flow rates, the feed velocity can be determined by assuming that the deposition velocity is given:

$$v_f = \frac{v_r WH}{\pi r_f^2} \tag{6.3}$$

Feed rate can be related to motor drive speed through [1, 2]:

$$v_f = \omega_p R_p \tag{6.4}$$

where ω_p is the angular velocity of the pinch rollers and R_p is the pinch roller radius, which can be rearranged to solve for ω_p and the drive motor speed.

The force required to push the filament through the extrusion head can be determined if the pressure drop, ΔP, through the liquefier can be estimated:

$$F = \Delta PA \tag{6.5}$$

where A is the cross-sectional area of the filament. Pressure drop through the nozzle can be modeled analytically; see [1]. With the force determined, the torque Γ and power P required to drive the filament can be computed:

$$\Gamma = FR_p \tag{6.6}$$

$$P_{mot} = \omega_p \Gamma \tag{6.7}$$

This model assumes that one motor drives the pinch rollers through a gear train.

If the force generated by the rollers exceeds the exit pressure, then buckling will occur in the filament feedstock, assuming there is no slippage between the material and the rollers. Another good analysis of the pinch roller feed system can be found in Turner et al. [3]. This analysis indicates that the feed forces are related to elastic modulus and that the more brittle the material, the more difficult it is to feed it through the nozzle. This would mean that composite materials that use ceramic fillers, for example, will require very precise control of feed rates. The increased modulus would lead to higher pressure and thus a higher force generated by the pinch rollers. The greater therefore is the chance of slippage or a mismatch between input and output that would result in non-flow or buckling at the liquefier entry point.

6.2.4   Solidification

Once the material is extruded, it should ideally remain the same shape and size. Gravity and surface tension, however, may cause the material to change shape, while size may vary according to cooling and drying effects. If the material is extruded in the form of a gel, the material may shrink upon drying, as well as possibly becoming porous. If the material is extruded in a molten state, it may also shrink when cooling. The cooling is also very likely to be nonlinear. If this nonlinear effect is significant, then it is possible the resulting part will distort upon cooling. This can be minimized by ensuring the temperature differential between the reservoir and the surrounding atmosphere is kept to a minimum (i.e., use of a controlled environmental chamber when building the part) and also by ensuring the cooling process is controlled with a gradual and slow profile.

It is reasonable to assume that a MEX AM system will extrude from a large reservoir to a small nozzle through the use of a conical interface. As mentioned before, the melt is generally expected to adhere to the walls of the liquefier and nozzle with zero velocity at these boundaries, subjecting the material to shear deformation during flow. The shear rate $\dot{\gamma}$ can be defined as [1]:

$$\dot{\gamma} = -\frac{dv}{dr} \tag{6.10}$$

and the shear stress as:

$$\tau = \left(\frac{\dot{\gamma}}{\phi}\right)^{\frac{1}{m}} \tag{6.11}$$

where m represents the flow exponent and ϕ represents the fluidity. The general flow characteristic of a material and its deviation from Newtonian behavior are reflected in the flow exponent m.

6.2.5   Positional Control

Like many AM technologies, MEX systems use a platform that indexes in the vertical direction to allow formation of individual layers. The extrusion head is typically carried on a plotting system that allows movement in the horizontal plane. This plotting must be coordinated with the extrusion rate to ensure smooth and consistent deposition.

Since the plotting head represents a mass and therefore contains an inertial element when moving in a specific direction, any change in direction must result in a deceleration followed by acceleration. The corresponding material flow rate must match this change in speed or else too much or too little material will be deposited in a particular region. For example, if the extrusion head is moving at a velocity v parallel to a nominal x direction and is then required to describe a right angle so that it then moves at the same velocity v in the perpendicular y direction, then, at some point, the instantaneous velocity will reach zero. If the extrusion rate is not zero at this point, then excess material will be deposited at the corner of this right-angled feature.

Since the requirement is to move a mechanical extrusion head in the horizontal plane, then the most appropriate mechanism to use would be a standard planar plotting system. This would involve two orthogonally mounted linear drive mechanisms like belt drives or leadscrews. Such drives need to be powerful enough to move the extrusion reservoir at the required velocity and be responsive enough to permit rapid changes in direction without backlash effects. The system must also be sufficiently reliable to permit constant movement over many hours without any loss in calibration. While cheaper systems often make use of belts driven by stepper motors, higher cost systems typically use servo drives with leadscrew technology.

Since rapid changes in direction can make it difficult to control material flow, a common strategy would be to draw the outline of the part to be built using a slower plotting speed to ensure that material flow is maintained at a constant rate. The internal fill pattern can be built more rapidly since the outline represents the external features of the part that corresponds to geometric precision. This outer shell also represents a constraining region that will prevent the filler material from affecting the overall precision. A typical fill pattern can be seen in Fig. 6.2. Determination of the outline and fill patterns will be covered in a later section of this chapter.

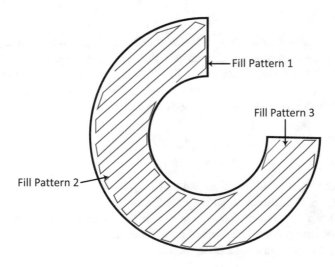

Fig. 6.2 A typical fill pattern using a MEX system, created in three stages (Adapted from [4])

6.2.6 Bonding

For heat-based systems, there must be sufficient residual heat energy to activate the surfaces of the adjacent regions, causing bonding. Alternatively, gel or paste-based systems must contain residual solvent or wetting agent in the extruded filament to ensure the new material will bond to the adjacent regions that have already been deposited. In both cases, we visualize the process in terms of energy supplied to the material by the extrusion head.

If there is insufficient energy, the regions may adhere, but there would be a distinct boundary between new and previously deposited materials. This can represent a fracture surface where the materials can be easily separated. Too much energy may cause the previously deposited material to flow, which in turn may result in a poorly defined part.

Once the material has been extruded, it must solidify and bond with adjacent material. Yardimci defined a set of governing equations that describe the thermal processes at work in a simple extruded road, laid down in a continuous, open-ended fashion along a direction x, based on various material properties [5]:

$$\rho \frac{\partial q}{\partial t} = k \frac{\partial^2 T}{\partial x^2} - S_c - S_l \tag{6.12}$$

where ρ is the material density, q is the specific enthalpy, and k is the effective thermal conductivity. T is the cross-sectional average road temperature. The term S_c is a sink term that describes convective losses:

$$S_c = \frac{h}{h_{\text{eff}}}(T - T_\infty) \tag{6.13}$$

h is the convective cooling heat transfer coefficient, and h_{eff} is a geometric term representing the ratio of the road element volume to surface for convective cooling. This would be somewhat dependent on the diameter of the nozzle. The temperature T_∞ is the steady-state value of the environment. The term S_l is a sink/source term that describes the thermal interaction between roads:

$$S_l = \frac{k}{\text{width}^2}(T - T_{\text{neigh}}) \tag{6.14}$$

where "width" is the width of the road and T_{neigh} is the temperature of the relevant neighboring road. If the material is laid adjacent to more material, this sink term will slow down the cooling rate. There is a critical temperature T_c above which a diffusive bonding process is activated and below which bonding is prohibited. On the basis of this, we can state a bonding potential φ as:

$$\varphi = \int_0^\tau (T - T_c) \, d\tau \tag{6.15}$$

6.2.7   Support Generation

All AM systems must have a means for supporting freestanding and disconnected features and for keeping all features of a part in place during the fabrication process. In polymer powder bed fusion, loose powder can be the support. But in Vat Photopolymerization (VPP), metal powder bed fusion, MEX, and other AM processes in subsequent chapters, such features must be kept in place by the additional fabrication of supports. Supports in such systems take two general forms:

- Similar material supports
- Secondary material supports

If a MEX system is built in the simplest possible way, then it will have only one extrusion reservoir. If it has only one reservoir, then supports must be made using the same material as the part. This may require parts and supports to be carefully designed and placed with respect to each other so that they can be separated at a later time. As mentioned earlier, adjustment of the temperature of the part material relative to the adjacent material can result in a fracture surface effect. This fracture surface can be used as a means of separating the supports from the part material. One possible way to achieve this may be to change the layer separation distance when depositing the part material on top of the support material or vice versa. The additional distance can affect the energy transfer sufficiently to result in this fracture phenomenon. Alternatively, adjustment of the reservoir or extrusion temperature when extruding supports might be an effective strategy. In all cases, however, the support material will be somewhat difficult to separate from the part.

The most effective way to remove supports from the part is to fabricate them in a different material. The variation in material properties can be exploited so that supports are easily distinguishable from part material, either visually (e.g., using a different color material), mechanically (e.g., using a weaker material for the supports), or chemically (e.g., using a material that can be removed using a solvent without affecting the part material). To do this, the MEX equipment should have a

Fig. 6.3 A medical model made using MEX AM technology from two different color materials, highlighting a bone tumor. (Photo courtesy of Stratasys)

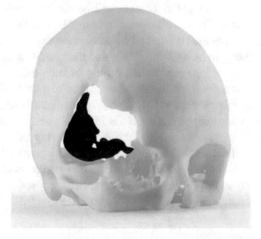

second extruder. In this way, the secondary material can be prepared with the correct build parameters and extruded in parallel with the current layer of build material, without delay. It may be interesting to note that a visually different material, when not used for supports, may also be used to highlight different features within a model, like the bone tumor shown in the medical model of Fig. 6.3.

6.3   Plotting and Path Control

As with nearly all Additive Manufacturing systems, MEX machines mostly take input from CAD systems using the generic STL file format. This file format enables easy extraction of the slice profile, giving the outline of each slice. As with most systems, the control software must also determine how to fill the material within the outline. This is particularly crucial to this type of system since extrusion heads physically deposit material that fills previously vacant space. There must be clear access for the extrusion head to deposit fill material within the outline without compromising the material that has already been laid down. Additionally, if the material is not laid down close enough to adjacent material, it will not bond effectively. In contrast, laser-based systems can permit, and in fact generally require, a significant amount of overlap from one scan to the next, and thus there are no head collision or overfilling-equivalent phenomena.

As mentioned earlier, part accuracy is maintained by plotting the outline material first, which will then act as a constraining region for the fill material. The outline would generally be plotted with a lower speed to ensure consistent material flow. The outline is determined by extracting intersections between a plane (representing the current cross-section of the build) and the triangles in the STL file. These intersections are then ordered so that they form a complete, continuous curve for each outline (there may be any number of these curves, either separate or nested inside of each other, depending upon the geometry of that cross-section). The only remaining thing for the software to do at this stage is to determine the start location for each outline. Since the extrusion nozzle is a finite diameter, this start location is defined by the center of the nozzle. The stop location will be the final point on this trajectory, located approximately one nozzle diameter from the start location. Since it is better to have a slight overlap than a gap and because it is very difficult to precisely control flow, there is likely to be a slight overfill and thus swelling in this start/stop region. If all the start/stop regions are stacked on top of each other, then there will be a "seam" running down the part. In most cases, it is best to have the start/stop regions randomly or evenly distributed around the part so that this seam is not obvious. However, a counter to this may be that a seam is inevitable and having it in an obvious region will make it more straightforward for removing during the post-processing stage.

Determining the fill pattern for the interior of the outlines is a much more difficult task for the control software. The first consideration is that there must be an offset inside the outline and that the extrusion nozzle must be placed inside this outline with minimal overlap. The software must then establish a start location for the fill and determine the trajectory according to a predefined fill pattern. This fill pattern is similar to those used in CNC planar pocket milling where a set amount of material must be removed with a cylindrical cutter [5]. As with CNC milling, there is no unique solution to achieving the filling pattern. Furthermore, the fill pattern may not be a continuous, unbroken trajectory for a particular shape. It is preferable to have as few individual paths as possible, but for complex patterns, an optimum value may be difficult to establish. As can be seen with even the relatively simple cross-section in Fig. 6.2, start and stop locations can be difficult to determine and are somewhat arbitrary. Even with a simpler geometry, like a circle that could be filled continuously using a spiral fill pattern, it is possible to fill from the outside–in or from the inside–out.

Spiral patterns in CNC are quite common, mainly because it is not quite so important as to how the material is removed from a pocket. However, they are less common as fill patterns for MEX, primarily for the following reason. Consider the example of building a simple solid cylinder. If a spiral pattern were used, every path on every layer would be directly above each other. This could severely compromise part strength and a weave pattern would be preferable. As with composite material weave patterning using material like carbon fiber, for example, it is better to cross the weave over each other at an angle so that there are no weakened regions due to the directionality in the fibers. Placing extrusion paths over each other in a crossing pattern can help to distribute the strength in each part more evenly.

Every additional weave pattern within a specific layer is going to cause a discontinuity that may result in a weakness within the corresponding part. For complex geometries, it is important to minimize the number of fill patterns used in a single layer. As mentioned earlier, and illustrated in Fig. 6.2, it is not possible to ensure that only one continuous fill pattern will successfully fill a single layer. Most outlines can be filled with a theoretically infinite number of fill pattern solutions. It is therefore unlikely that a software solution will provide the best or optimum solution in every case, but an efficient solution methodology should be designed to prevent too many separate patterns from being used in a single layer.

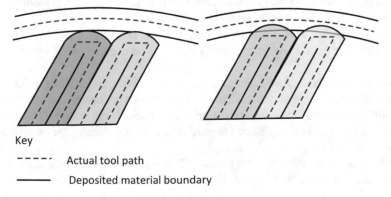

Key

- - - - -   Actual tool path

————   Deposited material boundary

Fig. 6.4 Extrusion of materials to maximize precision (left) or material strength (right) by controlling voids

Parts are weakened as a result of gaps between extruded roads. Since weave patterns achieve the best mechanical properties if they are extruded in a continuous path, there are many changes in direction. The curvature in the path for these changes in direction can result in gaps within the part as illustrated in Fig. 6.4. This figure illustrates two different ways to define the toolpath, one that will ensure no additional material will be applied to ensure no part swelling and good part accuracy. The second approach defines an overlap that will cause the material to flow into the void regions, increasing the strength but possibly decreasing accuracy if the part outline swells. However, in both cases, gaps are constrained within the outline material laid down at the perimeter. Additionally, by changing the flow rate at these directional change regions, less or more material can be extruded into these regions to compensate for gaps and swelling. This means that the material flow from the extrusion head should not be directly proportional to the instantaneous velocity of the head when the velocity is low, but rather should be increased or decreased slightly, depending on the toolpath strategy used. Furthermore, if the velocity is zero but the machine is known to be executing a directional change in a weave path, a small amount of flow should ideally be maintained. This will cause the affected region to swell slightly and thus help fill gaps. Obviously, care should be taken to ensure that excess material is not extruded to the extent that part geometry is compromised [6].

It can be seen that precise control of extrusion is a complex trade-off, dependent on a significant number of parameters, including:

- Input pressure: This variable is changed regularly during a build, as it is tightly coupled with other input control parameters. Changing the input pressure (or force applied to the material) results in a corresponding output flow rate change. A number of other parameters, however, also affect the flow to a lesser degree.
- Temperature: Maintaining a constant temperature within the melt inside the reservoir would be the ideal situation. However, small fluctuations are inevitable and will cause changes in the flow characteristics. Temperature sensing should be carried out somewhere within the reservoir, and therefore a loosely coupled parameter can be included in the control model for the input feed pressure to compensate for thermal variations. As the heat builds up, the pressure should drop slightly to maintain the same flow rate.
- Nozzle diameter: This is constant for a particular build, but many MEX systems do allow for interchangeable nozzles that can be used to offset speed against precision.
- Material characteristics: Ideally, control models should include information regarding the materials used. This would include viscosity information that would help in understanding the material flow through the nozzle. Since viscous flow, creep, etc. are very difficult to predict, accurately starting and stopping flow can be difficult.
- Gravity and other factors: If no pressure is applied to the reservoir, it is possible that material will still flow due to the mass of the molten material within the reservoir causing a pressure head. This may also be exacerbated by gaseous pressure buildup inside the reservoir if it is sealed. Surface tension of the melt and drag forces at the internal surfaces of the nozzle may retard this effect.

- Temperature buildup within the part: All parts will start to cool down as soon as the material has been extruded. However, different geometries will cool at different rates. Large, massive structures will hold their heat for longer times than smaller, thinner parts, due to the variation in surface to volume ratio. Since this may have an effect on the surrounding environment, it may also affect machine control.

Taking these and other factors into consideration can help one better control the flow of material from the nozzle and the corresponding precision of the final part. However, other uncontrollable or marginally controllable factors may still prove problematic to precisely control flow. Many MEX systems, for instance, resort to periodically cleaning the nozzles from time to time to prevent buildup of excess material adhered to the nozzle tip.

6.4   Material Extrusion Machine Types

By far the most common MEX AM technology is Fused Deposition Modeling (FDM), invented and developed by Stratasys, USA [2]. In this approach, a heating reservoir is used to liquefy the polymer that is fed into the system as a filament. The filament is pushed into the reservoir by a pinch roller arrangement, and it is this pushing that generates the extrusion pressure. A typical MEX machine of this type can be seen in Fig. 6.5, along with a picture of an extrusion head.

The initial MEX patent was awarded to Stratasys founder Scott Crump in 1992, and the company sold more MEX machines than any other AM machine type in the world for many subsequent years. As the original patents expired, hundreds of startups were formed worldwide to compete with Stratasys. In spite of tremendous com-

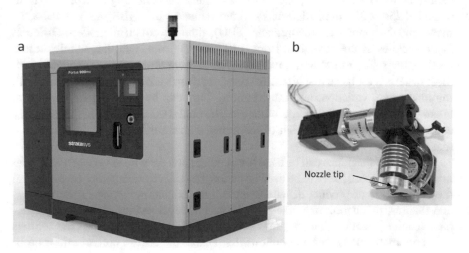

Fig. 6.5 (a) Typical Stratasys machine showing the outside and the extrusion head inside. (b) Nozzle tip. (Photo courtesy of Stratasys) [7, 8]

petition, Stratasys remains the market leader in MEX today. The major strength of MEX is the simplicity of machine design, low cost of machine components, range of materials, and the effective mechanical properties of resulting parts. Parts made using MEX can have among the best mechanical properties of any polymer-based Additive Manufacturing process, with only parts made from polymer PBF consistently performing at a comparable level.

The main drawback to using this technology is the build speed. As mentioned earlier, the inertia of the plotting heads means that the maximum speeds and accelerations that can be obtained are somewhat smaller than other systems. Furthermore, MEX requires material to be plotted in a point-wise, vector fashion that involves many changes in direction. To date, no line-wise or layer-wise MEX systems have been successfully commercialized.

6.4.1   MEX Machines from Stratasys

Stratasys sells a complete range of MEX equipment at various price points. As such, a review of Stratasys machines familiarizes the user with the breadth of commercial offerings available for MEX. Stratasys FDM machines range from low-cost, small-scale, minimal variable machines through to larger, more versatile, and more sophisticated machines that are inevitably more expensive. The company has separated its machines into several product lines.

The first product line, their F123 series, focuses on low-cost industrial machines currently starting around $12,000 USD. Stratasys promotes machines in this series as "true plug-and-play" which means that their operations are mostly automated and users have little control over process variables. The four models in the F123 series, the F120, F170, F270, and F370, have build volumes ranging from 250 mm^3 to 355 × 250 × 355 mm. Materials offered include PLA, ABS, ASA, a PC-ABS blend, and thermoplastic polyurethane (TPU), although not on all models. Different layer thicknesses are available, from 0.127 to 0.33 mm, with larger thicknesses used for their "fast draft" print mode. Soluble support structures are available with most materials for these machines. All these machines are designed to operate with minimal setup, variation, and intervention. They can be located without special attention to fume extraction and other environmental conditions. This means they can easily be placed in a design office rather than resorting to placing them in a machine shop.

While F123 machines can be used for making parts for a wide variety of applications, most parts are likely to be used as concept models by companies investigating the early stages of product development. More demanding applications, like jigs and fixtures, production manufacturing, and functional testing models would perhaps require machines that are more versatile, with more control over the settings and more material choices and options that enable the user to correct minor problems in the output model. Higher specification MEX machines are more expensive, not just because of the incorporated technology and increased range of materials but also because of the sales support, maintenance, and reliability. Stratasys has separated this higher-end technology through their FORTUS product line, with top-of-the-range models costing around $300,000 USD. The smallest FORTUS 380mc machine starts off roughly where the F123 machines end, with a slightly larger build envelope of 355 × 305 × 305 mm and a similar specification. Further up the range are machines with increases in size, accuracy, range of materials, and range of build speeds. The largest and most sophisticated machine is the F900, which has the highest productivity of all Stratasys FDM machines with a build envelope of 914 × 610 × 914 mm and at least 12 different material options.

Stratasys acquired the MakerBot company in 2013, which provided them with machines designed for the entry-level market. More about this class of machines will be covered in a later chapter. These entry-level machines are used extensively by educational institutions and individual hobbyists, as well as by companies who need cheap prototypes. MakerBot machines are available for purchase at prices ranging from $1500 to $5000.

It should be noted that MEX machines that operate with different layer thicknesses sometimes require the use of different nozzle diameters. These nozzles are manually changeable, and only one nozzle can be used for each reservoir/feeder system during a specific build. The nozzle diameter also controls the road width. Obviously, a larger diameter nozzle can extrude more material for a specific plotting speed and thus shorten the build time at the expense of lower precision.

FORTUS software options include the expected file preparation and build setup options. However, there are also software systems that allow the user to remotely monitor the build and schedule builds using a multiple machine setup. Stratasys has many customers who have purchased more than one machine, and their software is aimed at ensuring these customers can operate them with a minimum of user intervention. Much of this support was developed because of another Stratasys subsidiary called RedEye, which uses a large number of MEX machines as a service bureau (part-building manufacturing operation) for customers. Much of the operation of RedEye is based on customers logging into an Internet account and uploading STL files. Parts are scheduled for building and sent back to the customer within a few days, depending on part size, amount of finishing required, and order size.

Stratasys also recognizes that many parts coming off their machines will not be immediately suitable for the final application and that there may be an amount of finishing required. To assist in this, Stratasys provides a range of finishing stations that are designed to be compatible with various MEX materials. Finishing can be a mixture of chemically induced smoothing (using solvents that slightly melt the part surface) or burnishing using sodium bicarbonate as a light abrasive cleaning compound. Although there is a range of different material color filaments available, many applications require primers and coatings to achieve the right color and finish on a part. Figure 6.6 shows a wheel printed by 6Sixty Design and an identical machined alloy wheel.

Fig. 6.6   Machined alloy (gold) and polymer MEX (black) 20 inch wheel by 6Sixty Design

6.4.2   Other Material Extrusion Machines

Hundreds of companies have produced ME-type machines for sale. Because of stiff competition in the MEX market, and the fact that companies are being formed and going bankrupt at a fast rate, a comprehensive overview of companies is not attempted here. We will instead highlight two companies which provide variants of MEX machines that illustrate a portion of the breadth of this market. These companies are BigRep and Markforged.

BigRep is a German company that focuses on large-scale parts. They offer four MEX machine models that have build envelopes of 1 m³ to 1500 × 800 × 600 mm. Layer thickness is adjustable from 0.1 to 1.4 mm. The standard nozzle is 1 mm in diameter, with options for nozzles with diameters of 0.6 and 2 mm. Their feedstock is thermoplastic filament with a diameter of 2.85 mm, considerably larger than for smaller machines. The larger filament and layer thicknesses, compared to conventional MEX machines, enable much higher deposition and build rates. For example, their BigRep PRO machine offers a deposition rate of 115 cm³/hr for the 1 mm diameter nozzle, while their high-end EDGE machine offers a deposition speed of 1 m/s. Eight thermoplastic build materials are offered, along with two water-soluble support structures. In addition to the typical ABS and PLA build materials, polyamide, TPU, and carbon fiber-reinforced PET are also available.

The uniqueness of Markforged printers is their capability to print with continuous fibers to fabricate parts can be as strong as some aluminum alloys. Their continuous fiber machines have two deposition heads, one for thermoplastic and one for the fiber. Parts are fabricated by the typical thermoplastic deposition process, but, if reinforcement is desired, gaps are left in part cross-sections for the continuous fiber to be deposited. Resulting parts are selectively reinforced in each layer. Note that the continuous fibers are constrained to lie in one layer, thus strengthening properties particular in the x/y axes and less so in the z axis. As such, part orientation during build will dramatically affect part performance. Carbon fiber, fiberglass, and Kevlar fibers are available. Of their entry-level printers, the Mark Two offers all three fibers, a 0.1 mm layer thickness, and a build volume of $320 \times 132 \times 154$ mm. They also offer their Industrial Series printers, which are larger and more capable. Their X7 machine has the widest material availability, prints with 50 μm layers, and has a $330 \times 270 \times 200$ mm build volume.

6.4.3   Pellet-Fed Machines

An alternative to filaments is to use pellets as the feedstock material. In this case, an extruder is used in place of pinch rollers to force material through the heated reservoir and nozzle. Pellets are typically used as the feedstock for injection molding, so are readily available at significantly reduced cost, compared to MEX filaments. In these printers, two machine architectures are common: a stationary deposition head with the platform moving in X, Y, and Z directions and the conventional 2-axis deposition head (X and Y directions) with the platform indexing downward for each layer.

One of the first examples of a pellet-fed machine is the Big Area Additive Manufacturing (BAAM) machines developed by Cincinnati Incorporated and Oak Ridge National Laboratories. These machines have a gantry architecture, with the deposition head moving in X, Y, and Z directions and a stationary platform. They offer four BAAM models, with build volumes of $140 \times 65 \times 36$ inches to $240 \times 80 \times 72$ inches ($3.56 \times 1.65 \times 0.91$ m to $6.1 \times 2.29 \times 1.83$ m). A photo of their smallest machine is shown in Fig. 6.7. With nozzle diameters of 5–10 mm, their

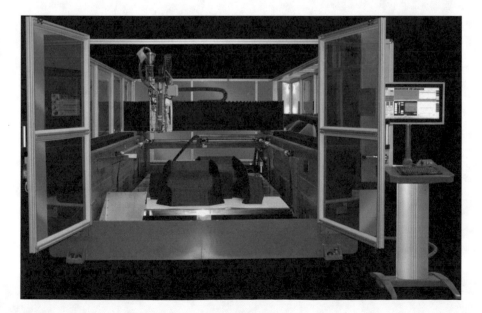

Fig. 6.7 BAAM machine from Cincinnati Incorporated

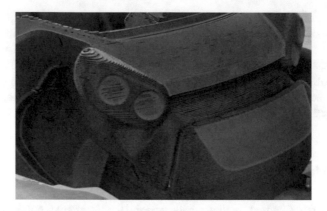

Fig. 6.8 Strati car front-end fabricated on a BAAM machine

machines can deposit 80 lb/hour of material (36.4 kg/hr). The developers have introduced some unique technologies to ensure part quality with such large parts and high deposition rates. Dynamic flow control was introduced into their deposition heads to precisely meter material flow. A unique tamping technology compresses the filament as it is deposited. Their process planning method must carefully consider temperature distributions in the build chamber so that filament bonding is achieved and high residual stresses are avoided. High deposition speed comes at the cost of reduced resolution. The large nozzle diameters mean that part surfaces are rough and require machining to yield smooth surfaces. This is evident in the photo, Fig. 6.8, of a Strati car body that was fabricated on one of their early machines. More information about this car design can be found in Chap. 18.

6.5   Materials

The most popular material is the ABSplus material, which can be used on all current Stratasys FDM machines. This is an updated version of the original ABS (acrylonitrile butadiene styrene) material that was developed for earlier MEX technology. Users interested in a translucent effect may opt for the ABSi material, which has similar properties to other materials in the ABS range. Some machines also have an option for ABS blended with polycarbonate (PC). Table 6.1 shows properties for various ABS materials and blends.

These properties are quite similar to many commonly used injection molding materials. It should be noted, however, that parts made using these materials on MEX machines may exhibit regions of lower strength than shown in this table because of interfacial regions in the layers and possible voids in the parts.

There are three other materials available for MEX technology that may be useful if the ABS materials cannot fulfill the requirements. A material that is predominantly PC-based can provide higher tensile properties, with a flexural strength of

Table 6.1 Variations in properties for the ABS range of MEX materials (compiled from Stratasys data sheets)

Property	ABS	ABSi	ABSplus	ABS/PC
Tensile strength	22 MPa	37 MPa	36 MPa	34.8 MPa
Tensile modulus	1627 MPa	1915 MPa	2265 MPa	1827 MPa
Elongation	6%	3.1%	4%	4.3%
Flexural strength	41 MPa	61 MPa	52 MPa	50 MPa
Flexural modulus	1834 MPa	1820 MPa	2198 MPa	1863 MPa
IZOD impact	106.78 J/m^2	101.4 J/m^2	96 J/m^2	123 J/m^2
Heat deflection @ 66 psi	90 °C	87 °C	96 °C	110 °C
Heat deflection @ 264 psi	76 °C	73 °C	82 °C	96 °C
Thermal expansion	5.60E-05 in/in/F	6.7E-6 in/in/F	4.90E-05 in/in/F	4.10E-5 in/in F
Specific gravity	1.05	1.08	1.04	1.2

104 MPa. A variation of this material is the PC-ISO, which is also PC-based, formulated to ISO 10993-1 and USP Class VI requirements. This material, while weaker than the normal PC with a flexural strength of 90 MPa, is certified for use in food and drug packaging and medical device manufacture. Another material that has been developed to suit industrial standards is the ULTEM 9085 material. This has particularly favorable flame, smoke, and toxicity (FST) ratings that makes it suitable for use in aircraft, marine, and ground vehicles. If applications require improved heat deflection, then an option would be to use the polyphenylsulfone (PPSF) material that has a heat deflection temperature at 264 psi of 189 °C. It should be noted that these last three materials can only be used in high-end machines and that they typically only work with breakaway support system, making their use somewhat difficult and specialized. The fact that they have numerous ASTM and similar standards attached to their materials indicates that Stratasys is seriously targeting final product manufacture (Direct Digital Manufacturing) as a key application for MEX.

Note that MEX works best with polymers that are amorphous in nature rather than the highly crystalline polymers that are more suitable for Powder Bed Fusion (PBF) processes. This is because the polymers that work best are those that are extruded in a viscous paste rather than in a lower viscosity form. As amorphous polymers, there is no distinct melting point, and the material increasingly softens and viscosity lowers with increasing temperature. The viscosity at which these amorphous polymers can be extruded under pressure is high enough that their shape will be largely maintained after extrusion, maintaining the extrusion shape and enabling them to solidify quickly and easily. Furthermore, when the material is added in an adjacent road or as a new layer, the previously extruded material can easily bond with it. This is different from Selective Laser Sintering, which relies on high crystallinity in the powdered material to ensure that there is a distinct material change from the powder state to a liquid state within a well-defined temperature region.

Many companies are introducing fiber-reinforced filaments suitable for MEX. A review of those provided by the Markforged company gives the readers an overview of the types of property improvements that are possible when using chopped fibers for MEX processes.

Markforged chopped fiber-reinforced filaments include Onyx, HSHT fiberglass, fiberglass, Onyx flame-retardant (FR), carbon fiber, and Kevlar filaments [9]. Onyx filament is nylon mixed with chopped carbon fiber that produces unique thermomechanical properties such as high strength, heat resistance, chemical resistance, and surface quality. Onyx FR is nylon-based filled with chopped carbon fiber designed for lightweight, strong, flame-resistance components. This plastic matrix is appropriate for aerospace, defense, automotive, and electronic housing applications. Mechanical properties of Onyx, Onyx FR, and nylon (white) are presented in Table 6.2.

Carbon fibers (CF) have very high stiffness and strength and can be made as thin filaments due to their crystalline structure. CF has the highest tensile strength and modulus among fibers used as reinforcement filaments. CF is also the stiffest and strongest fiber with the highest strength-to-weight ratio and, when combined with a polymer in composite form, can replace aluminum at half the weight. Carbon-reinforced composites (CRCs) have different applications such as robotic arms, forming tools, and fixtures. It is reported CRCs were used to provide a stiff and lightweight robot arm with 50 microns movement accuracy [9]. Mechanical properties for various fiber reinforcement materials are listed in Table 6.3.

Fiberglass filaments are broadly used as reinforcements. Fiberglass filaments are extremely thin strands of glass bundled together into a fiber. Fiberglass is about ten times stiffer than usual filaments for 3D printing and is suitable in composite form for functional prototyping, jigs, and fixtures. Fiberglass is 2–3 times more ductile than carbon fibers. This makes it possible, for instance, to print customized fiberglass-reinforced jaws that are strong enough to withstand clamping forces with no damage and marking of the components. High Strength High Temperature (HSHT) fiberglass has a higher impact resistance and heat deflection temperature

Table 6.2 Mechanical properties for plastic matrix filaments [9]

Plastic matrix	Onyx	Onyx FR	Nylon W
Tensile modulus (GPa)	1.4	1.3	1.7
Tensile stress at yield (MPa)	36	29	51
Tensile strain at yield (%)	25	33	4.5
Tensile stress at break (MPa)	30	31	36
Tensile strain at break (%)	58	58	150
Flexural strength (MPa)	81	79	50
Flexural modulus (GPa)	3.6	4.0	1.4
Heat deflection temp (°C)	145	145	41
Flame resistance	–	V-02	–
Izod impact – notched (J/m)	330	–	110
Density (g/cm³)	1.2	1.2	1.1

than standard fiberglass. Applications for HSHT fiberglass filaments include welding fixtures, thermoforms, mold inserts, and thermoset molds.

Kevlar is a synthetic, tough, and lightweight fiber that is conventionally used in different products such as sails, tires, and rope. Kevlar filament is 20% lighter than ABS with an eightfold higher impact resistance. Kevlar filament can be used for soft jaws, cradles, and end effectors since it exhibits high toughness and low density. The flexural stress–strain diagram for the introduced chopped filaments is shown in Fig. 6.9.

Table 6.3 Mechanical properties of fiber reinforcement [9]

Fiber reinforcement	Carbon	Kevlar®	Fiberglass	HSHT FG
Tensile strength (MPa)	800	610	590	600
Tensile modulus (GPa)	60	27	21	21
Tensile strain at break (%)	1.5	2.7	3.8	3.9
Flexural strength (MPa)	540	240	200	420
Flexural modulus (GPa)	51	26	22	21
Flexural strain at break (%)	1.2	2.1	1.1	2.2
Compressive strength (MPa)	320	97	140	192
Compressive modulus (MPa)	54	28	21	21
Compressive strain at break (%)	0.7	1.5	–	–
Heat deflection temp (°C)	105	105	105	150
Izod impact – notched (J/m)	960	2000	2600	3100
Density (g/cm³)	1.4	1.2	1.5	1.5

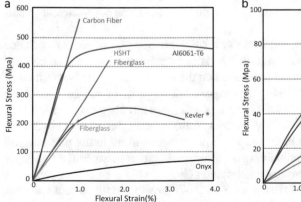
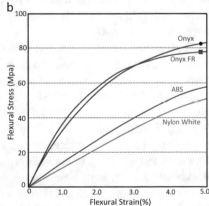

Fig. 6.9 Flexural stress–strain for (**a**) carbon fiber, HSHT fiberglass, Kevlar, fiberglass, Onyx and Al 9091. (**b**) Onyx, Onyx FR, ABS, and nylon filaments [9]

6.6 Limitations of MEX

MEX machines are very successful and meet the demands of many industrial as well as hobbyist users. This is partly because of the material properties and partly because of the low cost of the entry-level machines. There are, however, disadvantages when using this technology, mainly in terms of build speed, accuracy, and material density. Layer thickness is typically greater than 0.078 mm, so this limits feature sized and accuracy. Smaller nozzles with higher precision are possible but will lead to very long build times. Note also that all commercial nozzles are circular, and therefore it is impossible to draw sharp external corners; there will be a radius equivalent to that of the nozzle at any corner or edge. Internal corners and edges will also exhibit rounding. The actual shape produced is dependent on the nozzle, acceleration and deceleration characteristics, and the viscoelastic behavior of the material as it solidifies.

The speed of an MEX system is reliant on the feed rate and the plotting speed. Feed rate is also dependent on the ability to supply the material and the rate at which the liquefier can melt the material and feed it through the nozzle. If the liquefier were modified to increase the material flow rate, most likely it would result in an increase in mass. This in turn would make it more difficult to move the extrusion head faster. For precise movement, the plotting system is normally constructed using a lead-screw arrangement. Lower-cost systems can use belt drives, but flexing in the belts make it less accurate and there is also a lower torque reduction to the drive motor.

One method to improve the speed of motor drive systems is to reduce the corresponding friction. Stratasys used MagnaDrive technology to move the plotting head on early Quantum machines. By gliding the head on a cushion of air counterbalanced against magnetic forces attracting the head to a steel platen, friction was significantly reduced, making it easier to move the heads around at a higher speed. The fact that this system was replaced by conventional ball screw drives in the more recent FORTUS machines indicates that the improvement was not sufficient to balance against the cost.

One method to increase speed is to use a build strategy that attempts to balance the speed of using thick layers with the precision of using thin layers. The concept here is that thin layers only need to be used on the exterior of a part. The outline of a part can therefore be built using thin layers, but the interior can be built more quickly using thicker layers. Since most MEX machines have two extruder heads, it is possible that one head could have a thicker nozzle than the other. This thicker nozzle may be employed to build support structures and to fill in the part interior. However, the difficulty in maintaining a correct registration between the two layer thicknesses has prevented this approach from being broadly used commercially. A compromise on this solution is to use a honeycomb (or similar) fill pattern that uses less material and takes less time. This is only appropriate for applications where the reduced mass and strength of such a part is not an issue.

An important design consideration when using MEX is to account for the anisotropic nature of a part's properties. Additionally, different layering strategies result in different strengths. For instance, scanning strategy shown on the right in

Fig. 6.4 creates stronger parts than the left scanning strategy. Typically, properties are isotropic in the x–y plane since most software tools randomize raster orientation between layers; but if the raster fill pattern is set to preferentially deposit along a particular direction, then the properties in the x–y plane will also be anisotropic. In almost every case, the strength in the z-direction is measurably less than the strength in the x–y plane. Thus, for parts which undergo stress in a particular direction, it is best to build the part such that the major stress axes are aligned with the x–y plane rather than in the z-direction.

6.7 Bioextrusion

MEX technology has a large variety of materials that can be processed. If a material can be presented in a liquid form that can quickly solidify, then it is suitable to this process. As mentioned earlier, the creation of this liquid can be either through thermal processing of the material to create a melt or by using some form of chemical process where the material is in a gel form that can dry out or chemically harden quickly. These techniques are useful for bioextrusion. Bioextrusion is the process of creating biocompatible and/or biodegradable components that are used to generate frameworks, commonly referred to as "scaffolds," that play host to animal cells for the formation of tissue (tissue engineering). Such scaffolds should be porous, with micropores that allow cell adhesion and macro-pores that provide space for cells to grow.

There are a few commercial bioextrusion systems, like the modified MEX process used by Osteopore [10] to create scaffolds to assist in primarily head trauma recovery. This machine uses a conventional MEX-like process with settings for a proprietary material, based on the biocompatible polymer, polycaprolactone (PCL). Tissue engineering using MEX is seeing increased commercial interest; however, much of the work remains in a research stage, investigating many aspects of the process, including material choice, structural strength of scaffolds, coatings, biocompatibility, and effectiveness within various clinical scenarios. Many systems are custom developed in-house to match the specific interests of the researchers. There are, however, a small number of systems that are also available commercially.

6.7.1 Gel Formation

One common method of creating scaffolds is to use hydrogels. These are polymers that are water insoluble but can be dispersed in water. Hydrogels can therefore be extruded in a jellylike form. Following extrusion, the water can be removed and a solid, porous media remains. Such a media can be biocompatible and conducive to cell growth with low toxicity levels. Hydrogels can be based on naturally occurring polymers or synthetic polymers. The natural polymers are perhaps more biocompatible, whereas the synthetic ones are stronger. Synthetic hydrogels are rarely used in tissue engineering, however, because of the use of toxic reagents. Overall, use of hydrogels results in weak scaffolds that may be useful for soft tissue growth.

6.7.2   Melt Extrusion

Where stronger scaffolds are required, like when used to generate bony tissue, melt extrusion seems to be the process of choice. MEX can be used, but there are some difficulties in using this approach. In particular, MEX is somewhat unsuitable because of the expense of the materials. Biocompatible polymers suitable for tissue engineering are synthesized in relatively small quantities and are therefore only provided at high cost. Furthermore, the polymers often need to be mixed with other materials, like ceramics, that can seriously affect the flow characteristics, causing the material to behave in a non-Newtonian manner. Filaments fed by a pinch roller feed mechanism may not provide sufficient pressure at the nozzle tip for extrusion of biocompatible polymers, so many of the experimental systems use screw feed, similar to conventional injection molding and extrusion technology. Screw feed systems benefit from being able to feed small amounts of pellet-based feedstock, enabling one to work with a small material volume.

In addition to their layer-wise photopolymerization machines, EnvisionTEC [11] has also developed the 3D-Bioplotter system (see Fig. 6.10). This system is a MEX, screw feeding technology that is designed specifically for biopolymers. Lower temperature polymers can be extruded using a compressed gas feed, instead of a screw extruder, which results in a much simpler mechanism. Much of the system uses non-reactive stainless steel, and the machine itself has a small build envelope and software specifically aimed at scaffold fabrication. The melt reservoir is sealed apart from the nozzle, with a compressed air feed to assist the screw extrusion process.

Fig. 6.10 The
EnvisionTEC
3D-Bioplotter system.
(Photo courtesy of
EnvisionTEC) [12]

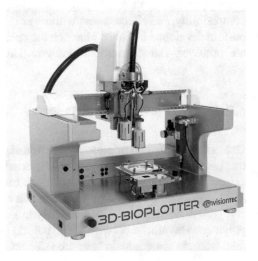

The system uses one extrusion head at a time, with a carousel feeder so that extruders can be swapped at any time during the process. This is particularly useful since most tissue engineering research focuses on building scaffolds with different regions made from different materials. Build parameters can be set for a variety of materials with control over the reservoir temperature, feed rate, and plotting speed to provide users with a versatile platform for tissue engineering research.

It should be noted that tissue engineering is an extremely complex research area, and the construction of physical scaffolds is just the starting point. This approach may result in scaffolds that are comparatively strong compared with hydrogel-based scaffolds, but they may fail in terms of biocompatibility and bio-toxicity. To overcome some of these shortcomings, a significant amount of post-processing is required.

6.7.3 Scaffold Architectures

One of the major limitations with MEX systems for conventional manufacturing applications relates to the diameter of the nozzle. For tissue engineering, however, this is not such a limitation. Scaffolds are generally built up so that roads are separated by a set distance so that the scaffold can have a specific macro porosity. In fact, the aim is to produce scaffolds that are as strong as possible but with as much porosity as possible. The greater the porosity, the more space there is for cells to grow. Scaffolds with greater than 66% porosity are common. Sometimes, therefore, it may be better to have a thicker nozzle to build stronger scaffold struts. The spacing between these struts can be used to determine the scaffold porosity.

The most effective geometry for scaffolds has yet to be determined. For many studies, scaffolds with a simple 0° and 90° orthogonal crossover pattern may be sufficient. More complex patterns vary the number of crossovers and their separation. Examples of typical patterns can be seen in Fig. 6.11. Many studies have investigated how cells proliferate in these different scaffold architectures and are usually carried out using bioreactors for in vitro (non-invasive) experiments. As such, samples are usually quite small and often cut from a larger scaffold structure. Using samples that are as large and complex in shape as the bones they are designed to replace and that are implanted in animal or human subjects is becoming more

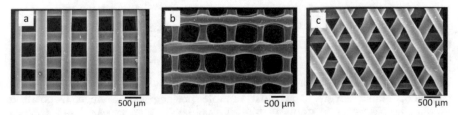

Fig. 6.11 Different scaffold designs showing a porous structure, with an actual image of a scaffold created using a bioextrusion system. (Elsevier license number 4602501409572) [13]

common as tissue engineering using AM is shifting from the research to the commercial realm. Nevertheless, many fundamental questions must be investigated to help make these procedures common.

6.8   Other Systems

Although the initial Stratasys patented designs for MEX remain the standard approach today, there are a number of other system designs which our readers should be familiar with. This section overviews a few of these variants.

6.8.1   Contour Crafting

In normal Additive Manufacturing, layers are considered as 2D shapes extruded linearly in the third dimension. Thicker layers result in lower part precision, particularly where there are slopes or curves in the vertical direction. A major innovative twist on the MEX approach can be found in the Contour Crafting technology developed by Prof. B. Khoshnevis and his team at the University of Southern California [14]. Taking the principle mentioned above that the exterior surface is the most critical in terms of meeting precision requirements, this research team has developed a method to smooth the surface with a scraping tool. This is similar to how artisans shape clay pottery and/or concrete using trowels. By contouring the layers as they are being deposited using the scraping tool to interpolate between these layers, very thick layers can be made that still replicate the intended geometry well.

Using this technique, it is conceptually possible to fabricate extremely large objects very quickly compared with other additive processes, since the exterior precision is no longer determined solely by the layer thickness. The scraper tool need not be a straight edge and can indeed be somewhat reconfigurable by positioning different parts of the tool in different regions or by using multiple passes. To illustrate this advantage, the team is in fact developing technology that can produce full-sized buildings using a mixture of the Contour Crafting process, fast-setting concrete, and robotic assembly (see Fig. 6.12).

6.8.2   Nonplanar Systems

There have been a few attempts at developing AM technology that does not use stratified, planar layers. The most notable projects are Shaped Deposition Manufacturing (SDM), Ballistic Particle Manufacturing (BPM), and Curved Laminated Object Manufacturing (Curved LOM) [16]. It is also possible to use MEX to extrude nonplanar layers that curve in the z-axis as well as the x–y axes. To do so, some type of 3D curved support system is required. This is primarily being investigated in MEX for long-fiber-reinforced filaments, so that the

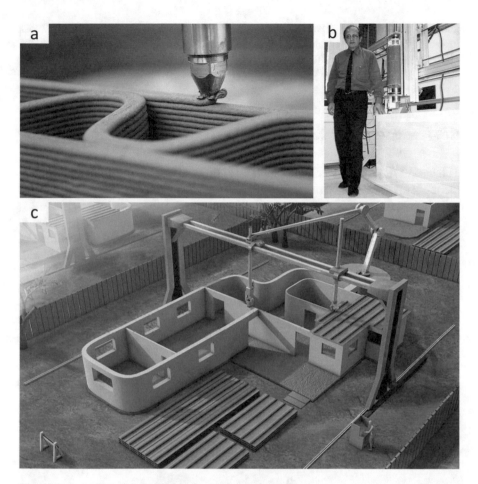

Fig. 6.12 (**a**, **b**) Contour Crafting technology, developed at USC, showing scraping device and large-scale machine. (**c**) Contour Crafting to build a house. (Photo courtesy of Contour Crafting Corporation) [15]

strength imparted by these filaments can be used in preferential directions based upon the needs of the part design. Process planning for nonplanar complex geometries becomes quite difficult, and there is not a simple software solution available to automate this process for arbitrary complex geometries. However, certain parts that require surface toughness can benefit from this nonplanar approach [17].

6.8.3   Material Extrusion of Ceramics

Another possible application of MEX is to develop ceramic part fabrication processes. In particular, MEX can be used to extrude ceramic pastes that can quickly solidify. The resulting parts can be fired using a high-temperature furnace to fuse and densify the ceramic particles. Resulting parts can have good properties with the geometric complexity characteristics of AM processes. The most prolific work investigating MEX to create ceramics came out of Rutgers University in the USA [18]. There has been very little commercial adoption of MEX for ceramics, but this area remains one of active research interest.

6.8.4   RepRap and Fab@Home

The basic MEX process is quite simple; and this can be illustrated by the development of two systems that are extremely low cost and capable of being constructed using minimal tools.

The RepRap project [19] is an experiment in open-source technology applied to AM. The initial idea was developed by a group at the University of Bath in the UK and designs and ideas continue to be developed by a number of enthusiasts worldwide. One concept that led to early interest is that a machine is capable of producing components for future machines, testing some of the theories of von Neumann on self-replicating machines. A large number of design variants exist, some using cold cure resins but most using a thermal extrusion head, but all are essentially variants of the MEX process, as illustrated by one of these designs shown in Fig. 6.13.

Fig. 6.13 The RepRap "Darwin" machine that is capable of making some of its own parts

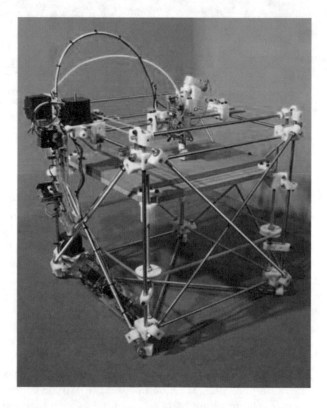

Another project that helped kick-start the availability of low-cost MEX technology was the Fab@Home project from Cornell University. The initial design used a frame constructed from laser-cut polymer sheets, assembled like a 3D jigsaw. Low-cost stepper motors and drives commonly found in inkjet printers were used for positioning, and the extrusion head was a compressed-air-fed syringe compatible with a variety of cold-cure materials. Fab@Home designs can be obtained free of charge and kits can be obtained for assembly at a very low cost.

Both of these approaches have inspired a variety of enthusiasts and start-up companies to develop their ideas. Some have focused on improving the designs so that they may be more robust or more versatile. Others have developed software routines that explore things like scanning patterns, more precise control, etc. Yet other enthusiasts have developed new potential applications for this technology, most notably using multiple materials that have unusual chemical or physical behavior. The Fab@Home technology has, for example, been used to develop 3D batteries and actuators. Some users have even experimented with chocolate to create edible sculptures.

This development of entry-level AM technology has sparked a 3D Printing revolution that has attracted huge amounts of media attention and brought it into the public domain. This explosion in interest is covered in more detail in Chap. 13 of this book.

6.9   Questions

1. Derive an expression for Q_T so that we can determine the flow through a circular extrusion nozzle.
2. The expression for total flow does not include a gravity coefficient. Derive an expression for Q_T that includes gravity, assuming there is a constant amount of material in the melt reservoir and the nozzle is pointing vertically downward.
3. Derive a pressure equation that connects the forces generated by the pinch rollers shown in Fig. 6.1 to the elastic modulus.
4. The expressions derived for solidification and bonding assume that a thermal process is being used. What do you think the terms will look like if a curing or drying process were used?
5. Why is MEX more suitable for medical scaffold architectures, compared with PBF-fabricated scaffolds made from a similar material?
6. In what ways is MEX similar to CNC pocket milling and in what ways is it different?
7. What are the differences between MEX and extrusion forming? Give an individual example to illustrate the differences.
8. Explain briefly what parameters would affect the precision of extrusion-based processes?
9. Do you think that Rapid Prototyping or medical scaffolding is a better application for MEX? Explain.

10. MEX machines, such as the MakerBot, are the Additive Manufacturing Technology that is strongly being marketed for personal use. What is it about MEX technology that makes it appealing for personal use, but not so much for commercial production applications?
11. Why does the MEX process work better with amorphous polymers compared to crystalline polymers?
12. Is it possible to use multiple materials with MEX, why or why not?
13. What are the two most significant benefits and drawbacks of MEX processes (one major benefit and one major drawback)?
14. Read online about the Echoviren Pavilion and the Canal House in Amsterdam. Each structure was made using a MEX process. What are the advantages and disadvantages to each construction process?
15. Read about the KamerMaker online. What type of structures could it potentially be used for?
16. Low-cost 3D printers typically use plastic PLA or ABS filament. What should a person consider before using one of these 3D printers for in-house repairs?
17. Why would Contour Crafting be strongly considered for constructing buildings on the moon?
18. Consider Fig. 6.4 and explain how the ideas shown can contribute to the dimensional accuracy of the part. Is it positively or negatively affected? Use your own words.
19. Why is MEX good at fabricating medical scaffolds? What are its limitations? Name some other AM technologies that are capable of scaffold fabrication.

References

1. Bellini, A., Guceri, S., & Bertoldi, M. (2004). Liquefier dynamics in fused deposition. *Journal of Manufacturing Science and Engineering, 126*(2), 237–246.
2. Turner, B. N., Strong, R., & Gold, S. A. (2014). A review of melt extrusion additive manufacturing processes: I. Process design and modeling. *Rapid Prototyping Journal, 20*(3), 192–204.
3. Smid, P. (2006). *CNC programming techniques: An insider's guide to effective methods and applications*. New York: Industrial Press Inc.
4. Atif Yardimci, M., & Güçeri, S. (1996). Conceptual framework for the thermal process modelling of fused deposition. *Rapid Prototyping Journal, 2*(2), 26–31.
5. Agarwala, M. K., et al. (1996). Structural quality of parts processed by fused deposition. *Rapid Prototyping Journal, 2*(4), 4–19.
6. Stratasys. (2020). *Fused deposition modelling*. http://www.stratasys.com
7. Stratasys. (2020). *FDM technology*. https://www.stratasys.com/fdm-technology
8. *Capabilities expand with next generation Fortus production 3D printer from Stratasys*. https://www.3dprintingmedia.network/capabilities-expand-next-generation-fortus-production-3d-printer-stratasys/. 2020.
9. Markforged. (2020). https://markforged.com/
10. Osteopore. (2020). *FDM produced tissue scaffolds*. http://www.osteopore.com/
11. Envisiontec. (2020). *Bioplotting system*. http://www.envisiontec.com
12. envisionTEC. (2020). https://www.flickr.com/photos/envisiontec/albums/72157672066966280

13. Zein, I., et al. (2002). Fused deposition modeling of novel scaffold architectures for tissue engineering applications. *Biomaterials, 23*(4), 1169–1185.
14. Contour Crafting. (2020). *Large scale extrusion-based process.* http://contourcrafting.com/
15. Contour Crafting. (2020). http://contourcrafting.com/building-construction/
16. Klosterman, D. A., et al. (1999). Development of a curved layer LOM process for monolithic ceramics and ceramic matrix composites. *Rapid Prototyping Journal, 5*(2), 61–71.
17. Yuan, L., & Gibson, I. (2007). A framework for development of a fiber-composite, curved FDM system. In *Proceedings of International Conference on Manufacturing Automation.* ICMA.
18. Jafari, M., et al. (2000). A novel system for fused deposition of advanced multiple ceramics. *Rapid Prototyping Journal, 6*(3), 161–175.
19. Reprap. (2020). *Self replicating AM machine concept.* http://www.reprap.org

Chapter 7
Material Jetting

Abstract Printing technologies progressed rapidly as the adoption of personal computers spread through offices and homes. Inkjet printing, in particular, is a huge market, and billions of dollars have been invested to make inkjet print heads reliable, inexpensive, and widely available. As a result of advances in inkjet printing technologies, many application areas beyond photos and text have been explored, including electronics packaging, optics, and Additive Manufacturing. Some of these applications have literally taken the technology into a new dimension. The employment of printing technologies in the creation of three-dimensional products quickly became a promising manufacturing practice, both widely studied and increasingly widely used.

This chapter will summarize how printing technologies are used in Additive Manufacturing to deposit build material layer-by-layer to produce 3D objects. The history of printing as a process to fabricate 3D parts is summarized, followed by a survey of commercial polymer printing machines. The focus of this chapter is on Material Jetting (MJT) in which all of the part material is dispensed from a print head. In Chap. 8 on Binder Jetting (BJT), printing of a binder onto a powder bed to form a part is covered, and in Chap. 5 on Powder Bed Fusion (PBF), printing sintering aids and inhibitors onto a powder bed prior to fusion is covered. In this chapter, some of the technical challenges of printing are introduced; material development for printing polymers, metals, and ceramics is investigated in some detail. Models of the MJT process are introduced that relate pressure required to fluid properties. Additionally, a printing indicator expression is derived and used to analyze printing conditions.

© Springer Nature Switzerland AG 2021
I. Gibson et al., *Additive Manufacturing Technologies*,
https://doi.org/10.1007/978-3-030-56127-7_7

7.1   Evolution of Printing as an Additive Manufacturing Process

Two-dimensional inkjet printing has been in existence since the 1960s, used for decades as a method for printing documents and images from computers and other digital devices. Inkjet printing is now widely used in the desktop printing industry, commercialized by companies such as HP and Canon. Le [1] provides a thorough review of the historical development of the inkjet printing industry.

Printing as a three-dimensional building method was first demonstrated in the 1980s with patents related to the development of Ballistic Particle Manufacturing, which involved simple deposition of "particles" of material onto an article [2]. The first commercially successful technology was the ModelMaker from Sanders Prototype (now Solidscape), introduced in 1994, which printed a basic wax material that was heated to liquid state [3]. These are considered the first MJT technologies, since they represent the first AM machines for which all of the part material is dispensed from a print head.

In 1996, 3D Systems joined the competition with the introduction of the Actua 2100, another wax-based printing machine. The Actua was revised in 1999 and marketed as the ThermoJet [3]. In 2001, Sanders Design International briefly entered the market with its Rapid ToolMaker but was quickly restrained due to intellectual property conflicts with Solidscape [3]. It is notable that all of these members of the first generation of printing machines relied on heated waxy thermoplastics as their build material; they are therefore most appropriate for concept modeling and investment casting patterns.

More recently, the focus of development has been on the deposition of acrylate photopolymers, wherein droplets of liquid monomer are formed and then exposed to ultraviolet light to promote polymerization. The reliance upon photopolymerization is similar to that in stereolithography, but other process challenges are significantly different. The leading edge of this second wave of machines arrived on the market with the Quadra from Objet Geometries of Israel in 2000, followed quickly by the revised QuadraTempo in 2001. Both machines jetted a photopolymer using print heads with over 1500 nozzles [3]. Objet merged with Stratasys in 2012 and continues to sell MJT machines under the tradename PolyJet.

In 2003, 3D Systems launched a competing technology with its InVision 3D printer. Multi-jet modeling, the printing system used in this machine, was actually an extension of the technology developed with the ThermoJet line [3], despite the change in material solidification strategy.

In the 2010s, several technologies have emerged that print metal or ceramic. Vader Systems developed a metal droplet printing technology for aluminum alloys, while XJet developed a technology for printing colloids containing metal or ceramic nanoparticles.

Companies continue to innovate, and MJT technologies remain a dynamic part of the AM industry, as will be discussed in the next sections.

7.2 Materials for Material Jetting

While industry players have introduced MJT machines that use waxy polymers and acrylic photopolymers, research groups have experimented with the potential for MJT machines that can build in many other materials. Among the materials most studied are polymers, ceramics, and metals. In addition to the commercially available materials, this section highlights achievements in related research areas.

For common droplet formation methods, the maximum printable viscosity threshold is generally considered to be in the range of 20–40 centipoise (cP) at the printing temperature [4–6]. An equivalent unit of measure is the millipascal-second, denoted mPa·s if SI units are preferred. To facilitate jetting, materials that are solid at room temperature must be heated so that they liquefy. For high-viscosity fluids, the viscosity of the fluid must be lowered to enable jetting. The most common practices are to use heat or solvents or other low-viscosity components in the fluid. In addition to these methods, it is also possible that in some polymer deposition cases shear thinning might occur, dependent upon the material or solution in use; drop-on-demand (DOD) printers produce strain rates of 10^3–10^4, which should be high enough to produce shear-thinning effects [4–7]. While other factors such as liquid density or surface tension and print head or nozzle design may affect the results, the limitation on viscosity quickly becomes the most problematic aspect for droplet formation in MJT.

7.2.1 Polymers

Polymers consist of an enormous class of materials, representing a wide range of mechanical properties and applications. And although polymers are the most common type of materials used in commercial AM machines, there is less discussion on polymer inkjet production of macro three-dimensional structures in the published scientific literature than for most other AM techniques.

Gao and Sonin [8] present the first notable academic study of the deposition and solidification of groups of molten polymer microdrops. They discuss findings related to three modes of deposition: columnar, sweep (linear), and repeated sweep (vertical walls). The two materials used in their investigations were a candelilla wax and a microcrystalline petroleum wax, deposited in droplets 50 μm in diameter from a print head 3–5 mm from a cooled substrate. The authors first consider the effects of droplet deposition frequency and cooling on columnar formation. As would be expected, if the drops are deposited rapidly ($\geq$50 Hz in this case), the substrate on which they impinge is still at an elevated temperature, reducing the solidification contact angle and resulting in ball-like depositions instead of columns (Fig. 7.1a). Numerical analyses of the relevant characteristic times of cooling are included. Gao and Sonin also consider horizontal deposition of droplets and the subsequent formation of lines. They propose that smooth solid lines will be formed only in a small

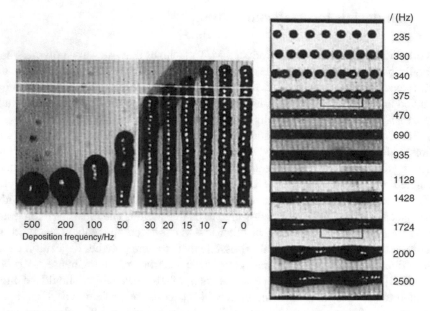

Fig. 7.1 (**a**) Columnar formation and (**b**) line formation as functions of droplet impingement frequency [8]

range of droplet frequencies, dependent upon the sweep speed, droplet size, and solidification contact angle (Fig. 7.1b). Finally, they propose that wall-like deposition will involve a combination of the relevant aspects from each of the above situations.

Reis et al. [9] also provide some discussion on the linear deposition of droplets. They deposited molten Mobilwax paraffin wax with a heated print head from Solidscape. They varied both the print head horizontal speed and the velocity of droplet flight from the nozzle. For low droplet speeds, low sweep speeds created discontinuous deposition, and high sweep speeds created continuous lines (Fig. 7.2a–c). High droplet impact speed led to splashing at high sweep speeds and line bulges at low sweep speeds (Fig. 7.2d–f).

From these studies, it is clear that process variables such as print head speed, droplet velocity, and droplet frequency affect the quality of the deposit. These process variables vary depending upon the characteristics of the fluid being printed, so some process development, or fine-tuning, is generally required when trying to print a new material or develop a new printing technology.

Feng et al. [10] present a full system, based on a print head from MicroFab Technologies Inc., that functions similar to the commercially available machines. It prints a wax material which is heated to 80 °C, more than ten degrees past its melting point, and deposits it in layers 13–60 μm thick. The deposition pattern is controlled by varying the droplet size and velocity, as well as the pitch and hatch spacing of the lines produced. An example of the result, a 2.5-dimensional gear, is presented in Fig. 7.3.

The earliest and most often used solution to the problem of high viscosity is to heat the material until its viscosity drops to an acceptable point. As discussed in Sect. 7.5, for example, commercial machines such as the Solidscape S300 printer series and 3D Systems' ProJet MJP 2500 print proprietary thermoplastics, which contain mixtures of various waxes and polymers that are solid at ambient temperatures but convert to a liquid phase at elevated printing temperatures [11]. In developing their hot-melt materials in the 2000s, for example, 3D Systems investigated various mixtures consisting of a low shrinkage polymer resin, a low-viscosity material such as paraffin wax, a microcrystalline wax, a toughening polymer, and a small amount of plasticizer, with the possible additions of antioxidants, coloring agents, or heat-dissipating filler [12]. These materials were formulated to have a viscosity of 18–25 cP and a surface tension of 24–29 dynes/cm at the printing temperature of 130 °C. De Gans et al. [13] contend that they have used a micropipette optimized for polymer printing applications that was able to print Newtonian fluids with viscosities up to 160 cP.

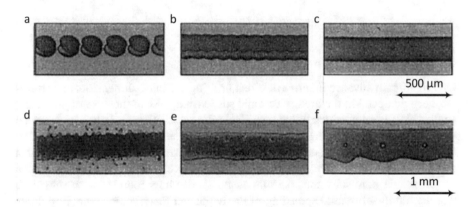

Fig. 7.2 Results of varying sweep and impact speeds [9]

Fig. 7.3 Wax gear [10]

For thermoset polymers, companies developed liquid photopolymerizable resins using prepolymers. For example, 3D Systems investigated a series of UV-curable printing materials, consisting of mixtures of high-molecular-weight monomers and oligomers such as urethane acrylate or methacrylate resins, urethane waxes, low-molecular-weight monomers and oligomers such as acrylates or methacrylates that function as diluents, a small amount of photoinitiators, and other additives such as stabilizers, surfactants, pigments, or fillers [14, 15]. These materials also benefited from the effects of hot-melt deposition, as they were printed at a temperature of 70–95 °C, with melting points between 45 and 65 °C. At the printing temperatures, these materials had a viscosity of about 10–16 cP.

One problem encountered, and the reason that the printing temperatures cannot be as high as those used in hot-melt deposition, is that when kept in the heated state for extended periods of time, the prepolymers begin to polymerize, raising the viscosity and possibly clogging the nozzles when they are finally printed [16]. Another complication is that the polymerization reaction, which occurs after printing, must be carefully controlled to assure dimensional accuracy.

7.2.2   Ceramics

One significant advance in terms of direct printing for three-dimensional structures has been achieved in the area of ceramic suspensions. As in the case of polymers, studies have been conducted that investigate the basic effects of modifying sweep speed, drop-to-drop spacing, substrate material, line spacing, and simple multilayer forms in the deposition of ceramics [16]. These experiments were conducted with a mixture of zirconia powder, solvent, and other additives, which were printed from a 62 μm nozzle onto substrates 6.5 mm away. The authors found that on substrates that permitted substantial spreading of the deposited materials, neighboring drops would merge to form single, larger shapes, whereas on other substrates the individual dots would remain independent (see Fig. 7.4). In examples where multiple layers were printed, the resulting deposition was uneven, with ridges and valleys throughout.

A sizable body of work has been amassed in which suspensions of alumina particles are printed via a wax carrier [4] which is melted by the print head. Suspensions of up to 40% solids loading have been successfully deposited at viscosities of 2.9–38.0 cP at a measurement temperature of 100 °C; higher concentrations of the suspended powder have resulted in prohibitively high viscosities. Because this deposition method results in a part with only partial ceramic density, the green part must be burnt out and sintered, resulting in a final product which is 80% dense but whose dimensions are subject to dramatic shrinkage [17]. A part created in this manner is shown in Fig. 7.5.

Similar attempts have been made with zirconia powder, using material with 14% ceramic content by volume [18], with an example shown in Fig. 7.6, as well as with PZT at up to 40% ceramic particles by volume [19].

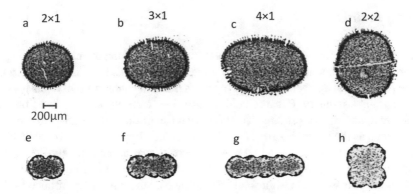

Fig. 7.4 Droplets on two different substrates [16]

Fig. 7.5 Sintered alumina impeller [17]

Fig. 7.6 Sintered zirconia vertical walls [18]

7.2.3   Metals

Much of the printing work related to metals has focused upon the use of printing for electronics applications – formation of traces, connections, and soldering. Liu and Orme [20] present an overview of the progress made in solder droplet deposition for the electronics industry. Because solder has a low melting point, it is an obvious choice as a material for printing. They reported use of droplets of 25–500 μm, with results such as the IC test board in Fig. 7.7, which has 70 μm droplets of Sn63/Pb37. In related work, a solder was jetted whose viscosity was approximately 1.3 cP, continuously jetted under a pressure of 138 kPa. Many of the results to which they refer are those of researchers at MicroFab Technologies, who have also produced solder forms such as 25 μm diameter columns.

There is, however, some work in true three-dimensional fabrication with metals. Priest et al. [21] provide a survey of liquid metal printing technologies and history, including alternative technologies employed and ongoing research initiatives. Metals that had been printed included copper, aluminum, tin, various solders, and mercury. One major challenge identified for depositing metals is that the melting point of the material is often high enough to significantly damage components of the printing system.

Orme et al. [22, 23] report on a process that uses droplets of Rose's metal (an alloy of bismuth, lead, and tin). They employ nozzles of diameter 25–150 μm with resulting droplets of 47–283 μm. In specific cases, parts with porosity as low as 0.03% were formed without post-processing, and the microstructure formed is more uniform than that of standard casting. In discussion of this technology, considerations of jet disturbance, aerodynamic travel, and thermal effects are all presented.

Yamaguchi et al. [24, 25] used a piezoelectrically driven actuator to deposit droplets of an alloy (Bi–Pb–Sn–Cd–In), whose melting point was 47 °C. They heated the material to 55 °C and ejected it from nozzles 200 μm, 50 μm, and less than 8 μm in diameter. As expected, the finer droplets created parts with better reso-

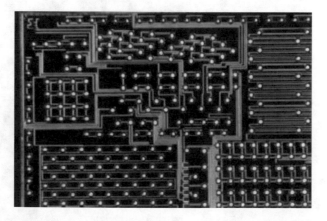

Fig. 7.7 IC test board with solder droplets [20]

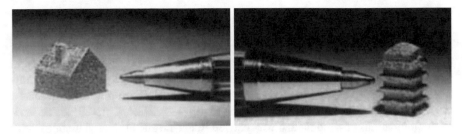

Fig. 7.8 Examples of parts fabricated with metal printing [25]

lution. The density, or "packing rate," of some parts reached 98%. Other examples of fabricated parts are shown in Fig. 7.8.

Several research groups have demonstrated aluminum deposition [26, 27]. In one example, fairly simple shapes have been formed from Al2024 alloy printed from a 100 μm orifice. In another example, pressure pulses of argon gas in the range of 20–100 KPa were used to eject droplets of molten aluminum at the rate of 1–5 drops per second. To achieve this, the aluminum was melted at 750 °C, and the substrate heated to 300 °C. The nozzle orifice used was 0.3 mm in diameter, with a resulting droplet size of 200–500 μm and a deposited line of width 1.00 mm and thickness 0.17 mm. The final product was a near-net shape part of density up to 92%.

As these examples have shown, metal MJT is well on its way to becoming a viable process for three-dimensional prototyping and manufacturing. While industry has only barely begun to use MJT for metals, the economic and efficiency advantages that printing provides ensure that it will be pursued in the future.

7.2.4   Solution- and Dispersion-Based Deposition

As hot-melt printing has very specific requirements for material properties, many current applications have turned to solution- or dispersion-based deposition. This allows the delivery of solids or high-molecular-weight polymers in a carrier liquid with viscosity low enough to be successfully printed. De Gans et al. [5] provided a review of a number of polymeric applications in which this strategy is employed.

A number of investigators have used solution and dispersion techniques in accurate deposition of very small amounts of polymer in thin layers for mesoscale applications, such as polymer light-emitting displays, electronic components, and surface coatings and masks. For example, Shimoda et al. [28] present a technique to develop light-emitting polymer diode displays using inkjet deposition of conductive polymers. Three different electroluminescent polymers (polyfluorine and two derivatives) were printed in organic solvents at 1–2 wt%. As another example, De Gans et al. [5] report a number of other results related to the creation of polymer light-emitting displays: a precursor of poly(*p*-phenylene vinylene) (PPV) was printed as a 0.3%wt solution; and PPV derivatives were printed in 0.5–2.0 wt% solutions in solvents such as tetraline, anisole, and o-xylene. Such low-weight percentages are typical.

In deposition of ceramics, the use of a low-viscosity carrier is also a popular approach. Tay and Edirisinghe [16], for example, used ceramic powder dispersed in industrial methylated spirit with dispersant, binder, and plasticizer additives resulting in a material that was 4.5% zirconia by volume. The resulting material had a viscosity of 3.0 cP at 20 °C and a shear rate of 1000 s^{-1}. Zhao et al. [29] tested various combinations of zirconia and wax carried in octane and isopropyl alcohol, with a dispersant added to reduce sedimentation. The viscosities of these materials were 0.6–2.9 cP at 25 °C; the one finally selected was 14.2% zirconia by volume.

Despite the success of solution and dispersion deposition for these specific applications, however, there are some serious drawbacks, especially in considering the potential for building complex, large 3D components. The low concentrations of polymer and solid used in the solutions and dispersions restrict the total amount of material that can be deposited. Additionally, it can be difficult to control the deposition pattern of this material within the area of the droplet's impact. Shimoda et al. [28], among others, report the formation of rings of deposited material around the edge of the droplet. They attribute this to the fact that the contact line of the drying drop is pinned on the substrate. As the liquid evaporates from the edges, it is replenished from the interior, carrying the solutes to the edge. They contend that this effect can be mitigated by control of the droplet drying conditions.

Another difficulty with solutions or dispersions, especially those based on volatile solvents, is that use of these materials can result in precipitations forming in the nozzle after a very short period of time [13], which can clog the nozzle, making deposition unreliable or impossible.

The company XJet [30] seems to have overcome many of these difficulties with the use of solution deposition. With their NanoParticle Jetting™ (NPJ) technology, parts are fabricated by jetting ceramic or metal nanoparticle colloid droplets from thousands of nozzles. The carrier fluid evaporates leaving behind the nanoparticles. Support structures are used, as in any other MJT process, which are removed in a solvent bath. The final part fabrication step is to sinter the part overnight. The company claims that parts are highly accurate and suffer very little shrinkage.

7.3 Material Processing Fundamentals

7.3.1 Technical Challenges of MJT

As evidenced by the industry and research applications discussed in the previous section, MJT already has a strong foothold in terms of becoming a successful AM technology. There are, however, some serious technical shortcomings that have prevented its development from further growth. To identify and address those problems, the relevant phenomena and strategic approaches taken by its developers must be understood. In the next two sections, the technical challenges of the printing process are outlined, the most important of its limitations relevant to the deposition of functional polymers is identified, and how those limitations are currently addressed is summarized.

MJT for three-dimensional fabrication is an extremely complex process, with challenging technical issues throughout. The first of these challenges is formulation of the liquid material. If the material is not in liquid form to begin with, this may mean suspending particles in a carrier liquid, dissolving materials in a solvent, melting a solid polymer, or mixing a formulation of monomer or prepolymer with a polymerization initiator. In many cases, other substances such as surfactants are added to the liquid to attain acceptable characteristics. Entire industries are devoted to the mixture of inks for two-dimensional printing, and it is reasonable to assume that in the future this will also be the case for three-dimensional fabrication.

The second hurdle to overcome is droplet formation. To use inkjet deposition methods, the material must be converted from a continuous volume of liquid into a number of small discrete droplets. This function is often dependent upon a finely tuned relationship between the material being printed, the hardware involved, and the process parameters; a number of methods of achieving droplet formation are discussed in this section. Small changes to the material, such as the addition of tiny particles [31], can dramatically change its droplet-forming behavior, as can changes to the physical setup.

A third challenge is control of the deposition of these droplets; this involves issues of droplet flight path, impact, and substrate wetting or interaction [32–36]. In printing processes, either the print head or the substrate is usually moving, so the calculation of the trajectory of the droplets must take this issue into account. In addition to the location of the droplets' arrival, droplet velocity and size will also affect the deposition characteristics and must be measured and controlled via nozzle design and operation [37]. The quality of the impacted droplet must also be controlled: if smaller droplets, called satellites, break off from the main droplet during flight, then the deposited material will be spread over a larger area than intended, and the deposition will not have well-defined boundaries. In the same way, if the droplet splashes on impact, forming what is called a "crown," similar results will occur [38]. All of the effects will negatively impact the print quality of the printed material.

Concurrently, the conversion of the liquid material droplets to solid geometry must be carefully controlled; as discussed in Sect. 7.2, MJT relies on a phase change of the printed material. Examples of phase change modes employed in existing printing technologies are solidification of a melted material (e.g., wax, solder), evaporation of the liquid portion of a solution (e.g., some ceramic approaches), and curing of a photopolymer (e.g., Objet, ProJet machines) or other chemical reactions. The phase change must occur either during droplet flight or soon after impact; the time and place of this conversion will also affect the droplet's interaction with the substrate [39, 40] and the final deposition created. To further complicate the matter, drops may solidify nonuniformly, creating warpage and other undesirable results [41].

An additional challenge is to control the deposition of droplets on top of previously deposited layers, rather than only upon the initial substrate [8, 16]. The droplets will interact differently, for example, with a metal plate substrate than with a surface of previously printed wax droplets. To create substantive three-dimensional parts, each layer deposited must be fully bound to the previous layer to prevent delamination, but must not damage that layer while being created. Commercially available machines tend to approach this problem by employing devices that plane or otherwise smooth the surface periodically [41–43].

Operational considerations also pose a challenge in process planning for MJT. For example, because nozzles are so small, they often clog, preventing droplets from exiting. Much attention has been given to monitoring and maintaining nozzle performance during operation [41]. Most machines currently in use go through purge and cleaning cycles during their builds to keep as many nozzles open as possible; they may also wipe the nozzles periodically [42]. Some machines may also employ complex sensing systems to identify and compensate for malfunctioning or inconsistent nozzles [44, 45]. In addition, many machines, including all commercial AM machines, have replaceable nozzles in case of permanent blockage.

Finally, to achieve the best print resolution, it is advantageous to produce many small droplets very close together. However, this requires high nozzle density in the print head, which is unattainable for many nozzle manufacturing processes. An alternative to nozzle density is to make multiple passes over the same area, effectively using process planning instead of hardware to create the desired effect [42]. Even in cases where high nozzle density is possible, however, problems arise due to crosstalk – basically an "overlapping" of the thermal or pressure differentials used to drive adjacent nozzles.

In approaching a printing process, these numerous challenges must in some sense be addressed sequentially: flight pattern cannot be studied until droplets are formed, and layering cannot be investigated until deposition of single droplets is controlled. In terms of functional polymer deposition, the challenge of material preparation has effectively been addressed; numerous polymer resins and mixtures already exist. It is the second challenge – droplet formation – that is therefore the current limiting factor in deposition of these materials. To understand these limitations, Sect. 7.3.2 reviews the dynamic processes that are currently used to form droplets.

7.3.2   Droplet Formation Technologies

Over the time that two-dimensional inkjet printing has evolved, a number of methods for creating and expelling droplets have been developed. The main distinction in categorizing the most common of technologies refers to the possible modes of expulsion: continuous stream (CS) and drop-on-demand (DOD). This distinction refers to the form in which the liquid exits the nozzle – as either a continuous col-

Fig. 7.9 Continuous (left) and drop-on-demand (right) deposition [46]

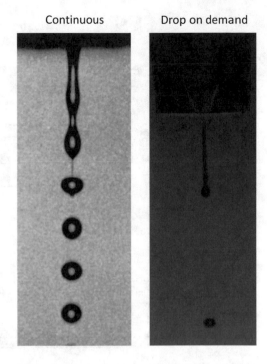

Continuous   Drop on demand

umn of liquid or as discrete droplets. Figure 7.9 shows the distinction between continuous (left) and DOD (right) formations.

7.3.3   Continuous Mode

In CS mode, a steady pressure is applied to the fluid reservoir, causing a pressurized column of fluid to be ejected from the nozzle. After departing the nozzle, this stream breaks into droplets due to Rayleigh instability. The breakup can be made more consistent by vibrating, perturbing, or modulating the jet at a fixed frequency close to the spontaneous droplet formation rate, in which case the droplet formation process is synchronized with the forced vibration, and ink droplets of uniform mass are ejected [47]. Because droplets are produced at constant intervals, their deposition must be controlled after they separate from the jet. To achieve this, they are introduced to a charging field and thus attain an electrostatic charge. These charged particles then pass through a deflection field, which directs the particles to their desired destinations – either a location on the substrate or a container of material to be recycled or disposed. Figure 7.10 shows a schematic of the function of this type of binary deflection continuous system.

An advantage of CS deposition is the high throughput rate; it has therefore seen widespread use in applications such as food and pharmaceutical labeling [3]. Two

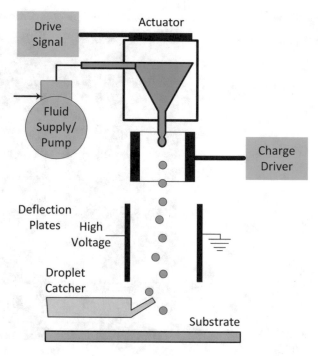

Fig. 7.10 Binary deflection continuous printing [46]

major constraints related to this method of droplet formation are, however, that the materials must be able to carry a charge and that the fluid deflected into the catcher must be either disposed of or reprocessed, causing problems in cases where the fluid is costly or where waste management is an issue.

In terms of droplets formed, commercially available systems typically generate droplets that are about 150 μm in diameter at a rate of 80–100 kHz, but frequencies of up to 1 MHz and droplet sizes ranging from 6 μm (10 fl) to 1 mm (0.5 μl) have been reported [46]. It has also been shown that, in general, droplets formed from continuous jets are almost twice the diameter of the undisturbed jet [48].

A few AM investigators have opted to use continuous printing methods. Blazdell et al. [49] used a continuous printer from BioDot, which was modulated at 66 kHz while ejecting ceramic ink from 50 and 75 μm nozzles. They used 280 kPa of air pressure. Blazdell [50] reports later results in which this BioDot system was modulated at 64 kHz, using a 60 μm nozzle that was also 60 μm in length. Early development of BJT processes used CS. At present commercial BJT machines (from 3D Systems and ExOne) use standard drop-on-demand print heads. In metal fabrication, Tseng et al. [51] used a continuous jet in depositing their solder alloy, which had a viscosity of about 2 cP at the printing temperature. Orme et al. [22, 23] also report the use of an unspecified continuous system in deposition of solders and metals.

7.3.4   Drop-on-Demand Mode

In DOD mode, in contrast, individual droplets are produced directly from the nozzle. Droplets are formed only when individual pressure pulses in the nozzle cause the fluid to be expelled; these pressure pulses are created at specific times by thermal, electrostatic, piezoelectric, acoustic, or other actuators [1]. Figure 7.11 shows the basic functions of a DOD setup. Liu and Orme [20] assert that DOD methods can deposit droplets of 25–120 μm at a rate of 0–2000 drops per second.

In the current DOD printing industry, thermal (bubble-jet) and piezoelectric actuator technologies dominate; these are shown in Fig. 7.12. Thermal actuators rely on a resistor to heat the liquid within a reservoir until a bubble expands in it, forcing a droplet out of the nozzle. Piezoelectric actuators rely upon the deformation of a piezoelectric element to reduce the volume of the liquid reservoir, which causes a droplet to be ejected. As noted by Basaran [52], the waveforms employed in piezoelectrically driven DOD systems can vary from simple positive square waves to complex negative–positive–negative waves in which the amplitude, duration, and other parameters are carefully modulated to create the droplets as desired.

In their review of polymer deposition, De Gans et al. [5] assert that DOD is the preferable method for all applications that they discuss due to its smaller drop size (often of diameter similar to the orifice) and higher placement accuracy in comparison to CS methods. They further argue that piezoelectric DOD is more widely applicable than thermal because it does not rely on the formation of a vapor bubble or on heating that can damage sensitive materials.

The preference for piezoelectrically driven DOD printing is reflected in the number of investigators who use and study such setups. For example, Gao and Sonin [8]

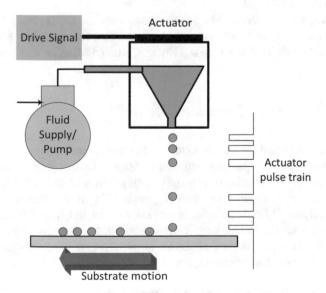

Fig. 7.11 Schematic of drop-on-demand printing system [46]

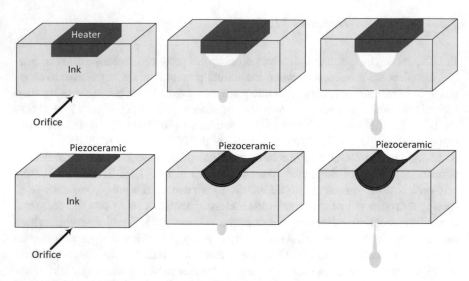

Fig. 7.12   Thermal (top) and piezoelectric (bottom) DOD ejection

use this technology to deposit 50 μm droplets of two waxes, whose viscosity at
100 °C is about 16 cP. Sirringhaus et al. [53] and Shimoda et al. [28] both use piezoelectric DOD deposition for polymer solutions, as discussed in Sect. 7.2.1. In
ceramic deposition, Reis et al. [9] print mixtures with viscosities 6.5 and 14.5 cP at
100 °C and frequencies of 6–20 kHz. Yamaguchi et al. [24, 25] also used a piezoelectrically driven DOD device at frequencies up to 20 Hz for the deposition of
metal droplets. Similarly, the solder droplets on the circuit board in Fig. 7.7 were
deposited with a DOD system.

At present, all commercial AM printing machines use DOD print heads, generally from a major manufacturer of printers or printing technologies. Such companies include Hewlett-Packard, Canon, Dimatix, Konica Minolta, and Xaar.

7.3.5   Other Droplet Formation Methods

Aside from the standard CS and DOD methods, other technologies have been experimentally investigated but have not enjoyed widespread use in industry applications.
Liquid spark jetting, a relative of thermal printing, relies on an electrical spark discharge instead of a resistor to form a gas bubble in the reservoir [47, 53].
Electrohydrodynamic inkjet techniques employ an extremely powerful electric field
to pull a meniscus and, under very specific conditions, droplets from a pressure-
controlled capillary tube; these droplets are significantly smaller than the tube from
which they emanate. Electrorheological fluid jetting uses an ink whose properties
change under high electric fields; the fluid flows only when the electric field is
turned off [54]. In their flextensional ultrasound droplet ejectors, Percin and

Khuri-Yakub [55] demonstrate both DOD and continuous droplet formation with a system in which a plate containing the nozzle orifice acts as the actuator, vibrating at resonant frequencies and forming droplets by creating capillary waves on the liquid surface as well as an increased pressure in the liquid. Focused acoustic beam ejection uses a lens to focus an ultrasound beam onto the free surface of a fluid, using the acoustic pressure transient generated by the focused tone burst to eject a fluid droplet [56]. Meacham et al. extended this work [57] to develop an inexpensive ultrasonic droplet generator and developed a fundamental understanding of its droplet formation mechanisms [58]. This work was extended and demonstrated ejection of thermoset resins with a viscosity of approximately 600 cP [59]. Fukumoto et al. [60] present a variant technology in which ultrasonic waves are focused onto the surface of the liquid, forming surface waves that eventually break off into a mist of small droplets. Overviews of these various droplet formation methods are given by Lee [54] and Basaran [52].

Beginning in the 2000s, Canon Océ developed a novel droplet generation technology based on electromagnetics [61] for metals with melting points up to 2000 °C. Metal feedstock is melted in a chamber that is surrounded by an electromagnetic coil that sets up a magnetic field in the chamber. If the molten metal is electrically conductive, the magnetic field induces an electrical current in the metal that creates a Lorentz force, given as the cross-product of the current and magnetic field. If configured correctly, the Lorentz force acts to drive the molten metal into the print head's orifice and helps to eject a droplet. Canon Océ partnered with Nottingham University to develop 3D printing machines that utilize their print heads [62]. The collaboration has succeeded in printing tin and silver materials, including silver–tin composites, with droplets around 40 μm in diameter, and fabricating a wide range of part shapes.

In the early 2010s, a father–son team developed a similar magnetic field enhanced print head for depositing metal droplets [63] and formed the company Vader Systems. In 2019, their company was purchased by Xerox Corp. The print head has been used to deposit several aluminum alloys, using wire feedstock. The print head melts the feedstock in a chamber that is surrounded by an electromagnetic coil, although with a significantly different configuration than the Canon Océ print head. A DC pulse is applied to the coil resulting in a radially inward Lorentz force on the molten metal, which helps eject a droplet from the orifice in conjunction with a pressurized inert gas. As such, the print head is capable of drop-on-demand printing at rates up to 1000 Hz.

More recently, a droplet generation approach called acoustophoretic printing was demonstrated that is capable of printing high-viscosity fluids [64]. The method combines a pressurized fluid nozzle with an external acoustic source. Specifically, a tapered glass nozzle was placed along the central axis of a Fabry–Perot emitter that induces an acoustic field focused near the nozzle tip. Computational models indicated that ejection forces of approximately 20 g were achieved at the nozzle tip which causes droplets to be detached from the nozzle. A wide variety of fluids were ejected by their experimental apparatus including optical resins, liquid metals, foods such as chocolate, and cell-laden biological matrices.

7.4   Cold Spray

Cold spray additive manufacturing (CSAM) is a type of cold spraying process that can produce freestanding components or fabricate features on existing components [65]. In this method no droplet is formed; however, because the materials are presurrized and ejected to the build plate or existing part, CSAM is categorized as MJT. This technology was developed as an AM process in the early 2000s, and recently several companies like TWI Additive Manufacturing and Spee3D have developed the technology into a viable manufacturing process.

In cold spray processes, a high velocity pressurized gas such as helium or nitrogen at 70 bar and 1100 °C accelerates small powder particles [66]. Upon impact on an existing part or build plate, the particles deform and bond together. The stream of powder particles creates a layer, and by continuing this process, a layered structure is obtained. Cold spray is commonly utilized to apply thermal barrier coatings on metal parts that will be subject to high temperatures, such as turbine blades, using ceramic materials.

The feedstock in this method is often fine powder of the materials. The "cold" term refers to the fact that particles are not melted. Part material is built up by a solid-state bonding mechanism that is activated by the severe plastic deformation that particles undergo upon impact. The spot size in cold spray is typically on the order of several millimeters, which is considerably larger than PBF processes, but is comparable to some Directed Energy Deposition (DED) processes. This method eliminates the heat-affected zone that is typical of thermally driven processes and avoids the need for a protective atmosphere.

In typical implementations, the spray gun is mounted on a motion platform that can be moved around the build plate or existing component. Often a robot arm is used as the motion platform, but other platforms, such as a gantry, could be used. In some implementations, both the spray gun and build plate can move, enabling a high degree of motion complexity and geometric freedom in part designs. Cold spray is an efficient process with low material waste and energy consumption. These characteristics give rise to several application types, including part repair, feature fabrication on existing parts, stand-alone part manufacture, and even fabrication of functionally graded materials by varying the powder composition. Figure 7.13 (a) shows the cold spray process on existing parts and (b) shows the cross-section of the system including pressurized gas–particle feeder and cooling system.

7.5   MJT Process Modeling

Conservation of energy concepts provides an appropriate context for investigating droplet generation mechanisms for MJT. Essentially, the energy imparted by the actuation method to the liquid must be sufficient to balance three requirements: fluid flow losses, surface energy, and kinetic energy. The losses originate from a

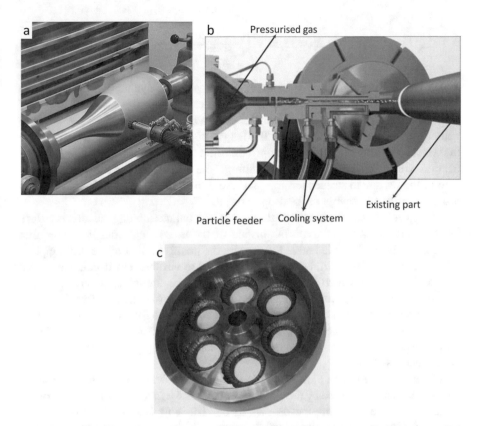

Fig. 7.13 (**a**) Cold spraying, (**b**) cross-section of the cold spray machine (photo courtesy of ABS Industries & Impact Innovations) [67], and (**c**) flywheel made by copper from Speed3D (Build time 179mins, weight 15.7 kg)

conversion of kinetic energy to thermal energy due to the viscosity of the fluid within the nozzle; this conversion can be thought of as a result of internal friction of the liquid. The surface energy requirement is the additional energy needed to form the free surface of the droplet or jet. Finally, the resulting droplet or jet must still retain enough kinetic energy to propel the liquid from the nozzle toward the substrate. This energy conservation can be summarized as

$$E_{\text{imparted}} = E_{\text{loss}} + E_{\text{surface}} + E_{\text{kinetic}} \tag{7.1}$$

The conservation law can be considered in the form of actual energy calculations or in the form of pressure, or energy per unit volume, calculations. For example, Sweet used the following approximation for the gauge pressure required in the reservoir of a continuous jetting system [68]:

$$\Delta p = 32\mu d_j^2 v_j \int_{l_1}^{l_2} \frac{dl}{d_n^4} + \frac{2\sigma}{d_j} + \frac{\rho v_j^2}{2} \tag{7.2}$$

where Δp is the total gauge pressure required, μ is the dynamic viscosity of the liquid, ρ is the liquid's density, σ is the liquid's surface tension, d_j is the diameter of the resultant jet, dn is the inner diameter of the nozzle or supply tubing, v_j is the velocity of the resultant jet, and l is the length of the nozzle or supply tubing. The first term on the right of (7.2) is an approximation of the pressure loss due to viscous friction within the nozzle and supply tubing. The second term is the internal pressure of the jet due to surface tension, and the third term is the pressure required to provide the kinetic energy of the droplet or jet.

Energy conservation can also be thought of as a balance among the effects before the fluid crosses a boundary at the orifice of the nozzle and after it crosses that boundary. Before the fluid leaves the nozzle, the positive effect of the driving pressure gradient accelerates it, but energy losses due to viscous flow decelerate it. The kinetic energy with which it leaves the nozzle must be enough to cover the kinetic energy of the traveling fluid as well as the surface energy of the new free surface.

As indicated earlier, actuation energy is typically in the form of heating (bubble-jet) or vibration of a piezoelectric actuator. Various electrical energy waveforms may be used for actuation. In any event, these are standard types of inputs and will not be discussed further.

While the liquid to be ejected travels through the nozzle, before forming droplets, its motion is governed by the standard equations for incompressible, Newtonian fluids, as we are assuming these flows to be. The flow is fully described by the Navier–Stokes and continuity equations; however, these equations are difficult to solve analytically, so we will proceed with a simplification. The first term on the right side of (7.2) takes advantage of one situation for which an analytical solution is possible that of steady, incompressible, laminar flow through a straight circular tube of constant cross-section. The solution is the Hagen–Poiseuille law [69], which reflects the viscous losses due to wall effects:

$$\Delta p = \frac{8Q\mu l}{\pi r^4 \sigma} \tag{7.3}$$

where Q is the flow rate and r is the tube radius. Note that this expression is most applicable when the nozzle is a long, narrow glass tube. However, it can also apply when the fluid is viscous, as we will see shortly.

Another assumption made by using the Hagen–Poiseuille equation is that the flow within the nozzle is fully developed. For the case of laminar flow in a cylindrical pipe, the length of the entry region l_e where flow is not yet fully developed is defined as 0.06 times the diameter of the pipe, multiplied by the Reynolds number [69]:

$$l_e = 0.06d \, \text{Re} = \frac{0.06\rho\bar{v}d^2}{\mu} \tag{7.4}$$

where $\bar{v}$ is the average flow velocity across the pipe. To appreciate the magnitude of this effect, consider printing with a 20 μm nozzle in a plate that is 0.1 mm thick, where the droplet ejection speed is 10 m/s. The entry lengths for a fluid with the density of water and varying viscosities are shown in Table 7.1.

Entry lengths are a small fraction of the nozzle length for fluids with viscosities of 40 cP or greater. As a result, we can conclude that flows are fully developed through most of a nozzle for fluids that are at the higher end of the range of printable viscosities.

Many readers will have encountered the primary concepts of fluid mechanics in an undergraduate course and may be familiar with the Navier–Stokes equation, viscosity, surface tension, etc. As a reminder, viscosity is a measure of the resistance of a fluid to being deformed by shear or extensional forces. We will restrict our attention to dynamic, or absolute, viscosity, which has units of pressure–time; in the SI system, units are typically Pa·s or mPa·s, for millipascals-seconds. Viscosity is also given in units of poise or centipoises, named after Jean Louis Marie Poiseuille. Centipoise is abbreviated cP, which conveniently has the same magnitude as mPa·s, that is, 1 cP is equal to 1 mPa s. Surface tension is given in units of force per length, or energy per unit area; in the SI system, surface tension often has units of N/m or J/m^2.

We can investigate the printing situation further by computing the pressures required for ejection. Equation (7.3) will be used to compute the pressure required to print droplets for various fluid viscosities and nozzle diameters. For many printing situations, wall friction dominates the forces required to print; hence we will only investigate the first term on the right of (7.2) and ignore the second and third terms (which are at least one order of magnitude smaller than wall friction).

Figure 7.14 shows how the pressure required to overcome wall friction varies with fluid viscosity and nozzle diameter. As can be seen, pressure needs to increase

Table 7.1 Entry lengths for "water" at various viscosities

Viscosity (cP)	Density (kg/m^3)	Entry length (μm)
1	1000	240
	1250	300
10	1000	24
	1250	30
40	1000	6
	1250	7.5
100	1000	2.4
	1250	3
200	1000	1.2
	1250	1.5

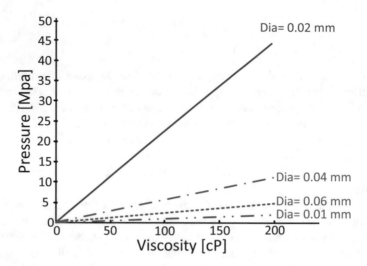

Fig. 7.14 Pressure required to overcome wall friction for printing through nozzles of different diameters

sharply as nozzles vary from 0.1 to 0.02 mm in diameter. This could be expected, given the quadratic dependence of pressure on diameter in (7.3). Pressure is seen to increase linearly with viscosity, which again can be expected from (7.3). As indicated, wall friction dominates for many printing conditions. However, as nozzle size increases, the surface tension of the fluid becomes more important. Also, as viscosity increases, viscous losses become important, as viscous fluids can absorb considerable acoustic energy. Regardless, this analysis provides good insight into pressure variations under many typical printing conditions.

Fluid flows when printing are almost always laminar; i.e., the Reynolds number is less than 2100. As a reminder, the Reynolds number is

$$\mathrm{Re} = \frac{\rho v r}{\mu} \tag{7.5}$$

Another dimensionless number of relevance in printing is the Weber number, which describes the relative importance of a fluid's inertia compared with its surface tension. The expression for the Weber number is

$$\mathrm{We} = \frac{\rho v^2 r}{\gamma} \tag{7.6}$$

Several research groups have determined that a combination of the Reynolds and Weber numbers is a particularly good indication of the potential for successful printing of a fluid [17]. Specifically, if the ratio of the Reynolds number to the square root of the Weber number has a value between 1 and 10, then it is likely that ejection of the fluid will be successful. This condition will be called the "printing indicator" and is

$$1 \le \frac{\text{Re}}{\text{We}^{1/2}} = \frac{\sqrt{\rho r \gamma}}{\mu} \le 10 \qquad (7.7)$$

Note that in the development of the printing indicator, the researchers used orifice radius as the characteristic length, not diameter, hence the change to r from d in Eqs. 7.4 and 7.5.

The inverse of the printing indicator is another dimensionless number called the Ohnesorge number that relates viscous and surface tension forces. Note that values of this ratio that are low indicate that flows are viscosity limited, while large values indicate flows that are dominated by surface tension. The low value of 1 for the printing indicator means that the maximum fluid viscosity should be between 20 and 40 cP.

Some examples of Reynolds numbers and printing indicators are given in Table 7.2. For these results, the surface tension is 0.072 N/m, the density is 1000 kg/m^3 (same as water at room temperature), and the droplet velocity is 1 m/s.

It is important to realize that the printing indicator is a guide, not a law to be followed. Water is usually easy to print through most print heads, regardless of the nozzle size. But the printing indicator predicts that water (with a viscosity of 1 cP) should not be ejectable since its surface tension is too high. We will see in the next section how materials can be modified in order to make printing feasible.

Table 7.2 Reynolds numbers and printing indicator values for some printing conditions

Nozzle radius [mm]	Viscosity [cP]	Reynolds no.	Printing indicator
0.02	1	20	37.9
	10	2	3.79
	40	0.5	0.949
	100	0.2	0.379
0.05	1	50	60.0
	10	5	6.00
	40	1.25	1.50
	100	0.5	0.600
0.1	1	100	84.9
	10	10	8.49
	40	2.5	2.12
	100	1	0.849

7.6   Material Jetting Machines

The three main companies involved in the development of early AM technologies are still the main players offering printing-based machines: Solidscape, 3D Systems, and Stratasys (after their merger with Objet Geometries). Solidscape sells the S300 series and S500 printer, both descendants of the previous ModelMaker line and based upon the melted wax technique. The machines deposit two materials, a thermoplastic part material and a waxy support material – to form layers 0.0005 in. thick [70]. It should be noted that these machines also fly-cut layers after deposition to ensure that the layer is flat for the subsequent layer. Because of their high precision and accuracy, these machines are often used to fabricate investment casting patterns for the jewelry and dentistry industries.

3D Systems and Stratasys offer machines using the ability to print and cure acrylic photopolymers. Stratasys markets PolyJet printers. These machines print a number of different acrylic-based photopolymer materials in layers of 0.013–0.032 mm thick from heads containing hundreds of individual nozzles, resulting in rapid, line-wise deposition efficiency. Each photopolymer layer is cured by ultraviolet light immediately as it is printed, producing fully cured models without post-curing. A planarizing step is used to ensure a planar surface after depositing one or more layers; the step consists of a compaction operation or a material removal operation similar to Solidscape's fly-cut approach. Support structures are built in a gel-like material, which is removed by hand and water jetting [71]. See Fig. 7.15 for an illustration of Stratasys' PolyJet approach, which is employed in all of their systems. The Connex line of machines provides multi-material capability using a technology called digital materials. In this approach, two or more materials are deposited in patterns that result in an effective "digital material" with variable mechanical properties or colors. Stratasys claims that their J750 and J850 machines

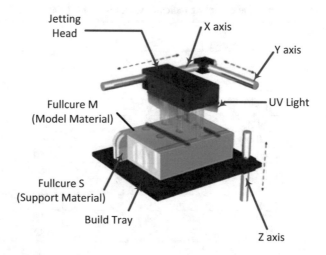

Fig. 7.15   Stratasys PolyJet approach [71]

Table 7.3 Sampling of commercially available MJT machines [11, 70, 71]

Company/ product	Build size X × Y × Z (mm)	Min. layer (mm)	Resolution X, Y (DPI)	Material	Support
Solidscape					
S325	152 × 152 × 102	0.0254	5000 × 5000	Wax	Wax
S350	152 × 152 × 1012	0.00635–0.0762	5000 × 5000	Wax	Wax
S390	152 × 152 × 102	0.0254–0.0508	5000 × 5000	Wax	Wax
3D systems					
MJP 2500	294 × 211 × 144	0.032	800 × 900 to 1600 × 900	Photo curable resin	Photo curable resin, wax option
ProJet MJP 3600	203 (to 298) × 185 × 203	0.016–0.032	375 × 450 or 750 × 750	Photo curable resin	Wax
ProJet MJP 5600	518 × 381 × 300	0.013–0.016	600 × 600 or 750 × 750	Photo curable resin	Wax
Stratasys					
Connex 3 objet 500	490 × 390 × 200	0.016	600 × 600	Photo curable resin	Soluble resin
J850	490 × 390 × 200	0.014	600 × 600	Photo curable resin	Soluble resin
Objet 1000 Plus	1000 × 800 × 500	0.016	300 × 300	Photo curable resin	Soluble resin

can produce 500,000 colors. A sampling of the machines currently available from these three vendors is presented in Table 7.3.

To date, all MJT machine vendors utilize standard inkjet print heads. For example, the Xaar 1003 AMx is a typical example of the heads used in MJT machines. This specific print head has 1000 nozzles in two rows that are interleaved to provide a nozzle pitch of 23.5 µm. Droplets of 6–42 pL (picoliter) can be generated by the print head. Large format MJT machines may have six or more of these 1000 nozzle print heads to deposit considerable material quickly over a relatively large area.

7.7   Process Parameters in Material Jetting

Like other AM techniques, process parameter values must be selected to achieve successful ejection and part fabrication. When a nozzle with a smaller diameter is used, more pressure is needed to overcome the wall friction. This pressure and the distance between nozzle and substrate should be maintained to achieve effective layering. Generally, these two parameters are highly dependent on each other.

Fig. 7.16 Effective process parameters in MJT

Another process parameter is viscosity which must be kept in proportion to other parameters. For instance, water with viscosity 1 cP may not be ejectable through small nozzles due to high surface tension.

Density of the material is also important in MJT. Higher density (considering the constant volume of droplets) leads to higher impact onto the substrate/previous layers. If the impact is too high, this can lead to delamination or increasing surface roughness.

Finally, surface tension is another influential factor that affects printing quality in MJT. Surface tension has the reverse effect from temperature, called thermocapillarity. Therefore, the temperature of the material should be maintained slightly above melting temperature. Figure 7.16 shows the effective process parameters in MJT AM.

7.8   Rotative Material Jetting

It is possible to use very large numbers of print heads along with unique machine architectures to enable MJT to produce parts very quickly. For instance, when multiple printing nozzles are aligned across an entire build area, rather than using a

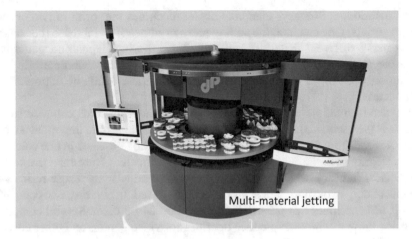

Multi-material jetting

Fig. 7.17 AM polar i2 multi-MJT. (Photo courtesy of dp polar)

single print head or smaller array of print heads, the entire build volume can be printed in one sweeping motion. In addition, if you design a machine that rotates a circular build plate underneath a set of nozzles, then you can enable continuous printing without any pauses to change machine/nozzle/build plate direction, resulting in a significant reduction in time and cost. The company dp polar GmbH patented the High-Speed Rotative (HSR) system for multi-material printing using a continuous rotating print platform. Their AM polar i2 system is equipped with three printing stations and provides the ability to produce parts with multiple materials in a 2 m^2 build area (700 L build volume). They claim their machine can achieve a 10 L/h fabrication rate in high-quality production mode. Figure 7.17 shows the AM polar i2 for multi-material printing.

7.9   Process Benefits and Drawbacks

Each AM process has advantages and disadvantages. The primary advantages of printing-based technologies, both MJT and binder printing, include low cost, high speed, scalability, ease of building parts in multiple materials, and the capability of printing colors. Printing machines are much lower in cost than AM machines which use expensive energy sources such as lasers and electron beams. In general, printing machines can be assembled from standard components (drives, stages, print heads), while other AM machines have many more machine-specific components. High speed and scalability are related: by using print heads with hundreds or thousands of nozzles, it is possible to deposit a lot of material quickly and over a considerable area. Scalability in this context means that printing speed can be increased by adding another print head to a machine, a relatively easy task, much easier than adding another laser to a VP or PBF machine.

As mentioned, Stratasys markets PolyJet machines that print in two or more part materials. One can imagine adding more print heads to increase the capability to many different materials and utilizing dithering deposition patterns to raise the number of effective materials into the hundreds. Compatibility and resolution need to be ensured, but it seems that these kinds of improvements will occur.

Related to multiple materials, colors can be printed by some commercial AM machines (see Sect. 8.3). The capability of printing in color is an important advance in the AM industry; for many years, parts could only be fabricated in one color. The only exception was the selectively colorable SL resins that Huntsman markets for the medical industry, which were developed in the mid-1990s. These resins were capable of only two colors, amber and either blue or red. In contrast, two companies market AM machines that print in high resolution 24-bit color. Several companies are using these machines to produce figurines for video gamers and other consumers (see Chaps. 3, 8, and 12).

For completeness, a few disadvantages of MJT will provide a more balanced presentation. The choice of materials to date is limited. Only waxes and photopolymers are commercially available polymers. Very few ceramic and metal materials are printable. Part accuracy for large parts is generally not as good as with some other processes, notably Vat Photopolymerization (VPP) and Material Extrusion (MEX) due to the need to use large droplets to print large components efficiently. In general, MJT accuracies have been improving across the industry and are starting to rival accuracies for all AM processes for small- and medium-sized components.

7.10   Summary

MJT takes advantage of inexpensive print heads developed for 2D printing applications. By combining these print heads with simple motions controllers, accurate deposition of wax, photopolymer, and other materials is possible. The primary advantages of MJT include low cost, high speed, scalability, ease of building parts in multiple materials, and the capability of printing colors. MJT machines are typically lower in cost than AM machines that use lasers or electron beams. Two primary mechanisms exist for droplet generation, continuous mode, and drop-on-demand. Most commercial MJT machines utilize drop-on-demand print heads. High speed and scalability are related: by using print heads with hundreds or thousands of nozzles, it is possible to deposit a lot of material quickly and over a considerable area. Scalability in this context means that printing speed can be increased by adding another print head to a machine, a relatively easy task, much easier than adding another laser to a VPP or PBF machine.

7.11   Questions

1. List five types of material that can be directly printed.
2. According to the printing indicator (7.7), what is the smallest diameter nozzle that could be used to print a ceramic–wax material that has the following properties:

   (a) Viscosity of 15 cP, density of 1800 kg/m^3, and surface tension of 0.025 N/m.
   (b) Viscosity of 7 cP, density of 1500 kg/m^3, and surface tension of 0.025 N/m.
   (c) Viscosity of 38 cP, density of 2100 kg/m^3, and surface tension of 0.025 N/m.

3. Develop a build time model for a printing machine. Assume that the part platform is to be filled with parts and the platform is L mm long and W mm wide. The print head width is H mm. Assume that a layer requires three passes of the print head, the print head can print in both directions of travel ($+X$ and $-X$), and the layer thickness is T mm. Figure Q 7.1 shows a schematic for the question. Assume that a delay of D seconds is required for cleaning the print heads every K layers. The height of the parts to be printed is P mm.

Fig. Q 7.1 Schematic for questions 4–5

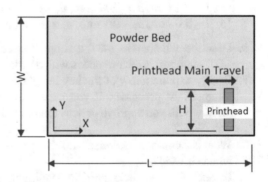

Table Q 7.1 Question 3

	L	W	H	T	D	K	P
(b)	300	185	50	0.04	10	20	60
(c)	300	185	50	0.028	12	25	85
(d)	260	250	60	0.015	12	25	60
(e)	340	340	60	0.015	12	25	60
(f)	490	390	60	0.015	12	25	80

Table Q 7.2 Question 5

	d_e [mm]	d_x [mm]	l [mm]	μ [cP]	ρ [kg/m^3]	γ [N/m]	v [m/s]
(b)	0.04	0.02	0.1	1	1000	0.072	10
(c)	0.04	0.02	0.1	40	1000	0.072	10
(d)	0.04	0.02	1.0	1	1000	0.072	10
(e)	0.1	0.04	5.0	1	1000	0.072	10
(f)	0.1	0.04	5.0	40	1000	0.025	10

(a) Develop a build time model using the variables listed in the problem statement.

Compute the build time for a layer of parts given the variable values in the following table.

4. The integral in (7.2) can be evaluated analytically for simple nozzle shapes. Assume that the nozzle is conical with the entrance diameter of d_e and the exit diameter d_x.

(a) Evaluate the integral analytically.

Use your integrated expression to compute pressure drop through the nozzle, instead of (7.3), for the following variable values:

5. Using the integral from Problem 5, develop a computer program to compute pressure drop through the nozzle for various nozzle sizes and fluid properties. Compute and plot the pressure drop for the printing conditions of Fig. 7.14 but using nozzles of the following dimensions:

   (a) $l = 0.1$ mm, $d_e = 0.06$ mm, $d_x = 0.02$ mm,
   (b) $l = 0.1$ mm, $d_e = 0.08$ mm, $d_x = 0.04$ mm,
   (c) $l = 0.1$ mm, $d_e = 0.12$ mm, $d_x = 0.05$ mm,
   (d) $l = 5.0$ mm, $d_e = 0.1$ mm, $d_x = 0.05$ mm.

6. Describe whether in MJT a support structure is necessary or not. Contrast MJT's support structure needs with a process that has the opposite need.

7. Briefly explain a few of the key benefits and the drawbacks of MJT compared to other AM processes. If you were a manager in an automotive design company, would you approve or deny the purchase of this process for your organization. Why?

8. What is a common chemical additive put into MJT photopolymer materials, and why is it used?

9. In Sect. 7.1, Ballistic Particle Manufacturing is mentioned in the historical development of 3D printing. Read about this online, and describe the key differences between this method and MJT as practiced today.

10. According to the printing indicator (7.7), estimate whether the following diameter nozzle could be working well. If not, choose a feasible nozzle for that one. According to the printing indicator (7.7), estimate whether the following diameter nozzle would work well. If not, choose a feasible nozzle for that one.

   (a) Viscosity of 20 c_P, density of 5500 kg/m³, surface tension of 0.025 N/m, and nozzle diameter 0.01 mm.
   (b) Viscosity of 40 c_P, density of 1200 kg/m³, surface tension of 0.025 N/m, and nozzle diameter 0.03 mm.
   (c) Viscosity of 20 c_P, density of 4500 kg/m³, surface tension of 0.045 N/m, and nozzle diameter 0.02 mm.
   (d) Viscosity of 40 c_P, density of 5500 kg/m³, surface tension of 0.045 N/m, and nozzle diameter 0.02 mm.

11. Use the printing indicator formula and (1) calculate the minimum nozzle diameter which could be used for printing the materials with the following information. (2) Also calculate the length of the entry region l_e using formula (7.4) provided in the text book.

   (a) $\mu = 10c_P$, $\rho = 1000\text{Kg/m}^3$, $\nu = 1$ m/s, $\gamma = 0.072$ N/M, surface tension is 0.072 N/m,
   (b) $\mu = 40c_P$, $\rho = 1000\text{Kg/m}^3$, $\nu = 8$ m/s, $\gamma = 0.025$ N/M, surface tension is 0.05 N/m.

References

1. Le, H. P. (1998). Progress and trends in ink-jet printing technology. *Journal of Imaging Science and Technology, 42*(1), 49–62.
2. *The rapid prototyping patent museum: basic technology patents.* (2013). US Patents.
3. Wohlers, T., (2004). *Wohlers Report*, Wohlers Associates.
4. Derby, B., & Reis, N. (2003). Inkjet printing of highly loaded particulate suspensions. *MRS Bulletin, 28*(11), 815–818.
5. De Gans, B. J., Duineveld, P. C., & Schubert, U. S. (2004). Inkjet printing of polymers: State of the art and future developments. *Advanced Materials, 16*(3), 203–213.
6. *MicroFab Technote 99–02 fluid properties effects on ink-jet device performance.* (1999). MicroFab Technologies, Inc.
7. Paton, A. D., & Kruse, J. M. (1995). *Reduced nozzle viscous impedance.* US Patents.
8. Gao, F., & Sonin, A. A. (1994). Precise deposition of molten microdrops: The physics of digital microfabrication. *Proceedings of the Royal Society A, 444*(1922), 533–554.
9. Reis, N., et al. (1998). Direct inkjet deposition of ceramic green bodies: II–jet behaviour and deposit formation. MRS Online Proceedings Library Archive. 542.
10. Feng, W., Fuh, J. Y., & Wong, Y. S. (2006). *Development of a drop-on-demand micro dispensing system.* In *Materials science forum.* Uetikon: Trans Tech Publ.
11. *3D Systems, 3D Printers.* (2020). http://www.3dsystems.com/3d-printers
12. Leyden, R. N., & Hull, C. W. (1999). *Method for selective deposition modeling.* US Patents.
13. de Gans, B. J., et al. (2004). Ink-jet printing polymers and polymer libraries using micropipettes. *Macromolecular Rapid Communications, 25*(1), 292–296.
14. Xu, P., et al. (2008). *Phase change support material composition.* US Patents.
15. Schmidt, K. A. (2005). *Selective deposition modeling with curable phase change materials.* US Patents.
16. Tay, B., & Edirisinghe, M. (2001). Investigation of some phenomena occurring during continuous ink-jet printing of ceramics. *Journal of Materials Research, 16*(2), 373–384.
17. Ainsley, C., Reis, N., & Derby, B. (2002). Freeform fabrication by controlled droplet deposition of powder filled melts. *Journal of Materials Science, 37*(15), 3155–3161.
18. Zhao, X., et al. (2002). Direct ink-jet printing of vertical walls. *Journal of the American Ceramic Society, 85*(8), 2113–2115.
19. Wang, T., & Derby, B. (2005). Ink-jet printing and sintering of PZT. *Journal of the American Ceramic Society, 88*(8), 2053–2058.
20. Liu, Q., & Orme, M. (2001). High precision solder droplet printing technology and the state-of-the-art. *Journal of Materials Processing Technology, 115*(3), 271–283.
21. Priest, J. W., Smith, C., & DuBois, P. (1997). Liquid metal jetting for printing metal parts. In *Solid Freeform Fabrication Proceedings. Austin: University of Texas at Austin.*

22. Orme, M. (1993). A novel technique of rapid solidification net-form materials synthesis. *Journal of Materials Engineering and Performance, 2*(3), 399–405.
23. Orme, M. E., Huang, C., & Courier, J. (1996). Precision droplet-based manufacturing and material synthesis: fluid dynamics and thermal control issues. *Atomization and Sprays, 6*(3), 305–329.
24. Yamaguchi, K. (2003). Generation of 3-dimensional microstructure by metal jet. *Microsystem Technologies, 9*(3), 215–219.
25. Yamaguchi, K., et al. (2000). Generation of three-dimensional micro structure using metal jet. *Precision Engineering, 24*(1), 2–8.
26. Cao, W., & Miyamoto, Y. (2006). Freeform fabrication of aluminum parts by direct deposition of molten aluminum. *Journal of Materials Processing Technology, 173*(2), 209–212.
27. Liu, Q., & Orme, M. (2001). On precision droplet-based net-form manufacturing technology. *Proceedings of the Institution of Mechanical Engineers, Part B: Journal of Engineering Manufacture, 215*(10), 1333–1355.
28. Shimoda, T., et al. (2003). Inkjet printing of light-emitting polymer displays. *MRS Bulletin, 28*(11), 821–827.
29. Zhao, X., et al. (2001). Ceramic freeforming using an advanced multinozzle ink-jet printer. *Journal of Materials Synthesis and Processing, 9*(6), 319–327.
30. XJet. (2020). https://startup-map.berlin/companies/xjet.
31. Furbank, R. J., & Morris, J. F. (2004). An experimental study of particle effects on drop formation. *Physics of Fluids, 16*(5), 1777–1790.
32. Bechtel, S., et al. (1981). Impact of a liquid drop against a flat surface. *IBM Journal of Research, 25*(6), 963–971.
33. Pasandideh-Fard, M., et al. (1996). Capillary effects during droplet impact on a solid surface. *Physics of Fluids, 8*(3), 650–659.
34. Thoroddsen, S., & Sakakibara, J. (1998). Evolution of the fingering pattern of an impacting drop. *Physics of Fluids, 10*(6), 1359–1374.
35. Bhola, R., & Chandra, S. (1999). Parameters controlling solidification of molten wax droplets falling on a solid surface. *Journal of Materials Science, 34*(19), 4883–4894.
36. Attinger, D., Zhao, Z., & Poulikakos, D. (2000). An experimental study of molten micro-droplet surface deposition and solidification: Transient behavior and wetting angle dynamics. *Journal of Heat Transfer, 122*(3), 544–556.
37. Zhou, W., et al. (2013). What controls dynamics of droplet shape evolution upon impingement on a solid surface? *AICHE Journal, 59*(8), 3071–3082.
38. Bussmann, M., Chandra, S., & Mostaghimi, J. (2000). Modeling the splash of a droplet impacting a solid surface. *Physics of Fluids, 12*(12), 3121–3132.
39. Orme, M., & Huang, C. (1997). Phase change manipulation for droplet-based solid freeform fabrication. *Journal of Heat Transfer, 119*(4), 818–823.
40. Schiaffino, S., & Sonin, A. A. (1997). Molten droplet deposition and solidification at low weber numbers. *Physics of Fluids, 9*(11), 3172–3187.
41. Sanders, R., Forsyth, L., & Philbrook, K. (1996). *3-D Model maker*. US Patents.
42. Thayer, J. S., et al. (2001). *Selective deposition modeling system and method*. US Patents.
43. Gothait, H. (2005). *System and method for three dimensional model printing*. US Patents.
44. Gothait, H. (2001). *Apparatus and method for three dimensional model printing*. US Patents.
45. Bedal, B. J., & Bui, L. V. (2002). *Method and apparatus for controlling the drop volume in a selective deposition modeling environment*. US Patents.
46. *MicroFab Technote 99–01 background on ink-jet technology*. (1999). MicroFab Technologies, Inc.
47. Tay, B., Evans, J., & Edirisinghe, M. (2003). Solid freeform fabrication of ceramics. *International Materials Reviews, 48*(6), 341–370.
48. Teng, W. D., & Edirisinghe, M. (1998). Development of continuous direct ink jet printing of ceramics. *BRIT CERAM T, 97*(4), 169–173.

49. Blazdell, P., et al. (1995). The computer aided manufacture of ceramics using multilayer jet printing. *Journal of Materials Science Letters, 14*(22), 1562–1565.
50. Blazdell, P. (2003). Solid free-forming of ceramics using a continuous jet printer. *Journal of Materials Processing Technology, 137*(1–3), 49–54.
51. Tseng, A. A., Lee, M., & Zhao, B. (2001). Design and operation of a droplet deposition system for freeform fabrication of metal parts. *Journal of Engineering Materials and Technology, 123*(1), 74–84.
52. Basaran, O. A. (2002). Small-scale free surface flows with breakup: Drop formation and emerging applications. *AICHE Journal, 48*(9), 1842–1848.
53. Sirringhaus, H., et al. (2000). High-resolution inkjet printing of all-polymer transistor circuits. *Science, 290*(5499), 2123–2126.
54. Lee, E. R. (2002). *Microdrop generation.* CRC press.
55. Percin, G., & Khuri-Yakub, B. T. (2002). Piezoelectrically actuated flextensional micromachined ultrasound droplet ejectors. *IEEE Transactions on Ultrasonics, Ferroelectrics, and Frequency Control, 49*(6), 756–766.
56. Elrod, S., et al. (1989). Nozzleless droplet formation with focused acoustic beams. *Journal of Applied Physics, 65*(9), 3441–3447.
57. Meacham, J., et al. (2004). Micromachined ultrasonic droplet generator based on a liquid horn structure. *Review of Scientific Instruments, 75*(5), 1347–1352.
58. Meacham, J., et al. (2005). Droplet formation and ejection from a micromachined ultrasonic droplet generator: Visualization and scaling. *Physics of Fluids, 17*(10), 100605.
59. Meacham, J. M., et al. (2010). Micromachined ultrasonic print-head for deposition of high-viscosity materials. *Journal of Manufacturing Science and Engineering, 132*(3), 030905.
60. Fukumoto, H., et al. (2000). Printing with ink mist ejected by ultrasonic waves. *Journal of Imaging Science and Technology, 44*(5), 398–405.
61. Rasa, M. V., et al. (2013). *Device for ejecting droplets of a fluid having a high temperature.* US Patent. p. 8,444,028.
62. Simonelli, M., et al. (2019). Towards digital metal additive manufacturing via high-temperature drop-on-demand jetting. *Additive Manufacturing, 30.*
63. Vader, S., & Vader, Z.. (2017). *Conductive liquid three dimensional printer.* US Patent. p. 9,616,494.
64. Foresti, D., et al. (2018). *Acoustophoretic printing. Science Advances, 4,* 1659.
65. Yin, S., et al. (2018). Cold spray additive manufacturing and repair: Fundamentals and applications. *Additive Manufacturing, 21,* 628–650.
66. Moridi, A., et al. (2014). Cold spray coating: Review of material systems and future perspectives. *Surface Engineering, 30*(6), 369–395.
67. ASB Industries & Impact Innovations. (2020). https://www.asbindustries.com/cold-spray-coatings
68. Sweet, R. G. (1964). *High-frequency oscillography with electrostatically deflected ink jets.* Stanford University, Stanford Electronics Labs.
69. Munson, B., Young, D., & Okiishi, T. (1998). Fundamentals of fluid mechanics.
70. Solidscape. (2020). *T66 Benchtop: Product Description.* http://www.solid-scape.com
71. Stratasys (2020). *Polyjet technology.* https://www.stratasys.com/polyjet-technology

Chapter 8
Binder Jetting

Abstract Binder Jetting (BJT) methods were developed in the early 1990s, primarily at MIT. They developed what they called the 3D Printing (3DP) process in which a binder is printed onto a powder bed to form part cross-sections. This concept can be compared to Powder Bed Fusion (PBF), where a laser melts powder particles to define a part cross-section. A wide range of polymer, composite, metal, and ceramic materials have been demonstrated, but only a subset of these are commercially available. Some BJT machines contain nozzles that print color, enabling the fabrication of parts with many colors. Several companies licensed the 3DP technology from MIT and became successful machine developers, including ExOne and Z Corp (purchased by 3D Systems in 2011). A novel continuous printing technology was developed by Voxeljet that can, in principle, fabricate parts of unlimited length. A resurgence of interest in BJT has occurred with several companies adopting the process for metal part fabrication.

8.1 Introduction

Binder Jetting was originally known as three-dimensional printing (3DP) and was invented at MIT and licensed to more than five companies for commercialization. In contrast to the Material Jetting (MJT) processes described in Chap. 7, BJT processes print a binder into a powder bed to fabricate a part. Hence, in BJT, only a small portion of the part material is delivered through the print head. Most of the part material is comprised of powder in the powder bed. Typically, binder droplets (~80 μm in diameter) form spherical agglomerates of binder liquid and powder particles as well as provide bonding to the previously printed layer. Once a layer is printed, the powder bed is lowered and a new layer of powder is spread onto it (typically via a counter-rotating rolling mechanism) [1], very similar to the recoating methods used in PBF processes, as presented in Chap. 5. This process (printing

© Springer Nature Switzerland AG 2021
I. Gibson et al., *Additive Manufacturing Technologies*,
https://doi.org/10.1007/978-3-030-56127-7_8

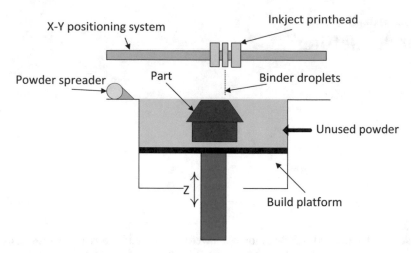

Fig. 8.1 Schematic of the BJT process

binder into bed; recoating bed with new layer of powder) is repeated until the part, or array of parts, is completed. A schematic of the BJT process is shown in Fig. 8.1.

Because a print head can contain many ejection nozzles, BJT typically features an array of parallel nozzles for rapid patterning. Since the process can be economically scaled by simply increasing the number of printer nozzles, the process is considered a scalable, line-wise patterning process. Such embodiments typically have a high deposition speed at a relatively low cost (due to the lack of a high-powered energy source) [1].

The printed part is sometimes left in the powder bed after its completion in order for the binder to fully set and for the green part to gain strength. Post-processing involves removing the part from the powder bed, removing unbound powder via pressurized air, and infiltrating the part with an infiltrant to make it stronger and possibly to impart other mechanical properties. For metal and ceramic BJT processes, a furnace operation to vaporize polymer constituents of the binder followed by powder sintering and/or infiltration is done.

The BJT process shares many of the same advantages of polymer PBF processes. Parts are self-supporting in the powder bed so that support structures are not needed. Similar to other processes, parts can be arrayed in one layer and stacked in the powder bed to greatly increase the number of parts that can be built at one time. Finally, assemblies of parts and kinematic joints can be fabricated since loose powder can be removed between the parts.

Applications of BJT processes are highly dependent upon the material being processed. Low-cost BJT machines use a plaster-based powder and a water-based binder to fabricate parts. Polymer powders are also available. Some machines have color print heads and can print visually attractive parts. With this capability, a market has developed for colorful figures from various computer games, as well as personal busts or sculptures, with images taken from cameras. Infiltrants are often used to strengthen parts after they are removed from the powder bed. With either the starch or polymer powders, parts are typically considered visual prototypes or light-duty functional prototypes. In some cases, particularly with elastomeric infiltrants, parts can be used for functional purposes. With polymer powders and wax-based infiltrants, parts can be used as patterns for investment casting, since the powder and wax can be burned out of the mold.

For metal powders, parts can be used as functional prototypes or for production purposes, provided that the parts have been designed specifically for the metal alloys available. Molds and cores for sand casting can be fabricated by some BJT machines that use silica or foundry sand as the powder. This is a sizable application in the automotive and heavy equipment industries.

8.2   Materials

8.2.1   Commercially Available Materials

When Z Corporation first started in the mid-1990s, their first material was starch-based and used a water-based binder similar to a standard household glue. At present, the commercially available powder from 3D Systems is plaster-based (calcium sulfate hemihydrate) and the binder is water-based [2]. Printed parts are fairly weak, so they are typically infiltrated with another material. 3D Systems provides three infiltrants, the ColorBond infiltrant, which is acrylate-based and is similar to super-glue, StrengthMax infiltrant which is a two-part infiltrant, and Salt Water Cure, an eco-friendly and hazard-free infiltrant. Strength, stiffness, and elongation data are given on 3D Systems' website for parts fabricated with these infiltrants. In general, parts with any of the infiltrants are much stiffer than typical thermoplastics or VP resins, but are less strong, and have very low elongation at break (0.04–0.23%).

Voxeljet [3], on the other hand, supplies a PMMA (poly-methyl methacrylate) powder and uses a liquid binder that reacts at room temperature. They recommend that parts stay in the powder bed for several hours to ensure that the binder is completely cured. For investment casting pattern fabrication, they offer a wax-based binder for use with PMMA powder that is somewhat larger in particle size than the powder used for parts. They claim excellent pattern burnout for investment casting.

For materials from both companies, unprinted powders are fully recyclable, meaning that they can be reused in subsequent builds. A desirable characteristic of powders is a high packing density so that printed parts have a high volume fraction of powder and are strong enough to survive depowdering and cleanup operations. High packing densities can be achieved by tailoring powder particle shape or by including a range of particle sizes so that small particles fill in gaps between larger particles. In practice, both approaches are used whenever possible. Quite a few othor infiltrant materials have been adopted, and many users have experimented with a variety of materials, so alternatives are possible that can produce parts with a wide range of mechanical properties.

Some as-built BJT parts have as much as 60% porosity, which may need to be filled in some circumstances. If a polymer is used as a filler, this can be achieved using a dipping process. Where the base material is metal (e.g., steel), the binder should be removed using a furnace, and, in some cases, a lower-melting-point material (e.g., bronze) can be used to infiltrate the voids.

ExOne markets machines that use either metal or sand powders for metal parts or sand casting molds and cores, respectively [4]. In the metals' area, they market several stainless steel materials and several other metals that are infiltrated with bronze. Included in the former category are 316L, 17-4PH, and 304L stainless steels, while 316 stainless steel, 420 stainless steel, and tungsten are metals in the latter category. They have many additional metals under development (2020), including Inconels, tool steels, cobalt chrome, and other tungsten alloys.

Desktop Metal is another MIT spin-off company that markets BJT systems for metals. Similar to ExOne, a polymer binder is used. They currently offer 316L and 17–4 stainless steels, H13 tool steel, and 4140 steel, with copper and Inconel 625 under development for their composite filament Material Extrusion (MEX) system. They claim to use metal injection molding feedstock materials in order to leverage decades of materials development efforts from that industry.

For metal and ceramic materials, polymer binders are used to produce a "green" part (a term from the ceramics industry which denotes a part before it has been sintered or fired). After fabrication, the "green" part is removed from the AM machine and then is subject to 2–3 furnace cycles. In the first cycle, low temperature is used for several hours to burn off the polymer binder. In the second cycle, high temperature is used to sinter the metal particles together. At this stage, the part is approximately 60 percent dense. In the third cycle, a bronze ingot (or other alloy with a lower melting temperature than the powder alloy) is placed in the furnace in contact with the part so that bronze infiltrates into the part's pores, resulting in parts that are greater than 90 percent dense. Other than how the green part is formed, this process is identical to the indirect processing approach for metal and ceramic part fabrication discussed in Chap. 5 and illustrated in Fig. 5.7.

In some metal MJT cases, the infiltration step is skipped and the part is sintered to near-full density. Almost all new materials being developed for BJT are designed such that rather than infiltrating with bronze (or another alloy) to form a composite structure, controlled sintering is utilized to form a dense single-material component. Sintering to nearly full density leads to considerable shrinkage and distortion, and there is always some porosity left in the part after sintering. Careful control of this distortion is critical for accurate parts to be built. This is very difficult for arbitrary complex geometries, and thus simulation of distortion and compensation for this distortion is a current major research area, and companies such as Ansys have recently introduced algorithms to help simulate and compensate for this distortion. Accurate distortion prediction and compensation is particularly critical for widespread adoption of metal BJT without infiltration. Both ExOne and Voxeljet market machines that use sand for the fabrication of molds and cores for sand casting.

ExOne offers a silica sand and two-part binder, where one part (binder catalyst) is coated on a layer and the second part is printed onto the layer, causing a polymerization reaction to occur and binding sand particles together. They claim that only standard foundry materials are used so that resulting molds and cores enable easy integration into existing manufacturing and foundry processes. Voxeljet also offers a silica sand with an inorganic binder and claims that their materials also integrate well into existing foundry processes.

Finally, ExOne markets several ceramic materials (beta stage), including silicon carbide, boron carbide, alumina, and zirconia. Similar to the metals' process, an organic binder is used that requires an elevated temperature curing cycle. Then, parts undergo furnace cycles to burn out the binder and sinter the ceramic particles to impart decent strength and stiffness.

8.2.2 Metal and Ceramic Materials Research

A wide range of materials has been developed for BJT. Printing into metal and ceramic powder beds was first demonstrated in the early 1990s. Various powder mixes, including compositions and size distributions, have been explored. The first report of using BJT for the fabrication of ceramics was in 1993; fired components were reported as typically greater than 99.2% dense [5]. Alumina, silica, and titanium dioxide have been made with this process [6].

Research involving the BJT of ceramics encountered early setbacks because of the use of dry powders. The fine powders needed for good powder bed density did not generally flow well enough to spread into defect-free layers [5]. Furthermore, since green part density was inadequate with the use of dry powders, isostatic pressing was implemented after the printing process. This extraneous requirement severely limits the types of part shapes capable of being processed.

To counteract the problems encountered with recoating a dry ceramic powder bed, research on ceramic BJT has shifted to the use of a slurry-based working material. In this approach, layers are first deposited by inkjet printing a layer of slurry over the build area. After the slurry dries, the binder is selectively printed to define the part shape. This is repeated for each individual layer, at the cost of significantly increased build time. Multiple jets containing different material composition or concentration could be employed to prepare components with composition and density variation on a fine scale (100 μm) [7]. Alumina and silicon nitride have been processed with this technique, improving green part density to 67% and utilizing layer thicknesses as small as 10 μm [8].

A variation of this method was developed to fabricate metal parts starting with metal oxide powders [9]. The ceramic BJT process was used until the furnace sintering step. While in the furnace, a hydrogen atmosphere is introduced, causing a reduction reaction to occur between the hydrogen and the oxygen atoms in the metal oxide. The reduction reaction converts the oxide to metal. After reduction, the metal particles are sintered to form a metal part. This process has been demonstrated for several material systems, including iron, steels, and copper. Unfortunately, reaction thermodynamics prevent alumina and titanium oxide from being reduced to aluminum and titanium, respectively. This metal oxide reduction 3DP (MO3DP) process was demonstrated using a Z Corp Z405 machine [10]. Metal oxide powders containing iron oxide, chromium oxide, and a small amount of molybdenum were prepared by spray drying the powder composition with polyvinyl alcohol (PVA) to form clusters of powder particles coated with PVA. Upon reduction, the material composition formed a maraging steel. Water was selectively printed into the powder bed to define part cross-sections, since the water will dissolve PVA, causing the clusters to stick together. A variety of shapes (trusses, channels, thin walls) were fabricated using the process to demonstrate the feasibility of producing cellular materials.

A key advantage of BJT is its economical fabrication of standard engineering materials. Simply put, the BJT process does not require high energy, does not involve lasers or any toxic materials, and is relatively inexpensive and fast. Part creation rate is limited to approximately twice the binder flow rate. A typical inkjet nozzle can deliver up to approximately 1 cm^3/min of binder; thus a machine with a 1000 nozzle print head could create up to 1000 cm^3/min of printed component. Because commercial BJT machines can have over 30,000 nozzles, BJT is being increasingly used as a production process.

8.3   Process Variations

Almost all commercially available BJT machines use the architecture shown in Fig. 8.1. An array of print heads is mounted on an XY translation mechanism. If the process is capable of printing colored parts, certain print heads are dedicated to printing binder material, while others are dedicated to printing color. Typically, the print heads are standard, off-the-shelf print heads found in machines for 2D printing of posters, banners, and similar applications.

Powder handling and recoating systems are similar to those used in PBF processes. Differences arise when comparing low-cost visual model printers (for plaster or polymer powders) to the metal or sand printers. For the low-cost printers, powder containers (vats) can be hand-carried. In the latter cases, however, powder beds can weigh hundreds or thousands of pounds, necessitating different material handling and powder bed manipulation methods. For large sand printers, the vats utilize a rail system for conveying powder beds to and from depowdering stations, and cranes are used for transporting parts or molds.

The capability of continuous printing or of fabricating parts that are larger than the AM machine fabricating them has been discussed in the research community. In recent years, two different approaches have been demonstrated for continuous printing of parts. One approach was introduced by Voxeljet in 2013 and is based on linear translation of the part being fabricated. The second approach is called spiral growth manufacturing and was developed by researchers at the University of Liverpool, UK.

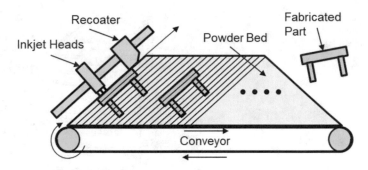

Fig. 8.2 Schematic of the continuous printing concept

The Voxeljet continuous printing process is a novel idea that utilizes an inclined build plane. That is, the build surface of the powder bed is inclined at an angle of 30°, less than the powder's critical angle of repose. This continuous printing BJT concept is illustrated in Fig. 8.2. Powder recoating and BJT are performed on this inclined build surface. The powder bed translates on a conveyor belt from the front toward the back of the machine. In contrast to typical batch fabrication, parts emerge continuously at the back of the machine. In principle, parts could be infinitely long, certainly much longer than the machine. Although this technology is no longer commercially available, this type of continuous part fabrication capability represents an important concept for achieving economical manufacture of moderate to high production volumes of parts.

The second continuous fabrication approach, spiral growth manufacturing (SGM), was invented by Chris Sutcliffe at the University of Liverpool in the early 2000s. The patent US2008109102 is a good reference for further information [11]. In the BJT variant of SGM, the powder bed is circular and rotates continuously. Binder printing and recoating are performed continuously. As the machine operates, the powder bed indexes downward continuously to accommodate the next layer of powder. As such, the top layer of powder forms a spiral in the powder bed. A machine schematic from the patent is shown in Fig. 8.3. Note that this technology preceded the development of the circular MJ machines from company dp polar highlighted in Sect. 7.8.

In the figure, object 2 is the cylindrical build chamber. Plates 10, 8, 14, and 23 do not rotate; plate 14 supports the build chamber and slides up and down on the pillars 12. The build chamber rotates, driven by the lead screw numbered 6. Four powder supply hoppers are shown by objects 24, so this indicates that the machine has four build stations, each with print heads and recoater mechanisms. As a consequence, for each rotation of the build chamber, effectively four layers are deposited and processed. So, for example, if the layer thickness is 0.1 mm, each rotation of the build chamber adds 0.4 mm to the powder bed height and plate 14, and the build chamber must translate downward by 0.4 mm per rotation to accommodate this increase in bed height.

Fig. 8.3 Schematic of a
spiral growth
manufacturing BJT
machine

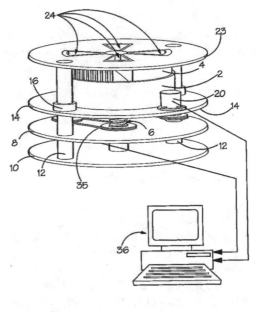

Fig. 8.4 Skewing an image for SGM printing

The linear velocity of the outer edge of the build chamber is greater than the linear velocity at the inner edge, which means that the powder along the outer edge passes the print head at a faster speed. This has important consequences for the printing conditions across the width of the chamber: more binder has to be deposited per unit time along the outer edge compared to the inner edge. Effectively this means that the images being printed have to be pre-skewed in order to compensate for the differences in speed. As an example, Fig. 8.4 shows how an image must be skewed so that the printed image is fabricated properly [12]. This image was printed on a SGM machine with two print heads, hence the two images on the right. Note that the inner edge is on the left side of the image, which is stretched, while the outer edge is compressed. This type of skewing is necessary for continuous rotative MJT as well.

8.4   BJT Machines

A wide variety of powder and binder materials can be used for BJT, which enables significant flexibility in the process. MIT licensed the BJT technology according to the type of material and application that each licensee was allowed to exploit. Z Corp (now owned by 3D Systems) marketed machines that build concept models in starch and plaster powder using a low-viscosity glue as binder. At the other end of the spectrum, ExOne markets machines that build in metal powder, with a strong polymer material that is used as the binder, as well as silica sand for sand casting applications. Therics developed a number of applications of BJT for tissue engineering scaffolding, drug delivery, and other medical applications, and was purchased by Integra LifeSciences in 2008.

Voxeljet is a German company that markets BJT machines that use polymer and sand powders for concept models, functional models, investment casting patterns, and sand casting applications. Desktop Metal markets metal machines, while HP has announced that they are developing metal machines based on their MultiJet Fusion technology, currently used for PBF machines.

When Z Corp was purchased by 3D Systems in 2012, their product line was merged into 3D Systems' ProJct line of printers. These printers are now branded as the ProJet X60 line of printers, with the smallest being the ProJet 160 and the largest being the ProJet 860Pro. Specifications for some of these machines are shown in Table 8.1. Machine names consisting only of numbers fabricate parts that are monochrome only, while suffixes of C, Plus, or Pro indicate that parts can be printed in color, up to the full CMYK color model (Cyan, Magenta, Yellow, Key (black)).

Voxeljet sold their first machine in 2005. They now offer a range of machines from the smallest VX200 to the huge VX4000, which has a 4-meter-long powder bed. The VX4000 processes foundry sand materials for the sand casting industry. They also marketed the VXC800, which is the infinitely continuous printer that was described earlier. A photo of the VXC800 is shown in Fig. 8.5. For this machine, the layer thickness was 150–400 microns. The surface area of the inclined plane was 500×850 mm.

The ExOne Corporation markets a line of BJT machines that fabricate metal parts and sand casting molds and cores in foundry sand. Strong polymer binders are required with these heavy powders. Stainless steel–bronze parts made with this technology [4] are typically accurate to ±0.125 mm. Several ExOne machine models are also listed in Table 8.1.

Desktop Metal has two BJT systems for metal parts, their Production System and their Shop System. Their BJT technology utilizes a large print head that spreads powder and prints binder in a single pass. Their system is reported to use more than 30,000 nozzles to jet up to 3 billion drops per second and fabricate up to 12,000 cm^3/hour. Support structures are printed, along with the parts, but anti-sintering agents can be printed between a part and its support to enable easy support removal after sintering. GE Additive recently introduced BJT technology to compete directly with ExOne and Desktop Metal.

Fig. 8.5 Voxeljet VXC800 machine. (Photo courtesy of Voxeljet)

Table 8.1 Machine specifications for a sampling of binder printing machines

Company/ models	Deposition rate L/hr	Build size ($l \times w \times h$) (mm)	Resolution DPI	Layer thickness (mm)	Number of nozzles	Materials
3D systems						
ProJet CJP 260Plus	0.87	$236 \times 185 \times 127$	300×450	0.10	604	Calcium sulfate hemihydrate (plaster of Paris)
ProJet CJP 460Plus	1.18	$203 \times 254 \times 203$	300×450	0.10	604	Calcium sulfate
ProJet CJP 860Pro	2.9	$508 \times 381 \times 229$	600×540	0.10	1520	Calcium sulfate
ExOne						
Innovent+	0.16	$160 \times 65 \times 65$	800×800	0.03	–	Metals
X1 25PRO	3.60	$400 \times 250 \times 250$	800×800	0.03	–	Metals
X1 160PRO	4.80	$800 \times 500 \times 400$	–	0.10	–	Metals
S-max pro	100–135	$1800 \times 1000 \times 700$	–	0.26	–	Casting sand
Voxeljet						
VX200	0.90	$300 \times 200 \times 150$	300×300	0.30	–	PMMA, sand
VX1000	9	$1000 \times 600 \times 500$	600×600	0.15	–	PMMA, sand
VX4000	123	$4000 \times 2000 \times 1000$	300×300	0.30	–	Sand

Applications for the metal material models include prototypes of metal parts, tooling, and more recently, direct digital manufacturing. As parts are fabricated in a powder bed, the surface finish of these parts is comparable to PBF parts. Finish machining is thus required for high tolerance and mating surfaces. ExOne developed a small machine that fabricates gold dental restorations, for example, copings for crowns. The materials and binder printing system were developed specifically for this application, since higher resolution is needed.

In the tooling area, ExOne promotes the advantages of conformal cooling in injection molds. In conformal cooling, cooling channels are routed close to the surfaces of the part cavity, particularly where hot spots are predicted. Using conventional machining processes, cooling channels are drilled as straight holes. With AM processes, however, cooling channels of virtually any shape and configuration can be designed into tools. Figure 8.6 illustrates one tool design with conformal cooling channels that was fabricated in an ExOne machine.

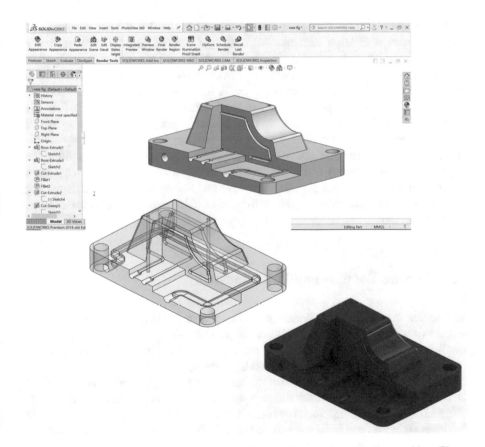

Fig. 8.6 Injection mold with conformal cooling channels fabricated in an ExOne machine. (Photo courtesy of ExOne Company)

Fig. 8.7 ExOne S-Max system. (Photo courtesy of ExOne)

The largest machines, the Voxeljet VX4000 and the ExOne S-Max, are intended for companies with large demands for castings, such as the automotive, heavy equipment, and oil and gas industries. A photo of the S-Max is shown in Fig. 8.7. Several dozen of these large machines (S-Max and its predecessor S15) have been installed. The machines print molds and cores for sand casting which, the companies claim, is much more economical than fabricating metal parts using PBF. In the Voxeljet VX4000, over 25,000 nozzles print binder to achieve build rates up to 120 liters per hour. Various metals can be cast into the printed molds, including aluminum, zinc, and even magnesium. Special equipment was developed for handling the large volumes of powders and heavy vats, including a silo and powder conveyor, conveyor track for transporting vats of powder and finished molds, and a debinding station. A typical installation with a S-Max or S15 machine occupies a room 40–50 m^2 in size.

8.5   Process Benefits and Drawbacks

BJT shares many of the advantages of MJT relative to other AM processes. With respect to MJT, BJT has some distinct advantages. First, it is faster since only a small fraction of the total part volume must be dispensed through the print heads. However, the need to recoat powder adds an extra step, slowing down BJT processes to some degree. Second, the combination of powder materials and additives in binders enables material compositions that are not possible, or not easily achieved, using MJT methods. Third, metal or ceramic slurries printed into powder beds result in higher solid loadings in BJT, compared with MJT, enabling better quality ceramic and metal parts to be produced.

As a general rule, however, parts fabricated using BJT processes tend to have poorer accuracies and surface finishes than parts made with MJT. Post-processing using infiltration steps and/or furnace operations are typically needed to fabricate dense parts or to ensure good mechanical properties, whereas MJT has lesser post-processing needs.

As with any set of manufacturing processes, the choice of manufacturing process and material depends largely on the requirements of the part or device. It is a matter of compromising on the best match between process capabilities and design requirements.

In metal BJT mechanical properties after furnace sintering are comparable with metal injection molding, which opens up a broad range of applications for direct replacement of powder metallurgy components by BJT.

Depending on the material and application, BJT requires costly post-processing steps compared to some AM techniques. For instance, metal BJT parts need to be sintered (or otherwise heat-treated) or infiltrated with a lower-melting-temperature metal. Full-color starch prototypes are also infiltrated with acrylic-coated materials to improve the vibrancy of colors. Sand casting cores and molds utilize a binder that imparts good mechanical properties, but these binders need curing.

BJT parts in the green state (as-built) typically have poor mechanical properties, are very brittle, and contain significant porosity. One main advantage of BJT over other 3D printing processes is that bonding occurs at room temperature. Therefore, dimensional deviations that are connected to thermal effects like distortion, shrinkage, cracks, and porosity that appear in other AM methods do not appear in BJT during the printing process. As a result, the build volumes of BJT printers are among the largest (up to 4000 mm in length). The bigger printers are normally used for manufacturing sand casting molds. Due to these bigger build sizes, it is common for these systems to be used for manufacture of multiple parts in series production lines.

Another important difference between BJT and most other AM techniques is no requirement for support structures during printing. The surrounding powder provides all the necessary support. This leads to decrease in the cost of supports, and removing and finishing of the printed samples, and allows for the creation of freeform structures with very few geometric restrictions. Since the parts in BJT do not need to be attached to the build platform, the whole build volume can be used. Therefore, BJT is suitable for low-to-medium batch production (Fig. 8.8 shows binder jet productions). However, metal parts may need support structures that are used during the sintering process so that the parts don't slump or warp, thus maintaining better dimensional stability through the furnace cycles.

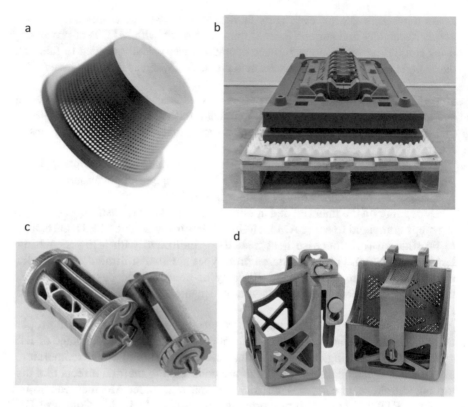

Fig. 8.8 Different parts produced by BJT from ExOne. (**a**) Filter, (**b**) casting mold block, (**c**) automobile part, (**d**) heavy equipment and machinery part [4, 13]

8.6   Summary

BJT processes share many of the advantages of MJT relative to other AM processes. Compared to MJT, BJT has some distinct advantages. First, it can be faster since only a small fraction of the total part volume must be dispensed through the print heads. Second, the combination of powder materials and additives in binders enables material compositions that are not possible, or not easily achieved, using direct methods. Third, slurries with higher solid loadings are possible with BJT, compared with MJ, enabling better quality ceramic and metal parts to be produced. As mentioned earlier, BJT processes lend themselves readily to printing colors onto parts. In addition, BJT variants have been commercialized for metal, polymer, ceramic, and composite applications, making BJT among the most flexible techniques material-wise. Some novel machine architectures have been demonstrated for BJT that enable continuous printing, including spiral growth manufacturing and an architecture with a slanted build surface. Recent growth in BJT is primarily tied to BJT of metal powders for Direct Digital Manufacturing of end-use components.

8.7 Questions

1. Explain why support structures are not needed in the BJT process for most materials.
2. List several characteristics of a good binder material.
3. Identify several methods for achieving a high packing density in the powder bed.
4. Develop a build time model for a conventional BJT machine. Assume that the part platform is to be filled with parts and the platform is L mm long and W mm wide. The print head width is H mm. Assume that a layer requires two passes of the print head, the print head can print in both directions of travel ($+X$ and $-X$), and the layer thickness is T mm. Figure Q8.1 shows a schematic for the question. Assume that a delay of D seconds is required for cleaning the print heads every K layers. The height of the parts to be printed is P mm. Assume the powder bed recoating time is 10s and the layer thickness is 0.1 mm.

   (a) Develop a build time model using the variables listed in the problem statement.
   Compute the build time for a layer of parts given the variable values in the following table.

5. How would you modify the build time model for the continuous printing Voxeljet machine? Discuss the time saved if this machine were successfully commercialized.
6. What are the similarities and differences between MJT and BJT processes? List at least three similarities and three differences in your discussion.

Fig. Q8.1 Schematic for questions 4–5

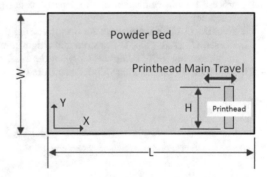

Table Q8.1 Question 4

	L	W	H	T	D	K	P
(b)	300	185	50	0.04	10	20	60
(c)	300	185	50	0.028	12	25	85
(d)	260	250	60	0.015	12	25	60
(e)	340	340	60	0.015	12	25	60
(f)	490	390	60	0.015	12	25	80

7. The Digital Grotesque in 2013 is the first entire closable room constructed using AM and uses BJT. Why was BJT used to create the Digital Grotesque?
8. Perform some background research on the Landscape House, by Universe Architects. Would you consider this a 3D printed building?
9. Can metal parts in three-dimensional printing be infiltrated by other material? If yes, explain the reasons, discuss how, and give examples of common material that may be used.

References

1. Sachs, E., et al. (1992). Three dimensional printing: Rapid tooling and prototypes directly from a CAD model. *Journal of Engineering for Industry, 114*(4), 481–488.
2. 3D Systems. (2020). *3D printers*. https://www.3dsystems.com/3d-printers
3. Voxeljet Corporation. (2020). www.voxeljet.com
4. ExOne. (2020). https://www.exone.com
5. Cima, M., et al. (1995). Structural ceramic components by 3D printing. In *1995 International Solid Freeform Fabrication Symposium.*
6. Uhland, S. A., et al. (2001). Strength of green ceramics with low binder content. *Journal of the American Ceramic Society, 84*(12), 2809–2818.
7. Cima, M., et al. (1992). Microstructural elements of components derived from 3D printing. In *1992 International Solid Freeform Fabrication Symposium.*
8. Grau, J., et al. (1997). High green density ceramic components fabricated by the slurry-based 3DP process. In *1997 International Solid Freeform Fabrication Symposium.*
9. Williams, C. B., Cochran, J. K., & Rosen, D. W. (2011). Additive manufacturing of metallic cellular materials via three-dimensional printing. *The International Journal of Advanced Manufacturing Technology, 53*(1–4), 231–239.
10. Williams, C. B. (2008). Design and development of a layer-based additive manufacturing process for the realization of metal parts of designed mesostructure. Georgia Institute of Technology.
11. Sutcliffe, C. (2008). Apparatus for manufacturing three dimensional items. US Patents.
12. Hauser, C., et al. (2008). Transformation algorithms for image preparation in spiral growth manufacturing (SGM). *Rapid Prototyping Journal, 14*(4), 188–196.
13. Digital metal. (2020). https://digitalmetal.tech/

Chapter 9
Sheet Lamination

Abstract Sheet Lamination (SHL) was one of the earliest commercialized AM techniques, but it has had only limited success in the marketplace. In SHL, sheets of materials are cut, stacked, and bonded (not always in that order) to form an object, and the material not used in the part cannot be easily reused and is typically discarded. Sheet material, however, can be some of the cheapest available and quite easy to handle. Metal variants exist as do paper, polymer, and ceramic variants. This process has been shown to be useful for the construction of very large, bulky objects. Although industrial applications for SHL are limited at the present time, continued research and technology development ensure these techniques will continue to serve niche applications for years to come.

9.1 Introduction

One of the first commercialized (1991) Additive Manufacturing techniques was Laminated Object Manufacturing (LOM). LOM involved layer-by-layer lamination of paper material sheets, cut using a CO_2 laser, each sheet representing one cross-sectional layer of the CAD model of the part. In LOM, the portion of the paper sheet which is not contained within the final part is sliced into cubes of material using a crosshatch cutting operation to ease subsequent removal. A schematic of the LOM process can be seen in Fig. 9.1.

A number of other processes have been developed based on Sheet Lamination (SHL) involving other build materials and cutting strategies. Because of the construction principle, only the outer contours of the parts are cut, and the sheets can be either cut and then stacked or stacked and then cut. These processes can be further categorized based on the mechanism employed to achieve bonding between layers: (a) gluing or adhesive bonding, (b) thermal bonding, (c) clamping, and (d) ultrasonic welding. As the use of ultrasonic welding involves unique solid-state bonding characteristics and can enable a wide range of applications, an extended discussion of this bonding approach is included at the end of this chapter.

© Springer Nature Switzerland AG 2021
I. Gibson et al., *Additive Manufacturing Technologies*,
https://doi.org/10.1007/978-3-030-56127-7_9

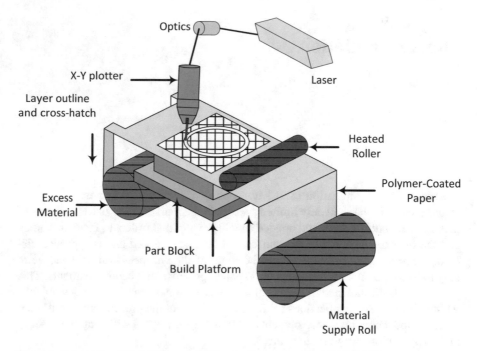

Fig. 9.1 Schematic of the LOM process (Elsevier license number 4635780060051) [1]

9.1.1 Gluing or Adhesive Bonding

The most popular SHL techniques have included a paper build material bonded using a polymer-based adhesive. Initially LOM was developed using adhesive-backed paper similar to the "butcher paper" used to wrap meat. Paper thicknesses range from 0.07 to 0.2 mm. Potentially any sheet material that can be precisely cut using a laser or mechanical cutter and that can be bonded can be utilized for part construction.

A further classification is possible within these processes based upon the order in which they bond and cut the sheet. In some processes the laminate is bonded first to the substrate and is then formed into the cross-sectional shape ("bond-then-form" processes). For other processes, the laminate is first cut and then bonded to the substrate ("form-then-bond" processes).

9.1.2 Bond-then-Form Processes

In "bond-then-form" processes, the building process typically consists of three steps in the following sequence: placing the laminate, bonding it to the substrate, and cutting it according to the slice contour. The original LOM machines used this process with adhesive-backed rolls of material. A heated roller passes across the sheet after placing it for each layer, melting the adhesive and producing a bond between layers. A laser (or in some cases a mechanical cutting knife) designed to cut to a depth of one layer thickness cuts the cross-sectional outline based on the slice information. The unused material is left in place as support material and is diced using a crosshatch pattern into small rectangular pieces called "tiles" or "cubes." This process of bonding and cutting is repeated until the complete part is built. After part construction, the part block is taken out and post-processed. The crosshatched pieces of excess material are separated from the part using typical wood carving tools (called decubing). It is relatively difficult to remove the part from the part block when it is cold; therefore, it is often put into an oven to warm before decubing or the part block is processed immediately after part buildup.

Although historically many people continue to associate paper SHL with the Laminated Object Manufacturing machines introduced in 1991 by Helisys Inc., USA, and subsequently supported by Cubic Technologies, USA (after Helisys' bankruptcy), many paper-based SHL machines have been introduced over many years by companies such as Mcor Technologies (Ireland) and Wuhan Binhu (China). These systems made use of plain paper as the build material, and selectively dispensed adhesive only where needed. Because the support material is not adhesively bonded, unlike in LOM, the support removal process is easier. The use of color inkjet printing onto paper by Mcor Technologies enabled the production of full-color paper parts directly from a CAD file.

Solidimension (Be'erot, Israel) took the concepts of LOM and further developed them in 1999 into a commercial prototyping system for laminating polyvinyl chloride (PVC) plastic sheets. Solidimension sold its own machines under the Solido name [2] and under other names via resellers. This machine utilized an x–y plotter for cutting the PVC sheets and for writing with "anti-glue" pens, which inhibit bonding in prescribed locations. This machine used a unique approach to support material removal. Support material was subdivided into regions, and unique patterns for cutting and bonding the excess material were used to enable easy support material removal. An example of this support material strategy can be seen in Fig. 9.2. Solido machines are no longer offered for sale; however if history is any guide, others may pick up this unique idea and offer similar machines someday.

Fig. 9.2 Support material removal for three golf balls made using a Solidimension machine, showing (**a**) the balls still encased in a central region, being separated from the larger block of bonded material; (**b**) the support material is glued in an accordion-like manner so that the excess material can be pulled out easily as a continuous piece; and (**c**) the balls after complete removal of excess support material. (Photo courtesy of 3D Systems)

Bond-then-form SHL principles have also been successfully applied to fabrication of parts from metal, ceramic, and composite materials. In this case, rather than paper or polymer sheets, ceramic or metal-filled tapes are used as the build material to form green parts, and high-temperature furnace post-processing is used to debind and sinter the structure. These tapes are used for part construction employing a standard SHL process.

Specific advantages of LOM-like bond-then-form adhesive-based processes include:

(a) Little shrinkage, residual stresses, and distortion problems within the process.
(b) When using paper feedstock, the end material is similar to plywood, a typical pattern making material amenable to common finishing operations.
(c) Large parts can be fabricated rapidly.
(d) A variety of build materials can be used, including paper and polymer sheets and metal- or ceramic-filled tapes.
(e) Nontoxic, stable, and easy-to-handle feedstock.
(f) Low material, machine, and process costs relative to other AM systems.

Paper-based SHL has several limitations, including the following: (a) most paper-based parts require coating to prevent moisture absorption and excessive wear; (b) the control of part accuracy in the Z-dimension is difficult (due to swelling or inconsistent sheet material thickness); (c) mechanical and thermal properties of the parts are inhomogeneous due to the glue used in the laminated structure; and (d) small part feature detail is difficult to maintain due to the manual decubing process.

In general, parts produced by paper-based SHL have been most successfully applied in industries where wooden patterns are often used or in applications where most features are upward-facing. Examples of good applications for paper SHL include patterns for sand casting and 3D topographical maps – where each layer represents a particular elevation of the map.

9.1.3 Form-then-Bond Processes

In form-then-bond processes, sheet material is cut to shape first and then bonded to the substrate. This approach is popular for construction of parts in metallic or ceramic materials that are thermally bonded (discussed in Sect. 9.3.1), but implementation has primarily been at the research level. One example of a glue-based form-then-bond process is the "Offset Fabbing" system patented by Ennex Corp., USA. In this process, a suitable sheet material with an adhesive backing is placed on a carrier and is cut to the outline of the desired cross-section using a two-dimensional plotting knife. Parting lines and outlines of support structures are also cut. The shaped laminate is then placed on top of the previously deposited layers and bonded to it. This process continues until the part is complete. A schematic of the process is shown in Fig. 9.3.

The form-then-bond approach facilitates construction of parts with internal features and channels. Internal features and small channels are difficult or impossible

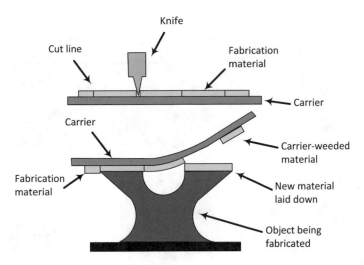

Fig. 9.3   Offset Fabbing system, Ennex Corp [3]

with a bond-then-form approach because the excess material is solid, and thus material inside internal features cannot be removed once bonded (unless the part is cut open). Another advantage of form-then-bond approaches is that there is no danger of cutting into the previous layers, unlike in bond-then-form processes where cutting occurs after placing the layer on the previous layer; thus, laser power control or knife pressure is less demanding. Also, the time-consuming and potentially damage-causing decubing step is eliminated. However, these processes require external supports for building overhanging features; some type of tooling or alignment system to ensure a newly bonded layer is registered properly with respect to the previous layers; or a flexible material carrier that can accurately place material regardless of geometry.

Computer-Aided Manufacturing of Laminated Engineering Materials (CAM-LEM, Inc., USA) was developed as a process for fabrication of functional ceramic parts using a form-then-bond method, as shown in Fig. 9.4. In this process, individual slices are laser cut from sheet stock of green ceramic or metal tape. These slices are precisely stacked one over another to create the part. After assembly the layers are bonded using heat and pressure or another adhesive method to ensure intimate contact between layers. The green part is then furnace processed in a manner identical to indirect processing of metal or ceramic green parts made using Powder Bed Fusion (PBF), as introduced in Chap. 5. The CL-100 machine produced parts from up to five types of materials, including materials of differing thickness, which were automatically incorporated into a build. One or more of these materials may act as secondary support materials to enable internal voids or channels and overhangs. These support materials were later removed using thermal or chemical means. An application for this technology is for the fabrication of microfluidic structures (structures with microscale internal cavities and channels). An example microfluidic structure made using CAM-LEM is shown in Fig. 9.5.

Another example of a form then bond process is the Stratoconception approach [4], where the model is sliced into thicker layers. These layers are machined and

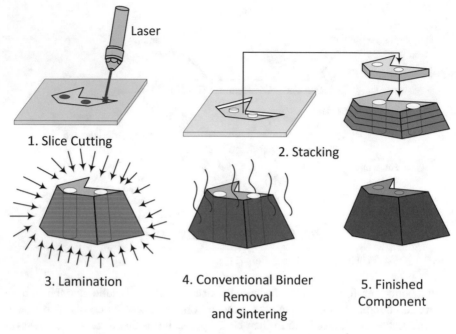

Laser

1. Slice Cutting

2. Stacking

3. Lamination

4. Conventional Binder
Removal
and Sintering

5. Finished
Component

Fig. 9.4 CAM-LEM process. (Photo courtesy of CAM-LEM, Inc.)

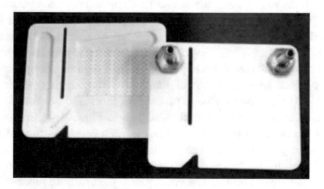

Fig. 9.5 A ceramic microfluidic distillation device cutaway view (left) and finished part (right).
(Photo courtesy of CAM-LEM, Inc.)

then glued together to form a part. The use of a multi-axis machining center enables
the edges of each layer to be contoured to better match the STL file, helping eliminate the stair-step effect that occurs with increasing layer thickness. This and similar
cutting techniques to eliminate vertical edges on each layer have been used by many
different researchers to build statues, large works of art, and other structures from
foam, wood, and other materials.

9.2 Materials

As covered in the previous section, a wide variety of materials has been processed using a variety of SHL processes, including plastics, metals, ceramics, and paper. A brief survey will be offered identifying the materials and their characteristics that facilitate SHL.

Butcher paper was the first material used in the original Helisys LOM process. Butcher paper is coated on one side with a thin layer of a thermoplastic polymer. It is this polymer coating that melts and ensures that one layer of paper bonds to the previous layer. Since butcher paper is fairly strong and heavy, it forms sturdy parts after a suitable thickness has been fabricated (> 5–6 mm typically). After part fabrication, parts are finished as if they were wood by sanding, filing, staining, and varnishing or sealing.

Mcor Technologies printers used standard copy paper in A4 or US letter sizes with weights of 20lb or 43lb. Either white or colored paper can be used. The water-based glue binds paper sheets and results in fairly rigid parts although, similar to the Helisys process, a minimum thickness of 5 or 6 mm is required to ensure good strength.

In the metals area, both bond-then-form and form-then-bond approaches have been pursued. Perhaps the most conceptually simple fabrication process is the sheet metal clamping approach, where sheet metal is cut to form part cross-sections, then simply clamped together. Other processes use several types of bonding methods. Some researchers were interested in demonstrating the feasibility of some metal SHL process advances, rather than fabricating functional devices, and simply used an adhesive to bond sheets together. In other cases, the adhesive bonded structures were meant to be functional prototypes, not just proof-of-concepts. Aluminum and low-carbon steel materials were most commonly used, unless functional molds or dies were desired, in which case tool steels were used.

Thermal and diffusion bonding approaches provide much stronger parts than adhesively bonded parts. Thermal bonding, to be discussed in the next section, has been demonstrated with a variety of aluminum and steel sheets and several types of bonding mechanisms, including brazing and welding. Diffusion bonding, to be covered in Sect. 9.4, has also been demonstrated on a variety of metals and is the important joining mechanism for Ultrasonic Additive Manufacturing, where aluminum, titanium, stainless steel, brass, Inconel, and copper materials have been demonstrated.

In SHL processes, ceramic materials are most often fabricated using bond-then-form processes using ceramic-filled tapes. Tape casting methods form sheets of material composed of powdered ceramics, such as SiC, TiC-Ni composite, or alumina, and a polymer binder. Metal powder tapes can also be used to fabricate metal parts. These tapes are then used for part construction employing a standard SHL process. Various SiC, alumina, TiC–Ni composite, and other material tapes have been used to build parts. A challenge with this process is that thermal post-processing to consolidate metal or ceramic powders results in a large amount of shrinkage (12–18%) which can lead to dimensional inaccuracies and distortion. This is typical of many conventional powder-based processes, such as powder injection molding, and strategies have been developed to address the effects of shrinkage, although limitations exist.

For polymer materials, the Solidimension example is the most well-known and used PVC sheets. Foam blocks have also been used in some research machines, as well as by sculptors who create large sculptures by stacking blocks cut by hot wire or CNC milling. Additionally, some research efforts have successfully demonstrated the automated lay-up of polymer composite sheets. The area of polymer SHL is broad and not very well defined, since it stretches from sculpture to composites manufacturing.

9.3 Material Processing Fundamentals

As indicated, several types of processes are evident under the general category of SHL. Thermal bonding and sheet metal clamping are covered in this section. In the next section, a more in-depth coverage is provided for the Ultrasonic Additive Manufacturing process.

9.3.1 Thermal Bonding

Many organizations around the world have successfully applied thermal bonding to SHL of functional metal parts and tooling. A few examples will be mentioned to demonstrate the flexibility of this approach. Yi et al. [5] have successfully fabricated 3D metallic parts using precut 1-mm-thick steel sheets that are then diffusion bonded. They demonstrated continuity in grain structure across sheet interfaces without any physical discontinuities. Himmer et al. [6] produced aluminum injection molding dies with intricate cooling channels using Al 3003 sheets coated with 0.1-mm-thick low-melting point Al 4343 (total sheet thickness 2.5 mm). The sheets were laser cut to an approximate, oversized cross-section, assembled using mechanical fasteners, bonded together by heating the assembly in a nitrogen atmosphere just above the melting point of the Al 4343 coating material, and then finish machined to the prescribed part dimensions and surface finish. Himmer et al. [7] also demonstrated satisfactory layer bonding using brazing and laser spot welding processes. Obikawa [8] manufactured metal parts employing a similar process from thinner steel sheets (0.2 mm thick), with their top and bottom surface coated with a low-melting-point alloy. Wimpenny et al. [1] produced laminated steel tooling with conformal cooling channels by brazing laser-cut steel sheets. Similarly, Yamasaki [9] manufactured dies for automobile body manufacturing using 0.5-mm-thick steel sheets. Each of these, and other investigators, have shown that thermally bonding metal sheets is an effective method for forming complex metal parts and tools, particularly those which have internal cavities and/or cooling channels.

Although extensively studied, sheet metal lamination approaches have gained little traction commercially. This is primarily due to the fact that bond-then-form processes require extensive post-processing to remove support materials, and form-then-bond processes are difficult to automate for arbitrary, complex geometries. In the case of form-then-bond processes, particularly if a cross-section has geometry that is disconnected from the remaining geometry (e.g., an island within a layer), accurate registration of laminates is difficult to achieve and may require a part-specific solution. Thus, upward-facing features where each cross-section's geometry is contiguously interconnected are the easiest to handle. Commercial interest in SHL is primarily in the area of inexpensive paper parts and large tooling, where internal, conformal cooling channels can provide significant benefits over traditional cooling strategies.

Another process that combined SHL with other forms of AM (including beam deposition, extrusion, and subtractive machining) was Shape Deposition Manufacturing (SDM) [10]. With SDM, the geometry of the part is subdivided into nonplanar segments. Each segment is deposited as an oversized, near-net shape region and then finished machined. Sequential deposition and machining of segments (rather than planar layers) forms the part. A decision is made concerning how each segment should be manufactured dependent on such factors as the accuracy, material, geometrical features, functional requirements, etc. Secondary support materials were commonly used to enable complex geometry to be made and for clearance between mechanisms that required differential motion after manufacture. A completely automated subdivision routine for arbitrary geometries, however, was never developed and intervention from a human "expert" is required for many types of geometries. As a result, though interesting and useful for certain complex multi-material structures, such a system was never commercially introduced.

9.3.2 Sheet Metal Clamping

In the case of assembling rigid metal laminates into simple shapes, it may be advantageous to simply clamp the sheets together using bolts and/or a clamping mechanism rather than using an adhesive or thermal bonding method. Clamping is quick and inexpensive and enables the laminates to be disassembled in order to modify a particular laminate's cross-section and/or for easy recycling of the materials. In addition, the clamping or bolting mechanism can act as a reference point to register each laminate with respect to one another.

When clamping, it is often advantageous to simply cut a profile into one edge of a laminate, leaving three edges of the rectangular sheet uncut. An example of such

Fig. 9.6 Profiled edge
laminate tool. (Photo
courtesy of Fraunhofer
CCL)

a "profiled edge laminate" construction is shown in Fig. 9.6. Of course, this type of profiled edge can also be utilized with adhesive and thermally bonded layers as well. The major benefit of this approach is the ease with which the layers can be clamped (i.e., bolting the laminates together through a set of holes, as could be done using the through-holes visible on the right edge of Fig. 9.6). The drawbacks of a profile approach are that clamping forces for most tools would then be perpendicular to the laminate interface, and the laminates might separate from one another (leaving gaps) under certain conditions, such as when pressurized polymers are injected into a mold made from such a tool.

9.4   Ultrasonic Additive Manufacturing

Ultrasonic Additive Manufacturing (UAM), originally known as Ultrasonic Consolidation, is a hybrid SHL process combining ultrasonic metal seam welding and CNC milling and first commercialized by Solidica Inc., USA, in 2000. In UAM, the object is built up on a rigidly held base plate bolted onto a heated platen, with temperatures ranging from room temperature to approximately 200 °C. Parts are built from bottom to top, and each layer is composed of several metal foils laid side-by-side and then trimmed using CNC milling.

Early UAM systems were equipped with 1 kW power systems suitable for processing softer material such as aluminum and copper. The power system was enhanced in 2008 by Edison Welding Institute (EWI) and increased to 9 kW, enabling UAM to process stainless steel and Inconel. Solidica Inc. and EWI created a joint venture known as Fabrisonic LLC in 2011 to produce, develop, and distribute new equipment. UAM machines available from Fabrisonic have build chamber

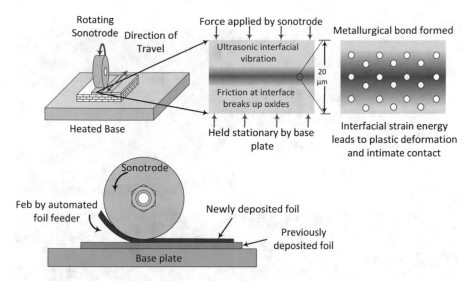

Fig. 9.7 Schematic of Ultrasonic Additive Manufacturing

dimensions up to 1829 × 1829 × 914 mm, which is significantly bigger than most AM systems.

During UAM, a rotating sonotrode travels along the length of a thin metal foil (typically 100–150 μm thick). The foil is held closely in contact with the base plate or previous layer by applying a normal force via the rotating sonotrode, as shown schematically in Fig. 9.7. The sonotrode oscillates transversely to the direction of motion, at a constant 20 kHz frequency and user-set oscillation amplitude. After depositing a foil, another foil is deposited adjacent to it. This procedure is repeated until a complete layer is placed. The next layer is bonded to the previously deposited layer using the same procedure. Typically four layers of deposited metal foils are termed one level in UAM. After deposition of one level, the CNC milling head shapes the deposited foils/layers to their slice contour (the contour does not need to be vertical, but can be a curved or angled surface, based on the local part geometry). This additive–subtractive process continues until the final geometry of the part is achieved. Thus, UAM is a bond-then-form process, where the forming can occur after each layer or after a number of layers, depending on the settings chosen by the user. Additionally, each layer is typically deposited as a combination of foils laid side-by-side rather than a single large sheet, as is typically practiced in SHL processes.

By the introduction of CNC machining, the dimensional accuracy and surface finish of UAM end products is not dependent on the foil thickness, but on the CNC milling approach that is used. This eliminates the stair-stepping effects and layer-thickness-dependent accuracy aspects of other AM processes. Due to the combination of low-temperature ultrasonic bonding, and additive-plus-subtractive processing, the UAM process is capable of creating complex, multifunctional 3D parts, including objects with complex internal features, objects made up of multiple

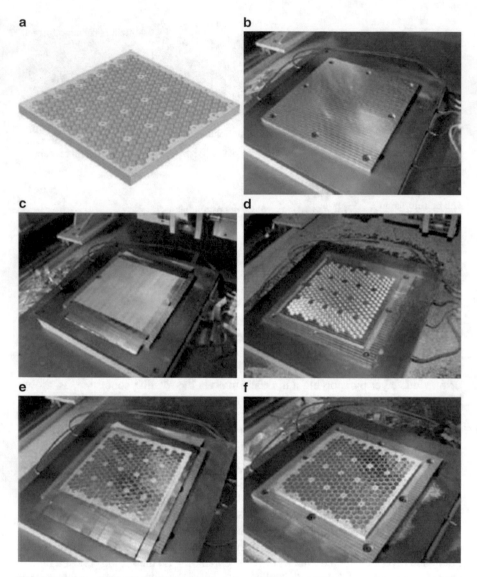

Fig. 9.8 Fabrication procedure for a honeycomb structure using UAM

materials, and objects integrated with wiring, fiber optics, sensors, and instruments. The lack of an automated support material in commercial systems, however, means that many types of complex overhanging geometries cannot be built using UAM.

To better illustrate the UAM process, Fig. 9.8a–f illustrates the steps utilized to fabricate a honeycomb panel (270 mm by 240 mm by 10 mm). The cutaway CAD model showing the internal honeycomb features is shown in Fig. 9.8a. The part is fabricated on a 350 mm by 350 mm by 13 mm Al 3003 base plate, which is firmly bolted to a heated platen, as shown in Fig. 9.8b. Metal foils used for this part are Al

3003 foils 25 mm wide and 0.15 mm thick. The first layer of deposited foils is shown in Fig. 9.8c. Since the width of one layer is much larger than the width of the individual metal foils, multiple foils are deposited side-by-side for one layer. After the deposition of the first layer, a second layer is deposited on the first layer and so on, as seen in Fig. 9.8d. After every four layers of deposition, the UAM machine trims the excess tape ends and machines internal and external features based on the CAD geometry. After every 40 layers, the machine does a surface machining pass at the exact height of that layer (in this case the z-height of the 40th layer is 0.15 mm per layer times 40 layers, or 6 mm) to compensate for any excess z-height that may occur due to variability in foil thicknesses. A surface machining pass can occur at any point in the process if, for instance, a build interruption or failure occurs (enabling the build to be continued from any user-specific z-height). After a series of repetitive bonding and machining operations the face sheet layers are deposited to enclose the internal features, as shown Fig. 9.8e. Four layers are deposited, and the final panel is shown in Fig. 9.8f.

9.4.1   UAM Bond Quality

There are two widely accepted quality parameters for evaluating UAM-made structures, which are linear welding density (LWD) and part strength. LWD is defined as the percentage of interface which is bonded divided by the total length of the interface between two ultrasonically consolidated foils, determined metallographically. An example of a microstructure sample made from four layers of Al 3003 tapes by UAM is shown in Fig. 9.9. The black areas represent the unbonded regions along the interfaces. In this microstructure, a LWD of 100% occurred only between Layer 1 and the base plate.

Fig. 9.9 A UAM part made from four layers of Al 3003 foils. LWD is determined by calculating the bonded interface divided by the total interface (arrows show the sonotrode traveling direction for each layer). (Elsevier license number 4841900933578) [11]

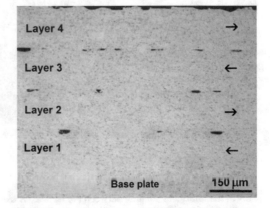

9.4.2 UAM Process Fundamentals

Ultrasonic metal welding (UMW) is a versatile joining technology for various industries, including in electronics, automotive, and aerospace. Compared to other metal fusion processes, the UMW solid-state joining approach does not require high-temperature diffusion or metal melting; and the maximum processing temperature is generally no higher than 50% of the melting point of the joined metals. Therefore, thermal residual stresses and thermally induced deformation due to reso-lidification of molten metals, which are important considerations in thermal welding processes and many AM processes (such as PBF, Directed Energy Deposition (DED), and thermal bonding-based SHL processes), are not a major consideration in UAM.

Bonding in UMW can be by (a) mechanical interlocking, (b) melting of interface materials, (c) diffusion bonding, and (d) atomic forces across nascent metal surfaces (e.g., solid-state metallurgical bonding). In UAM, bonding of foils to one another appears to be almost exclusively by nascent metal forces (metallurgical bonding), whereas bonding between foils and embedded structures, such as reinforcement fibers, is primarily by mechanical interlocking. An example of a stainless steel 304 wire mesh embedded between Al 3003 foils using the UAM process is shown as Fig. 9.10. This figure illustrates that the mesh is mechanically interlocked with the Al 3003 matrix, whereas the SS mesh metallurgically bonded to itself and the Al 3003 layers metallurgically bonded to each other. Mechanical interlocking between the Al and SS mesh was due to plastic deformation of Al around and through the mesh. Thus, mechanical interlocking can take place for material combinations between dissimilar metals or between materials with significant hardness differences. For material combinations of similar materials or materials with similar hardness values, metallurgical bonding appears to be the dominant bond formation mechanism.

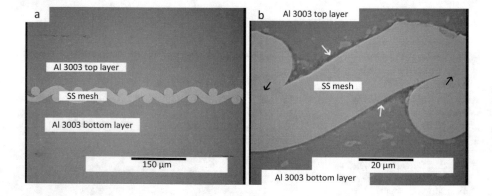

Fig. 9.10 SEM microstructures of Al 3003/SS mesh: (**a**) SS mesh embedded between Al 3003 layers, (**b**) Al 3003/SS mesh interface at a higher magnification. The white arrows illustrate the lack of metallurgical bonding between the Al and SS materials. The black arrows indicate areas of metallurgical bonding between SS mesh elements. (Emerald license number 4841921140237)

Two conditions must be fulfilled for establishment of solid-state bonding during UAM: (a) generation of atomically clean metal surfaces and (b) intimate contact between clean metal surfaces. As all engineering metals contain surface oxides, the oxides must be displaced in order to achieve atomically clean metal surfaces in intimate contact. The ease with which oxide layers can be displaced depends on the ratio of metal oxide hardness to base metal hardness, where higher ratios facilitate easier removal. Due to the significant hardness differences between aluminum and aluminum oxide, Al 3003 alloys are one of the best-suited materials for ultrasonic welding. Nonstructural noble metals, such as gold which do not have surface oxide layers, are quite amenable to ultrasonic welding. Materials with difficult-to-remove oxide layers are problematic for ultrasonic welding. However, difficult-to-weld materials have been shown to be UAM-compatible when employing chemical or mechanical techniques for removing the surface oxide layers just prior to welding.

Plastic deformation at the foil interfaces is critical for UAM, to break up surface oxides and overcome surface roughness. The magnitude of plastic deformation necessary to achieve bonding can be reduced by decreasing the surface roughness of the interface materials prior to welding, such as by surface machining (which occurred between Layer 1 and the base plate in Fig. 9.9) and/or by removing the surface oxides by chemical stripping or surface finishing. In addition, factors which enhance plastic deformation are also beneficial for bonding, such as using more ductile materials and/or by thermally or acoustically softening the materials during bonding.

Metallic materials experience property changes when subjected to ultrasonic excitations, including acoustic softening, increase in crystallographic defects, and enhanced diffusivities. In particular, metal softening in the presence of ultrasonic excitations, known as the "Blaha effect" or "acoustic softening," means that the magnitude of stresses necessary to initiate plastic deformation are significantly lower [12]. The softening effect of ultrasonic energy on metals is similar to the effect of heating and can in fact reduce the flow stress of a metallic material more effectively than heating. Thus, acoustic softening results in plastic deformation at strains much less than would otherwise be needed to achieve plastic deformation.

UAM processes also involve metal deformation at high strain rates. High strain rate deformation facilitates formation of vacancies within welded metals, and thus excess vacancy concentration grows rapidly. As a result, the ductility and diffusivity of the metal are enhanced. Both of these characteristics aid in UAM bonding.

9.4.3 UAM Process Parameters and Process Optimization

The important controllable process parameters of UAM are (a) oscillation amplitude, (b) normal force, (c) travel speed, and (d) temperature. It has been found that the quality of bonding in UAM is significantly affected by each of these process parameters. A brief discussion of each of these parameters and how they affect bonding in UAM follows

9.4.3.1   Oscillation Amplitude

Energy input directly affects the degree of elastic/plastic deformation between mating metal interfaces and consequently affects bond formation. Oscillation amplitude and frequency of the sonotrode determine the amount of ultrasonic energy available for bond formation. In commercial UAM machines, the frequency of oscillation is not adjustable, as it is preset based on sonotrode geometry, transducer and booster hardware, and the machine power supply. In UAM, the directly controllable parameter for ultrasonic energy input is oscillation amplitude [13].

Generally speaking, the higher the oscillation amplitude, the greater the ultrasonic energy delivered. Consequently, for greater energy, more elastic/plastic deformation occurs at the mating metal interface and therefore better welding quality is achieved. However, there is an optimum oscillation amplitude level for a particular foil thickness, geometry, and material combination. A sufficient amount of ultrasonic energy input is needed to achieve plastic deformation, to help fill the voids due to surface roughness that are inherently present at the interface. However, when energy input exceeds a critical level, bonding deteriorates as excess plastic deformation can damage previously formed bonds at the welding interface due to excessive stress and/or fatigue.

9.4.3.2   Normal Force

Normal force is the load applied on the foil by the sonotrode, pressing the layers together. Sufficient normal force is required to ensure that the ultrasonic energy in the sonotrode is delivered to the foils to establish metallurgical bonds across the interface. This process parameter also has an optimized level for best bonding. A normal force higher or lower than the optimum level degrades the quality of bonds and lowers the LWD obtained. When normal force increases beyond the optimum level, the stress condition at the mating interface may be so severe that the formed bonds are damaged, just as when oscillation amplitude exceeds its optimum level.

9.4.3.3   Sonotrode Travel Speed

Welding exposure time has a direct effect on bond strength during ultrasonic welding. In UAM welding, exposure time is determined by the travel speed of the sonotrode. Higher speeds result in shorter welding exposure times for a given area. Over-input of ultrasonic energy may cause destruction of previously formed metal bonds and metal fatigue. Thus, to avoid bond damage caused by excess ultrasonic energy, an optimum travel speed is important for strong bonds.

9.4.3.4   Preheat Temperature

Metallurgical bonds can be established at ambient temperature during UAM processing. However, for many materials an increased preheat temperature facilitates bond formation. Heating directly benefits bond formation by reducing the flow stress of metals. However, excess heating can have deleterious effects. High levels of metal foil softening can result in pieces of the metal foil sticking to the sonotrode. In addition, in the case of fabrication of structures with embedded electronics, excess temperature may damage embedded electronics. For certain materials, such as Cu, enhanced oxide formation at elevated temperatures will impede oxide removal. Finally, for some materials elevated temperatures cause metallurgical "aging" phenomena such as precipitation hardening, which can embrittle the material and cause premature part failure.

9.4.3.5   Other Parameters

Metal foil thickness is another important factor to be considered in UAM. The most common metal foils used in UAM are on the order of ~150 μm. Generally speaking, bonds are more easily formed between thin metal foils than between thick ones. However, foil damage is a major concern for UAM of thinner metal foils, as they are easily scratched or bent; and thus metal foils between 100 and 200 μm are most often used in UAM.

In addition to material-related constants, process optimization is influenced by the surface condition of the sonotrode, particularly the sonotrode surface roughness. A typical sonotrode in UAM is made of titanium or tool steel. The surface of the sonotrode is EDM roughened to enhance friction between the sonotrode and foil being deposited. However, surface roughness of the sonotrode decreases significantly after extended use. Thus, optimized parameters change along with the condition of the sonotrode surface. It is necessary to practice regular sonotrode roughness measurements and modify process parameters accordingly. Also, the sonotrode surface roughness is imprinted onto the upper-most surface of the just-deposited foil (see upper surface of Fig. 9.9). As a result, this surface roughness must be overcome by plastic deformation during deposition of the next layer. An optimum surface roughness condition would be one which involves no slip between the sonotrode and the foil being deposited, without significantly increasing the surface roughness of the deposited foil. As slip often increases with decreasing roughness, sonotrode surface roughness is inherently difficult to optimize.

9.4.4   Microstructures and Mechanical Properties of UAM Parts

9.4.4.1   Defects

The most common defects in UAM-made parts are voids. Voids occur either along the interfaces between layers or between the foils that are laid side-by-side to form each layer. For ease of discussions, defects are classified into three types according to defect origin. Type 1 defects are the voids along layer/layer interfaces due to foil surface roughness and/or insufficient input energy. Type 2 defects are damaged areas, also at the layer/layer interface, that are created when excess energy input during UAM results in the breaking of previously formed bonds. Type 3 defects are found between adjacent foils within a layer.

One can identify defect types by observing the existence of oxide layers on the surfaces of the defects or by looking at the defect morphology. For Type 1 defects, since the metal surfaces have not bonded, oxide layers are not damaged and removed and can be observed. In addition, Type 1 defects typically have a flat upper surface and a rounded lower surface (where the flat upper surface is the newly deposited, smooth foil and the rounded lower surface is the unbonded upper surface of the previously deposited foil, as seen in Fig. 9.11). For Type 2 defects, since bonding has occurred, oxide layers have been disturbed and are difficult to locate. Type 2 defects thus have a different morphology than Type 1 defects, as they represent voids where the interface has been torn apart after bonding, rather than regions which have never bonded.

Type 3 defects are the physical gaps between adjacent metal foils, as shown in Fig. 9.12. In UAM, the foil width setting within the software determines the offset distance the sonotrode and foil placement mechanism are moved between depositions of adjacent foils within a layer. If the setting value is larger than the actual metal foil width, there will always be gaps between adjacent foils. The larger the width setting above the foil width, the larger the average physical gap. If the width

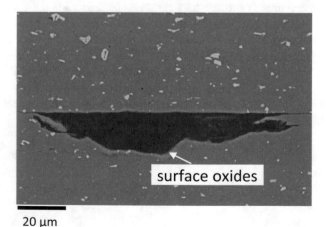

20 µm

Fig. 9.11 Type 1 UAM defect (arrow indicates location of surface oxides)

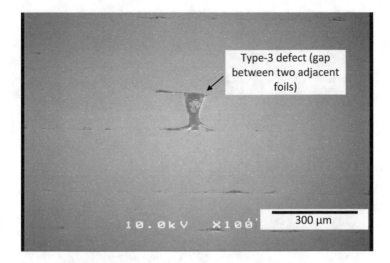

Fig. 9.12 Type 3 defect observed between adjacent foils. (Note the morphology of the Type 1 defects between layers indicate that this micrograph is upside down with respect to build orientation)

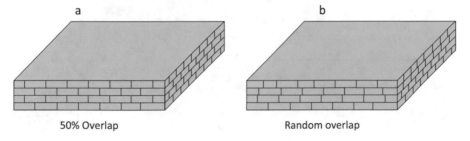

Fig. 9.13 Schematic illustrating (**a**) 50% foil overlap and (**b**) random foil overlap in UAM

setting is smaller than the actual width of the foil, gaps will be minimized. However, excess overlap results in surface unevenness at the overlapping areas and difficulty with welding. Thus, positioning inaccuracies of the foil placement mechanism in a UAM machine, combined with improper width settings, cause Type 3 defects.

Defects strongly affect the strength of UAM parts. Process parameter optimization (including optimization of width settings) to maximize LWD and minimize Type 3 defects is the most effective means to increase bond strength. With optimized parameters, Type 1 and Type 3 defects are minimized and Type 2 defects do not occur.

Type 1 defects can be reduced by surface machining a small amount of metal (~10 μm, or the largest roughness observed at the upper-most deposited surface, as in Fig. 9.9) after depositing each layer. Post-process heat treatment can also be used to significantly reduce all types of defects.

The degradation of part mechanical properties due to Type 3 defects can be reduced by designed arrangement of successive layers. Successive layers in a UAM part can be arranged so that 50% overlap across layers is obtained, as shown in Fig. 9.13. Although somewhat counterintuitive, it has been shown that better tensile properties result from a 50% overlap than when random foil arrangements are used.

Fig. 9.14   Ultrasonically consolidated Ni 201 foils

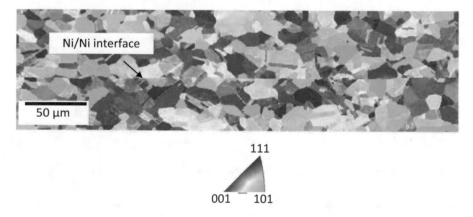

Fig. 9.15   An image of several inverse pole figures of contiguous areas along a well-bonded Ni–Ni interface stitched together. The grains in the image are color coded to reflect their orientation (Elsevier license number 4841930784673) [14]

9.4.4.2   UAM Microstructures

Typical microstructures from Al 3003 tapes with representative defects were shown in Figs. 9.9 and 9.12. Figure 9.14 shows the microstructure of two Ni 201 foils deposited on an Al 3003 substrate. Plastic deformation of Ni foils near the foil surfaces can be experimentally visualized using orientation imaging microscopy, as shown in Fig. 9.15. Smooth intragrain color transition within a few grains at the surface indicates the foil interfaces undergo some plastic deformation during UAM processing, whereas the absence of intragranular color transitions away from the foil surfaces indicates that the original microstructure is retained in the bulk of the foil.

In addition to UAM of similar materials, UAM of dissimilar materials is quite effective. Many dissimilar metal foils can be bonded with distinct interfaces, with a high degree of LWD and without intermetallic formation [15].

9.4.4.3 Mechanical Properties

Mechanical properties of UAM parts are highly anisotropic due to the anisotropic properties of metal foils, the presence of defects, and the alignment of grain boundaries along the foil-to-foil interfaces. Most metal foils used in UAM are prepared via rolling. Grains within the foils are often elongated along the rolling direction. As a result, foils are typically stronger along the rolling direction, and thus UAM parts are typically stronger in the x-axis than in the y- or z-axes. A typical transverse y-axis strength for a UAM part is about 85% of the published bulk strength value for a particular material, whereas the longitudinal x-axis strength typically exceeds published values for a material. In the z direction, perpendicular to the layer interfaces, UAM parts are much weaker than the x and y properties. This is primarily due to the fact that the bond formed across the foil interfaces, even at 100% LWD, is not as strong as the more isotropic inter-granular bonding within the foils. Thus, z direction strength values are often 50% of the published value for a particular material, with very little ductility.

Thus, when considering UAM for part fabrication, it is important to consider the anisotropic aspects of UAM parts with respect to their design. Heat treatment can be used to normalize these properties if this anisotropy results in unacceptable properties.

Another factor which affects mechanical properties is the interfacial plastic deformation which foils undergo during UAM. This plastic deformation increases the hardness of the metal as a result of work hardening effects. Although this work hardening improves the strength, it has a negative effect on ductility.

9.4.5 UAM Applications

UAM provides unique opportunities for manufacture of structures with complex internal geometries, manufacture of structures from multiple materials, fiber embedment during manufacture, and embedding of electronics and other features to form smart structures. Each of these application areas is discussed below.

9.4.5.1   Internal Features

As with other AM techniques, UAM is capable of producing complex internal features within metallic materials. These include honeycomb structures, internal pipes or channels, and enclosed cavities. During UAM, internal geometrical features of a part are fabricated via CNC trimming before depositing the next layer (see Fig. 9.8). Not all internal feature types are possible, and all of the "top" surface of internal features will have a stair-step geometry and not a CNC-milled surface, as the CNC can only mill the upward-facing surfaces of internal geometries. After fabrication of an internal feature is completed, metal foils are placed over the cavities or channels and welded, thus enclosing the internal features.

It has been shown that it becomes quite difficult to bond parts using UAM when their height-to-width ratio is near 1:1 [16]. In order to achieve higher ratios, support materials or other restraints are necessary to make the part rigid enough such that there is differential motion between the existing part and the foils that are being added. The development of an effective support material dispensing system for UAM would dramatically increase its ability to make more freeform shapes and larger internal features. Without support materials, internal features must be designed and oriented in such a way that the sonotrode is always supported by an existing, rigid feature while depositing a subsequent layer. As a result, for instance, internal cooling channels cannot be perpendicular to the sonotrode traveling direction, and honeycomb structures must be small enough that there are always at least two ribs supporting the deposition of the foil face sheets.

9.4.5.2   Material Flexibility

A wide range of metallic materials has been used with UAM. Theoretically, any metal which can be ultrasonically welded is a candidate material for the UAM process. Materials which have been successfully bonded using UAM include Al 3003 (H18 and O condition), Al 6061, Al 2024, Inconel® 600, brass, SS 316, SS 347, Ni 201, and high purity copper. Ultrasonic weldabilities of a number of other metallic materials have been widely demonstrated [15, 17–20]. Thus, there is significant material flexibility for UAM processes. In addition to metal foils, other materials have been used, including MetPreg® (an alumina fiber-reinforced Al matrix composite tape) and prewoven stainless steel AISI 304 wire meshes (see Fig. 9.10), which both have been bonded to Al 3003 using UAM.

By depositing various metal foils at different desired layers or locations during UAM, multi-material structures or functionally gradient materials can be produced. Composition variation and resultant property changes can be designed to meet various application needs. For instance, by changing materials it is possible to optimize thermal conductivity, wear resistance, strength, ductility, and other properties at specific locations within a part.

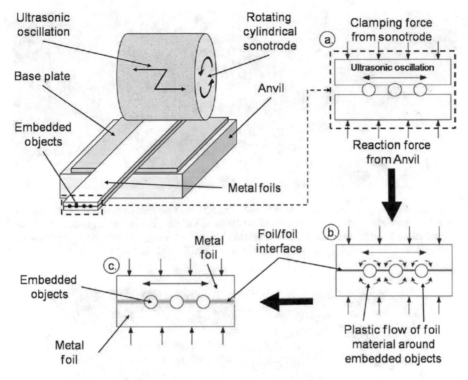

Fig. 9.16   Schematic diagram of the Ultrasonic Additive Manufacturing process for object embedment (Elsevier license number 4571741199346) [21]

9.4.5.3   Fiber/Object Embedment

A significant advantage of UAM over other metal AM methods is its low operating temperature. UAM also induces high plastic flow in the metal foils as they are being deposited. This combination enables UAM to embed objects inside the metal matrix without melting the objects. Figure 9.16 shows how UAM embeds small objects that are placed between adjacent layers during deposition. To embed larger objects, or to ensure precise placement of smaller features, a pocket or groove must first be milled into the prior layer(s) before placement of the embedded object.

UAM is particularly suited to fiber embedment. As can be seen in Fig. 9.17, bonding near an embedded fiber is much better than bonding away from the fiber for a particular set of process parameter conditions. Plastic flow predicted by modeling done at Sheffield University by Mariani and Ghassemieh (2009) has shown that in some cases there can be one hundred times the degree of interfacial metal flow in the presence of a fiber when compared to bonding of foils without a fiber.

The most commonly embedded fibers are silicon carbide structural fibers within Al matrices (thus forming an Al/SiC metal matrix composite) and optical fibers within Al matrices. Fibers can also be placed and embedded between dissimilar materials, as seen in Fig. 9.18. In the case of dissimilar materials, the presence of a

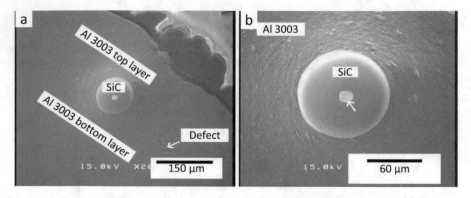

Fig. 9.17 SEM microstructures of Al 3003/SiC: (**a**) SiC fiber embedded between Al 3003 layers showing a lack of defects near the fiber; and (**b**) the same SiC fiber at a higher magnification showing excellent bonding near the fiber. © Emerald Group Publishing Limited, "Use of Ultrasonic Consolidation for Fabrication of Multi-Material Structures," Janaki Ram et al. [15]

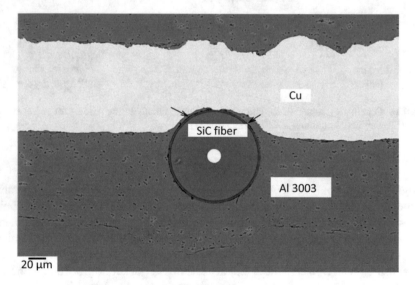

Fig. 9.18 SiC Fiber embedded between copper and aluminum using UAM. Black arrows denote regions where the softer Al extruded around the fiber during embedment, resulting in displacement of the fiber away from the interface into the Al base material

stiff fiber exacerbates the plastic deformation between the stiffer and less stiff material, causing the material with a lower flow stress to deform more than the higher flow stress material. In addition, in contrast to the case of embedment between similar materials where the fiber center is typically aligned with the foil interfaces, the fiber is offset into the softer material (compare Figs. 9.17 and 9.18).

Embedded ceramic fibers are typically mechanically entrapped within metal matrices, without any chemical bonding between fiber and matrix materials. As a result of this mechanical entrapment, friction aids in the transfer of tensile loads from the matrix to the fiber, thus strengthening the part, whereas the lack of chemical bonding means that there is little resistance to shear loading at the fiber/matrix interface, thus weakening the structure for this failure mode.

During plastic deformation grain boundaries prevent dislocation movements which leads to "work hardening" (an increase in mechanical properties such as strength, stiffness, and yield stress). To enhance dislocation motion and alleviate these effects, energy is often added during deformation, in the form of heat. Langenecker [23] showed how both heat and ultrasonic excitation can provide dislocation enhancing energy, by generating stress-strain curves for static tensile tests in which samples were subject to ultrasonic excitation at 20 kHz and compared to samples without ultrasound energy applied at various temperatures. Figure 9.19 shows that ultrasonic excitation at various energy densities at room temperature can produce softening identical to samples formed at a higher temperature without application of ultrasound energy.

UAM is a candidate manufacturing process for fabrication of long-fiber-reinforced metal matrix composites (MMC). However, to utilize UAM to make end-use MMC parts, several technical difficulties need to be overcome, including automatic fiber feeding and alignment mechanisms, and the ability to change the fiber/foil direction between layers.

Optical fibers have been successfully embedded by many researchers worldwide. Since UAM operates at relatively low processing temperatures, many types of opti-

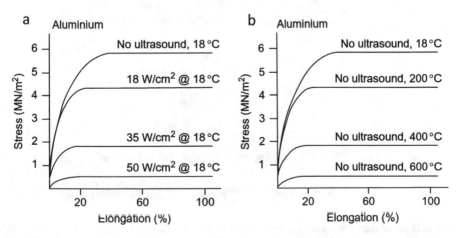

Fig. 9.19 Comparison of the equivalent energy requirements for ultrasonic vs. thermal energy (Elsevier license number 4571741199346) [21, 23]

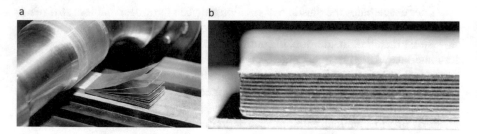

Fig. 9.20 UAM manufactured functional graded laminate that is created through the alternate layering of Cu (darker layer) and Al (lighter layer) foil materials bonded in the solid state. (**a**) The laminate during UAM and (**b**) a close-up of the final Cu/Al functionally graded laminate. (Elsevier license number 4571741199346) [21]

cal fibers can be deposited without damage, thus enabling data and energy to be optically transported through the metal structure. Fig. 9.20 shows a UAM functionally graded laminate comprising Cu and Al sheets.

9.4.5.4 Smart Structures

Smart structures are structures which can sense, transmit, control, and/or react to data, such as environmental conditions. In a smart structure, sensors, actuators, processors, thermal management devices, and more can be integrated to achieve a desired functionality (see Fig. 9.21). Fabrication of smart structures is difficult for

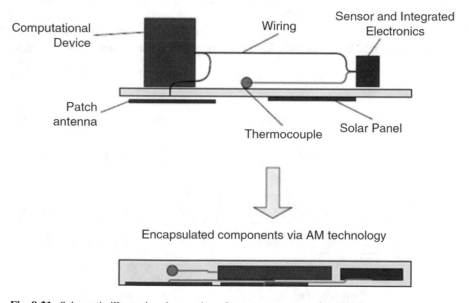

Fig. 9.21 Schematic illustrating the creation of a smart structure using UAM

conventional manufacturing processes, as they do not enable full three-dimensional control over geometry, composition, and/or placement of components. AM processes are inherently suited to the fabrication of smart structures, and UAM, in particular, offers several advantages. Since UAM is the only AM process whereby metal structures can be formed at low temperatures, UAM offers excellent processing capability for fabrication of smart structures. In addition to traditional internal self-supporting features (honeycomb structures, cooling channels, etc.), larger internal cavities can designed to enable placement of electronics, actuators, heat pipes, or other features at optimum location within a structure [24]. Many types of embedded electronics, sensors, and thermal management devices have been inserted into UAM cavities. Sensors for recording temperature, acceleration, stress, strain, magnetism, and other environmental factors have been fully encapsulated and have remained functional after UAM embedment. In addition to prefabricated electronics, it is feasible to fabricate customized electronics in UAM with the integration of Direct Write (DW) technologies (see Chap. 11). By combining UAM with DW, electronic features (conductors, insulators, batteries, capacitors, etc.) can be directly created within or on UAM-made structures in an automated manner.

9.5   Sheet Lamination Benefits and Drawbacks

SHL has some benefits compared to other AM processes, such as speed, low cost, and ease of material handling. In most AM processes, the entire layer must be formed by motion of an extrusion nozzle, laser, or other deposition/energy source everywhere within the layer. In SHL, layer formation can be very fast due to the fact that cutting of only the outline and not the entire cross-sectional area is done.

Sheet Lamination of paper can produce full-color prints inexpensively since the raw material is normal office paper and color is added via normal inkjet printing. Excess paper can be recycled with low environmental issues.

UAM can create metal parts and has the ability to create functionally gradient parts in separate layers. UAM and other SHL are well-suited to embed artifacts into the part like wires, sensors, fibers, or computational devices.

Bonding between sheets of material is an area of needed research. The strength of parts made via SHL are highly dependent upon the strength of the bond between sheets. Better adhesives, more effective fusion, and other bonding techniques are areas which require significant research to further advance the process into a more mainstream technology.

A disadvantage of SHL is that an entire sheet is consumed per layer, so material waste can be high if parts don't make full use of the build volume. When layers are bonded before the 2D layer is cut out, the removal of excess material can be quite labor-intensive. Certain geometries, such as lattice structures, may be impossible to build since excess material may not be removable. Mechanical properties are highly anisotropic, with weak interlayer bonding normal to the build plane often resulting in delamination. Furthermore, many reports of safety concerns due to a chance of fire when cutting paper and other combustible materials using lasers have been reported [25].

9.6 Commercial Trends

At the time of this publication (2020) companies which produce SHL machines are having trouble with market growth. For instance, although Fabrisonic is successfully deploying UAM services and selling some machines, market adoption of UAM benefits of dissimilar material bonding, object embedding, etc. has been slow. Future development of UAM technology may involve benefits of micromachining and ultrasonic bonding to provide fast production with good surface quality and dimensional accuracy. In addition, it appears like the layer-by-layer fabrication and encapsulation of fully integrated functional materials enabling production of components with circuitry and sensors included within a dissimilar metal matrix are continuing to gain traction, although slowly [21].

Most companies which were formed to commercialize SHL techniques have gone out of business or abandoned this technology. Although researchers have illustrated that many sheet types, including paper, ceramic, metal, polymer, and composite materials, can be successfully made into 3D objects using SHL, the commercial lack of success makes SHL primarily a niche AM technology today.

9.7 Summary

As illustrated in this chapter, a broad range of SHL techniques exist. From the initial LOM paper-based technology to UAM, SHL processes have shown themselves to be robust, flexible, and capable of producing many types of applications and materials. The basic method of trimming a sheet of material to form a cross-sectional layer is inherently fast, as trimming only occurs at the layer's outline rather than needing to melt or cure the entire cross-sectional area to form a layer. This means that SHL approaches exhibit the speed benefits of a layer-wise process while still utilizing a point-wise energy source. Many variations of SHL processes have been demonstrated, which have proved to be suitable for many different types of metal, ceramic, polymer, and paper materials.

Future variations of SHL techniques will likely include better materials, new bonding methods, novel support material strategies, new sheet placement mechanisms, and new forming/cutting techniques. As these developments occur, SHL techniques will likely move from the fringe of AM to a more central role in the future for an increasing number of products.

9.8   Questions

1. Discuss the benefits and drawbacks of bond-then-form versus form-then-bond approaches. In your discussion, include discussion of processes which can use secondary support material and those which do not.
2. Find four papers not mentioned in the references to this chapter which discuss the creation of tooling from laminated sheets of metal. Discuss the primary benefits and drawbacks identified in these papers to this approach to tooling. Based upon this, what do you think about the commercial viability of this approach?
3. Find three examples where SDM was used to make a complex component. What about this approach proved to be useful for these components? How might these beneficial principles be better applied to AM today?
4. What are the primary benefits and drawbacks of UAM compared to other metal AM processes? Discuss UAM and at least three other metal AM processes in your comparison.
5. Compare and contrast several different machine architectures for paper SHL processes, including new architectures you brainstorm. Start with the Helisys and Mcor Technologies examples. Investigate form-then-bond and bond-then-form approaches. Include inkjet printing capability for color part fabrication. Evaluate the pros and cons of each technology and compare with the machine architectures of commercial machines, if you can find them.
6. What is an application for Ultrasonic Additive Manufacturing that is not possible using other metal AM techniques?
7. Describe three different types of bonding in SHL processes and when these types of bonding would be used.
8. Why do UAM structures have a lower residual stress compared to metal PBF and DED processes?
9. Describe the unique microstructures and mechanical properties of UAM parts. In your explanation, also note the primary disadvantages caused by defects in UAM.
10. Find a recent reference (within the past 2 years) where a manufacturer used metal Sheet Lamination for molds. What was their reason for doing so?
11. What are the two conditions that must be fulfilled for solid-state bonding to occur during UAM?
12. If we use UAM to make a part from Al3003 foils, which kind of post-processing might be useful to improve its mechanical properties?
13. Why is UAM research commonly combined with embedded electronics research, whereas this combination is less common in other metal AM processes?
14. What are design limitation for fabricating internal structures via UAM?
15. Use a ruler to estimate the linear weld density (LWD) in Figure Q9.1 below:

$$\%LWD = \frac{\text{Bonded interface length}}{\text{Total interface length}} \times 100$$

Fig. Q 9.1 A UAM part made from four layers of Al 3003 foils. LWD is determined by calculating the bonded interface divided by the total interface (arrows show the sonotrode traveling direction for each layer). (Elsevier license number 4841900933578) [11]

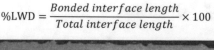

$$\%LWD = \frac{Bonded\ interface\ length}{Total\ interface\ length} \times 100$$

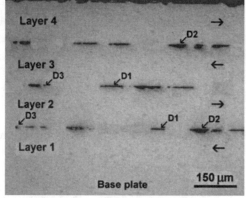

References

1. Wimpenny, D. I., Bryden, B., & Pashby, I. R. (2003). Rapid laminated tooling. *Journal of Materials Processing Technology, 138*(1–3), 214–218.
2. Solido. (2020). http://www.solido3d.com/
3. Ennex Corp. (2020). https://www.ennex.com/
4. Stratoconception. (2020). www.stratoconception.com
5. Yi, S., et al. (2004). Study of the key technologies of LOM for functional metal parts. *Journal of Materials Processing Technology, 150*(1–2), 175–181.
6. Himmer, T., Nakagawa, T., & Anzai, M. (1999). Lamination of metal sheets. *Computers in Industry, 39*(1), 27–33.
7. Himmer, T., et al. (2004). Metal laminated tooling–a quick and flexible tooling concept. In *Proceedings of the solid freeform fabrication symposium*. Austin, TX.
8. Obikawa, T. (1998). Rapid manufacturing system by sheet steel lamination. In *Proceeding of International Conference on Computer aided Production Engineering*.
9. Yamasaki, H. (2000). Applying laminated die to manufacture automobile part in large size. *Die Mould Technology, 15*(7), 36–45.
10. Prinz, F. B., & Weiss, L. E. (1998). Novel applications and implementations of shape deposition manufacturing. *Naval Research Reviews, 50*, 19–26.
11. Ram, G. J., Yang, Y., & Stucker, B. (2006). Effect of process parameters on bond formation during ultrasonic consolidation of aluminum alloy 3003. *Journal of Manufacturing Systems, 25*(3), 221.
12. Blaha, F., & Langenecker, B. (1966). Plasticity test on metal crystals in an ultrasonic field. *Acta Metallurgica, 7*, 93–100.
13. Barber, J. R. (2002). *Elasticity*. Springer.
14. Adams, B. L., et al. (2008). Accessing the elastic–plastic properties closure by rotation and lamination. *Acta Materialia, 56*(1), 128–139.
15. Janaki Ram, G., et al. (2007). Use of ultrasonic consolidation for fabrication of multi-material structures. *Rapid Prototyping Journal, 13*(4), 226–235.
16. Robinson, C., et al. (2006). Maximum height to width ratio of freestanding structures built using ultrasonic consolidation. In *Seventeeth Solid Freeform Fabrication Proceedings*.

17. Flood, G. (1997). Ultrasonic energy welds copper to aluminium. *Welding Journal, 76*(1), 43–47.
18. Gunduz, I. E., et al. (2005). Enhanced diffusion and phase transformations during ultrasonic welding of zinc and aluminum. *Scripta Materialia, 52*(9), 939–943.
19. Joshi, K. C. (1971). The formation of ultrasonic bonds between metals. *Welding Journal, 50*(12), 840–848.
20. Weare, N. (1960). *Fundamental studies of ultrasonic welding. Welding Journal, 37*(84), 331s–341s.
21. Friel, R. J. (2015). 13 - Power ultrasonics for additive manufacturing and consolidating of materials. In J. A. Gallego-Juárez & K. F. Graff (Eds.), *Power Ultrasonics* (pp. 313–335). Oxford: Woodhead Publishing.
22. Yang, Y., Janaki Ram, G., & Stucker, B. E. (2010). An analytical energy model for metal foil deposition in ultrasonic consolidation. *Rapid Prototyping Journal, 16*(1), 20–28.
23. Langenecker, B. (1966). Effects of ultrasound on deformation characteristics of metals. *IEEE Transactions on Sonics and Ultrasonics, 13*(1), 1–8.
24. Robinson, C., et al. (2007). *Fabrication of a Mini-SAR Antenna Array using Ultrasonic Consolidation and Direct-Write*. Albuquerque: Sandia National Lab.(SNL-NM).
25. Sheet Lamination. (2020). http://canadamakes.ca/what-is-sheet-lamination/

Chapter 10
Directed Energy Deposition

Abstract Directed Energy Deposition (DED) is a method for melting material as it is being deposited layer-by-layer. Material in wire or powder form is delivered along with the energy required to melt it. Although it has been shown that a number of material types can be processed this way, DED is almost exclusively applied to metals in both research and commercialized instantiations. DED presents unique advantages and disadvantages that make it particularly suited for repair and feature addition to an existing part. DED is gaining industrial interest because of its ability to create near-net-shape, large freeform structures more quickly and inexpensively than traditionally made near-net-shape castings and forgings.

10.1 Introduction

Directed Energy Deposition processes enable the creation of parts by melting material as it is being deposited. Although this basic approach can work for polymers, ceramics, and metal matrix composites, it is predominantly used for metal powders. Thus, this technology is often referred to as "metal deposition" technology.

DED processes direct energy into a narrow, focused region to heat a substrate, melting the substrate and simultaneously melting material that is being deposited into the substrate's melt pool. Unlike Powder Bed Fusion (PBF) techniques (see Chap. 5), DED processes are *not* used to melt a material that is pre-laid in a powder bed but are used to *melt materials as they are being deposited*.

DED processes use a focused heat source (typically a laser or electron beam) to melt the feedstock material and build up three-dimensional objects in a manner similar to the Material Extrusion (MEX) processes from Chap. 6. Each pass of the DED head creates a track of solidified material, and adjacent lines of material make up layers. Complex three-dimensional geometry requires either support material or a multi-axis deposition head. A schematic representation of a DED process using powder feedstock material and a laser is shown in Fig. 10.1.

Commercial DED processes include using a laser or electron beam to melt powders or wires. In many ways, DED techniques can be used in an identical manner to

I. Gibson et al., *Additive Manufacturing Technologies*,
https://doi.org/10.1007/978-3-030-56127-7_10

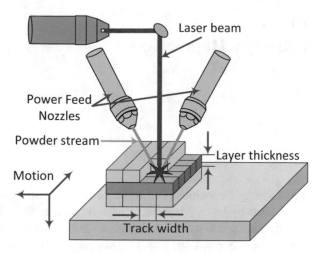

Fig. 10.1 Schematic of a typical laser powder DED process

laser cladding and plasma welding machines. For the purposes of this chapter, however, DED machines are considered those which are designed to create complex 3D shapes directly from CAD files, rather than the traditional welding and cladding technologies, which were designed for joining or to apply coatings and do not typically use 3D CAD data as an input format.

A number of organizations have developed DED machines using lasers and powder feeders. These machines have been referred to as Laser Engineered Net Shaping (LENS) [1], Directed Light Fabrication (DLF) [2], Direct Metal Deposition (DMD), 3D Laser Cladding, Laser Generation, Laser-Based Metal Deposition (LBMD), Laser Freeform Fabrication (LFF), Laser Direct Casting, LaserCast [3], Laser Consolidation, LASFORM, and others. Although the general approach is the same, differences between these machines commonly include changes in laser power, laser spot size, laser type, powder delivery method, inert gas delivery method, feedback control scheme, and/or the type of motion control utilized. Because these processes all involve deposition, melting, and solidification of powdered material or wire feedstock using a traveling melt pool, the resulting parts attain a high density during the build process. The microstructure of parts made from DED processes (Figs. 10.2 and 10.3) are similar to PBF processes (see Fig. 5.18), wherein each pass of the laser or heat source creates a track of rapidly solidified material.

As can be seen from Figs. 10.2 and 10.3, the microstructure of a DED part can be different between layers and even within layers. In the Ti/TiC deposit shown in Fig. 10.2, the larger particles present in the microstructure are unmelted carbides. The presence of fewer unmelted carbides in a particular region is due to a higher overall heat input for that region of the melt pool. By changing process parameters, it is possible to create fewer or more unmelted carbides within a layer, and by increasing laser power, for instance, a greater amount of the previously deposited layer (or substrate for the first layer) will be remelted. By comparing the thickness of the last-deposited layer with the first- or second-deposited layer (such as in

Fig. 10.2 LENS-deposited
Ti/TiC metal matrix
composite structure
(four layers on top of a Ti
substrate)

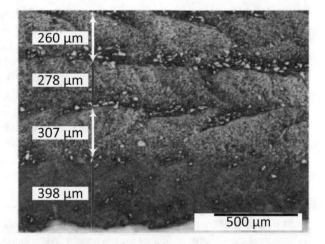

Fig. 10.3 CoCrMo deposit on CoCrMo: (**a**) side view (every other layer is deposited perpendicular to the previous layer using a 0,90,0 pattern); and (**b**) top view of deposit

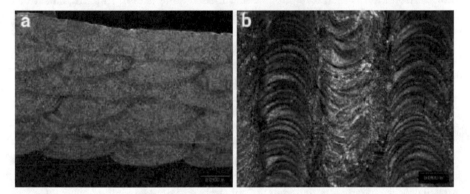

Fig. 10.3a), an estimate of the proportion of a layer that is remelted during subsequent deposition can be made. Each of these issues is discussed in the following section.

10.2   General Directed Energy Deposition Process Description

As the most common type of DED system is a powder-based laser deposition system optimized for metals, we will use a typical laser powder deposition (LPD) process as the paradigm process against which other DED processes will be compared. In LPD, a "deposition head" is utilized to deposit material onto the substrate. A deposition head is typically an integrated collection of laser optics, powder nozzle(s), inert gas tubing, and, in some cases, sensors. The substrate can be either a flat plate on which a new part will be fabricated or an existing part onto which additional geometry will be added. Deposition is controlled by relative differential motion between the substrate and deposition head. This differential motion is accomplished by moving the deposition head, by moving the substrate, or by a combination of substrate and deposition head motion. 3-axis systems, whereby the deposition occurs in a vertical manner, are typical. However, 4- or 5-axis systems using either rotary tables or robotic arms are also available. In addition, several companies sell LPD deposition heads as "tools" for inclusion in multi-tool-changer CNC milling machines. By integration into a CNC milling machine, a LPD head can enable additive plus subtractive capabilities in one apparatus. This is particularly useful for overhaul and repair (as discussed below).

The kinetic energy of powder particles being fed from a powder nozzle into the melt pool is greater than the effect of gravity on powders during flight. As a result, nonvertical deposition is just as effective as vertical deposition. Multi-axis deposition head motion is therefore possible and indeed quite useful. In particular, if the substrate is very large and/or heavy, it is easier to accurately control the motion of the deposition head than the substrate. Conversely, if the substrate is a simple flat plate, it is easier to move the substrate than the deposition head. Thus, depending on the geometries desired and whether new parts will be fabricated onto flat plates or new geometry will be added to existing parts, the optimum design of a LPD apparatus will change.

In LPD, the laser generates a small molten pool (typically 0.25–1 mm in diameter and 0.1–0.75 mm in depth) on the substrate as powder is injected into the pool. The powder is melted as it enters the pool and solidifies as the laser beam moves away. Under some conditions, the powder can be melted during flight and arrive at the substrate in a molten state; however, this is atypical and the normal procedure is to use process parameters that melt the substrate and powder as they enter the molten pool.

The typical small molten pool and relatively rapid traverse speed combine to produce very high cooling rates (typically 10^3–10^5 °C/s) and large thermal gradients. Depending upon the material or alloy being deposited, these high cooling rates can produce unique solidification grain structures and/or nonequilibrium grain structures which are not possible using traditional processing. At lower cooling rates, such as when using higher beam powers or lower traverse speeds – which is typical when using electron beams rather than lasers for DED – the grain features grow and look more like cast grain structures.

The passing of the beam creates a thin track of solidified metal deposited on and welded to the layer below. A layer is generated by a number of consecutive overlapping tracks. The amount of track overlap is typically 25% of the track width (which results in remelting of previously deposited material), and typical layer thicknesses employed are 0.05–0.5 mm. After each layer is formed, the deposition head moves away from the substrate by one layer thickness.

10.3   Material Delivery

DED processes can utilize both powder and wire feedstock material. Each has limitations and drawbacks with respect to each other.

10.3.1   Powder Feeding

Powder is the most versatile feedstock, and most metal and ceramic materials are readily available in powder form. However, not all powder is captured in the melt pool (e.g., less than 100% powder capture efficiency), so excess powder is utilized. Care must be taken to ensure excess powder is recaptured in a clean state if recycling is desired.

Excess powder feeding is not necessarily a negative attribute, as it makes DED processes geometrically flexible and forgiving. This is due to the fact that excess powder flow enables the melt pool size to dynamically change. As described below, DED processes using powder feeding can enable overlapping scan lines to be used without the swelling or overfeeding problems inherent in MEX (discussed in Sect. 6.3).

In DED, the energy density of the beam must be above a critical amount to form a melt pool on the substrate. When a laser is focused to a small spot size, there is a region above and below the focal plane where the laser energy density is high enough to form a melt pool. This region is labeled in Fig. 10.4. If the substrate surface is either too far above or too far below the focal plane, no melt pool will form.

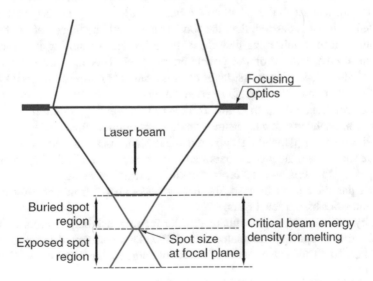

Fig. 10.4 Schematic illustrating laser optics and energy density terminology for DED

Similarly, the melt pool will not grow to a height that moves the surface of the melt pool outside this region.

Within this critical beam energy density region, the height and volume of the deposit melt pool is dependent upon melt pool location with respect to the focal plane, scan rate, laser power, powder flow rate, and surface morphology. Thus, for a given set of parameters, the deposit height approaches the layer thickness offset value only after a number of layers of deposition. This is evident, for instance, in Fig. 10.2, where a constant layer thickness of 200 μm was used as the deposition head z-offset for each layer. The substrate was initially located within the buried spot region, but not far enough within it to achieve the desired thickness for the layers shown (i.e., the laser power, scan rate, and powder flow settings caused the deposit to be thicker than the layer thickness specified). Thus, deposit thickness approached the layer thickness z-offset as the spot became effectively more "buried" during each subsequent layer addition. In Fig. 10.2, however, too few layers were deposited to reach the steady-state layer thickness value.

If the laser and scanning parameters settings used are inherently incapable of producing a deposit thickness at least as thick as the layer thickness z-offset value, subsequent layers will become thinner and thinner. Eventually, no deposit will occur when scanning for the next layer starts outside the critical energy density region (i.e., when the substrate starts out below the exposed spot region, there is insufficient energy density to form a melt pool on the substrate).

In practice, when the first layer is formed on a substrate, the laser focal plane is typically buried below the surface of the substrate approximately 1 mm. In this way, a portion of the substrate material is melted and becomes a part of the melt pool. The first layer, in this case, will be made up of a mixture of melted substrate combined with material from the powder feeders, and the amount of material added to the surface for the first layer is dependent upon process parameters and focal plane location with respect to the substrate surface. If little mixing of the substrate and deposited material is desired, then the focal plane should be placed at or above the substrate surface to minimize melting of the substrate – resulting in a melt pool made up almost entirely of the powdered material. This may be desirable, for instance, when depositing a first layer of "material A" on top of "material B" that might form "intermetallic AB" if mixed in a molten state. In order to suppress intermetallic formation, a sharp transition from A to B is typically required.

In summary, the first few layers may be thicker or thinner than the layer thickness set by the operator, depending upon the focal plane location with respect to the substrate surface and the process parameters chosen. As a result, the layer thickness converges to the steady-state layer thickness setting after several layers or, if improper parameters are utilized, the laser "walks away" from the substrate and deposition stops after a few layers.

The dynamic thickness benefits of powder feeding also help overcome the corrugated surface topology associated with DED. This corrugated topology can be seen in Fig. 10.3b and is a remnant of the set of parallel, deposited tracks (beads) of material which make up a layer. As in MEX, in DED a subsequent layer is typically deposited in a different orientation than the previous layer. Common scan patterns from layer to layer are typically multiples of 30, 45, and 90 degrees (e.g., 0, 90, 0,

90...; 0, 90, 180, 270, 360...; 0, 45, 90...315, 360...; and 0, 30, 60...330, 360...).
Layer orientations can also be randomized between layers at pre-set multiples. The
main benefits of changing orientation from layer-to-layer are the elimination of
preferential grain growth (which otherwise makes the properties anisotropic) and
minimization of residual stresses.

Changing orientation between layers can be accomplished easily when using
powders, as the presence of excess powder flow provides for dynamic leveling of
the deposit thickness and melt pool at each region of the deposited layer. This means
that powdered material feedstock allows the melt pool size to dynamically change
to fill the bottoms of the corrugated texture without growing too thick at the top of
each corrugation. This is not as easy for wire feeding.

Powder is typically fed by first fluidizing a container of powder material (by bub-
bling up a gas through the powder and/or applying ultrasonic vibration) and then
using a pressure drop to transfer the fluidized powder from the container to the laser
head through tubing. Powder is focused at the substrate/laser interaction zone using
either co-axial feeding, 4-nozzle feeding, or single nozzle feeding. In the case of
co-axial feeding, the powder is introduced as a toroid surrounding the laser beam,
which is focused to a small spot size using shielding gas flow, as illustrated in
Fig. 10.5a. The two main benefits of co-axial feeding are that it enables a higher
capture efficiency of powder, and the focusing shielding gas can protect the melt
pool from oxidation when depositing in the presence of air. Single nozzle feeding
involves a single nozzle pointed at the interaction zone between the laser and substrate. The main benefits of single nozzle feeding are the apparatus simplicity (and
thus lower cost), a better powder capture efficiency than 4-nozzle feeding, and the
ability to deposit material into tight locations (such as when adding material to the
inside of a channel or tube). The main drawback of single nozzle feeding is that the
melt pool geometry is direction-specific (i.e. the melt pool is different when feeding
toward the nozzle, versus away from the nozzle or at right angles to the nozzle).
4-nozzle feeding involves 4 separate nozzle heads equally spaced at 90 degree
increments around the laser beam, focused to intersect at the melt pool. The
main benefit of a 4-nozzle feeding system is that the flow characteristics of 4-nozzle

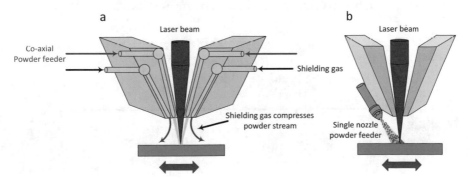

Fig. 10.5 Illustration of powder nozzle configurations: (**a**) co-axial nozzle feeding; and (**b**) single-
nozzle feeding

feeding gives more consistency in build height for complex and arbitrary 3D geometries that involve combinations of thick and thin regions.

10.3.2 Wire Feeding

In the case of wire feeding, the volume of the deposit is always the volume of the wire that has been fed, and there is 100% feedstock capture efficiency (minus a little "splatter" from the melt pool). Wires are most effective for simple geometries, "blocky" geometries without many thin/thick transitions, or for coating of surfaces. When complex, large, and/or fully dense parts are desired, geometry-related process parameters (such as hatch width, layer thickness, wire diameter, and wire feed rate) must be carefully controlled to achieve a proper deposit size and shape. Just as in MEX, large deposits with geometric complexity must have porosity designed into them to remain geometrically accurate.

For certain geometries, it is not possible to control the geometry-related process parameters accurately enough to achieve both high accuracy and low porosity with a wire feeder unless periodic subtractive processing (such as CNC machining) is done to reset the geometry to a known state. For most applications of DED low porosity is more important than geometric accuracy. Thus, wire-based DED scan processes are designed to be pore-free, at the sacrifice of dimensional accuracy. Thus, the selection of a wire feeding system versus a powder feeding system is best done after determining what type of deposit geometries is required, whether dimensional accuracy is critical, and whether a subtractive milling system will be integrated with the additive deposition head.

10.4 DED Systems

Based on the source of energy, DED systems are divided into four main categories comprising laser, friction stir, electron beam, and arc as shown in Fig. 10.6.

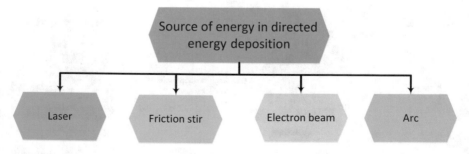

Fig. 10.6 Source of energy for DED processes

10.4.1 Laser Powder Deposition Processes

Low volume powder feed systems are typified by the Laser Engineered Net Shaping (LENS) process where parts are built on a substrate with the possibility of multiple powder feed systems (to create multi-material components). High volume powder feed systems aim to deliver larger volumes of material producing thick layers, which increases deposition rate but negatively affects the accuracy of parts. Much of DED technology is based around laser cladding and all of these technologies are suitable for multi-axis approaches where layers need not be planar, either by mounting on a robot arm or providing a tiltable rotating platform.

10.4.1.1 Laser Engineered Net Shaping

One of the first commercialized DED processes, Laser Engineered Net Shaping, was developed by Sandia National Laboratories, USA, and commercialized by Optomec, USA. Optomec's "LENS 750" machine was launched in 1997. Subsequently, the company launched numerous improvements, offering machines with build volumes up to 900 x 1500 x 900 mm and multi-laser head capability. Optomec's machines originally used an Nd-YAG laser, but more recent machines utilize fiber lasers.

LENS machines process materials in an enclosed inert gas chamber (*see* Fig. 10.7). An oxygen removal, gas recirculation system is used to keep the oxygen concentration in the gas (typically argon) near or below 10 ppm oxygen. This is several orders of magnitude cleaner than the inert gas systems used in PBF machines. The inert gas chamber, laser type, and 4-nozzle feeder design utilized by Optomec make their LENS machines some of the most flexible platforms for DED, as many materials can be effectively processed with this combination of laser type and atmospheric conditions. Most LENS machines are 3-axis and do not use closed-loop feedback control;

Fig. 10.7 Optomec
LENS® CS 800 system.
(Photo courtesy of
Optomec)

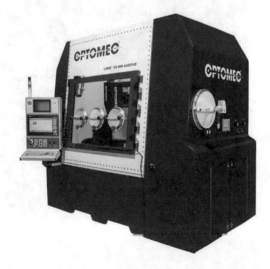

however 5-axis "laser wrist" systems can enable deposition from any orientation, and systems for monitoring build height and melt pool area can be used to dynamically change process parameters to maintain constant deposition characteristics.

POM Group, USA, was an early company building LPD machines, subsequently acquired by DM3D Technology. Their DED machines with 5-axis, co-axial powder feed capability build parts using a shielding gas approach. A key feature of DM3D machines has always been the integrated closed-loop control system (*see* Fig. 10.8). The feedback control system adjusts process variables such as powder flow rate, deposition velocity, and laser power to maintain deposit conditions. Closed loop control of DED systems has been shown to be effective at maintaining build quality. Thus, not only do DM3D Technology and Optomec machines offer this option, but so do competing LPD machine manufacturers as well as companies building electron beam DED machines that utilize wire feeders. DM3D machines have historically utilized CO_2 lasers, which have the benefit of being an economical, high-powered heat source. But the absorptivity of most materials is much less at CO_2 laser wavelengths than for Nd-YAG or fiber lasers (as discussed in Chap. 5 and shown in Fig. 5.14), and thus almost all new DED machines now utilize fiber, diode, or Nd-YAG lasers. For machines where CO_2 lasers are still used, in order to compensate for their lower absorptivity, a larger amount of laser energy is applied, resulting in a larger heat affected zone and overall heat input.

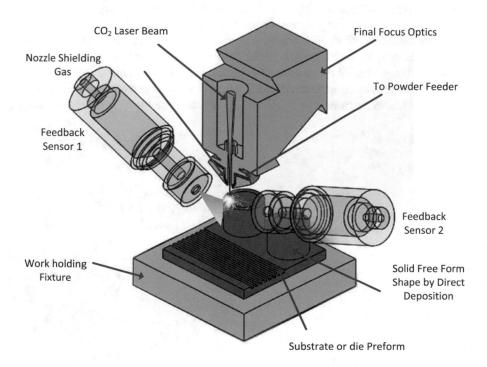

Fig. 10.8 POM DED machine schematic. (Photo courtesy of POM)

Another company which was involved early in the development of DED machines was AeroMet Inc., USA – until the division was closed in 2005. The AeroMet machine was specifically developed for producing large aerospace "rib-on-plate" components using prealloyed titanium powders and an 18 kW CO_2 laser (*see* Fig. 10.9). Although they were able to demonstrate the effectiveness of building rib-on-plate structures cost-effectively, the division was not sustainable financially and was closed. The characteristics of using such a high-powered laser are that large deposits can be made quite quickly, but at the cost of geometric precision and a much larger heat affected zone. Companies which today are interested in the high-deposition-rate characteristics of the AeroMet machine typically choose to use electron beam energy sources with wire feed, and thus there are no commercially available LPD machines similar to AeroMet's machines. It is reported [4] that China has created a similar machine to make structures for its fighter jets.

The benefits behind adding features to simple shapes to form aerospace and other structures with an otherwise poor "buy-to-fly" ratio are compelling. The term buy-to-fly refers to the amount of wrought material that is purchased as a block that is required to form a complex part. In many cases, 80% or more of the material is machined away to provide a stiff, lightweight frame for aerospace structures. By building ribs onto flat plates using DED, the amount of waste material can be reduced significantly. This has both significant cost and environmental benefits. This is also true for other geometries where small features protrude from a large object, thus requiring a significant waste of material when machined from a block. This benefit is illustrated in the electronics housing deposited using LENS on the hemispherical plate shown in Fig. 10.10.

Another example of LPD is the Laser Consolidation process from Accufusion, Canada. The key features of this process are the small spot-size laser, accurate

Fig. 10.9 AeroMet System. (Photo courtesy of MTS Systems Corp)

Fig. 10.10 Electronics
Housing in 316SS. (Photo
courtesy of Optomec and
Sandia National
Laboratories)

motion control, and single-nozzle powder feeding. This enables the creation of
small parts with much better accuracy and surface finish than other DED processes,
but with the drawback of a significantly lower deposition rate.

Controlled Metal Buildup (CMB) is a hybrid metal deposition process developed
by the Fraunhofer Institute for Production Technology, Germany. It illustrates an
integrated additive and subtractive manufacturing approach that a number of
research organizations and companies are developing around the globe. In CMB, a
diode laser beam is used, and the build material is introduced in the form of a wire.
After depositing a layer, it is shaped to the corresponding slice contour by a high-
speed milling cutter. The use of milling after each deposited layer eliminates the
geometric drawbacks of a wire feeder and enables highly accurate parts to be built.
The process has been applied primarily to weld repairs and modifications to tools
and dies. Several DED machine manufacturers now offer their LPD deposition
heads for integration into subtractive machine tools, a market pioneered by Hybrid
Manufacturing Technologies, which designs, builds, and integrates LPD heads into
existing CNC machine tool platforms.

10.4.1.2   DED Process Terminology

Many research organizations and companies have created their own terminology for
their DED machines. Table 10.1 shows many of these names.

10.4.1.3   Laser Cladding

Cladding is bonding of dissimilar or similar metals to achieve different surface and
mechanical properties. Bonding processes in cladding include roll bonding, explosive welding, and laser bonding. All of these processes are difficult to control when
creating a 3D structure. However, they do "add" material, and the flexibility in

Table 10.1 Different
powder-based laser
deposition DED techniques

Terminology	Acronym
Direct Laser Deposition [5]	DLD
Directed light Fabrication [6, 7]	DLF
Direct Metal Deposition [8]	DMD
Laser-based multi-directional metal deposition [9, 10]	LBMDMD
Laser consolidation/ freeform laser consolidation [11–13]	LC
Laser direct casting [14, 15]	LDC
Laser metal deposition [16, 17]	LMD
Laser Forming [18]	LASFORM
Laser Powder Fusion [15]	LPF
Shape Deposition Manufacturing [19]	SDM

Fig. 10.11 Cladding on shaft. (Photo courtesy of Hardchrome Engineering) [20]

material delivery makes them AM processes if they are computer-controlled systems which are driven directly by CAD data.

Laser cladding uses wire or powder as feedstock in order to coat the part or substrate to repair damaged components (due to wear, corrosion, abrasion, or breakage) or to produce a new part. A key usage of laser cladding is to improve the tribological properties of surfaces.

Hard materials such as Tungsten carbide parts with over 2000 Vickers can be used for part repair or coating. Figure 10.11 shows a typical cladding AM process which uses a large shaft as the substrate [20].

Laser cladding AM is used for different applications such as high erosion protection from liquid or gas flows, high-impact conditions, and part-to-part wear and abrasion resistance. The benefits of laser cladding are:

• Limited dilution of coating for maximum purity and performance.

- Precise, fully fused deposited layer.
- Short processing time minimizes heat spread.
- Improved component wear performance.
- Cost-effective component life extension.
- Can be used to reclaim worn or fatigued components.
- Reduces maintenance and part replacement costs.

In laser cladding AM, the build rate is typically higher than LENS, but less than Electron Beam DED due to higher energy delivery. Typical laser power, build rate, and layer thickness for laser cladding are 1–10 kW, 1–12 kg/h, and 0.2–4 mm, respectively. Due to higher layer thickness compared to LENS, the process has lower cooling rate. As a result, less brittleness, martensitic structure, etc. are observed.

10.4.2   Electron Beam Based Metal Deposition Processes

Around 2007 engineers at NASA's Langley Research Center developed Electron Beam Freeform Fabrication (EBF[3]) [21] using an electron beam gun, a dual wire feed, and computer controls to manufacture metallic structures for building parts on Earth, Mars, or the international space station (Fig. 10.12).

Using an electron beam as a thermal source and a wire feeder, EBF[3] is capable of rapid deposition under high current flows or better accuracy using slower deposition rates. The primary considerations which led to the development of EBF[3] for space-

Fig. 10.12 NASA engineers test the EBF[3] system during a parabolic flight in 2007 [22]

based applications include the following: electron beams are much more efficient at converting electrical energy into a beam than most lasers, which conserves scarce electrical resources; electron beams work effectively in a vacuum but not in the presence of inert gases and thus are well suited for the space environment; and powders are inherently difficult to contain safely in low-gravity environments, and thus wire feeding is preferred.

NASA Langley Research Center has two EBF^3 machines. One is a ground-based system that is a commercially available from Sciaky. This machine is equipped with a 42 kW, 60 kV accelerating voltage electron beam gun, a vacuum system, a positioning system, and dual wire feeders capable of independent, simultaneous operation [23]. This system can use fine or coarse wire depending on the parts to be printed or two different alloys to produce parts with compositional gradients. The gun and positioning table can be tilted in six directions which provides good degree of freedom to produce freeform surfaces. This electron beam system is within a vacuum chamber capable of 5×10^{-5} torr and measuring 2.5 m by 2 m by 2.7 m. Figure 10.13 shows ground-based and portable EBF^3 systems.

The second **EBF^3** system is portable and with a chamber size of $300 \times 300 \times 150$ mm. It includes 4-axis motion and a single wire feeder [24]. NASA Langley produced different parts by this system with complex shape transitions, unsupported overhangs, and the ability to control the process with varied wire feed angles relative to the molten pool (Fig. 10.14). However, due to surface quality of as-built EBF^3 samples, post-processing such as machining and surface treatment is normally required [25].

Sciaky, USA, has developed a number of electron beam-based DED machines which utilize wire feedstock. These Sciaky machines are built inside very large vacuum chambers and enable depositions within a build volume exceeding 6 meters in their largest dimension. These machines are excellent at building large, bulky deposits very quickly, enabling large rib-on-plate structures and other deposits to be produced (typically for aerospace applications), eliminating the long lead times

Fig. 10.13 (a) Ground based EBF^3 system at NASA Langley Research Center (b) Portable EBF^3 system at NASA Langley Research Center [23]

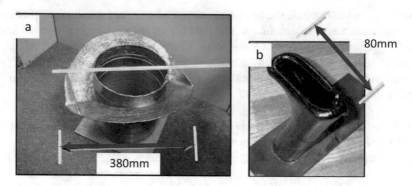

Fig. 10.14 Examples of parts fabricated at NASA Langley using the EBF process [23]

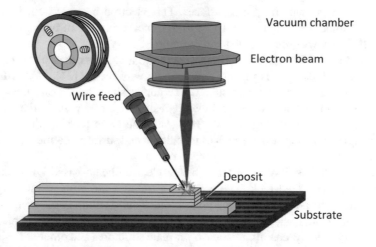

Fig. 10.15 Schematic of Electron Beam Freeform Fabrication (EBF³) system components

needed for the forged components they replace. These machines are also being investigated to produce tools for automobile manufacturing.

Figure 10.15 shows a schematic of the EBF³ process. Metal wire feedstock is moved to a molten pool that is created and sustained using a focused electron beam in a high vacuum environment (1×10^{-4} torr or lower). EBF³ has almost 100% efficiency in feedstock consumption and 95% efficiency in power usage. Another significant advantage of EBF³ is effective processing of highly reflective materials such as aluminum and copper. Many weldable and non-weldable alloys can be deposited. EBF³ has an extremely high production rate compared to laser-based systems with up to 2500 cm³/hr. (150 in³/hr). This method is limited by positioning precision and wire feed capabilities. The control factor for the smallest detail is wire feedstock, so fine diameter wires can be used for adding fine details, and larger diameter wires are suitable for higher deposition rates. EBF³ provides reliable solutions to issues of deposition rate, process efficiency, and material compatibility [23].

EBF^3 is reportedly used to produce Ti parts for vertical tails of the F-35 Strike Fighter. This DED application gives shorter lead times and less wasted materials than traditional processes for partners Lockheed Martin and Brisbane, Australia,-based Ferra Engineering.

10.4.3   Wire Arc Additive Manufacturing (WAAM)

Several research groups have investigated the use of welding and/or plasma-based technologies as a heat source for DED. Pioneering work in this area included work at Southern Methodist University, USA, utilizing gas metal arc welding combined with 4 ½-axis milling to produce three-dimensional structures. Similar work was also been demonstrated by the Korea Institute of Science and Technology, which demonstrated combined CO_2 arc welding and 5-axis milling for part production. These approaches showed arc processes to be lower-cost alternatives to laser and electron beam approaches; however, the larger heat-affected zone and other process control issues kept these approaches from widespread commercialization until recently.

Most metal AM systems, including PBF systems, have relatively low deposition rates, and part size is limited by the enclosed working envelope. These methods are best suited to small components with high complexity. The combination of an electric arc as a heat source and wire as feedstock, referred to as Wire Arc Additive Manufacturing (WAAM), is increasingly being researched as it overcomes these limitations. The first patent in this area was registered in 1925, and the first WAAM system was developed in early 1990. WAAM hardware comprises standard, off-the-shelf welding equipment: welding power source, torches, and wire feeding systems. CNC gantries or robotic systems provide the motion using MIG as the preferred process of choice because it enables wire as the consumable electrode and coaxiality with the welding torch for easier toolpath generation. MIG is capable of processing many materials, including Al and Steel. However, due to arc wandering, it is not useful for processing Ti parts. To process Ti parts, tungsten inert gas and plasma arc are better choices. Similar to other DED processes, WAAM has been shown to reduce fabrication time by 40–60% and post-machining time by 15–20% compared to traditional manufacturing processes [26].

Depending on the heat source, there are three common types of WAAM processes: Gas Metal Arc Welding (GMAW) [27], Gas Tungsten Arc Welding (GTAW) [28], and Plasma Arc Welding (PAW) [29]. A good review of various WAAM techniques are available in [26].

WAAM has similar defect characteristics to other DED techniques, including porosity, residual stress, delamination, oxidation, cracking, poor surface finish, and deformation issues [26]. To solve these problems, post-processes such as heat treatment, interpass cold rolling, peening, ultrasonic impact treatment, and interpass cooling are suggested.

10.4.3.1   Post-Process (Heat Treatment) for DED

Post-process heat, deformation, and other residual stress treatments are widely used in metal AM techniques to tailor mechanical properties and reduce residual stresses. The selection of a proper treatment process depends on the target material, Additive Manufacturing methods, working temperature, and heat treatment conditions. Heat treatment has been shown to improve the mechanical strength and enable grain refinement of WAAM components [26, 30].

Interpass Cold Rolling

Interpass cold rolling was developed by Cranfield University to reduce residual stresses and improve homogeneity without heat treatment. Thermal-based AM processes generally result in the produced part having anisotropic microstructural evolution and mechanical properties. Figure 10.16 shows an interpass cold rolling apparatus used in conjunction with a MIG torch. This apparatus significantly reduces anisotropy through plastic deformation, refines the microstructure, and enhances tensile strength in the longitudinal direction. However, interpass cold rolling does not work well for complex geometries and is not able to reduce overall part distortion in these cases. As such interpass cold rolling is well-suited for rib-on-plate structures where the deposited ribs are single pass rib buildups.

WAAM of large 3D metal Ti parts measuring around 1.2 m in length have been produced at Cranfield for articles like aircraft wing struts. An even larger Al wing component measuring 2.5 m in size was also printed in less than a day, and even bigger Ti parts using local shielding are now being produced [14].

Interpass Cooling

Researchers at the University of Wollongong in Australia developed and evaluated interpass cooling as a post-processing step for WAAM. A movable gas nozzle supplies argon, nitrogen or CO_2, which is used to provide active, forced cooling to the

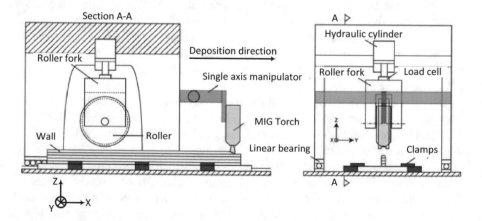

Fig. 10.16 Schematic diagram of WAAM with cold rolling process (Elsevier license number 4577940537820) [14]

printed components during and/or after layer deposition. With this cooling control system, the temperature history of each layer can be better controlled and custom-ized toward specific mechanical properties. Interpass cooling also improves the effi-ciency and reduces the dwell time between deposited layers. More investigation regarding residual stress and distortion is in progress. Fig. 10.17 shows an interpass cooling process [26].

Peening and Ultrasonic Impact Treatment
Peening and ultrasonic impact are cold-working post-processes that can be used for improving surface quality of WAAM parts. These techniques use high energy media to release tensile and compressive stresses within the components. Mechanical peening can result in increasing the surface hardness to a depth of 1–2 mm in carbon steels. Ultrasonic impact treatment leads to grain refinement and randomizes orien-tation and improves mechanical properties. It is reported that residual stresses of WAAM Ti6Al4V parts reduced by 58% and the microhardness increased by 28% after ultrasonic impact treatment. Also, the surface-treated layers undergo plastic deformation with significant grain refinement and dense dislocations [31]. The limi-tation of both peening and ultrasonic impact is low penetration depth which may not be appropriate for large metal WAAM parts. Both peening and ultrasonic impact treatments are described in more detail in later chapters.

10.4.4 Friction Stir Additive Manufacturing (FSAM)

Friction Stir Additive Manufacturing FSAM applies friction stir welding to Additive Manufacturing. Commercialized by MELD Manufacturing [32], it is a solid-state process which works below the melting temperature and offers various applications

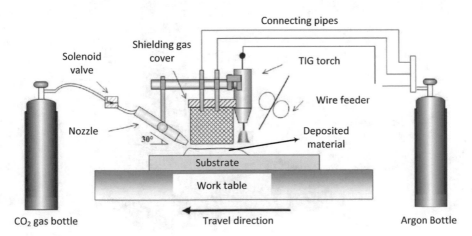

Fig. 10.17 Schematic diagram of the combined WAAM gas cooling process (Elsevier license number 4577970927636) [26]

such as alloying, component repair, metal joining, and more. Although FSAM is not technically a DED process, since it does not melt material as it is being deposited, it does apply energy (via plastic deformation) to bond materials as they are being deposited. As such, it is more similar to DED than other AM techniques, and thus it is included here.

FSAM plastically deforms or softens metal feedstock while simultaneously depositing that feedstock in a layer-wise fashion. This process can be performed in open atmosphere. Unlike wire and powder DED systems, FSAM can more readily process un-weldable materials. Due to the low processing temperatures, components have low residual stresses and can have high density with significantly low process energy. The low working temperature also reduces the chance of thermally induced porosity. There is also lower demand for expensive and time-consuming post-processes such as hot or cold isostatic processing. Figure 10.18 shows the schematic of FSAM with wire feed.

FSAM breaks up the granular structure of the feedstock, producing small grains, thus improving mechanical properties such as hardness, corrosion, and wear resistance as well as tensile strength. Figure 10.19 shows finer microstructure when processing Inconel 625 with FSAM (decreased by over 5%). It produces equiaxed grain structure and orientation, so more isotropic properties are obtained. This can create a usable part with good surface and bulk qualities.

The deposition rate for FSAM is comparable to DED wire systems, reported at up to 9 kg/h. In addition to the wire feedstock shown in Fig. 10.18, it is also possible to feed powders, flakes, and other material forms through the friction stir tool to achieve successful deposition. By mixing powders prior to deposition, real-time alloying, and composite depositions are achieved.

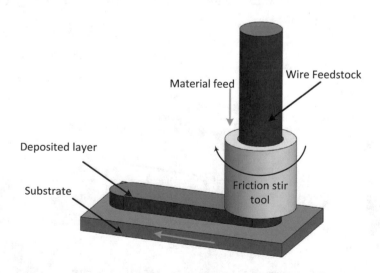

Fig. 10.18 Schematic of FSAM [32]

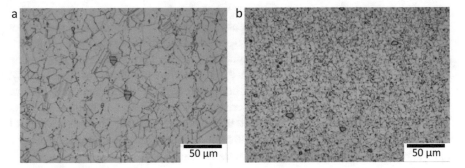

Fig. 10.19 Microstructure of Inconel 625 (**a**) before and (**b**) after FSAM. (Photo courtesy of Meld)

10.4.5 Other DED Materials and Processes

A number of materials have been explored for DED. In addition to multiple investigators who have demonstrated the processing of ceramics using a standard LENS process, other researchers have investigated almost any type of powder which can theoretically be melted using thermal energy sources. Plastic powder or even table sugar can be blown into a melt pool produced by a hot air gun to produce plastic or sugar parts. Although to date only metal-focused systems have been commercialized, it is likely only a matter of time before different material systems and machine architectures for DED become commercially viable.

10.5 Process Parameters

Most AM machines come pre-programmed with optimized process parameters for materials sold by the machine vendors, but DED machines are typically sold as flexible platforms; and thus DED users must identify the correct process parameters for their application and material. Optimum process parameters are material dependent and application/geometry dependent. Important process parameters include track scan spacing, powder feed rate, beam traverse speed, beam power, and beam spot size. Powder feed rate, beam power, and traverse speed are all interrelated; for instance, an increase in feed rate has a similar effect to lowering the beam power. Likewise, increasing beam power or powder feed rate and decreasing traverse speed all increase deposit thickness. From an energy standpoint, as the scan speed is increased, the input beam energy decreases because of the shorter dwell time, resulting in a smaller melt pool on the substrate and more rapid cooling.

Scan pattern also plays an important role in part quality. As mentioned previously, it may be desirable to change the scan orientation from layer to layer to minimize residual stress buildup. Track width hatch spacing must be set so that adjacent beads overlap, and layer thickness settings must be less than the melt pool depth to produce a fully dense product. Sophisticated accessory equipment for melt pool imaging and real-time deposit height measurement for accurately monitoring the melt pool and deposit characteristics are worthwhile additions for repeatability, as it is possible to use melt pool size, shape, and temperature as feedback control inputs to maintain desired pool characteristics. To control deposit thickness, travel speed can be dynamically changed based upon sensor feedback. Similarly to control solidification rate, and thus microstructure and properties, the melt pool size can be monitored and then controlled by dynamically changing laser power.

10.6   Typical Materials and Microstructure

DED processes aim to produce fully dense functional parts. Any powder material or powder mixture which is stable in a molten pool can be used for construction of parts. In general, metals with high reflectivities and thermal conductivities are difficult to process, such as gold and some alloys of aluminum and copper. Most other metals are quite straightforward to process, unless there is improper atmospheric preparation and bonding is inhibited by oxide formation. Generally, metallic materials that exhibit reasonably good weldability are easy to process.

Ceramics are more difficult to process, as few can be heated to form a molten pool. Even in the event that a ceramic material can be melted, cracking often occurs during cooling due to thermal shock. Thus, most ceramics that are processed using DED are processed as part of a ceramic matrix or metal matrix composite.

For powder feedstock, the powder size typically ranges from approximately 20–150 μm. It is within this range that powder particles can be most easily fluidized and delivered using a flowing gas. Blended elemental powders can be used to produce an infinite number of alloy combinations, or prealloyed powders can be used. Elemental powders can be delivered in precise amounts to the melt zone using separate feeders to generate various alloys and/or composite materials in situ. When using elemental powders for generation of an alloy in situ, the enthalpy of mixing plays an important role in determining the homogeneity of the deposited alloy. A negative enthalpy of mixing (heat release) promotes homogeneous mixing of constituent elements, and, therefore, such alloy systems are quite suitable for processing using elemental powders.

The fruitfulness of creating multi-material or gradient material combinations to investigate material properties quickly is illustrated in Figs. 10.20 and 10.21. Figure 10.20 illustrates a tensile bar made with a smooth 1D transition between Ti-6-4 and Ti-22-23, where Fig. 10.21 illustrates the yield strength of various combinations of these alloys. Using optical methods, localized stress and strain fields can be calculated during a tensile test and correlated back to the alloy combination for that location. Using this methodology, the properties of a wide range of alloy combinations can be investigated in a single experiment. Creating larger samples

Fig. 10.20 Smooth transition between a 100% Ti-6-4- and 100% Ti-22-23 alloy in the gage section of a tensile bar. The transition region is shown at higher magnification. (Photo courtesy of Optomec)

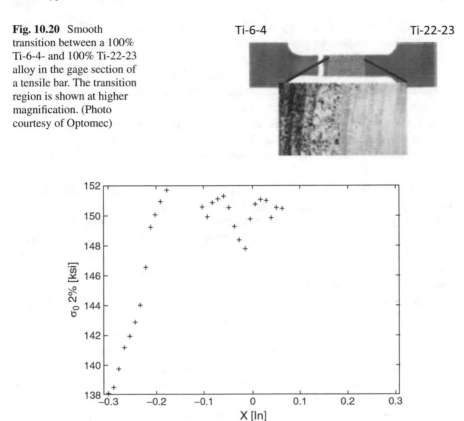

Fig. 10.21 Yield strength at various locations along the tensile bar from Fig. 10.20 representing the mechanical properties for different combinations of Ti-6-4 and Ti-22-23. (Photo courtesy of Optomec)

with 2D transitions of alloys (alloy transitions both longitudinally and transversely to the test axis using 3 or 4 powder feeders) can enable even greater numbers of alloy combinations to be investigated simultaneously.

DED processes can involve extremely high solidification cooling rates, from 10^3 to as high as 10^5 °C/s. (This is also true for metal PBF processes, and thus the following discussion is also relevant to parts made using metal PBF.) High cooling rates can lead to several microstructural advantages, including (a) suppression of diffusion controlled solid-state phase transformations; (b) formation of supersaturated solutions and nonequilibrium phases; (c) formation of extremely fine microstructures with dramatically reduced elemental segregation; and (d) formation of very fine secondary phase particles (inclusions, carbides, etc.). Parts produced using DED experience a complex thermal history in a manner very similar to multi-pass weld deposits. Changes in cooling rate during part construction can occur due to heat buildup, especially in thin-wall sections. Also, energy introduced during deposition of subsequent layers can reheat previously deposited material, changing the microstructure of previously deposited layers. The thermal history, including peak temperatures, time at peak temperature, and cooling rates, can be different at each point in a part, leading to phase transformations and a variety of microstructures within a single component.

As shown in Figs. 10.2, 10.3, and 10.20, parts made using DED typically exhibit a layered microstructure with an extremely fine solidification substructure. The interface region generally shows no visible porosity and a thin heat-affected zone (HAZ), as can be seen, for example, on the microstructure at the interface region of a LENS deposited medical-grade CoCrMo alloy onto a CoCrMo wrought substrate of the same composition (Fig. 10.22). Some materials exhibit pronounced columnar grain structures aligned in the laser scan direction, while some materials exhibit fine equiaxed structures. The deposited material generally shows no visible porosity, although gas evolution during melting due to excess moisture in the powder or from entrapped gases in gas-atomized powders can cause pores in the deposit. Pores can also result if excess energy is utilized, resulting in material vaporization and "key-holing." Parts generally show excellent layer-to-layer bonding, although lack-of-fusion defects can form at layer interfaces when the process parameters are not properly optimized and insufficient energy density is utilized.

Residual stresses are generated as a result of solidification, which can lead to cracking during or after part construction. For example, LENS-deposited TiC ceramic structures are prone to cracking as a result of residual stresses (Fig. 10.23). Residual stresses pose a particularly significant problem when dealing with metal-lurgically incompatible dissimilar material combinations.

Formation of brittle intermetallic phases formed at the interface of dissimilar materials in combination with residual stresses can also lead to cracking. This can be overcome by suppressing the formation of intermetallics using appropriate

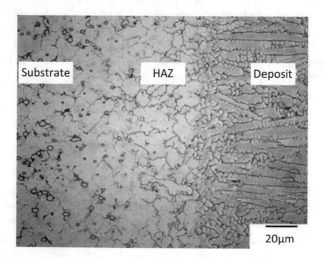

Fig. 10.22 CoCrMo LENS deposit on a wrought CoCrMo substrate of the same composition (deposit occurred from the right of the picture) (Springer nature license agreement number 4635810376596) [33]

Fig. 10.23 Cracks in a TiC LENS deposit due to residual stresses (Elsevier license number 4635800416420) [34]

processing parameters or by the use of a suitable interlayer. For instance, in several research projects, it has been demonstrated that it is possible to suppress the formation of brittle intermetallics when depositing Ti on CoCrMo by placing the focal plane above the CoCrMo substrate during deposition of the first layer and depositing a thin coating of Ti using a low laser power and rapid scan rate. Subsequent layers are likewise deposited using relatively thin deposits at high scan rates and low laser power to avoid reheating of the Ti/CoCrMo interface. Once a sufficient Ti deposit is accumulated, normal process parameters for higher deposition rate can be utilized. However, if excess heat is introduced either during the deposition of subsequent layers or in subsequent heat treatment, equilibrium intermetallics will form, and cracking and delamination occurs. In other work, CoCrMo has been successfully deposited on a porous Ta substrate when employing Zr as an interlayer material, a combination that is otherwise prone to cracking and delamination.

It is common for laser-deposited parts to exhibit superior yield and tensile strengths because of their fine grain structure. Ductility of DED parts, however, is generally considered to be inferior to wrought or cast equivalents. Layer orientation can have a great influence on % elongation, with the worst being the z direction. However, in many alloys ductility can be recovered and anisotropy minimized by heat treatment – without significant loss of strength in most cases.

10.7 Processing–Structure–Properties Relationships

Parts produced in DED processes exhibit high cooling rate cast microstructures. Processing conditions influence the solidification microstructure in ways that can be predicted in part by rapid solidification theory. For a specific material, solidification microstructure depends on the local solidification conditions, specifically the solidification rate and temperature gradient at the solid/liquid interface. By calculating the solidification rate and thermal gradient, the microstructure can be predicted based upon calibrated "solidification maps" from the literature.

To better understand solidification microstructures in DED processes, Beuth and Klingbeil [35] have developed procedures for calculating thermal gradients, G, and solidification rates, R, analytically and numerically. These calculated G and R values can then be plotted on solidification maps to determine the types of microstructures which can be achieved with different DED equipment, process parameters, and material combinations. Solutions for both thin-walls [35] and bulky deposits [36] have been described. For brevity's sake, the latter work by Bontha et al. [36] based upon the 3D Rosenthal solution for a moving point heat source on an infinite substrate will be introduced here (Fig. 10.24). This 3D Rosenthal solution also has been applied to PBF techniques in an identical fashion.

In this simplified model, material deposition is ignored. The model considers only heat conduction within the melt pool and substrate due to a traveling heat source moving at velocity, V. The fraction of impinging energy absorbed is αQ, which is a simplification of the physically complex temperature-dependent absorption of the beam by regions of the melt pool and solid, absorption of energy by powder in flight, and other factors. Thus a single parameter, α, represents the fraction of impinging beam energy power absorbed.

It is assumed the beam moves only in the x direction, and thus the beam's relative coordinates (x_0, y_0, z_0) from Fig. 10.24 are related to the fixed coordinates (x, y, z) at any time t as $(x_0, y_0, z_0) = (x - Vt, y, z)$.

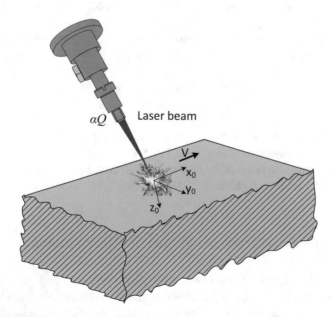

Fig. 10.24 3D Rosenthal geometry considered

With the above conditions, the Rosenthal solution for temperature T at time t for any location in an infinite half-space can be expressed in dimensionless form as:

$$\bar{T} = \frac{e^{-\left(\bar{x}_0 + \sqrt{\bar{x}_0^2 + \bar{y}_0^2 + \bar{z}_0^2}\right)}}{2\sqrt{\bar{x}_0^2 + \bar{y}_0^2 + \bar{z}_0^2}} \tag{10.1}$$

where

$$\bar{T} = \frac{T - T_0}{(\alpha Q / \pi k)(\rho c V / 2k)},$$

$$\bar{x}_0 = \frac{x_0}{(2k / \rho c V)}, \bar{y}_0 = \frac{y_0}{(2k / \rho c V)} \text{ and } \bar{z}_0 = \frac{z_0}{(2k / \rho c V)}. \tag{10.2}$$

In these equations, T_0 is the initial temperature, and ρ, c, and k are density, specific heat, and thermal conductivity of the substrate, respectively. In this simplified model, the thermophysical properties are assumed to be temperature independent and are often selected at the melting temperature, since cooling rate and thermal gradient at the solid/liquid interface is of greatest interest.

The parameters of interest are solidification cooling rate and thermal gradient. The dimensionless expression for cooling rate becomes:

$$\frac{\partial \bar{T}}{\partial \bar{t}} = \frac{1}{2} \frac{e^{-\left((\bar{x}-\bar{t}) + \sqrt{(\bar{x}-\bar{t})^2 + \bar{y}_0^2 + \bar{z}_0^2}\right)}}{\sqrt{(\bar{x}-\bar{t})^2 + \bar{y}_0^2 + \bar{z}_0^2}}$$

$$\times \left\{ 1 + \frac{(\bar{x}-\bar{t})}{\left(\sqrt{(\bar{x}-\bar{t})^2 + \bar{y}_0^2 + \bar{z}_0^2}\right)} + \frac{(\bar{x}-\bar{t})}{\left((\bar{x}-\bar{t})^2 + \bar{y}_0^2 + \bar{z}_0^2\right)} \right\}. \tag{10.3}$$

where the dimensionless x coordinate is related to the dimensionless x_0 by $\bar{x} = \bar{x}_0 + \bar{t}$

where $\bar{t} = \left(t / \left(2k / \rho c V^2\right)\right)$ and the dimensionless cooling rate is related to the actual cooling rate by:

$$\frac{\partial \bar{T}}{\partial \bar{t}} = \left(\frac{2k}{\rho c V}\right)^2 \left(\frac{\pi k}{\alpha Q V}\right) \frac{\partial T}{\partial t}. \tag{10.4}$$

The dimensionless thermal gradient is obtained by differentiating (10.1) with respect to the dimensionless spatial coordinates, giving

$$\left|\overline{\nabla T}\right| = \sqrt{\left(\frac{\partial \overline{T}}{\partial \overline{x}_0}\right)^2 + \left(\frac{\partial \overline{T}}{\partial \overline{y}_0}\right)^2 + \left(\frac{\partial \overline{T}}{\partial \overline{z}_0}\right)^2}, \tag{10.5}$$

where

$$\frac{\partial \overline{T}}{\partial \overline{x}_0} = -\frac{1}{2} \frac{e^{-\left(\overline{x}_0 + \sqrt{\overline{x}_0^2 + \overline{y}_0^2 + \overline{z}_0^2}\right)}}{\sqrt{\overline{x}_0^2 + \overline{y}_0^2 + \overline{z}_0^2}}$$

$$\times \left\{ 1 + \frac{\overline{x}_0}{\left(\sqrt{\overline{x}_0^2 + \overline{y}_0^2 + \overline{z}_0^2}\right)} + \frac{\overline{x}_0}{\left(\overline{x}_0^2 + \overline{y}_0^2 + \overline{z}_0^2\right)} \right\}, \tag{10.6}$$

$$\frac{\partial \overline{T}}{\partial \overline{y}_0} = -\frac{1}{2} \frac{\overline{y}_0 e^{-\left(\overline{x}_0 + \sqrt{\overline{x}_0^2 + \overline{y}_0^2 + \overline{z}_0^2}\right)}}{\left(\overline{x}_0^2 + \overline{y}_0^2 + \overline{z}_0^2\right)} \left\{ 1 + \frac{1}{\left(\sqrt{\overline{x}_0^2 + \overline{y}_0^2 + \overline{z}_0^2}\right)} \right\}, \tag{10.7}$$

and

$$\frac{\partial \overline{T}}{\partial \overline{z}_0} = -\frac{1}{2} \frac{\overline{z}_0 e^{-\left(\overline{x}_0 + \sqrt{\overline{x}_0^2 + \overline{y}_0^2 + \overline{z}_0^2}\right)}}{\left(\overline{x}_0^2 + \overline{y}_0^2 + \overline{z}_0^2\right)} \left\{ 1 + \frac{1}{\left(\sqrt{\overline{x}_0^2 + \overline{y}_0^2 + \overline{z}_0^2}\right)} \right\}. \tag{10.8}$$

As defined above, the relationship between the dimensionless thermal gradient $\left|\overline{\nabla T}\right|$ and the actual thermal gradient $|\nabla T|$ is given by

$$\left|\overline{\nabla T}\right| = \left(\frac{2k}{\rho c V}\right)^2 \left(\frac{\pi k}{\alpha Q}\right) |\nabla T|. \tag{10.9}$$

Using this formulation, temperature, cooling rates, and thermal gradients can be solved for any location (x,y,z) and time (t).

For microstructure prediction purposes, solidification characteristics are of interest; and thus we need to know the cooling rate and thermal gradients at the boundary of the melt pool. The roots of (10.1) can be solved numerically for temperature T equal to melting temperature T_m to find the dimensions of the melt pool. Similarly to (10.2) for normalized temperature, normalized melting temperature can be represented by:

$$\overline{T}_m = \frac{T_m - T_0}{(\alpha Q / \pi k)(\rho c V / 2k)}. \tag{10.10}$$

Given cooling rate $\dfrac{\partial T}{\partial t}$ from (10.4) and thermal gradient, G, defined as $=|\nabla T|$, we can define the solidification velocity, R, as

$$R = \frac{1}{G}\frac{\partial T}{\partial t}. \tag{10.11}$$

We can now solve these sets of equations for specific process parameters (i.e., laser power, velocity, material properties, etc.) for a machine/material combination of interest. After this derivation, Bontha et al. [36] used this analytical model to demonstrate the difference between solidification microstructures which can be achieved using a small scale DED process with a lower-powered laser beam, such as utilized in a LENS machine, compared to a high-powered laser beam system, such as practiced by AeroMet for Ti–6Al–4 V. Assumptions included the thermo-physical properties of Ti–6Al–4 V at $T_m = 1654\ °C$, a room temperature initial substrate temperature $T_0 = 25\ °C$, fraction of energy absorbed $\alpha = 35$, laser power from 350 to 850 W, and beam velocity ranging from 2.12 to 10.6 mm/s. For the high-powered beam system, a laser power range from 5 to 30 kW was selected. A set of graphs representing microstructures with low-powered systems is shown in Fig. 10.25. Microstructures from high-powered systems are shown in Fig. 10.26 for comparison.

As can be seen from Fig. 10.25, lower-powered DED systems cannot create mixed or equiaxed Ti–6Al–4 V microstructures, as the lower overall heat input

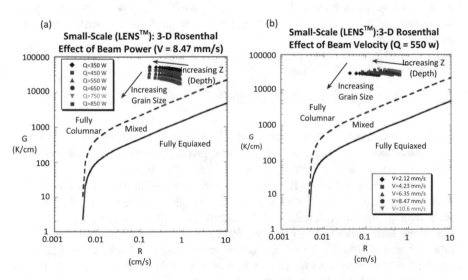

Fig. 10.25 Process maps showing microstructures predicted by the 3D Rosenthal solution for a lower powered (LENS like) DED system for Ti 6Al 4 V (Elsevier License number 4635810822810) [36]

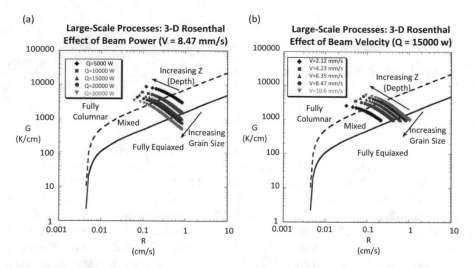

Fig. 10.26 Process maps showing microstructures predicted by the 3D Rosenthal solution for a higher-powered (AeroMet-like) DED system for Ti–6Al–4 V (Elsevier License number 4635810822810 [36]

means that there are very large thermal gradients. For higher-powered DED systems (relevant to AeroMet-like processes and most electron beam DED process), it is possible to create dendritic, mixed, or fully equiaxed microstructures depending upon the process parameter combinations used. As a result, without the need for extensive experimentation, process maps such as these, when combined with appropriate modeling, can be used to predict the type of DED system (specifically the scan rates and laser power) needed to achieve a desired microstructure type for a particular alloy.

10.8   DED Benefits and Drawbacks

DED processes are capable of producing fully dense parts with highly controllable microstructural features. These processes can produce functionally graded components with composition variations in the x, y, and z directions.

The main limitations of DED processes are poor resolution and surface finish. An accuracy better than 0.25 mm and a surface roughness of less than 25 μm (arithmetic average) are difficult with most DED processes. Slower build speed is another limitation. Build times can be very long for these processes, with typical deposition rates as low as 25–40 g/h. To achieve better accuracies, small beam sizes and deposition rates are required. Conversely, to achieve rapid deposition rates, degradation of resolution and surface finish result. Changes in laser power and scan rate to achieve better accuracies or deposition rates may also affect the microstructures of the deposited components, and thus finding an optimum deposition condition necessitates trade-offs between build speed, accuracy, and microstructure.

Examples of the unique capabilities of DED include:

- DED offers the capability for unparalleled control of microstructure. The ability to change material composition and solidification rate by simply changing powder feeder mixtures and process parameters gives designers and researchers tremendous freedom. This design freedom is further explored in Chap. 19.
- DED is capable of producing directionally solidified and single crystal structures.
- DED can be utilized for effectively repairing and refurbishing defective and service damaged high-technology components such as turbine blades.
- DED processes are capable of producing in situ generated composite and heterogeneous material parts. For example, Banerjee et al. [37] have successfully produced Ti–6Al–4 V/TiB composite parts using the LENS process employing a blend of pure pre-alloyed Ti–6Al–4 V and elemental B powders (98 wt% Ti–6Al–4 V + 2 wt% B). The deposited material exhibited a homogeneous refined dispersion of nano-scale TiB precipitates within the Ti–6Al–4 V α/β matrix.
- DED can be used to deposit thin layers of dense, corrosion-resistant, and wear-resistant metals on components to improve their performance and lifetime. One example includes deposition of dense Ti/TiC coatings as bearing surfaces on Ti biomedical implants, as illustrated in Fig. 10.2.

When contrasted with other AM processes, DED processes cannot produce as complex of structures as PBF processes. This is due to the need for more dense support structures (or multi-axis deposition) for complex geometries and the fact that the larger melt pools in DED result in a reduced ability to produce small-scale features, greater surface roughness, and less accuracy.

Post-processing of parts made using DED typically involves removal of support structures or the substrate, if the substrate is not intended to be a part of the final component. Finish machining operations because of relatively poor part accuracy and surface finish are commonly needed. Stress relief heat treatment may be required to relieve residual stresses. In addition, depending upon the material, heat treatment may be necessary to produce the desired microstructure(s). For instance, parts constructed in age-hardenable materials will require either a direct aging treatment or solution treatment followed by an aging treatment to achieve precipitation of strengthening phases.

DED processes are uniquely suited amongst AM process for repair and feature addition. As this AM process is formulated around deposition, there is no need to deposit on a featureless plate or substrate. Instead, DED is often most successful when used to add value to other components by repairing features, adding new features to an existing component, and/or coating a component with material which is optimized for the service conditions of that component in a particular location.

As a result of the combined strengths of DED processes, practitioners of DED primarily fall into one of several categories. First, DED has been highly utilized by research organizations interested in the development of new material alloys and the application of new or advanced materials to new industries. Second, DED has found great success in facilities that focus on repair, overhaul, and modernization of metallic structures. Third, DED is useful for adding features and/or material to existing structures to improve their performance characteristics. In this third category, DED can be used to improve the life of injection molding or die casting dies by depositing wear-resistant alloys in high-wear locations; it is being actively researched by multiple biomedical companies for improving the characteristics of biomedical implants; and it is used to extend the wear characteristics of everything from drive shafts to motorcycle engine components. Fourth, DED is increasingly used to produce near net structures in place of wrought billets, particularly for applications where conventional manufacturing results in a large buy-to-fly ratio.

10.9   Questions

1. Discuss three characteristics where DED is similar to MEX and three characteristics where DED is different than MEX.
2. Read the following reference related to thin-wall structures made using DED. What are the main differences between modeling thin wall and bulky structures? What ramifications does this have for processing?
   Vasinonta, Aditad, Jack L. Beuth, and Raymond Ong. "Melt pool size control in thin-walled and bulky parts via process maps." Solid Freeform Fabrication Proceedings. Proc. 2001 Solid Freeform Fabrication Symposium, Austin, 2001.
3. Why is solidification rate considered the key characteristic to control in DED processing?
4. From the literature, determine how solidification rate is monitored. From this information, describe an effective, simple closed-loop control methodology for solidification rate.
5. Why are DED processes particularly suitable for repair?
6. In DED, what are the two main methods of material delivery? Compare and contrast these two methods.
7. When DED is used for repair, describe how the size of the melt pool compared to the size of the substrate affects the properties of the resulting component.
8. Compare and contrast the speed at which the energy source is moved in DED versus VPP, MEX, and PBF processes. Explain why this is the case.
9. Selective Laser Sintering (SLS) and Laser Engineered Net Shaping (LENS) were both developed in the early days of AM. Which turned out to be a more successful commercial process, and why? Describe at least three differences between them in your justification for why.
10. Hard tooling is manufacturing tools from hardened metals. How does DED compare to other AM technologies for the production of hard tooling? How would your answer change if you're focused on the repair of hard tooling?

References

1. Keicher, D. M., & Miller, W. D. (1998). LENSTM moves beyond RP to direct fabrication. *Metal Powder Report, 12*(53), 26–28.
2. Lewis, G.K., et al. (1994). Directed light fabrication. In *International Congress on Applications of Lasers & Electro-Optics*. LIA.
3. House, M., et al. (1996). Rapid laser forming of titanium near shape articles: LaserCast. in 1996 *International Solid Freeform Fabrication Symposium.*
4. 3ders. https://www.3ders.org/articles/20130529-china-shows-off-world-largest-3d-printed-titanium-fighter-component.html. (2013).
5. Conduit, B., et al. (2019). Probabilistic neural network identification of an alloy for direct laser deposition. *Materials & Design, 168*, 107644.
6. Shamsaei, N., et al. (2015). An overview of direct laser deposition for additive manufacturing; part II: Mechanical behavior, process parameter optimization and control. *Additive Manufacturing, 8*, 12–35.
7. Wu, X., et al. (2004). Microstructures of laser-deposited Ti–6Al–4V. *Materials & Design, 25*(2), 137–144.
8. Mazumder, J., et al. (1997). The direct metal deposition of H13 tool steel for 3-D components. *JOM, 49*(5), 55–60.
9. Dwivedi, R., & Kovacevic, R. (2005). *Process Planning for Multi-Directional Laser-Based Direct Metal Deposition., 219*(7), 695–707.
10. Dwivedi, R., & Kovacevic, R. (2006). An expert system for generation of machine inputs for laser-based multi-directional metal deposition. *International Journal of Machine Tools and Manufacture, 46*(14), 1811–1822.
11. Xue, L. (2018). Chapter 16 – Laser Consolidation—A Rapid Manufacturing Process for Making Net-Shape Functional Components, in *Advances in Laser Materials Processing (Second Edition)* J. Lawrence, Editor. Woodhead Publishing. p. 461–505.
12. Xue, L. & M. Islam. (1998). Free-form laser consolidation for producing functional metallic components. In *International Congress on Applications of Lasers & Electro-Optics*. LIA.
13. Xue, L., & Islam, M. (2000). Free-form laser consolidation for producing metallurgically sound and functional components. *Journal of Laser Applications, 12*(4), 160–165.
14. McLean, M. (1997). Laser direct casting high nickel alloy components. *Advances in Powder Metallurgy Particulate Materials, 3*, 21.
15. Sears, J. W. (1999). *Direct laser powder deposition-'State of the Art', Knolls Atomic Power Lab*. Niskayuna.
16. Prakash, V. J., et al. (2019). Laser Metal Deposition of Titanium Parts with Increased Productivity. In *3D Printing and Additive Manufacturing Technologies* (pp. 297–311). Singapore: Springer.
17. Azarniya, A., et al. (2019). Additive manufacturing of Ti–6Al–4V parts through laser metal deposition (LMD): Process, microstructure, and mechanical properties. *Journal of Alloys and Compounds, 804*, 163–191.
18. Arcella, F. G., & Froes, F. H. (2000). Producing titanium aerospace components from powder using laser forming. *JOM, 52*(5), 28–30.
19. Fessler, J.R., et al., *Laser deposition of metals for shape deposition manufacturing.* Proceedings of the Solid Freeform Fabrication Symposium, 1996: p. 117–124.
20. Hardchrome Engineering. 2020. http://www.hardchrome.com.au/?menuty=1
21. Electron Beam Freeform Fabrication, NASA Report. (2020) https://www.nasa.gov/topics/technology/features/ebf3.html.
22. *NASA engineers test the EBF3 system during a parabolic flight in 2007.* (2007). https://en.wikipedia.org/wiki/Electron-beam_freeform_fabrication#/media/File:NASA_EBF3_2007_test.jpg
23. Taminger, K. M., & Hafley, R. A. (2006). Electron beam freeform fabrication for cost effective near-net shape manufacturing.

24. Watson, J., et al. 2002. Development of a prototype low-voltage electron beam freeform fabrication system. .
25. Taminger, K., et al. (2004). Effect of surface treatments on electron beam freeform fabricated aluminum structures.
26. Wu, B., et al. (2018). A review of the wire arc additive manufacturing of metals: Properties, defects and quality improvement. *Journal of Manufacturing Processes, 35*, 127–139.
27. Ding, J., et al. (2011). Thermo-mechanical analysis of wire and arc additive layer Manufacturing process on large multi-layer parts. *Computational Materials Science, 50*(12), 3315–3322.
28. Dickens, P., et al. (1992). *Rapid prototyping using 3-D welding.* in 1992 *International Solid Freeform Fabrication Symposium.*
29. Spencer, J., Dickens, P., & Wykes, C. (1998). Rapid prototyping of metal parts by three-dimensional welding. *Proceedings of the Institution of Mechanical Engineers, Part B: Journal of Engineering Manufacture, 212*(3), 175–182.
30. Baufeld, B., Brandl, E., & Van der Biest, O. (2011). Wire based additive layer manufacturing: Comparison of microstructure and mechanical properties of Ti–6Al–4V components fabricated by laser-beam deposition and shaped metal deposition. *Journal of Materials Processing Technology, 211*(6), 1146–1158.
31. Li, G., et al. (2017). Effect of ultrasonic surface rolling at low temperatures on surface layer microstructure and properties of HIP Ti-6Al-4V alloy. *Surface and Coatings Technology, 316*, 75–84.
32. MELD Manufacturing. (2020). http://meldmanufacturing.com/
33. Janaki Ram, G. D., Esplin, C. K., & Stucker, B. E. (2008). Microstructure and wear properties of LENS® deposited medical grade CoCrMo. *Journal of Materials Science: Materials in Medicine, 19*(5), 2105–2111.
34. Liu, W., & DuPont, J. (2003). Fabrication of functionally graded TiC/Ti composites by laser engineered net shaping. *Scripta Materialia, 48*(9), 1337–1342.
35. Beuth, J., & Klingbeil, N. (2001). The role of process variables in laser-based direct metal solid freeform fabrication. *JOM, 53*(9), 36–39.
36. Bontha, S., et al. (2009). Effects of process variables and size-scale on solidification microstructure in beam-based fabrication of bulky 3D structures. *Materials Science and Engineering: A, 513*, 311–318.
37. Banerjee, R., et al. (2003). Direct laser deposition of in situ Ti–6Al–4V–TiB composites. *Materials Science and Engineering: A, 358*(1–2), 343–349.

Chapter 11
Direct Write Technologies

Abstract Direct Write (DW) technologies are a collection of AM and related approaches for creation of very small components. Most of these approaches were covered in prior chapters. However, since feature creation at very small scales brings unique challenges and benefits, it is appropriate to address this topic separately. In particular, there are increasing uses of DW technologies combined with compatible AM techniques to create unique devices that include electrical, electronic, and other functionality all in an integrated component. Approaches to DW and its applications are thus explored in this chapter.

11.1 Direct Write Technologies

Direct Write technologies are not specifically mentioned in ASTM or ISO standards. Instead they make use of the seven categories of AM processes described in the previous seven chapters. The term "Direct Write" in its broadest sense can mean any technology which can create two- or three-dimensional functional structures directly onto flat or conformal surfaces in complex shapes, without any tooling or masks [1]. Although Directed Energy Deposition (DED), Material Jetting (MJT), Material Extrusion (MEX), and other AM processes fit this definition; for the purposes of distinguishing between the technologies discussed in this chapter and the technologies discussed elsewhere in this book, we will limit our definition of DW to those technologies which are designed to build freeform structures or electronics with feature resolution in one or more dimensions below 50 µm. This "small-scale" interpretation is how the term DW is typically understood in the Additive Manufacturing community. Thus, for the purposes of this chapter, DW technologies are those processes which create meso-, micro-, and nano-scale structures using a freeform deposition tool.

Although freeform surface modification using lasers and other treatments in some cases can be referred to as DW [2], we will only discuss those technologies which add material to a surface. A more complete treatment of DW technologies can be found in books dedicated to this topic [3].

© Springer Nature Switzerland AG 2021   319
I. Gibson et al., *Additive Manufacturing Technologies*,
https://doi.org/10.1007/978-3-030-56127-7_11

11.2 Background

Although the initial use of some DW technologies predate the advent of AM, the development of DW technologies was dramatically accelerated in the 1990s by funding from the US Defense Advanced Research Projects Agency (DARPA) and its Mesoscopic Integrated Conformal Electronics (MICE) program. DARPA recognized the potential for creating novel components and devices if material deposition technologies could be further developed to enable manufacture of complex electronic circuitry and mesoscale devices onto or within flexible or complex three-dimensional objects. Many different DW technologies were developed or improved following funding from DARPA, including Matrix-Assisted Pulsed Laser Evaporation (MAPLE), nScrypt 3De, Maskless Mesoscale Materials Deposition (M^3D, now known as Aerosol Jet), and DW Thermal Spraying. As a result, most people have come to consider DW technologies as those devices which are designed to write or print passive or active electronic components (conductors, insulators, batteries, capacitors, antennas, etc.) directly from a computer file without any tooling or masks. However, DW devices have found broad applicability outside the direct production of circuitry and are now used to fabricate structures with tailored thermal, electrical, chemical, and biological responses, among other applications.

DW processes can be subdivided into five categories, including ink-based, laser transfer, thermal spray, beam deposition, liquid-phase, and beam tracing processes. Most of these use a 3D programmable dispensing or deposition head to accurately apply small amounts of material automatically to form circuitry or other useful devices on planar or complex geometries. The following sections of this chapter describe these basic approaches to DW processing and commercial examples, where appropriate.

11.3 Materials in Direct Write Technology

As discussed in the following sections, the large variation in available DW technologies means there are a large number of material types which can be used for DW. Materials in DW include inks, organic precursors, binders, slurries, pastes, flakes, powders, nanoparticles, and combinations thereof, commonly with specific customized chemical and rheological properties. Many of these materials have been developed for low-temperature deposition from 200–400 °C [4, 5]. Powder morphology and packing efficiency are important characteristics for many DW processes [6, 7]. When powders are used in DW processes, often significant porosity remains in the structure. In the electronic industry, this value of porosity has a logarithmic effect on conductivity and electrical performance of devices. To compensate for these problems, polydispersed nanopowders may be added to the original powder to reduce porosity and improve properties.

Table 11.1 List of some materials used in direct writing applications [5–8]

Material	Examples
Alloys	Al alloys, Pt–Ru, Ag–Pd, Cu–Ni, Sn–Pb solders
Biomaterials	Living neural cell, peptides, proteins, antibodies, DNA oligomers, simple eukaryotic cell, mammalian cells, *Escherichia coli* bacteria, enzymes
Ceramics	Al2O3, SiO2, PZT, TiO2, ZrO2
Dielectrics	Barium titanate, barium strontium titanate, PMN/glass ferrites, polyimide
Food	Chocolate
Metals	Ag, Al, Au, Cu, In, Pt, Sn
Nanoparticle materials	Ag, Au, Cu, BaTiO3, Carbon nanotubes, Fe3O4
Optical materials	Silicate glass, Epoxy resin
Oxides	YBaCuO, BiSrCaCuO, InOx
Polymers	PTF-epoxy + carbon + titanate, biodegradable polymer (PLGA), organic light emitting polymer, conductive polymers (e.g., PEDOT:PSS, polyaniline), organic semiconductors, UV-curing polymers
Resistors	Ag/Pd/glass, ruthenate
Speciality	Diamond
Composites	Mix of above materials
Other materials	Adhesives, electrolytes, lubricants, pharmaceuticals, radio-opaque materials, and surfactants

DW processes are compatible with many types of substrates, including ceramics, plastics (such as PVC, mylar, nylon, polyimide, PEEK, polystyrene), and metals (such as 304 stainless steel, aluminum, copper, and nitinol). However, not all DW processes are compatible with all build materials and/or substrates. Table 11.1 shows different types of DW materials that are available.

11.4 Ink-Based DW

The most varied, least expensive, and most simple approaches to DW involve the use of liquid inks. These inks are deposited on a surface and contain the basic materials which become the desired structure. A significant number of ink types are available including among others:

- Colloidal inks
- Nanoparticle-filled inks
- Fugitive organic inks
- Polyelectrolyte inks
- Sol–gel inks

After deposition, these inks solidify due to evaporation, gelation, solvent-driven reactions, or thermal energy to leave a deposit of the desired properties. A large number of research organizations, corporations, and universities worldwide are involved in the development of new and improved DW inks.

DW inks are typically either extruded as a continuous filament through a nozzle (see Chap. 6) or deposited as droplets using a print head (see Chap. 7). Important rheological properties of DW inks include their ability to (1) flow through the deposition apparatus, (2) retain shape after deposition, and (3) either span voids/gaps or fill voids/gaps, as the case may be. To build three-dimensional DW structures, it is highly desirable for the deposited inks to be able to form a predictable and stable 3D deposition shape and to bridge small gaps. For 2D electronic structures built onto a surface, it is highly desirable for the deposited inks to maintain a constant and controllable cross-section, as this will determine the material properties (e.g., conductivity, capacitance, etc.). In general, this means that viscoelastic materials which flow freely under shear through a nozzle but become rigid and set up quickly after that shear stress is released are preferred for DW inks.

In contrast to most MEX and many MJT materials, DW inks must be transformed after deposition to achieve the desired properties. This transformation may be due to the natural environment surrounding the deposit (such as during evaporation or gelation), but in many cases, external heating using a thermal source or some other post-processing step is required.

Figure 11.1 illustrates the two most common methodologies for DW ink dispensing. Continuous dispensing, as in (a), has the merits of a continuous cross-sectional area and a wider range of ink rheologies. Droplet dispensing, as in (b), can be parallelized and done in a very rapid fashion; however, the deposit cross-sections are discontinuous, as the building blocks are basically overlapping hemispherical droplet splats, and the rheological properties must be within a tighter range (as discussed in Sect. 7.5). Nozzle dispensing and quill processes both create continuous deposits from DW inks. Printing and aerosol jet processes both create droplets from DW inks. These four approaches are discussed in more detail below.

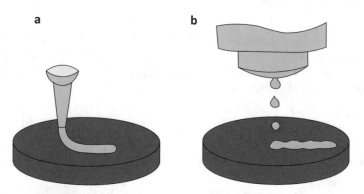

Fig. 11.1 Schematic illustration of direct ink writing techniques: (**a**) continuous filament writing and (**b**) droplet jetting [9]

11.5   Nozzle Dispensing Processes

Nozzle DW processes are technologies which use a pump or syringe mechanism to push DW inks through an orifice for deposition onto a substrate. A 3-axis motion control system is typically used with these nozzle systems to enable deposition onto complex surfaces or to build up scaffolds or other 3D geometry, as illustrated in Fig. 11.2. Some nozzle devices are packaged with a scanning system that first determines the topology of the substrate on which the deposit is to be made and then deposits material conformally over that substrate surface based on the scan data.

For nozzle processes, the main differentiating factors between devices are the (1) nozzle design, (2) motion control system, and (3) pump design. Nozzle design determines the size and shape of the deposit, directly influences the smallest feature size, and has a large effect on the types of inks which can be used (i.e., the viscosity of the ink and the size and type of fillers which can be used in the inks). The motion controller determines the dimensional accuracy and repeatability of the deposit, the maximum size of the deposit which can be made, and the speed at which deposition can occur. The pump design determines the volumetric control and repeatability of dispensing, how accurately the deposits can be started and stopped, and the speed at which deposition can occur. The difference between these three factors for different manufacturers and designs determines the price and performance of a nozzle-based DW process.

Micropen and nScrypt are two companies with well-developed extrusion nozzles and deposition systems for DW. Micropen stopped selling machines, subsequently merged with Exxelia, and currently sells DW services and solutions. nScrypt markets and sells nozzles, pumps, and integrated scanning, dispensing, and motion control systems for DW. A wide range of nozzle designs are available, and feedback systems help ensure that the stand-off distance between the nozzle and the substrate remains substantially constant to enable repeatable and accurate deposition of traces across conformal surfaces.

Fig. 11.2 A schematic drawing showing the deposition of a scaffold using a nozzle process [9]. (Photo courtesy of nScrypt)

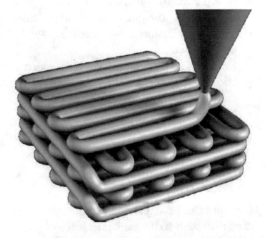

One characteristic of nScrypt systems is their aspirating function, causing the material to be pulled back into the print nozzle at the end of a deposition path. This aspiration function enables precise starts and stops. In addition, a conical nozzle design enables a large range of viscous materials to be dispensed. The pump and nozzle design, when combined, enable viscosities which are processable over six orders of magnitude, from 1 cp to 1000,000 cp (the equivalent of processing materials ranging from water to peanut butter). This means that virtually any electronic ink or paste can be utilized. Materials ranging from electronic inks to quick setting concrete have been deposited using nScrypt systems.

Simple DW nozzle devices can be built using off-the-shelf syringes, pumps, and 3-axis motion controllers for a few thousand dollars, such as by using an inexpensive MEX system such as the Fab@Home system developed at Cornell University [10]. These enable one to experiment with nozzle-based DW processes for a relatively low capital investment. Fully integrated devices with multiple nozzles capable of higher dimensional accuracy, dispensing repeatability, and wider range of material viscosities can cost significantly more than $250,000.

Nozzle DW processes have successfully been used to fabricate devices such as integrated RC filters, multilayer voltage transformers, resistor networks, porous chemical sensors, biological scaffolds, and other components. Three aspects of nozzle-based processes make them interesting candidates for DW practitioners: (1) these processes can deposit fine line traces on nonplanar substrates, (2) they work with the largest variety of inks of any DW technology, and (3) they can be built up from interchangeable low-cost components and integrated easily onto various types of multi-axis motion control systems. The main drawback of nozzle-based systems is that the inks must typically be thermally post-processed to achieve the robust properties desired for most end-use applications. Thus, a thermal or laser post-deposition-processing system is highly beneficial. Although the types of materials which have been deposited successfully using nozzle-based processes are too numerous to list, examples include [11]:

- Electronic materials – metal powders (silver, copper, gold, etc.) or ceramic powders (alumina, silica, etc.) suspended in a liquid precursor that after deposition and thermal post-processing form resistors, conductors, antennas, dielectrics, etc.
- Thermoset materials – adhesives, epoxies, etc. for encapsulation, insulation, adhesion, etc.
- Solders – lead-free, leaded, etc. as electrical connections
- Biological materials – synthetic polymers and natural polymers, including living cells
- Nanomaterials – nanoparticles suspended in gels, slurries, etc.

11.5.1   Quill-Type Processes

DW inks can be deposited using a quill-type device, much like a quill pen can be used to deposit writing ink on a piece of paper. These processes work by dipping the pen into a container of ink. The ink adheres to the surface of the pen, and then, when the pen is put near the substrate, the ink is transferred from the pen to the substrate. By controlling the pen motion, an accurate pattern can be produced. The primary DW method for doing this is the dip-pen nanolithography (DPN) technique developed by a number of universities and sold by NanoInk. This process works by dipping an atomic force microscope (AFM) tip into an inkwell of specially formulated DW ink. The ink adheres to the AFM tip and then is used to write a pattern onto a substrate, as illustrated in Fig. 11.3a. NanoInk ceased operations in 2013, but a number of organizations continue to use DPN for nano-scale DW research and development.

Dip-pen nanolithography is capable of producing 14 nm line widths with 5 nm spatial resolution. It is typically used to produce features on flat surfaces (although uneven topography at the nm scale is unavoidable even on so-called flat surfaces). Various 1D and 2D arrays of pen tips are available, with some 1D 8-pen designs capable of individual tip actuation (either "on" or "off" with respect to each other by lifting individual AFM tips using a thermal bimorph approach) so that not all print heads produce the deposition pattern being traced by the motion controller or so that unused pens can be used for AFM scanning. The scalability of DPN was demonstrated using the PrintArray™, which had 55,000 AFM quills in a square centimeter, enabling 55,000 identical patterns to be made at one time. This array, however, did not enable individual tip actuation.

One use of dip-pen nanolithography is the placement of DNA molecules in specific patterns. DNA is inherently viscous, so the pens used for these materials must be stiffer than for most NanoInks. Also, unique inkwell arrays have been developed to enable multiple tips to be charged with the same ink or different inks for different tips. When combining a multi-material inkwell with an actuated pen array, multi-material nano-scale features can be produced. Based on the physics of adhesion to AFM tips at very small length scales, most inks developed for other DW processes cannot be used with dip-pen nanolithography.

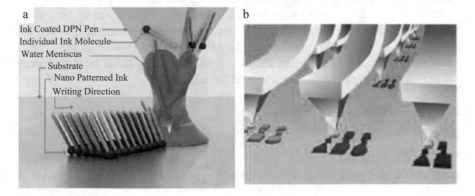

a
Ink Coated DPN Pen
Individual Ink Molecule
Water Meniscus
Substrate
Nano Patterned Ink
Writing Direction

b

Fig. 11.3 (a) A schematic showing how an AFM tip is used to write a pattern on a substrate. (b) An illustration of a 2D array of print heads (55,000 per cm2) [12]. (Photo courtesy of NanoInk)

11.5.2 Inkjet Printing Processes

Thousands of organizations around the world practice the deposition of DW precursor inks using inkjet printing [3, 13]. This is primarily done to form complex electronic circuitry on flat surfaces, as deposition onto a conformal substrate is difficult. The inkjet printing approach to DW fabrication is comparable to MJT, as discussed in Chap. 7. In the case of DW, the print heads and motion control systems are optimized for printing high accuracy electronic traces from DW inks onto relatively flat substrates in one or just a few layers rather than the buildup of three-dimensional objects from low-melting-point polymers or photopolymers.

The primary benefits of inkjet approaches to DW are their speed and low cost. Parallel sets of inkjet print heads can be used to very rapidly deposit DW inks. By setting up arrays of print heads, very large areas can be printed rapidly. In addition, there are numerous suppliers for inkjet print heads.

The primary drawbacks of inkjet approaches to DW are the difficulty inherent with printing on conformal surfaces, the use of droplets as building blocks which can affect material continuity, more stringent requirements on ink rheology, and a limited droplet size range. Since inkjet print heads deposit material in a droplet by droplet manner, the fundamental building block is hemispherical (see Fig. 11.1b). In order to produce consistent conductive paths, for instance, the droplets need a repeatable degree of overlap. This overlap is relatively easy to control between droplets that are aligned with the print head motion, but for deposits that are at an angle with respect to the print head motion, there will be a classic "stair-step" effect, resulting in a change in cross-sectional area at locations in the deposit. This can be overcome by using only a single droplet print head and controlling its motion so that it follows the desired traces (similar to the Solidscape approach to MJT). It can also be overcome using a material removal system (such as a laser) to trim the deposits after their deposition to a highly accurate, repeatable cross-section, giving consistent conductivity, resistivity, or other properties throughout the deposit. However, these solutions to stair stepping mean that one cannot take advantage of the parallel nature of inkjet printing, or a more complicated apparatus is needed.

Most inkjet print heads work best with inks of low viscosity at or near room temperature. However, the rheological properties (primarily viscosity) which are needed to print a DW ink can often only be achieved when printing is done at elevated temperatures. The modeling introduced in Chap. 7 is useful for determining the material types which can be considered for inkjet DW.

11.5.3 Aerosol DW

Aerosol DW processes make deposits from inks or ink-like materials suspended as an aerosol mist. The commercialized version of this approach is the Aerosol Jet process developed by Optomec. The Aerosol Jet process begins with atomization of a liquid molecular precursor or a colloidal suspension of materials, which can include metal, dielectric, ferrite, resistor, or biological materials. The aerosol stream is delivered to a deposition head using a carrier gas. The stream is surrounded by a coaxial sheath air flow and exits the chamber through an orifice directed at the substrate. This coaxial flow focuses the aerosol stream onto the substrate and allows for deposition of features with dimensions as small as 5 μm. Typically either laser chemical decomposition or thermal treatment is used to process the deposit to the desired state.

The Aerosol Jet process can be controlled to be gentle enough to deposit living cells. A schematic illustration of the Aerosol Jet process is shown in Fig. 11.4.

The Aerosol Jet process was initially conceived as a process which made use of the physics of laser guidance. When photons of light interact with free-floating or suspended small particles, there is a slight amount of "force" applied to these particles, and these particles move in the direction of photon motion. When applied to aerosol DW, a laser is transmitted through the mist into a hollow fiber optic. The laser propels tiny droplets from the mist into and through the hollow fiber, depositing the droplets onto a substrate where the fiber ends [14]. Laser guidance, however, entrains and moves droplets slowly and inefficiently. To overcome this drawback, further iterations with the technology involved pressurizing the atomizer and using a pressure drop and flow of gas through the tube between the atomizer and the deposition head as the primary means of droplet propulsion. Lasers can still be used, however, to provide in-flight energy to the droplets or to modify them thermally or chemically. The ability to laser-process the aerosol droplets in-flight and/or on the substrate enables the deposition of a wider variety of materials, as both untreated and coaxially laser-treated materials can be considered.

One benefit of a collimated aerosol spraying process is its high stand-off distance and large working distance. The nozzle can be between 1 and 5 mm from the

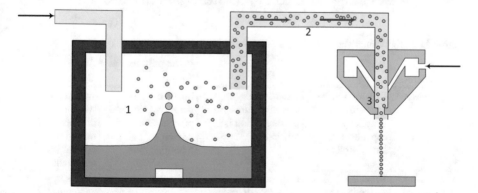

Fig. 11.4 Aerosol Jet System. (1) Liquid material is placed into an atomizer, creating a dense aerosol of tiny droplets 1–5 μm in size. (2) The aerosol is carried by a gas flow to the deposition head (with optional in-flight laser processing). (3) Within the deposition head, the aerosol is focused by a second gas flow, and the resulting high-velocity stream is jetted onto the substrate creating features as small as 10 μm in size. (Courtesy of Optomec)

substrate with little variation in deposit shape and size within that range. This means that repeatable deposits are possible on substrates which have steps or other geometrical features on their surface. A major application for Aerosol Jet is creating interconnects for solar panels, which makes use of its unique ability to deposit conductive traces on a substrate with widely varying stand-off distances.

The Aerosol Jet process is also more flexible than inkjet printing processes, as it can process a wide range of material viscosities (0.7–2500 cPs); it has variable line widths from 5 to 5000 μm and layer thicknesses between 0.025 and 10 μm. The main drawback of the Aerosol Jet process is its complexity compared to other ink-based processes. The Aerosol Jet process has been parallelized to include large numbers of print heads in an array, so the process can be made quite fast and flexible, in spite of its complexity.

Table 11.2 summarizes the key benefits and drawbacks of ink-based approaches to DW.

11.6   Laser Transfer DW

When a focused high-energy laser beam is absorbed by a material, that material may be heated, melted, ablated, or some combination thereof. In the case of ablation, there is direct evaporation (or transformation to plasma) of material. During ablation, a gas or plasma is formed, which expands rapidly as further laser energy is added. This rapid expansion can create a shock wave within a material, or it can propel a material. By focusing the expansion of the material during ablation (utilizing shock waves produced by laser ablation) or taking advantage of rapid thermal expansion inherent with laser heating, materials can be accurately transferred in a very repeatable and accurate manner from one location to another. Laser transfer DW makes use of these phenomena by transferring material from a foil, tape, or

Table 11.2 Key benefits and drawbacks of ink-based approaches to DW

	Nozzle	Quill	Inkjet printing	Aerosol
Manufacturer	nScrypt	NanoInk	Various	Optomec
Key benefits	Greatest range of viscosities, simplicity, capable of 3D lattice structures	Nano-scale structures, massive parallelization is possible	Speed due to parallelization of print heads, numerous manufacturers	Widest range of working distances and line widths, coaxial laser treatment
Key drawbacks	Knowledge of surface topography needed to maintain constant stand-off distance	Only relevant at very small length scales, requires precise motion controllers and custom inks	Need flat plates or low-curvature substrates, limited ink viscosity ranges	Complex apparatus. Requires inks which can be aerosolized

plate onto a substrate. By carefully controlling the energy and location of the impinging laser, complex patterns of transferred material can be formed on a substrate.

Two different mechanisms for laser transfer are illustrated in Fig. 11.5. Figure 11.5a illustrates a laser transfer process where a transparent carrier (a foil or plate donor substrate which is transparent to the laser wavelength) is coated with a sacrificial transfer material and the dynamic release layer (the build material). The impinging laser energy ablates the transfer material (forming a plasma or gas), which propels the build material toward the substrate. The material impacts the substrate and adheres, forming a coating on the substrate. When using a pulsed laser, a precise amount of material can be deposited per pulse.

Figure 11.5b shows a slightly different mechanism for material transfer. In this case the laser pulse ablates a portion of the surface of a foil. This ablation and absorption of thermal energy create thermal waves and shock waves in the material. These waves are transmitted through the material and cause a portion of the material on the opposing side to fracture from the surface in a brittle manner (known as spallation). The fractured material is propelled toward the substrate, forming a deposit coating on the substrate (not shown).

The Matrix-Assisted Pulsed Laser Evaporation DW (MAPLE DW) process was developed to make use of these laser transfer mechanisms [17]. A schematic of the MAPLE DW process is shown in Fig. 11.6. In this process, a laser transparent quartz disk or polymer tape is coated on one side with a film (a few microns thick), which consists of a powdered material that is to be deposited and a polymer binder. The coated disk or tape is placed in close proximity and parallel to the substrate. A laser is focused onto the coated film. When a laser pulse strikes the coating, the polymer is evaporated, and the powdered material is deposited on the substrate, firmly adhering to it. By appropriate control of the positions of both the ribbon and the substrate, complex patterns can be deposited. By changing the type of ribbon, multi-material structures can be produced.

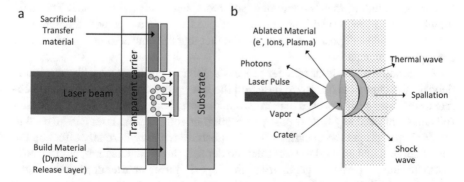

Fig. 11.5 (**a**) Mechanism for laser transfer using a sacrificial transfer material (based on [15]), (**b**) Mechanism for laser transfer using thermal shock and spallation (based on [16]). (Courtesy of Douglas B. Chrisey)

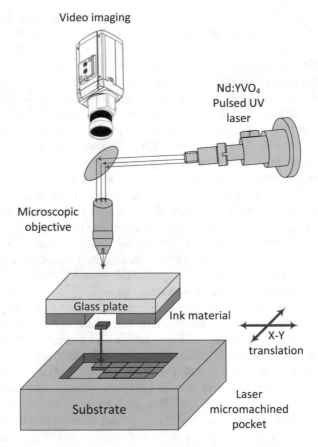

Video imaging

Nd:YVO$_4$
Pulsed UV
laser

Microscopic
objective

Glass plate

Ink material

X-Y
translation

Laser
micromachined
pocket

Substrate

Fig. 11.6 Matrix-Assisted Pulsed Laser Evaporation DW (MAPLE DW) process [18]. (Photo courtesy of PennWell Corp., Laser Focus World)

Laser transfer processes have been used to create deposits of a wide variety of materials, including metals, ceramics, polymers, and even living tissue. The main drawbacks of a laser transfer process are the need to form a tape with the appropriate transfer and/or deposit materials and the fact that the unused portions of the tape are typically wasted.

The benefits of the laser transfer process are that it produces a highly repeatable deposit (the deposit is quantized based on the laser pulse energy), it can be as accurate as the laser scanners used to manipulate the laser beam, and the deposited materials may not need any further post-processing. In addition, the laser can be used to either simply propel the material onto the substrate without thermally affecting the substrate, or it can be used to laser treat the deposit (including heating, cutting, etc.) to modify the properties or geometry of the deposit during or after deposition. In the case of a rigid tape (such as when using a glass plate), the plate is typically mechanically suspended above the substrate. When a flexible polymer tape is used, it can be laid directly onto the substrate before laser processing and then peeled from the substrate after laser processing, leaving behind the desired pattern.

11.7   Thermal Spray DW

Thermal spray is a process that accelerates material to high velocities and deposits them on a substrate, as shown in Fig. 11.7. Material is introduced into a combustion or plasma flame (plume) in powder or wire form. The plume melts and imparts thermal and kinetic energy to the material, creating high-velocity droplets. By controlling the plume characteristics and material state (e.g., molten or softened), it is possible to deposit a wide range of metals, ceramics, polymers, or composites thereof. Particles can be deposited in a solid or semisolid state, which enables the creation of useful deposits at or near room temperature. Thermal spray techniques for DW have been commercialized by MesoScribe Technologies, Inc. [19].

A deposit is built up by successive impingement of droplets, which yield flattened, solidified platelets, referred to as splats. The deposit microstructure, and thus its properties, strongly depends on the processing parameters utilized. Key characteristics of thermal spray DW include (1) a high volumetric deposition rate, (2)

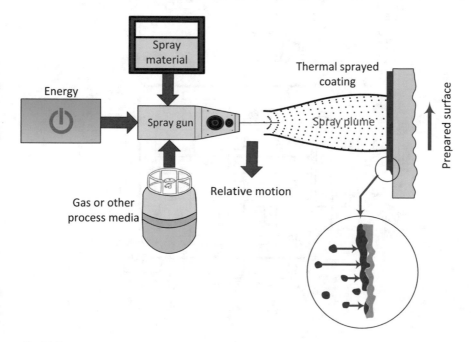

Fig. 11.7   General apparatus for thermal spray [20]. (Courtesy of and (C) Copyright Sulzer Metco All rights reserved)

material flexibility, (3) useful material properties in the as-deposited state (without thermal treatment or curing), and (4) moderate thermal input during processing, allowing for deposition on a variety of substrates.

DW thermal spray differs from traditional thermal spray in that the size and shape of the deposit are controlled by a unique aperture system. A schematic aperture system from an issued patent is shown in Fig. 11.8. This aperture is made up of adjustable, moving metal foils (702a and 702b moving horizontally and 708a and 708b moving vertically) which constrain the plume to desired dimensions (region 720). The distance between the moving foils determines the amount of spray which reaches the substrate. The foils are in constant motion to avoid overheating and buildup of the material being sprayed. The used foils become a waste product of the system.

Because the temperature of the substrate is kept low and no post treatment is typically required, DW thermal spray is well-suited to produce multilayer devices formed from different materials. It is possible to create insulating layers, conductive/electronic layers, and further insulating layers stacked one on top of the other (including vias for signal transmission between layers) by changing between various metal, ceramic, and polymer materials. DW thermal spray has been used to successfully fabricate thermocouples, strain gages, antennas, and other devices for harsh environments directly from precursor metal and ceramic powders. In addition, DW thermal spray, combined with ultrafast laser micromachining, has been shown to be capable of fabricating thermopiles for power generation [21].

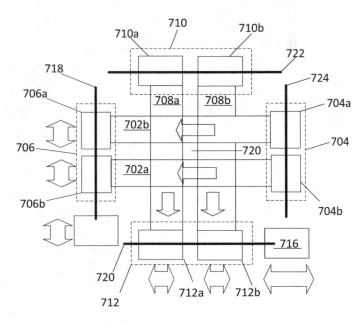

Fig. 11.8 Schematic aperture apparatus for DW thermal spray (US patent 6576861). Foils 702a, b and 708a, b are in constant motion and are adjusted to allow different amounts of spray to reach the substrate through hole 720 in the center

11.8 Electroforming

Electroforming is a technique for producing metal components with high aspect ratios. It is based on electrodeposition and photo-curing processes. Electroforming is a series of steps including cleaning, coating, exposing, developing, deposition, and harvesting as shown in Fig. 11.9.

In the first step, (a) the metal substrate is cleaned and degreased from any contamination. Then according to the designed pattern/part, the substrate is coated by a light-sensitive coating/photoresist mask (b). The masked substrate is then exposed to ultraviolet light that hardens the photoresist layer (c). In this stage laser direct imaging, like that introduced by VECO [22], produces a pattern with high accuracy features. In laser direct imaging, the photo-masks are reusable, thus saving the costs of production. After transferring the image by exposing, the substrate is developed (d), rinsed, and dried and is ready for deposition. In the next step, (e) the part is placed in an electrolytic bath to deposit metal on the patterned area. Electroforming includes electrolytic solvent, two electrodes, and DC current. The current is passed through the solution and transfers metal ions from the anode to cathode by continuous deposition. In this case the materials are accurately deposited on a precise microscale. At the end, the produced parts are harvested from the substrate. The accuracy in electroforming is related to the stress in the material and exposure method. The highest accuracy is obtained when using laser direct imaging. Feature accuracy is ±2 μm with standard deviation ±0.75 μm for 50 μm thickness in geometry. The geometric accuracy of thicker samples drops to ±0.25% of material thickness.

Electroforming has some key benefits:

- Possibility to produce large numbers of micro components
- Ultra-precision and repeatability and ease in controlling the process

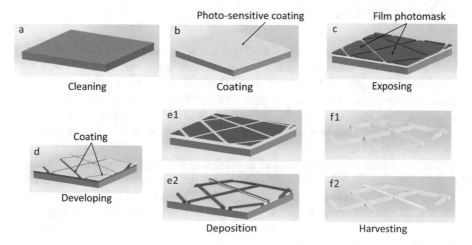

Fig. 11.9 Different steps in electroforming

- Cost-effective for mass production/customization
- Using advanced photo tooling that provides good flexibility in design
- Consistent material properties such as hardness, grain structure, etc.
- Production of sharp edges

Electroforming is typically only feasible for nickel, copper, and iron. Electrodeposited parts have high internal stresses, which can be controlled by stress-reducing additives. Finally, the part must be removed from the substrate, normally by chemical, thermal, or mechanical means.

11.9 Beam Deposition DW

Several DW procedures have been developed based upon vapor deposition technologies using, primarily, thermal decomposition of precursor gases. Vapor deposition technologies produce solid material by condensation, chemical reaction, or conversion of material from a vapor state. In the case of chemical vapor deposition (CVD), thermal energy is utilized to convert a reactant gas to a solid at a substrate. In the regions where a heat source has raised the temperature above a certain threshold, solid material is formed from the surrounding gaseous precursor reactants. The chemical composition and properties of the deposit are related to the thermal history during material deposition. By moving a localized heat source across a substrate (such as by scanning a laser), a complex geometry can be formed. A large number of research groups over 30 years has investigated the use of vapor deposition technologies for Additive Manufacturing purposes [23]. A few examples of these technologies are described below.

11.9.1 Laser CVD

Laser chemical vapor deposition (LCVD) is a DW process which uses heat from a laser to selectively transform gaseous reactants into solid materials. In some systems, multiple gases can be fed into a small reactant chamber at different times to form multi-material structures, or mixtures of gases with varying concentrations can be used to form gradient structures. Sometimes flowing jets of gas are used to create a localized gaseous atmosphere, rather than filling a chamber with the gaseous precursor materials.

The resolution of a LCVD deposit is related to the laser beam diameter, energy density, and wavelength (which directly impact the size of the heated zone on the substrate) as well as substrate thermal properties. Depending on the gases present at the heated reactive zone, many different metals and ceramics can be deposited, including composites. LCVD has been used to deposit carbon fibers and multilayered carbon structures in addition to numerous types of metal and ceramic structures.

A LCVD system developed at the Georgia Institute of Technology is displayed in Fig. 11.10. This design constrains the reactant gas (which is often highly corrosive and/or biologically harmful) to a small chamber that is separated from the motion controllers and other mechanisms. This small, separated reaction chamber has multiple benefits, including an ability to quickly change between reagent gas materials for multi-material deposition, and better protection of the hardware from corrosion. By monitoring the thermal and dimensional characteristics of the deposit, process parameters can be controlled to create deposits of desired geometry and material properties.

LCVD systems are capable of depositing many types of materials, including carbon, silicon carbide, boron, boron nitride, and molybdenum onto various substrates including graphite, grafoil, zirconia, alumina, tungsten, and silicon [25]. DW patterns as well as fibers have been successfully deposited. A wide variety of materials and deposit geometries makes LCVD a viable technology for further DW developments. LCVD is most comparable to microthermal spray, in that deposits of metals and ceramics are directly formed without posttreatment but without the "splat" geometry inherent in thermal spray. The benefits of LCVD are the unique materials and geometries it can deposit. However, LCVD has a very low deposition rate and a relatively high system complexity and cost compared to most DW approaches (particularly ink-based technologies). High-temperature deposition can be another disadvantage of the process. In addition, the need to deposit on surfaces

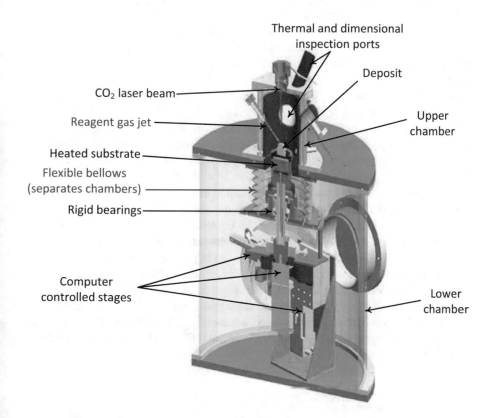

Fig. 11.10 The LCVD system developed at Georgia Tech [24]

that are contained within a controlled-atmosphere chamber limits its ability to make deposits on larger pre-existing structures.

LCVD can be combined with layer-wise deposition of powders (similar to the Binder Jetting (BJT) techniques in Chap. 8) to more rapidly fabricate structures than when using LCVD alone. In this case the solid material created from the vapor phase is used to bind the powdered material together in regions where the laser has heated the powder bed. This process is known as Selective Area Laser Deposition Vapor Infiltration (SALDVI). In SALDVI, the build rates are much higher than when the entire structure is fabricated from LCVD-deposited materials only; but the resultant structures may be porous and are composite in nature. The build rate difference between LCVD and SALDVI is analogous to the difference between BJT and MJT.

11.9.2   Focused Ion Beam CVD

A focused ion beam (FIB) is a beam of ionized gallium atoms created when a gallium metal source is placed in contact with a tungsten needle and heated. Liquid gallium wets the needle, and the imposition of a strong electric field causes ionization and emission of gallium atoms. These ions are accelerated and focused in a small stream (with a spot size as low as a few nanometers) using electrostatic lenses. A FIB is similar in conceptualization to an electron beam source, and thus FIB is often combined with electron beams, such as in a dual-beam FIB-scanning electron microscope system.

FIB processing, when done by itself, can be destructive, as high-energy gallium ions striking a substrate will cause sputtering and removal of atoms. This enables FIB to be used as a nanomachining tool. However, due to sputtering effects and implantation of gallium atoms, surfaces near the machining zone will be changed by deposition of the removed material and ion implantation. This sputtering and ion implantation, if properly controlled, can also be a benefit for certain applications.

Direct write deposition using FIB is possible in a manner similar to LCVD. By scanning the FIB source over a substrate in the presence of CVD gaseous precursors, solid materials are deposited onto the substrate (and/or implanted within the surface of the substrate) [26, 27]. These deposits can be submicron in size and feature resolution. FIB CVD for DW has been used to produce combinations of metallic and dielectric materials to create three-dimensional structures and circuitry. In addition, FIB CVD is being used in the integrated circuits (IC) industry to repair faulty circuitry. Both the machining and deposition features of FIB are used for IC repairs. In the case of short-circuits, excess material can be removed using a FIB. In the case of improperly formed electrical contacts, FIB CVD can be used to draw conductive traces to connect electrical circuitry.

11.9.3   Electron Beam CVD

Electron beams can be used to induce CVD in a manner similar to FIB CVD and LCVD. Electron beam CVD is slower than laser or FIB CVD; however, FIB CVD and electron beam CVD both have a better resolution than LCVD [28].

11.10   Liquid-Phase Deposition

Similar to the vapor techniques described above, thermal or electrical energy can be used to convert liquid-phase materials into solid materials. These thermochemical and electrochemical techniques can be applied in a localized manner to create pre-scribed patterns of solid material.

Drexel University illustrated the use of thermochemical means for DW traces using ThermoChemical Liquid Deposition (TCLD). In TCLD, liquid reactants are sprayed through a nozzle onto a hot substrate. The reactants thermally decompose or react with one another on the hot surface to form a solid deposit on the substrate. By controlling the motion of the nozzle and the spraying parameters, a 3D shape of deposited material can be formed. This is conceptually similar to the ink-based DW approaches discussed above but requires a high-temperature substrate during deposition.

A second electrochemical liquid deposition (ECLD) approach was also tested at Drexel. In ECLD, a conductive substrate is submerged in a plating bath and connected to a DC power source as the cathode, as in Fig. 11.11. A pin made up of the material to be deposited is used as the anode. By submerging the pin in the bath near the substrate and applying an appropriate voltage and current, electrochemical decomposition and ion transfer result in a deposit of the pin material onto the substrate. By moving the pin, a prescribed geometry can be traced. As electrochemical plating is a slow process, the volumetric rate of deposition for ELCD can be increased by putting a thin layer of metal powder in the plating bath on the surface of the substrate (similar conceptually to SALDVI described above). In this case, the deposited material acts as a binder for the powdered materials, and the volumetric rate of deposition is significantly increased [29].

Thermochemical and electrochemical techniques can be used to produce complex geometry solids at small length scales from any metal compatible with thermochemical or electrochemical deposition, respectively. These processes are also compatible with some ceramics. However, these approaches are not available commercially and may have few benefits over the other DW techniques described above. Drawbacks of TCLD-based approaches are the need for a heated substrate and the use of chemical precursors which may be toxic or corrosive. Drawbacks of ECLD-based approaches include the slow deposition rate of electrochemical processes and the fact that, when used as a binder for powders, the resultant product is porous and requires further processing (such as sintering or infiltration) to achieve desirable properties.

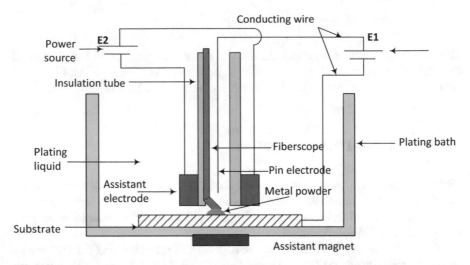

Fig. 11.11 Schematic of an electrochemical liquid deposition system [29]. (MATERIALS & DESIGN by Zongyan He, Jack G. Zhou, and Ampere A. Tseng. Copyright 2000 by Elsevier Science & Technology Journals. Reproduced with permission of Elsevier Science & Technology Journals in the format Textbook via Copyright Clearance Center)

11.11   Beam Tracing Approaches to Additive/Subtractive DW

By combining layer-wise additive approaches with freeform beam (electron, FIB, or laser) subtractive approaches, it is possible to create DW features. Many coating techniques exist to add a thin layer of material to a substrate. These include physical vapor deposition, electrochemical or thermochemical deposition, chemical vapor deposition, and other thin film techniques used in the fabrication of integrated circuits. Once these layers are added across the surface of a substrate, a beam can be used to trim each layer into the prescribed cross-sectional geometry. These micro- or nanodiameter beams are used to selectively cure or remove materials deposited in a layer-by-layer fashion. This approach is conceptually similar to the bond-then-form Sheet Lamination (SHL) techniques discussed in Chap. 9.

11.11.1   Electron Beam Tracing

Electron beams can be used to either cure or remove materials for DW. Standard spin-on deposition coating equipment can be used to produce thin films between 30 and 80 nm. These films are then exposed to a prescribed pattern using an electron beam. Following exposure, the uncured material is removed using standard IC fabrication techniques. This methodology can produce line-edge definition down to 3.3 nm. A converse approach can also be used, whereby the exposed material is removed and the unexposed material remains behind. In the case of curing, low-energy electrons can be utilized (and are often considered more desirable, to reduce the occurrence of secondary electron scattering). These techniques fit well within existing IC fabrication methodologies and enable maskless IC fabrication.

Another variant of electron beam tracing is to produce a thin layer of the desired material using physical vapor deposition or a similar approach and then to use high-powered electron beams to remove portions of the coating to form the desired pattern.

Electron beams are not particularly efficient for either curing or removing layers of material, however. Thus, electron beam tracing techniques for DW are quite slow.

11.11.2 Focused Ion Beam Tracing

As discussed above, a focused ion beam can be utilized to machine materials in a prescribed pattern. By combining the steps of layer-wise deposition with FIB machining, a multilayer structure can be formed. If the deposited material is changed layer-by-layer, then a multi-material or gradient structure can be formed.

11.11.3 Laser Beam Tracing

Short-wavelength lasers can be utilized to either cure layers of deposited materials or ablatively remove materials to form micro and nano-scale DW features. To overcome the diffraction limit of traditional focusing optics, a number of nanopatterning techniques have been developed to create features that are smaller than half the optical wavelength of the laser [30]. These techniques include multiphoton absorption, near-field effects, and Bessel beam optical traps. Although these techniques can be used to cure features at the nano-scale, inherent problems with alignment and positioning at these length scales make it difficult to perform subwavelength nanopatterning in practice.

These laser approaches are conceptually identical to the electron beam and FIB-based additive plus subtractive approaches just mentioned. Some of the benefits of lasers for beam tracing DW are that they can process materials much more rapidly than electron beams, and they can do so without introducing FIB gallium ions. The main drawback of lasers is their relatively large spot size compared to electron and focused ion beams.

11.12 Hybrid Direct Write Technologies

As in most Additive Manufacturing techniques, there is an inherent trade-off between material deposition speed and accuracy for most DW processes. This will remain true until techniques are developed for line-wise or layer-wise deposition (such as is done with mask projection stereolithography using a DLP system, as described in Chap. 4). Thus, to achieve a good combination of deposition speed and accuracy, hybrid technologies are often necessary. Some examples of hybrid technologies have already been mentioned. These include the additive/subtractive beam tracing methods described above and the use of a laser in the Aerosol Jet aerosol system.

The primary form of hybrid technology used in DW is to form deposits quickly and inexpensively using an ink-based technique and then trim those deposits using a short-wavelength laser. This results in a good combination of build speed, accuracy, and overall cost for a wide variety of materials. In addition, the laser used to trim the ink-based deposits has the added benefit of being an energy source for curing the deposited inks, when used in a lower power or more diffuse manner. If DW is integrated with an AM process that includes a laser, such as stereolithography, the laser can be used to modify the DW traces [31].

One of the fastest hybrid DW approaches is the "roll-to-roll" approach. As the name suggests, roll-to-roll printing is analogous to high-speed 2D printing presses. In roll-to-roll DW, a paper or plastic sheet of material is used as a substrate material and moved through printing rolls that deposit patterns of DW ink that are subsequently thermally processed, such as with a flash lamp. Inkjet printing and other DW techniques can be added into the roll-to-roll facility to add ink patterns in a more flexible manner than the repeated patterns printed by a printing roll. When DW techniques and flexible laser systems are integrated into a roll-to-roll facility, this gives the combination of the speed of line-wise AM via the rolls plus the flexibility of point-wise AM via other DW techniques.

11.13 Applications of Direct Write

The applications of DW processes are growing rapidly [32]. There is a growing variety of materials which are available, including semiconductors, dielectric polymers, conductive metals, resistive materials, piezoelectrics, battery elements, capacitors, biological materials, and others. These can be deposited onto various substrate materials including plastic, metal, ceramic, glass, and even cloth. The combination of these types of materials and substrates means that the applications for DW are extremely broad.

The most often cited applications for DW techniques are related to the fabrication of sensors. DW approaches have been used to fabricate thermocouples, thermistors, magnetic flux sensors, strain gages, capacitive gages, crack detection sensors, accelerometers, pressure gages, and more [3, 19, 21, 33].

A second area of substantial interest is in antenna fabrication. Since DW, like other AM techniques, enables fabrication of complex geometries directly from CAD data, antenna designs of arbitrary complexity can be made on the surface of freeform objects including, for instance, fractal antennas on conformal surfaces. Figure 11.12 illustrates the fabrication of a fractal antenna on the abdomen of a worker honeybee using MAPLE DW.

Another area of interest for DW is as a freeform tool to connect combinations of electronic components on freeform surfaces. One area where this is particularly useful is in harsh environments, as shown in Fig. 11.13. In this example, DW thermal spray is used as a method for producing and connecting a series of electronic components that monitor and feedback information about the state of a turbine blade. A thermocouple, labeled high-temperature sensor in the figure is deposited on the hot region of the blade, whereas the supporting electronics are deposited on the cold regions of the blade. DW-produced conductors are used to transmit signals between regions and components.

Although most DW processes can produce thermal and strain sensors, there is still opportunity for improved conductors, insulators, antennas, batteries, capacitors, resistors, and other electronic circuitry. In addition, in every case where a conductive path is not possible between a power source and a deposited sensor, some form of local power generation is necessary. Several researchers have demonstrated the ability to create systems which use electromagnetic or thermopile power generation schemes using DW [21]. If designs for energy harvesting devices can be made robustly using DW, then remote monitoring and sensing of components and parts is possible. For instance, the ability to create a thermal sensor with integrated power harvesting and antenna directly onto an internally rotating component (such as a bearing) within a transmission could provide feedback to help optimize performance of systems from power plants to motor vehicles to jet engines. In addition,

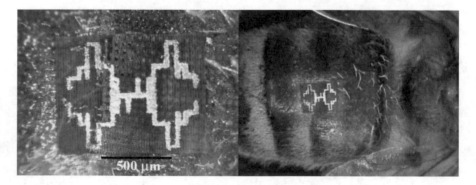

Fig. 11.12 35-GHz fractal antenna design (left) and MAPLE DW printed antenna on the abdomen of a dead drone honeybee (right). (Photo courtesy of Douglas B. Chrisey) [5]

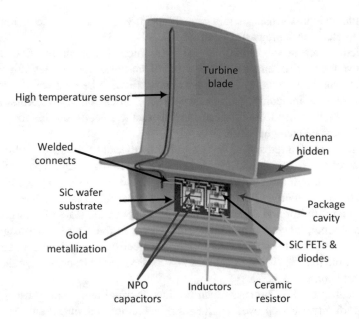

High temperature sensor

Turbine
blade

Antenna
hidden

Welded
connects

SiC wafer
substrate

Package
cavity

Gold
metallization

SiC FETs &
diodes

NPO
capacitors

Inductors

Ceramic
resistor

Fig. 11.13 A DW sensor and associated wiring on a turbine blade structure. Signal conditioning electronics are positioned on a more shielded spot. (Photo courtesy of MesoScribe Technologies, Inc. and Arkansas Power Electronics Int.)

this type of remote sensing could notify the operator of thermal spikes before catastrophic system failure, thus saving time and money.

DW techniques are rapidly growing. Ongoing investments in DW R&D indicate these technologies will continue to expand their application potential to become common methods for creating nano-, micro-, and mesoscale features and devices.

11.14   Technical Challenges in Direct Write

Although inkjet DW is widely used and inexpensive, much research is still needed to expand the process to more applications and materials. Most materials do not have a liquid precursor that can be inkjet printed. Research on jet and drop formation, drop impact, drying, and curing as well as intrinsic physical properties of fluids such as density, heat capacity, heat transfer rate, and their temperature dependencies are needed. To achieve high concentrations of solid particles in liquid-phase polymers, better understanding of the printing of non-Newtonian fluids would be helpful [4].

A further set of challenges involves the creation of low-cost inks that can create stable and long-lasting parts. Such inks would benefit from having optoelectronic, semiconductor, magnetic, and other functional properties. Accuracy is also important, thus limiting the volume of material per droplet, adding further demands to hardware.

Gas-phase deposition is limited to volatile meta-organic or inorganic materials and is sensitive to contamination. Although deposition can be achieved with 2D and 3D substrates, these techniques face challenges for deposition onto organic substrates like animal cell structures.

Laser-based DW processes generally suffer from low build rate. Also, laser-based processes have lower dimensional accuracy (>100 nm) compared to electron beam and focused ion beam [4, 34].

Additional technical challenges of DW that need to be overcome include:

- Lack of accurate and reliable process modeling and control
- A better understanding of rheological characteristics of deposited materials
- Determination of the physical properties of materials at small length scales
- A limited range of available materials
- Control of edge formation and elimination of necking at micro and nano-scales
- Complexity and nonlinearity of many DW processes

As these and other technical hurdles are overcome, the range of applications using DW will continue to grow.

11.15   Questions

1. From an internet search, identify two DW inks for conductive traces, one ink for resistors and one for dielectric traces that are commonly used in nozzle-based systems. Make a table which lists their room temperature properties and their primary benefits and drawbacks.
2. For the inks identified in problem 1, estimate the printing number Sect. 7.7. List all of your assumptions. Can any of the inks from problem 1 be used in an inkjet printing system?
3. Would you argue that DW techniques are a subset of AM technologies (like Powder Bed Fusion (PBF) or DED), or are they more an application of AM technologies? Why?
4. Two techniques for accelerating DW were discussed in this chapter where DW deposition was used to bind powders to form an object. What other DW techniques might be accelerated by the use of a similar approach? How would you go about doing this? What type of machine architecture would you propose?
5. Research thermocouple types that can withstand 1000 °C. Based on the materials that are needed, which DW techniques could be used to make these thermocouples and which could not?
6. What are hydrogels used for, made of, and how do they work?
7. Name one AM technology from a previous chapter that was not discussed in this chapter that could potentially be used as a Direct Write technology. How could you use this technology for DW purposes?

8. Satellite panels often are metal structures that hold electronics, wiring, heat pipes, fibers, sensors, antennas, and other satellite-related capabilities. If you wanted to fabricate advanced satellite structures using the combination of UAM and DW technologies, what type of DW technologies would you use, and what might the benefits of this combination offer compared to traditional satellite panel design?

9. A small college just received permission to purchase one direct write machine. Their objective is to allow students to learn by hands on experience. If you were in charge of deciding which DW machine to purchase and use, what would you choose and why?

References

1. Mortara, L., et al. (2009). Proposed classification scheme for direct writing technologies. *Rapid Prototyping Journal, 15*(4), 299–309.
2. Abraham, M. H., & Helvajian, H. (2004). Laser direct write for release of SiO 2 MEMS and nano-scale devices. In *Fifth International Symposium on Laser Precision Microfabrication*. International Society for Optics and Photonics.
3. Piqué, A., & Chrisey, D. B. (2001). *Direct-write technologies for rapid prototyping applications: sensors, electronics, and integrated power sources*. Elsevier.
4. Hon, K., Li, L., & Hutchings, I. (2008). Direct writing technology—Advances and developments. *CIRP Annals, 57*(2), 601–620.
5. Pique, A. & Chrisey, D. B. (2001). *Direct-write technologies for rapid prototyping applications: sensors, electronics, and integrated power sources*. Elsevier.
6. Pouliquen, O., Nicolas, M., & Weidman, P. (1997). Crystallization of non-Brownian spheres under horizontal shaking. *Physical Review Letters, 79*(19), 3640.
7. Torquato, S., Truskett, T. M., & Debenedetti, P. G. (2000). Is random close packing of spheres well defined? *Physical Review Letters, 84*(10), 2064.
8. Puippe, J. C., Acosta, R., & Von Gutfeld, R. (1981). Investigation of laser-enhanced electroplating mechanisms. *Journal of the Electrochemical Society, 128*(12), 2539–2545.
9. Li, B., et al. (2007). A robust true direct-print technology for tissue engineering. In *ASME 2007 International Manufacturing Science and Engineering Conference*. American Society of Mechanical Engineers.
10. FabatHome. (2020). www.fabathome.org
11. nScrypt designs and manufactures. (2020). www.nscryptinc.com
12. NanoInk Inc. (2020). https://scitech.com.au/nanotechnology-surface-metrology/nanoink/
13. Szczech, J., et al. (2000). Manufacture of microelectronic circuitry by drop-on-demand dispensing of nano-particle liquid suspensions. *MRS Online Proceedings Library Archive, 624*.
14. Essien, M., Renn, M., & Keicher, D. (2002). Development of mesoscale processes for direct write fabrication of electronic components. In *Proceedings of the Conference on Metal powder deposition for rapid prototyping*.
15. Paul Scherrer Institute (PSI). (2020). https://www.psi.ch/lmx-interfaces/
16. Young, D., & Chrisey, D. B. (2000). *Issues for Tissue Engineering by Direct Issues for Tissue Engineering by DirectWrite Technologies Write Technologies*. U.S. Naval Research Laboratory, Washington, DC.
17. Fitz-Gerald, J., et al. (2000). Matrix assisted pulsed laser evaporation direct write (MAPLE DW): A new method to rapidly prototype active and passive electronic circuit elements. MRS Online Proceedings Library Archive. 625.
18. Laser Focus World. (2020). https://www.laserfocusworld.com/

19. Sampath, S., et al. (2000). Thermal spray techniques for fabrication of meso-electronics and sensors. MRS Online Proceedings Library Archive. 624.
20. Oerlikon. (2020). https://www.oerlikon.com/en/
21. Chen, Q., et al. (2004). Ultrafast laser micromachining and patterning of thermal spray multilayers for thermopile fabrication. *Journal of Micromechanics and Microengineering, 14*(4), 506.
22. VECO, Precision metal. (2020). https://www.vecoprecision.com/
23. Kadekar, V., Fang, W., & Liou, F. (2004). Deposition technologies for micromanufacturing: A review. *Journal of Manufacturing Science and Engineering, 126*(4), 787–795.
24. Bondi, S., et al. (2006). Laser assisted chemical vapor deposition synthesis of carbon nanotubes and their characterization. *Carbon, 44*(8), 1393–1403.
25. Duty, C., Jean, D., & Lackey, W. (2001). Laser chemical vapour deposition: Materials, modelling, and process control. *International Materials Reviews, 46*(6), 271–287.
26. Hoffmann, P., et al. (2000). Focused electron beam induced deposition of gold and rhodium. MRS Online Proceedings Library Archive. 624.
27. Longo, D. M., & Hull, R. Direct focused ion beam writing of printheads for pattern transfer utilizing microcontact printing. *MRS Online Proceedings Library Archive, 624.*
28. Bhushan, B. (2017). *Springer handbook of nanotechnology.* Springer.
29. He, Z., Zhou, J. G., & Tseng, A. A. (2000). Feasibility study of chemical liquid deposition based solid freeform fabrication. *Materials and Design, 21*(2), 83–92.
30. McLeod, E., & Arnold, C. B. (2008) Laser direct write near-field nanopatterning using optically trapped microspheres. In *Lasers and Electro-Optics, 2008 and 2008 Conference on Quantum Electronics and Laser Science. CLEO/QELS 2008. Conference on.* IEEE.
31. Palmer, J., et al. (2005). Stereolithography: a basis for integrated meso manufacturing. In *16th Annual Solid Freeform Fabrication Symposium.*
32. Church, K. H., Fore, C., & Feeley, T. (2000). Commercial applications and review for direct write technologies. *MRS Online Proceedings Library Archive, 624.*
33. Lewis, J. A. (2006). Direct ink writing of 3D functional materials. *Advanced Functional Materials, 16*(17), 2193–2204.
34. Nagel, D. (2002). Technologies for micrometer and nanometer pattern and material transfer. In *Direct write technologies for rapid prototyping applications* (pp. 557–701). New York: Academic Press.

Chapter 12
Hybrid Additive Manufacturing

Abstract As manufacturing systems have evolved from manual to powered to automated processes over the past 200 years, it is common to find examples of multiple manufacturing systems combined into a single, hybrid machine to increase manufacturing efficiencies for certain categories of parts. In additive manufacturing, this trend is also accelerating. The most common hybrid AM systems involve the inclusion of a material removal (e.g., machining or cutting) step into the AM process chain. But hybrid AM systems go far beyond this one instantiation to include multiple AM processes in a single machine, combinations of traditional and AM manufacturing in a single machine, and more. In this chapter we explore various types of hybrid AM approaches and the unique benefits these hybrid machines enable.

12.1 Hybrid Manufacturing

Today's manufacturing systems are evolving quickly to include the flexibility for dynamic changes, high intelligence for autonomy, and sustainability for environmental benefits. Over the last few decades, many new manufacturing techniques have developed, but it is difficult to find a single, stand-alone manufacturing process which satisfies the desired flexibility. Hybrid Manufacturing (HM) has emerged to make use of the benefits of two or more processes to obtain desired aspects within a single workstation to produce parts with better quality or lower lead time [1, 2]. In HM the relative merits of each process can be harnessed, resulting in robust geometrical accuracy and mechanical properties. The ability to work, measure, and then rework material provides a higher quality of component with fewer defects. HM revolves around the improvement in the whole process rather than the detail of each process stage [3, 4].

The basic premise in a hybrid process is that the result of the combined hybrid process is greater than the sum of the individual processes [5]. An alternative definition of hybrid processes revolves around simultaneous control of the interaction between various mechanisms and operations to achieve a more effective result [6, 7].

© Springer Nature Switzerland AG 2021

I. Gibson et al., *Additive Manufacturing Technologies*,

https://doi.org/10.1007/978-3-030-56127-7_12

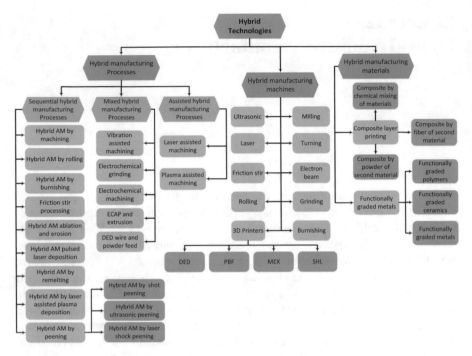

Fig. 12.1 Examples of hybrid technologies used in manufacturing

For instance, in laser-assisted milling, a laser is directed toward the contact point between the cutting tool and workpiece to ease the material removal process. HM technologies are divided into three main areas including machines, material, and processes, as shown in Fig. 12.1. Some of the AM processes discussed previously, such as Ultrasonic Additive Manufacturing, could be considered hybrid processes. However, those that were introduced in prior chapters will not be presented in detail here.

12.2   Hybrid Manufacturing Processes

Based on CIRP [6], Hybrid Manufacturing is classified into assisted and mixed processes. In assisted HM a secondary process assists the primary process in order to improve the overall manufacturing results. In a mixed HM process, the processes occur simultaneously. In some instances the exact same combination of processes occurs, and the difference between assisted and mixed processes is just the timing.

In thermally assisted machining, a heat source elevates the temperature of the workpiece to provide softening, which decreases the cutting force and tool damage, making the material easier to cut. When a heat source heats the part simultaneously with machining, it is a mixed process. When, for instance, a laser is applied before the cutting tool, it is an assisted process. Lasers are commonly used in hybrid

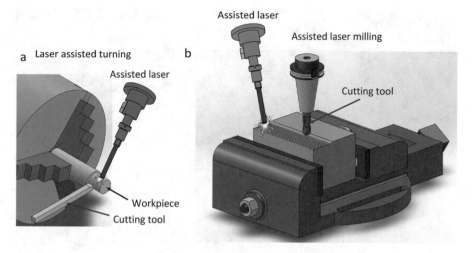

Fig. 12.2 Hybrid assisted manufacturing, (**a**) laser-assisted turning and (**b**) laser-assisted milling [8–12]

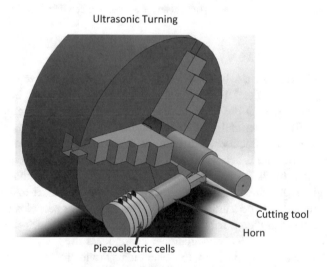

Fig. 12.3 Assisted hybrid manufacturing Ultrasonic Machining (turning)

processes as a heat source because of the beam controllability and easy delivery. Figure 12.2 shows thermally assisted turning and milling.

In mixed HM, the primary and secondary processes happen simultaneously [5, 6, 13–17]. Vibration-assisted machining, electrochemical machining, electrochemical grinding, equal channel angular pressing (ECAP) with extrusion, and dual powder and wire Directed Energy Deposition (DED) are all mixed HM methods. Vibration-assisted machining uses either acoustic softening or small reciprocating motions which periodically loses contact with the surface to aid the cutting process, thus improving surface finish and accuracy with near-zero burr [18] (Fig. 12.3).

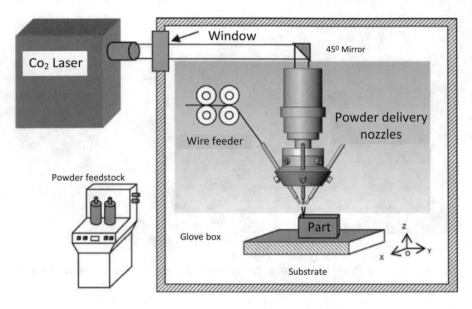

Fig. 12.4 Dual DED powder and wire feed (Elsevier license number: 4687471305451)

It is possible to create a hybrid AM process from a single AM type. For instance, the dual wire and powder DED process is considered a mixed hybrid DED AM method. As can be seen in Fig. 12.4, powder and wire can be fed simultaneously to create a part [19]. Powder and wire feed rates can be controlled separately to produce different volume fractions of a composite material. Furthermore, when fine detail is needed, powder is deposited, and when coarse features are required, wire is deposited. This exhibits the potential to produce different feature characteristics in different regions. As an example, processing TiC powder and Ti-6Al-4V wire by this hybrid method showed a uniform distribution of powder particles in the bulk of the material, and significant tribological and mechanical changes were observed when the fraction of powder reinforcement changed. If the volume ratios of these materials change during deposition, the laser energy reflection and absorption characteristics change, which can cause control difficulties. The simultaneous processing of multiple materials, particularly if in different forms, introduces challenges in process planning and control that need to be overcome before such hybrid processes can achieve wide applicability.

The most common hybrid AM approach is to combine an AM process for creation of a near-net-shape part with machining to ensure geometric accuracy. The DMG MORI Lasertec process [20, 21] applies this approach to combine DED, milling, and turning in a single platform.

12.3   Hybrid Additive Manufacturing Principles

In hybrid AM, subtractive or surface enhancement processes are commonly used as a secondary process to improve accuracy, physical properties, or microstructure of the printed parts. The secondary process can be subtractive like milling and turning or a surface enhancement method such as shot peening, remelting, laser shock peening, rolling, or burnishing that are sequentially applied following all or part of the AM (primary) process. Hybrid AM is defined by concepts that are explained in the next section.

12.3.1   Inseparable Hybrid Processes

Inseparable means the secondary process cannot be separated from the primary AM process. For instance, an AM process with a milling process to remove material following the deposition of each layer is an inseparable hybrid AM if milling is required to form the layer cross-section. However, post-processing such as machining after the AM part is completed does not constitute an inseparable hybrid process. Pre-processing such as preheating is also not included [22, 23].

12.3.2   Synergy in Hybrid AM

Synergy states that the secondary process can be applied either simultaneously or in a cyclic manner after deposition of one or more layers to improve the process and part quality. Most of the hybrid AM processes are cyclic-based, and the secondary process is applied after the AM process. A simultaneous process example is where laser assistance is utilized during Plasma Arc Additive Manufacturing (PAAM).

12.3.3   Hybrid Materials

Hybrid materials are a combination of two or more materials, which have different physical, chemical, and/or mechanical properties. AM has exceptional potential to produce parts with hybrid materials such as composites and functionally gradient materials. Hybrid materials produce enhanced functionality performance compared to single materials. Hybrid materials have been produced using every AM process but are easiest to achieve in Sheet Lamination (SHL), DED, Material Extrusion (MEX), Material Jetting (MJT), and Binder Jetting (BJT). Hybrid material components can be useful, for instance, in high-speed transmission shafts, where the contact surfaces should have higher wear resistance, and other parts of the shaft are

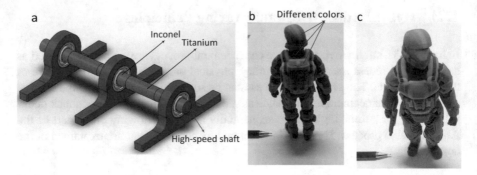

Fig. 12.5 Application for hybrid material (**a**) high-speed shaft made by Inconel and titanium (**b**) figurine made using different material mixtures

preferably made from the materials with lower weight and better thermal conductivity. Figure 12.5 shows examples of hybrid materials using metals and polymers.

12.3.4   Part Quality and Process Efficiency

Part quality and the process efficiency are often improved using hybrid AM. These improvements can be related to tailoring the mechanical properties, surface enhancements, and reducing dimensional deviations. However, in most hybrid AM (except in simultaneous hybrid AM), the secondary process does not assist the primary build process. For instance, in hybrid AM using subtractive machining, the material removal process has no effect on the build rate, but it does affect the overall time needed to produce a component.

12.4   Sequential Hybrid AM Classification Based on Secondary Processes

Hybrid AM processes are categorized based upon the secondary process. In general these processes are designed to provide surface enhancements. These surface enhancements can be used to increase accuracy or change residual stress by material removal or to enhance properties such as hardness, corrosion resistance, and more by processes such as rolling and peening. Hybrid AM can be further classified by the physical state of the material that is utilized during the secondary process. This classification can involve the same secondary processes but for different purposes. Properties can be changed in the (I) solid-state such as by conventional machining or shot peening [24–26], in a (II) semi-solid-state such as by friction stir processing [27], and in a (III) molten state such as by laser shock peening and laser-assisted PAAM [28, 29]. Figure 12.6 shows many of the secondary processes that are used in hybrid AM.

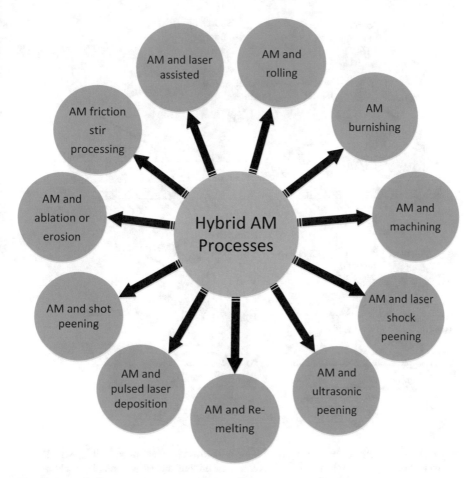

Fig. 12.6 Secondary hybrid AM processes

12.4.1 Hybrid AM by Machining

When a material removal process is used after deposition of each layer, the process is called hybrid AM by machining. Different AM processes such as Powder Bed Fusion (PBF), DED, and SHL can be coupled with machining to make a hybrid process. In hybrid AM by machining, both sidewalls and top surface can be machined after each layer deposition [30–32]. When sidewalls are machined, the objective is typically to improve final surface finish and dimensional accuracy. When face machining is used, the objective is typically to make a proper surface for the next layer.

Hybrid AM can introduce problems into the AM process that may affect the part's final properties and the ability of layers to be joined to one another. For instance, secondary thermal processes which create an oxide layer can reduce the performance of the final part, such as when using wire arc AM [31]. Hybrid AM by machining can be applied to many AM techniques that use solid feedstock such as

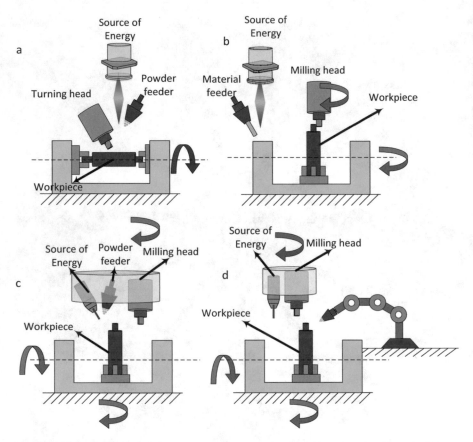

Fig. 12.7 Hybrid DED by machining, (**a**) turning with PAAM, (**b**) multi-axis milling with PAAM (**c**) integrated laser powder head DED with multi-axis milling, and (**d**) separate laser and powder head DED with multi-axis milling

PBF, MEX, MJT, BJT, and SHL, but the chips produced during the machine operation may interfere with and/or become trapped within subsequent layers.

DED is the most commonly reported process for hybrid systems. For instance, DED processes involving laser cladding can be performed without a working chamber, and therefore the installation of DED into machining centers is easy and cost-effective. DMG Mori has introduced two large-scale machines for hybrid AM. The LASERTEC 65 3D hybrid combines DED with a 5-axis machining center for part fabrication, repair, or coating. The next machine is their LASERTEC 125 3D that is larger and has a maximum table load of 2000 kg and 14,000 rpm spindle speed [20, 21]. Figure 12.7 shows different combinations of hybrid DED and machining.

Several companies are combining PBF with machining. Coherent Laser's Creator hybrid PBF utilizes 3-axis milling for small components, and the machining process is used cyclically, typically after each ten layers [33]. Matsuura has launched two types of medium-sized hybrid PBF and milling machines including the Avance-25 and Avance-60. A spindle speed up to 45,000 rpm provides high-speed machining for faster processing and finer surface finish [34].

12.4.2   Hybrid AM by Rolling

AM combined with rolling improves the surface of as-built components between layers and/or after completion of the entire component. Rolling can improve bonding and reduce surface waviness between deposits within a layer without producing chips (unlike machining). Smoothing overlapping regions reduces defects such as keyholes, porosity, and cracks. This secondary process can relax surface residual stresses, especially for thermal-based AM processes. Rolling is not thermal-based, so the risk for thermal degradation during the secondary process decreases. In addition, controlling the rolling process causes tailored grain structure and mechanical properties, decreasing distortion and increasing dimensional accuracy. Figure 12.8a

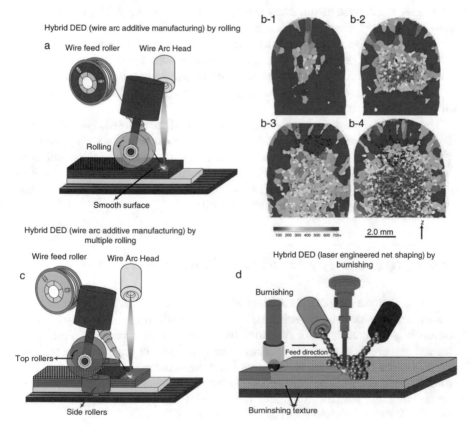

Fig. 12.8 (**a**) Hybrid AM by rolling; (**b**) EBSD reconstructed β grain size maps from Ti-6Al-4V rolled WAAM walls, (**b**-1, **b**-2) single roll pass with 50kN and 75kN loads, and (**b**-3, **b**-4) multiple rolling pass after deposition of each layer with 50kN and 75kN loads [35]; (**c**) multiple pass rolling; and (**d**) hybrid AM by burnishing

shows a schematic of a hybrid DED process with rolling. Figure 12.8b shows the EBSD reconstructed β grain size maps for a Ti-6Al-4V rolled wall; the view is a transverse, YZ cross-section close to the top of the wall. The effect of a single rolling pass with 50kN and 75kN rolling force shows that increasing the load produces finer grains (Fig. 12.8b-1 and b-2, respectively). When rolling is applied after deposition of each layer, a finer microstructure is obtained (b-3 and b-4, respectively) [35]. This secondary process can utilize single or multiple rolling passes (Fig. 12.8c). When using a single pass, rolling occurs on only the top surface of deposited layers. In the case of using multi-rolling (for instance, three rolls), side surfaces are also rolled, and better mechanical properties are obtained compared to single rolling. Hybrid AM by using rolling can be performed at high temperature, producing refined grain structure and improved mechanical properties.

12.4.3 Hybrid AM by Burnishing

Burnishing is a surface treatment by plastic deformation using sliding contacts such as by balls and cylinder (Fig. 12.8d) by inducing material flow on the surface. This improves the surface quality of AM parts by removing surface features such as peaks and valleys. Burnishing also refines microstructure and improves residual stress and hardness. Due to the high pressure of burnishing, the affected depth is higher than rolling. However, higher pressure in burnishing can produce rougher surfaces [7, 24].

The number of passes in burnishing is a function of the material. For soft materials, multi-pass burnishing is not suitable leading to increased surface roughness and creation of fracture points. For harder materials multi-pass burnishing improves surface characteristics, microstructure, and mechanical properties. When post heat treatment is carried out on burnished samples, fine microstructure is produced and cracks, pores and keyholes can be removed [36].

12.4.4 Hybrid AM by Friction Stir Processing

Friction stir processing can be applied after deposition of each layer to improve bonding and mechanical properties. Hybrid AM by friction stir processing has been successfully demonstrated for AM composites, metals, and polymers. Finer grain size and higher hardness can be obtained. The refinement in grain size is a function of the distance between the bulk material and the surfaces. In hybrid AM of hard materials by friction stir processing, due to the higher mechanical forces, tool wear can be observed, which can damage the material and produce rougher surfaces. Figure 12.9 shows a hybrid EBM and FSP process [37].

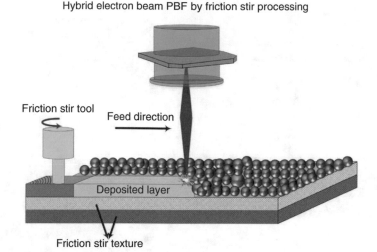

Hybrid electron beam PBF by friction stir processing

Fig. 12.9 Hybrid Electron Beam Powder Bed fusion (EB-PBF) by friction stir processing

12.4.5 Hybrid AM by Ablation or Erosion

In this process, the deposited layer is eroded or ablated by electron beam or laser. The secondary process subtracts a thin layer from the primary deposited layer to produce a smooth and precise cut. This improves the surface quality of outer surfaces, increases density, and can produce features with 50–100 μm accuracy. The advantage of this method over hybrid AM by conventional machining is that it is a noncontact subtractive process which reduces the chance of breaking and scratching the surface. However, ablation and erosion by electron beam increases the thermal residual stress, which can be a problem in materials with low thermal conductivity such as Ti alloys [38]. Figure 12.10a shows the schematic of Hybrid Laser Beam Powder Bed Fusion (LB-PBF) by ablation and (Fig. 12.10b) the surface before and after ablation.

12.4.6 Hybrid AM by Peening

12.4.6.1 Hybrid AM by Shot Peening

Shot peening uses a continuous stream of impacts from hard materials (beads) directed onto the part surface (Fig. 12.11). In shot peening, beads with high kinetic energy induce plastic deformation and increase work hardening when they bounce off the surface of an object. Bead material and hardness are selected according to the properties of the printed object, which can be metal, ceramic, or glass. Shot peening improves microstructure, mechanical properties, and surface quality. In

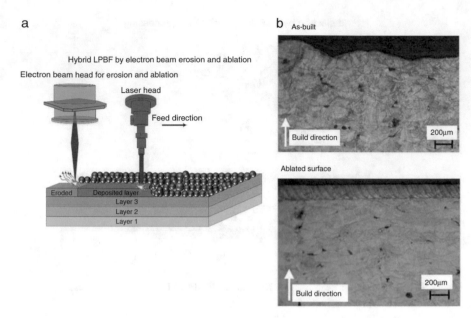

Fig. 12.10 Hybrid LB-PBF by electron beam erosion and ablation (**b**) SEM image of as-built and ablated material (Elsevier license number 4630720259893) [38]

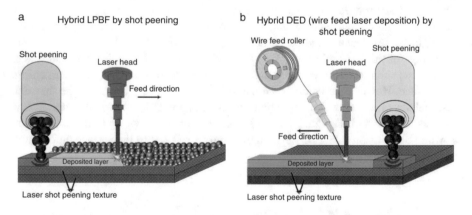

Fig. 12.11 Hybrid LB-PBF and DED by shot peening

thermal-based AM processes, high residual stresses are normally produced. Peening in hybrid AM helps to reduce these residual stresses and the likelihood of distortion [39]. Shot peening is an excellent secondary process for DED, MEX, cold spray, and SHL AM methods. In cold spray AM, shot peening is beneficial for the formation of metallurgical bonding. The strengthened peening effect improves bonding and reduces the porosity of cold sprayed AM parts [40].

Due to small bead size, when using shot peening for powder feed AM processes, chips and particles are produced, especially for soft materials like polymers. This results in powder contamination, so sifting and other manual operations are needed when recycling beads, which increases the cost of production. When using shot peening with powder AM processes, to avoid powder contamination, an alternative option is to use the same powder for peening as is used in the AM process [36]. Since AM powders are typically smaller than standard beads, this results in fine particle shot peening, which has a lower penetration depth. Fine particle shot peening is not a proper choice for soft materials such as polymers because of the high risk of bead deformation. Deformed beads produce rougher surfaces, so further post-processing may be necessary which further adds to the cost.

12.4.6.2   Hybrid AM by Ultrasonic Peening

In ultrasonic peening, ultrasonic energy (over 20 kHz) is used rather than kinetic energy from flying beads. An ultrasonic tool vibrates against the deposited material and refines the microstructure and mechanical properties. Hybrid AM by ultrasonic peening is a low-cost and rapid process compared with many other methods. This surface treatment enhances surface quality, fatigue resistance, corrosion resistance, tribological performance, and microstructure. This approach has been shown to produce large improvements in average global and local mechanical properties by producing columnar grain refinement as well as relieving induced residual stresses. Ultrasonic peening can be used in situ within the AM process for dynamically changing mechanical behavior and residual stress in a tunable fashion. Figure 12.12 shows the schematic of hybrid AM by ultrasonic peening [41].

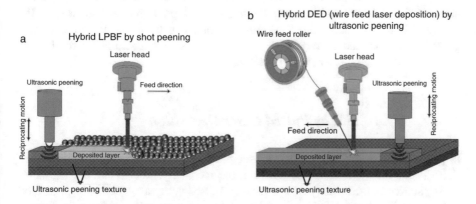

Fig. 12.12 Hybrid LB-PBF and DED by ultrasonic peening

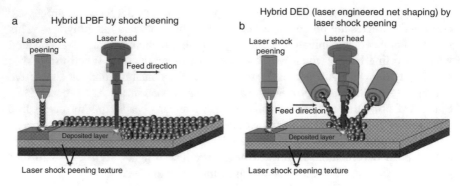

Fig. 12.13 Hybrid LB-PBF and DED by laser shock peening

12.4.6.3 Hybrid AM by Laser Shock Peening

In laser shock peening, shock waves are produced by recoil pressure generated by plasma formed from laser-material interactions. The shock can be amplified through the use of one or two overlays. A transparent overlay material, typically water, can be used to confine the plasma. A second overlay, an opaque material such as tape, can be used to further confine the plasma and protect the part surface. Overlays, however, complicate integration with AM and are often skipped. The advantage of laser shock peening, over other peening methods, is that compressive residual stresses are formed deeper in the part material. Laser shock peening in Hybrid AM is typically applied after deposition of each layer but can also be applied after the deposition of multiple layers. It is reported [42, 43] that in LB-PBF of stainless steel 316 L to obtain lower compressive residual stress, the laser shock peening has to be applied after each layer. In contrast, when laser shock peening is performed after multiple layers, a higher compressive residual stress is obtained. The higher the layer overlap in laser shock peening, the higher the depth of residual stress [24]. By changing how often laser shock peening is applied between layers, functionally graded properties can be produced. Figure 12.13 shows a schematic for hybrid LB-PBF and DED by laser shock peening.

12.4.7 Hybrid AM by Pulsed Laser Deposition

In pulsed laser deposition, a pulsed laser melts the deposited material, and laser shock peens the material. In this method, the secondary process can use the same energy source as the primary AM process. A pulsed laser can generate high surface energy densities up to 10^{18} W/m^2, and vaporization occurs that leads to the formation of plasma and recoil pressure and a protected layer. This deforms the deposited layer, and the microstructure and mechanical properties are enhanced. This method

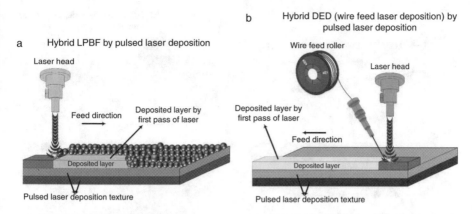

Fig. 12.14 (**a**) Hybrid LB-PBF by pulsed laser deposition and (**b**) Hybrid DED (wire feed laser deposition) by pulsed laser deposition (both processes used a single laser for deposition and secondary processing)

can control and reduce the residual stress, but controlling the temperature of the heat source is very complex and needs extensive knowledge of melting, solidification, phase change, thermomechanical properties of the material, and more. Figure 12.14 shows the hybrid LB-PBF and DED by pulsed laser deposition.

12.4.8   Hybrid AM by Remelting

Similar to hybrid AM by ablation and erosion, for hybrid AM by remelting, the energy source can be a laser or electron beam. The key feature of this method is using low laser energy, which prevents any vaporization or generation of plasma. In general, remelting increases the surface temperature to the material's melting point to melt a portion of the surface. Pores, keyholes, and inclusions are filled during remelting, leading to increased density and part quality. Remelting can be applied to each layer, after several layers, or only to final part surfaces. Remelting changes the microstructure in the remelted regions and provides the ability to tailor the mechanical properties. Remelting can improve fatigue and toughness properties. A disadvantage of remelting is an increase in total printing time [7, 24].

As-built LB-PBF parts have surface finishes in the range of 15–20 µm Ra, and it is reported that hybrid AM by remelting improves the surface finish to 2–8 µm Ra [38, 44–46]. If the remelting operation heats the surface to slightly below the melting temperature, increasing the number of passes has no effect on mechanical properties. Remelting also can be applied on DED system and improves the relative density, mechanical properties, and surface quality. Figure 12.15 shows the schematic of hybrid AM (LB-PBF and DED) by laser remelting.

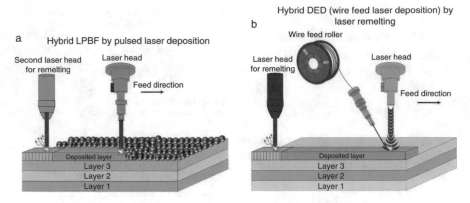

Fig. 12.15 (**a**) Hybrid LB-PBF by laser remelting and (**b**) Hybrid DED by remelting

12.4.9   Hybrid AM by Laser-Assisted Plasma Deposition

In Plasma Arc Additive Manufacturing (PAAM), due to high recoil pressure and cooling rate, rough surfaces are produced, and so the quality of the surface is lower than for many AM processes. To improve the quality of the AM process in PAAM, laser assistance as a secondary source of energy can be used simultaneously with plasma deposition. The assisted laser provides extra energy that helps in the formation of the plasma and improves the deposition process. The assisting laser can increase the plasma's energy density and reduce the plasma diameter to increase precision. These characteristics lead to a deeper melt pool, improved microstructure, and decreased porosity, compared to plasma deposition without the assisting laser [7, 24, 47]. Figure 12.16 shows the schematic of laser-assisted PAAM.

12.5   Summary

Hybrid AM uses two or more processes synergistically to improve the performance of printing or the properties of the produced part. The secondary processes are normally aimed at improvement of surface quality and microstructure of printed components. Tailored mechanical properties such as strain to failure, ultimate stress, hardness, residual stress, etc. are possible. Many materials, including ceramics, composites, polymers, and metals, can be processed by hybrid AM.

The secondary process in Hybrid AM often eliminates common defects in AM processes such as cracks, keyholes, pores, residual stress, the effect of overlapping, etc. Each secondary process is suitable for a specific application that must be selected according to these requirements. Secondary processes in hybrid AM can be based on cold working such as rolling, burnishing, machining, and shot peening. The secondary process can also utilize or generate heat, like in laser-assisted,

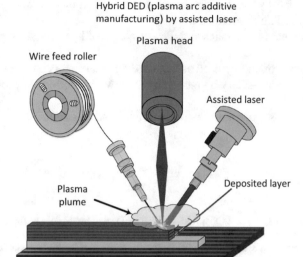

Fig. 12.16 Hybrid PAAM by assisted laser

friction stir, electron beam assisted, pulsed laser deposition, and remelting Hybrid AM. Hybrid AM offers significant technology development potential for next-generation AM machines.

12.6 Questions

1. What is a hybrid process in manufacturing? Why should a hybrid process be used?
2. What are the three mixed hybrid manufacturing processes? Please provide an example that shows how secondary processes operate and help to assist the primary process?
3. What are the benefits and drawbacks of hybrid AM?
4. In medical applications, the surface characteristics of an implant are among the most important factors. Which types of hybrid AM processes do you recommend to produce parts with the best surface quality for sliding part in knee joint?
5. For an application where a smooth surface and high surface hardness is required, which sort of hybrid AM is more suitable for these parts?
6. What are some advantages of hybrid AM by ultrasonic peening?
7. A group of researchers is looking to make a jet engine component made from titanium. The part is very complex and needs to have a good surface finish. Researchers wanted to use hybrid PBF by ablation or erosion because they think this method is cheap and suitable. However, an adviser recommends they

use hybrid PBF by machining. Discuss these recommendations, and explain which one you think should be used.

8. What are the main advantages and disadvantages of hybrid AM by pulsed laser deposition?
9. Why are lasers commonly used as assisted/secondary processes in hybrid AM?
10. What are the main issues when using hybrid AM by friction stir processing for hard materials?
11. What are the most important features of hybrid AM by rolling?

References

1. Klocke, F., Roderburg, A., & Zeppenfeld, C. (2011). Design methodology for hybrid production processes. *Procedia Engineering, 9*, 417–430.
2. Nau, B., Roderburg, A., & Klocke, F. (2011). Ramp-up of hybrid manufacturing technologies. *CIRP Journal of Manufacturing Science and Technology, 4*(3), 313–316.
3. Flynn, J. M., et al. (2016). Hybrid additive and subtractive machine tools–Research and industrial developments. *International Journal of Machine Tools and Manufacture, 101*, 79–101.
4. Merklein, M., et al. (2016). Hybrid additive manufacturing technologies–an analysis regarding potentials and applications. *Physics Procedia, 83*, 549–559.
5. Kozak, J., & Rajurkar, K. P. (2000). Hybrid machining process evaluation and development. In *Proceedings of 2nd international conference on machining and measurements of sculptured surfaces*, Keynote Paper, Krakow.
6. Lauwers, B., et al. (2014). Hybrid processes in manufacturing. *CIRP Annals, 63*(2), 561–583.
7. Sealy, M. P., et al. (2018). Hybrid processes in additive manufacturing. *Journal of Manufacturing Science and Engineering, 140*(6), 060801.
8. Brehl, D., & Dow, T. (2008). Review of vibration-assisted machining. *Precision Engineering, 32*(3), 153–172.
9. Lauwers, B., et al., (2010). Investigation of the process-material interaction in ultrasonic assisted grinding of ZrO2 based ceramic materials. In *Proceedings of the 4th CIRP International Conference on High Performance Cutting*.
10. Wang, Z., & Rajurkar, K. P. (2000). Cryogenic machining of hard-to-cut materials. *Wear, 239*(2), 168–175.
11. De Lacalle, L. L., et al. (2000). Using high pressure coolant in the drilling and turning of low machinability alloys. *The International Journal of Advanced Manufacturing Technology, 16*(2), 85–91.
12. Brecher, C., et al. (2011). Laser-assisted milling of advanced materials. *Physics Procedia, 12*, 599–606.
13. Rajurkar, K. P., et al. (1999). New developments in electro-chemical machining. *CIRP Annals, 48*(2), 567–579.
14. Kozak, J., Zybura-Skrabalak, M., & Skrabalak, G. (2016). Development of advanced abrasive electrical discharge grinding (AEDG) system for machining difficult-to-cut materials. *Procedia CIRP, 42*, 872–877.
15. Zhu, D., et al. (2011). Precision machining of small holes by the hybrid process of electrochemical removal and grinding. *CIRP Annals, 60*(1), 247–250.
16. Golabczak, A., & Swiecik, R. (2010, April). Electro-discharge grinding: Energy consumption and internal stresses in the surface layer. In *16th International Symposium on Electromachining (ISEM)*, Shanghai.
17. Koshy, P., Jain, V., & Lal, G. (1997). Grinding of cemented carbide with electrical spark assistance. *Journal of Materials Processing Technology, 72*(1), 61–68.

18. Alao, A.R. (2011). A fundamental study of vibration assisted machining. *Advanced Materials Research, 264*, 997–1002. Trans Tech Publ.
19. Wang, F., et al. (2007). Laser fabrication of Ti6Al4V/TiC composites using simultaneous powder and wire feed. *Materials science and Engineering: A, 445*, 461–466.
20. DMG MORI LESERTEC 125. https://au.dmgmori.com/products/machines/additive-manufacturing/powder-nozzle/lasertec-125-3d-hybrid. 2020.
21. DMG MORI, LASERTEC 65. https://www.dmgmori.co.jp/en/products/machine/id=3592. 2020.
22. Khorasani, A. M., et al. (2018). A comprehensive study on surface quality in 5-axis milling of SLM Ti-6Al-4V spherical components. *The International Journal of Advanced Manufacturing Technology, 94*(9–12), 3765–3784.
23. Khorasani, A. M., et al. (2018). Characterizing the effect of cutting condition, tool path, and heat treatment on cutting forces of selective laser melting spherical component in five-axis milling. *Journal of Manufacturing Science and Engineering, 140*(5), 051011.
24. Sealy, M., et al. (2016, August). Finite element modeling of hybrid additive manufacturing by laser shock peening. In *Solid Freeform Fabrication Symposium (SFF)*, Austin.
25. Gale, J., Achuthan, A., & Don, A. (2016, August). Material property enhancement in additive manufactured materials using an ultrasonic peening technique. In *Solid Freeform Fabrication Symposium (SFF)*, Austin.
26. El-Wardany, T. I., et al. (2014). Turbine disk fabrication with in situ material property variation, U. Patent, Editor. Google Patents.
27. Palanivel, S., et al. (2015). Friction stir additive manufacturing for high structural performance through microstructural control in an Mg based WE43 alloy. *Materials & Design (1980–2015), 65*, 934–952.
28. Lamikiz, A., et al. (2007). Laser polishing of parts built up by selective laser sintering. *International Journal of Machine Tools and Manufacture, 47*(12), 2040–2050.
29. Campanelli, S., et al. (2013). Taguchi optimization of the surface finish obtained by laser ablation on selective laser molten steel parts. *Procedia CIRP, 12*, 462–467.
30. Akula, S., & Karunakaran, K. (2006). Hybrid adaptive layer manufacturing: An intelligent art of direct metal rapid tooling process. *Robotics and Computer-Integrated Manufacturing, 22*(2), 113–123.
31. Karunakaran, K., Sreenathbabu, A., & Pushpa, V. (2004). Hybrid layered manufacturing: Direct rapid metal tool-making process. *Proceedings of the Institution of Mechanical Engineers, Part B: Journal of Engineering Manufacture, 218*(12), 1657–1665.
32. Lorenz, K., et al. (2015). A review of hybrid manufacturing. In *Solid Freeform Fabrication Conference Proceedings*.
33. Creator hybrid. https://creator.or-laser.com/en/. 2020.
34. Matsuura LUMEX. https://www.lumex-matsuura.com/english/LUMEX-Avance-25. 2020.
35. Donoghue, J., et al. (2016). The effectiveness of combining rolling deformation with Wire–Arc Additive Manufacture on β-grain refinement and texture modification in Ti–6Al–4V. *Materials Characterization, 114*, 103–114.
36. Book, T. A., & Sangid, M. D. (2016). Evaluation of select surface processing techniques for in situ application during the additive manufacturing build process. *JOM, 68*(7), 1780–1792.
37. Francis, R., Newkirk, J., & Liou, F. (2016). Investigation of forged-like microstructure produced by a hybrid manufacturing process. *Rapid Prototyping Journal, 22*(4), 717–726.
38. Yasa, E., Kruth, J.P., & Deckers, J. (2011). Manufacturing by combining selective laser melting and selective laser erosion/laser re-melting. *CIRP Annals, 60*(1), 263–266.
39. Hartmann, K., et al. (1994). Robot-assisted shape deposition manufacturing. In *Proceedings of the 1994 IEEE international conference on robotics and automation*. IEEE.
40. Xie, Y., et al. (2019). Strengthened peening effect on metallurgical bonding formation in cold spray additive manufacturing. *Journal of Thermal Spray Technology, 28*(4), 769–779.
41. Gale, J. (2017). Application of ultrasonic peening during DMLS production of 316L stainless steel and its effect on material behavior. *Rapid Prototyping Journal, 23*(6), 1185–1194.

42. Kalentics, N., Logé, R., & Boillat, E. (2017). *Method and device for implementing laser shock peening or warm laser shock peening during selective laser melting.* European Patent EP3147048A1.
43. Kalentics, N., et al. (2017). 3D Laser Shock Peening–A new method for the 3D control of residual stresses in Selective Laser Melting. *Materials & Design, 130,* 350–356.
44. Yasa, E., & Kruth, J.P. (2011). Application of laser re-melting on selective laser melting parts. *Advances in Production Engineering and Management, 6*(4), 259–270.
45. Yasa, E., & Kruth, J.P. (2010). Investigation of laser and process parameters for Selective Laser Erosion. *Precision Engineering, 34*(1), 101–112.
46. Yasa, E., Deckers, J., & Kruth, J.-P. (2011). The investigation of the influence of laser re-melting on density, surface quality and microstructure of selective laser melting parts. *Rapid Prototyping Journal, 17*(5), 312–327.
47. Qian, Y.-P., et al. (2008). Direct rapid high-temperature alloy prototyping by hybrid plasma-laser technology. *Journal of Materials Processing Technology, 208*(1), 99–104.

Chapter 13
The Impact of Low-Cost AM Systems

Abstract Since 2010, Additive Manufacturing has gone from a niche prototyping technology primarily known to the manufacturing, design, and engineering communities to a technology that is widely known to the general population as "3D Printing." Much of this growth in awareness is due to an increase in the adoption of the technology due to massive reductions in the cost of Material Extrusion (MEX) machines. By making it possible for individuals to afford AM machines for their own personal use, the potential for AM has become known to the masses. This chapter will discuss some of the issues surrounding low-cost AM technologies, including new machine developments due to patent expirations, the rise of the Maker movement, and some of the new business models that have resulted.

13.1 Introduction

When the first Additive Manufacturing machines came on to the market for the purposes of Rapid Prototyping, they were, not surprisingly, very expensive. The fact that they were aimed at early adopters, based around complex and new technologies, like lasers, and only produced in small volumes meant that the purchase price of such machines was commonly a quarter of a million US dollars or even more. Furthermore, the perceived value of these machines to these early adopters was also very high. Even at these prices, the return on investment (ROI) was often only a matter of months or attributable to a small selection of high-value projects. An automotive manufacturer could, for example, achieve the ROI just by proving the AM technology ensured a new vehicle was launched on or ahead of schedule. Such perceived value did little to bring the prices of these machines down.

As the technology became more popular, the market became more competitive. However, demand for these new machines was also high, particularly from the traditional market drivers of automotive, aerospace, and medicine as mentioned in previous chapters. Vendors did find themselves in competition with each other, but there were many different customers. Furthermore, different machines were exhibiting different strengths and weaknesses that the vendors exploited to develop

© Springer Nature Switzerland AG 2021
I. Gibson et al., *Additive Manufacturing Technologies*,
https://doi.org/10.1007/978-3-030-56127-7_13

differing markets. For most of these markets, the more successful vendors also ensured they had excellent intellectual property (IP) protection. All of this served to maintain the technology at a high cost.

Ultimately it was the issues surrounding IP that led to high prices for two decades. Many of the original technologies were protected by patents that prevented other companies from copying them. Machine manufacturers were very aggressive defending their IP as well as buying up related IP and the companies that owned the IP. The cost of IP protection and litigation had the effect of slowly eroding profit margins. But as patents started expiring, a huge shift in market dynamics occurred, and a low-cost AM technology marketplace was born.

As key patents expired and prominent players in the AM marketplace changed, a huge amount of interest by the financial and technical media ensued. Soon thereafter, the "3D printing" story was taken up by the mass media. Much of this interest can be associated with what is becoming known as disruptive innovation. AM certainly fits with a number of other technologies to form the basis for disruptive business models, which we will discuss. AM is also an enabling technology that has enabled many people to solve their own technical problems at home. Sharing these experiences and even profiting from them has spearheaded what is being commonly called the Maker movement, which we shall also examine. We will go on to consider how this branch of AM may develop in the future.

13.2   Intellectual Property

As mentioned in the introduction and in other parts of this book, the key patents with the most protection originated in the USA. While there was activity in other countries, 3D Systems, Stratasys, DTM, and Helisys were the principal vendor names for much of the world in the early days. Other companies were also present, like EOS, Sony, Sanders, and Objet, but they either came along at a later date or were in close IP conflict with these American vendors.

European and US patent law are similar, in that they both offer exclusivity for approximately 20 years from the initial filing date of the patent, in most cases. Charles Hull filed his first stereolithography patent in 1984 [1]. Scott Crump patented the Fused Deposition Modeling process in 1989 [2]. The major difference between these technologies, in this context, is that the MEX process is much easier to replicate at a low cost compared with Vat Photopolymerization (VPP) systems. In addition, subsequent patents on stereolithography were found to be quite critical to successful machine design and operation, and as such IP protection was effectively extended beyond initial expiration of the first patent. Thus, although the door for competition started to open due to expiration of applicable AM patents around 2004, it was not opened very wide until several years later, as other patents started to expire. This is evident from the figures quoted by Terry Wohlers from 2007 to 2017, shown in Fig. 13.1, concerning the number of low-cost AM machines purchased over that period [3].

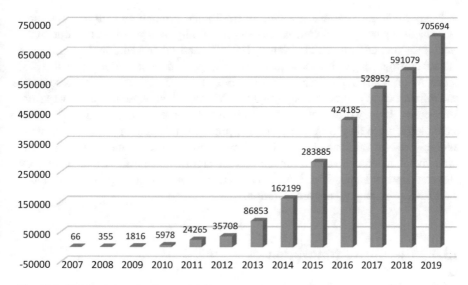

Fig. 13.1 Wohlers' data showing how the numbers of low-cost AM machines (purchase price less than $5000) has increased dramatically since 2007 [3]

MEX patent expiration enabled inexpensive machines to enter the market. But although there have been many copies of the original FDM patent, it took many years before competitors implemented an environmental control apparatus. This is because the patent for this development was not filed by Crump until a few years after the initial FDM patent.

There is an excellent review of expiring patents carried out by Hornick and Roland [4]. In it they make a number of interesting observations:

- There are a huge number of patents involved, and it is a very difficult to navigate through them. Many patents make multiple claims across numerous platforms. They often do not adhere to a single process.
- Earlier patents discussed broad-based approaches that are easy to defend and difficult to find ways around. It was quite easy to distinguish one process from another since they did not have the benefits that were gained through experience of actual application of the technology. Later patents resulting from discoveries resulting from use of these base technologies, which discuss subtle features like soluble supports or fill patterns, were claimed broadly but harder to defend.
- While 3D Systems patents were among the earliest to expire, there are obvious technical complexities in them that would make them difficult to replicate without significant industrial backing. An example relates to the formulation of resins to speed up the curing and build process or another to facilitate even spreading of resins.
- Some key droplet deposition and laser sintering patents expired in late 2014, early 2015. Expiration of these patents have helped increase competition in the high-end, high-quality AM marketplace by bringing machine costs down.

Many large corporations involved in traditional printing as well as equipment manufacturing companies view AM as an opportunity for growth. At the time of the writing of this edition (2020), Xerox, HP, GE, and other traditional printing and manufacturing companies have made very large, recent investments in the AM space. As one can imagine, a large printer company possesses knowledge, resources, and infrastructure to produce high-quality AM equipment in volume at very reasonable costs. It is worth noting that although low-cost AM started with MEX technology, since they were the easiest to produce and some of the first patents to expire, low-cost versions of VPP (Formlabs), polymer laser sintering (Norge), and even metal machines such as metal laser sintering (MatterFab) and lower-cost Directed Energy Deposition (DED) heads (Hybrid Manufacturing Technologies or the LENS print engine) are proliferating.

13.3   Disruptive Innovation

13.3.1   Disruptive Business Opportunities

Disruptive innovation and disruptive technology are terms that were originally defined by Christensen to describe activities or technology that create new markets [5]. A very obvious example of this is how the Internet made it possible to create online businesses which could not have existed before. However, the effects may be much more subtle, and it is possible to create a disruptive business merely by using existing technology in a different way. Often this process can be achieved by early adopters of technology or by those who have skills that are either difficult to learn or not commonly used in a disruptive way. Here we can say that although there are many musicians, there is only one David Bowie or Lady Gaga, who made use of their artistic talents to generate additional business opportunities.

There is no doubt that AM is a disruptive technology, which can be combined with other technologies to generate new businesses. Improvements and more widespread use of CAD technology has made it possible for individuals to design products with minimal cost and training. Google SketchUp and Tinkercad are online design tools that are basically free to use. While they are not as powerful and versatile as paid CAD software, these accessible and simple to use tools have opened up a new market for home designers, who would then like to find outlets for these designs. While many futurists opined this may mean a majority of people will have machines in their homes, we are not there yet, and current market adoption trends indicate home manufacturing may continue to be a niche/hobbyist undertaking. Some people may have a low-cost AM machine at home, but even those that do may not have a machine that can meet the functional requirements for their designs, and so they will look to outside services to build and supply parts based upon their designs. This has led to the establishment of companies who provide online services where designers cannot only have their models made but also find an outlet where their designs can be sold.

Fig. 13.2 Website image for i.materialise, showing how designers can post their ideas on the web and have them built and made available for others to buy [6]

Shapeways and i.materialise (see Fig. 13.2) both operate in this space, using techniques developed for social media platforms, where viewers can "like" other people's designs as well as discuss, share, and promote them. For those wishing to share designs but not concerned with the commercial aspects, there are also portals like the Thingiverse website. Issues surrounding copyright of designs are regularly discussed around these websites. Replicas or models inspired by merchandising for TV and movie shows, for example, can be regularly found on these sites, and some sharing sites will have more or less control than others.

Companies such as Shapeways, Sculpteo, Objective 3D, Zeal, 3D Printing Studio, and more offer online services such as 3D printing, 3D designing, 3D modeling, and part delivery logistics. These services provide a wide variety of items including jewelry, multicolor gift printing, metal car replacement parts, and more.

13.3.2   Media Attention

Since disruptive innovation is always going to attract media attention, it is worth considering some of the more prominent stories that have attracted interest and therefore structured how the general public view AM.

An example that has received attention for several years is the use of AM to create firearms. A great deal of attention was directed toward AM when it was realized that it was possible to create firearms using the technology, the most well-known of these being the Liberator, single-shot pistol [7]. Here, it was discovered that this gun could be largely manufactured using an AM machine and that the plans for its manufacture were posted online. A primary reason for this attracting so much attention was the obvious contentious nature of the topic, backed up by the fact that this design can be easily shared on the Internet. It was particularly interesting to note that the attention was focused on the negative impact of AM rather than that of the Internet. There are a number of issues that are worth discussing here:

– While the gun was indeed built, certain items like the firing pin and obviously the ammunition would need to be added to the design.
– The original gun was created using a relatively high-end AM machine by a skilled operator. Sharing of the design online does not include the build parameters, and a study by the New South Wales Police in Australia revealed that the user is at great risk if the gun is not built correctly [8].
– Improvised firearms have been possible for many years and can be constructed by anyone with a small amount of technical knowledge and access to simple manufacturing equipment [9].

Admittedly, AM can be used like any technology, for good or for bad. However, we can see here how the media can latch on to one example and confuse the public image. By bringing the technology into a much wider public domain than previously, this allowed specialists in the field the opportunity to more properly explain the true impact of AM.

In stark contrast to the use of AM for destructive purposes, there have been numerous articles that describe how AM can be used to create replacement body parts [10]. As described in (Chap. 21) in this book, there is a huge potential for AM to contribute in this direction. For example, AM Magazine reported on improvements in joint replacement when using an AM-produced metal device in ACL knee surgery with high accuracy. Another example is the use of AM to print medical drills that keep the bone cool during surgery (Fig. 13.3). The problem in the media coverage however is the timeframe attached to many of these solutions. Some applications have been implemented where AM has made significant improvements in medical and health care. However, this cannot be easily generalized into a conclusion that all medical problems can be solved this way. We must expect significant developments in the technology before we can make that conclusion, and the AM machines of the future may look nothing like the machines we have today. Furthermore, we need parallel efforts in the biomedical sciences as well because they are far from being ready to plug in directly to AM devices. Specialists in these fields need to understand that, while it is reasonable to speculate that AM for body parts is on the horizon, they must be wary of that message being misconstrued.

CNN released news about using 3D printing systems in the automobile industry entitled "General Motors is gone. Now come 3D printers and robots in Youngstown, Ohio." This report also showed the Silicon Valley transformation to include 3D

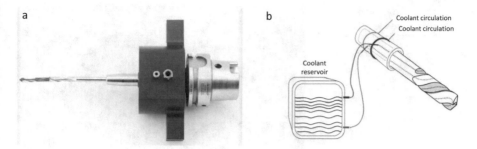

Fig. 13.3 (a) Researchers created an internally cooled bone drill that compensates for heat in drilling, protecting the bone from damage (b). The bone drill is connected to a *coolant* reservoir, so that water flows into and out of the drill without exiting into the patient's body. (Photo courtesy of Toolcraft) [11]

printing and automation with fewer person-hours and more robotic systems. This type of article is one of many focused on "Industry 4.0" (I4.0), which has become a popular idea in technology and business circles. I4.0 refers to the 4th Industrial Revolution and refers to the transformation of product design and manufacturing using new and emerging technologies. I4.0 discussions almost always discuss AM as an enabling technology to improve the time to market, performance, and quality of the parts and services that are provided for customers. I4.0 is often discussed in light of an advanced Internet-of-things (IoT), to help make AME ubiquitous to achieve fully online services. For instance, research was carried out in Taiwan [12] that showed online 3D printing can be successfully performed with high customer satisfaction. The system receives orders from different customers and distributes the required components to the nearest AM center. After AM, delivery vehicles visit the printing facilities sequentially to collect the different orders. To minimize the cycle time for delivering the order, a specific optimization process was designed and the workload balanced to suit different AM centers. Cycle time of printing to delivery can be reduced by around 33%.

Another sector that has attracted the media attention is the "cool gadget" area. One early example that elevated general public awareness was The Economist magazine front-page article titled "Print me a Stradivarius" (see Fig. 13.4). This was probably the first mass market article that highlighted how AM could be used for truly functional applications. While other media sites have included similar articles somewhere in their portfolios, the fact that this was on the front page of an international magazine certainly had an impact. It is also worthwhile taking note that the BBC have published regular articles about AM over the years, averaging around one every 2 months in a wide variety of areas.

It is important to note that the general public is often confused about the capabilities of AM as a result of media attention. Although much of the media hype around AM was driven by low-cost 3D printers, the general public often thinks these low-cost printers are the only 3D printers available. Media attention and resulting articles have one thing in common: they imply that you will eventually be able to create

Fig. 13.4 The Economist
magazine front cover

solutions for yourself. In some sense this has become a self-fulfilling prophecy as
entrepreneurs as well as existing companies that previously had no knowledge of
AM have been intrigued, studied the opportunities, and invested their time and
money to make these ideas a reality as a direct response to this media attention.

13.4   The Maker Movement

As alluded to earlier in this chapter, the number of users who have their own AM
machines has increased dramatically, and they are driving significant innovation.
These innovations are often aimed at providing solutions for problems that the user
has experienced around his or her home or workplace. Social media and other outlets have allowed these users to demonstrate these solutions and thus inspire other
users to do the same. While social media does help in dissemination, physical
demonstration is usually a much more effective way of presenting designs. To this
effect, Maker Faires® have almost literally taken the world by storm, as can be
seen in Fig. 13.5.

Fig. 13.5 Maker Faire web page

Maker Faires® are events where "Makers" congregate to display, demonstrate, and trade in items that they have designed and built themselves. While there is a lot of emphasis on technology, there is equal emphasis on design, environment, engagement, and fun in these fairs. Additive Manufacturing is certainly a component in this, but so are Arduino and Raspberry Pi microcontrollers, laser cutting, conventional machining and hand tools, and home crafting skills like carving, sewing, and knitting. Many Makers merge technology with more conventional craft and design to come up with personalized, often quirky systems that they would like others to see. The low-cost AM technologies have found a huge following within this Maker movement that is fueling a large amount of innovation and even spinning off into new commercial ventures.

Perhaps not surprisingly this Maker movement originated in the USA, originally promoted by Make magazine [13] for the first event in California in 2006. The USA has a long and distinguished culture of innovation, perhaps spawning from the original frontier mentality of having to make do with whatever is around you, and this can also be seen in terms of how innovation is an accepted part of everyday life. However, the concept of the fair is now a worldwide phenomenon with dozens of events held yearly attracting thousands of exhibitors and hundreds of thousands of attendees. The Maker Faire® has become the home for the MakerBot and other similar low-cost AM technologies, including RepRap designs [14], and for design sharing portals like Thingiverse.

A huge number of commercial entry-level machines can thank the RepRap project for their origins. It would be almost impossible to identify every company, there are so many with varying levels of success. Designs have evolved considerably from the original RepRap machines, but the basic principle remains very much the same, exploiting the hot-melt extrusion process that was facilitated by the expiration of the original FDM patents.

The Maker movement along with the FabLab [15] and Idea to Product (I2P [16]) concepts has done much to highlight the benefits of AM and associated technologies to the general public. FabLab and I2P labs are walk-in facilities aimed at providing an environment that encourages people to experiment with accessible manufacturing technologies and develop their ideas. Often the costs are at least partially absorbed by local authorities or donations. These are different from what is referred to as hardware incubators, where the costs are sourced from an investor network. All of these recognize in some way that there is a need to foster the creative processes.

Although the Wohlers Report does a great of estimating the number of low-cost machines sold each year (see Fig. 13.1), sales of low-cost AM machines often come from nontraditional sources and are notoriously difficult to track. New start-ups and small companies selling these small machines come and go on a regular basis. In addition, some of the parts and subassemblies are available in online shops as kits. The average selling price for low-cost MEX machines in 2017 was about $1150. By comparison, the average cost for industrial AM machines at that time was more than 81 times more expensive. While the number of low-cost machines sold per year is much more than the number of industrial machines sold per year [3], these low-cost machines represent a small fraction of the financial investment in AM since industrial machines cost much more to purchase and their consumable materials are significantly more expensive. Nevertheless, extensive growth in the number of sales of low-cost machine (528,952 in 2017 [3]) illustrates that Makers have a huge influence on the future of this technology.

13.5   The Future of Low-Cost AM

Low-cost AM has done much to bring AM technology into the public domain. It took more than 20 years to reach this stage, where many of the technological hurdles to AM were overcome by early adopters, and patent expiration opened the market to the masses. As cuh, there has never been a better time to get involved in AM. A lot of confusion still surrounds what AM can and cannot do, but even some of the negative press coverage has helped to promote the technology in some manner. While the majority of AM machines have exploited the Stratasys FDM and MEX process, subsequent patent expirations of other AM technologies have started to open the doors to some interesting new low-cost technologies, including in metals by Desktop Metal and others [17].

13.6   Questions

1. What low-cost technologies are available in your area? Are there any local vendors and are their machines different in any way to the standard MakerBot and RepRap variants?
2. Examine articles about AM in the press. Is the information presented accurate? Would you write the article in a different way? Does inaccurate reporting have any impact on how the general public views AM technology?

3. What other technologies are represented at Maker Faires?
4. Consider the copyright infringement issues that surrounded the development of YouTube. Could similar things happen with respect to model sharing sites in the future? How can this be regulated?
5. Do you think low-cost 3D printers will be as commonly used as 2D printers in the future? Why or why not?
6. Search online and research several types of 3D printers with costs between $500 and $5000. What are the main driving factors for these price differences? What are the primary benefits when going from an ultra-low-cost design to a more expensive design?

References

1. Hull, C. W. (1986). *Apparatus for production of three-dimensional objects by stereolithography.* US Patents.
2. Crump, S. S. (1992). *Apparatus and method for creating three-dimensional objects.* US Patents.
3. Wohlers, T. (2018). *Wohlers report. 3D printing and additive manufacturing state of the industry, annual worldwide progress report.* Wohlers Associates.
4. Hornick, J., & Roland, D. (2013). *Many 3D printing patents are expiring soon: here's a roundup & overview of them.* 3D Printing Industry.
5. Christensen, C. (2013). *The innovator's dilemma: When new technologies cause great firms to fail.* Harvard Business Review Press.
6. i.materialise. (2020). https://i.materialise.com/en/Getting-Started
7. Liberator (gun). (2020). https://en.wikipedia.org/wiki/Liberator_(gun)
8. The Guardian. (2020). https://www.theguardian.com/uk-news/2013/oct/25/3d-printed-guns-risk-user
9. Improvised firearm. (2020). https://en.wikipedia.org/wiki/Improvised_firearm
10. The Telegraph. (2020). https://www.telegraph.co.uk/technology/news/10629531/The-next-step-3D-printing-the-human-body.html
11. Toolcraft. (2020). https://www.toolcraft.de/en/cases/bone-drill.html
12. Chen, T., & Wang, Y.-C. (2019). An advanced IoT system for assisting ubiquitous manufacturing with 3D printing. *The International Journal of Advanced Manufacturing Technology, 103*, 1721–1733.
13. Make Magazine. (2020). https://makezine.com/
14. RepRap. (2020). https://reprap.org/wiki/RepRap
15. Gershenfeld, N. (2008). *Fab: The coming revolution on your desktop – from personal computers to personal fabrication.* Basic Books, New York (USA)
16. Idea 2 Product 3D Printing and Scanning. (2020). https://idea2product.net/
17. Desktop Metal. (2020). https://www.desktopmetal.com/products/production

Chapter 14
Materials for Additive Manufacturing

Abstract Good materials are crucial for effective AM, and different processes require these materials to be prepared in different ways. Some AM processes are capable of processing a wider range of materials than others. In this chapter we look at metals, ceramics, polymers, and composites and in particular how they change throughout AM processing. Ceramics have not been dealt with widely in other chapters so we cover a range of ways in which we can produce ceramic parts using AM. We also discuss problems that may occur when using different AM materials.

14.1 Introduction

All AM processes share the use of a computer to draw, store, and process geometric data and to drive the machine, including placement of material to fabricate parts. Further to this, post-processing is used to improve mechanical, tribological, surface, and geometrical properties. In AM the minimum deposited volume element (e.g., voxel) is preferably smaller than the smallest critical features of a component [1].

Materials added to an AM machine prior to building a part are often referred to as "feedstock" materials. Feedstock physical phase, shape, mechanical properties, chemical composition, etc. are critical factors in AM processes and drive the method and quality of produced parts. Early AM technologies focused on plastics, but material development has continued to grow far beyond that. Today, AM systems have been developed to use bio-inks, ceramics, metals, composites, glass, paper, graphene-embedded plastics, food, concrete, yarn, etc. to print 3D shapes. Materials for AM have shown extensive growth in recent years in terms of both developing new types of materials and sales volume. Between 2017 and 2019, spending on AM materials increased from \$1.13 to \$1.877 billion across all sectors, showing growth of 25.5, 31.9, and 28.3% in 2017–2019, respectively [2, 3]. More broadly, the market for AM feedstock has seen exponential growth between 2010 and 2019 [4].

The markets for all material feedstock types have increased over the years. Figure 14.1 shows the expenditures for 2017–2019 for different feedstock types. Although photopolymers comprise the largest feedstock segment, the polymer

© Springer Nature Switzerland AG 2021

I. Gibson et al., *Additive Manufacturing Technologies*,

https://doi.org/10.1007/978-3-030-56127-7_14

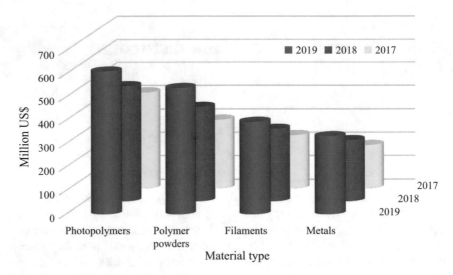

Fig. 14.1 Estimated expenditures for AM feedstock in 2017 and 2018 for different materials. (Source: *Wohlers Report 2020*) [4]

powders segment has grown at the highest rate, 85%, during this period, with metals a close second at 81.4%, filaments (for Material Extrusion (MEX)) at 75%, and photopolymers at 50%.

Generally, AM is used to produce both accurate net shape and near net shape components that can reduce production steps and costs.

Mechanical, chemical, physical, and thermal properties and subsequently the overall quality of printed components are directly related to the quality of the raw material put into an AM machine. Feedstock quality also affects production success and overall cost. As such, material feedstock considerations must be taken into account when selecting an AM process.

This chapter discusses advances and challenges in AM materials. We will overview common materials used in AM processes, give users guidance on selecting the appropriate feedstock materials, and discuss general properties of printed components. We will also discuss issues related to storage and recycling of feedstock materials.

AM feedstock type falls into three main categories comprising liquid, powder, and solid. Figure 14.2 shows relationships between feedstock type and the principle for most commercially available AM processes. Note that it is possible to include more than one material for some processes. For example, it is possible to include powder mixed with resin during Vat Photopolymerization (VPP). Similarly, Material Jetting (MJT) and MEX may include powder, but not as an essential item to carry the process out. Therefore, this diagram focuses on what material types are essential to ensure the AM process will work.

Fabricating multi-material structures has been an area of research since the early days of AM. However, relatively few applications have been commercialized until recently. The successful testing of an igniter made of multi-materials was announced by NASA in 2017. This example demonstrated that AM can reduce the cost of

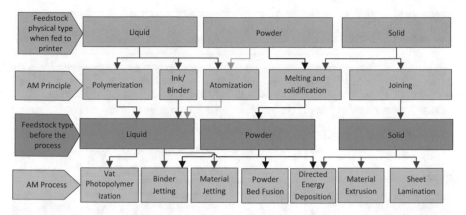

Fig. 14.2 A schematic diagram of the relative feedstock for AM processes

aerospace components such as rocket engines by up to one third and manufacturing time by half [5]. Also, chopped fibers have been developed by Markforged and others for MEX to provide improvements to thermal and mechanical properties. Researchers at Lawrence Livermore National Laboratory have printed composite silicon objects with shape memory characteristics.

AM process chains can be described in terms of single-step and multi-step operations. In the single-step case, the total build process is carried out in a single operation, and the basic geometry and properties of the final components are achieved simultaneously. Multi-step operations are most often used for metallic, ceramic and composite parts. The first step provides the basic geometry, and the following steps improve the strength, density, and other properties, usually using one or more furnace cycles. Figure 14.3 shows multi-step AM processes and principles.

14.2   Feedstock for AM Processes

Each AM process is designed to use one or more feedstock materials of a specific type, including liquid, filament, powder, strip, sheet, or wire feedstock. Materials also must deliver acceptable service properties for the intended application. Feedstock in different AM processes should be selected based on various properties such as mechanical properties, biocompatibility, transparency, color, moisture resistance, fire retardancy, toxicity emissions, sterilization, cost, and more.

A common classification for polymers for use in AM processes, as presented by Bourell et al. [1], differentiates between amorphous and semi-crystalline thermoplastics and thermosetting polymers. VPP uses the photolithographic cross-linking of liquid photosensitive thermoset polymer resins in order to form a solid. These liquid thermoset resins may have various additives. Additive particulates such as silica nanoparticles are used primarily to provide stiff, strong materials for structural applications. If particles are small enough or chemically bound to resin components, the particles do not settle, and the resin does not require periodic stirring;

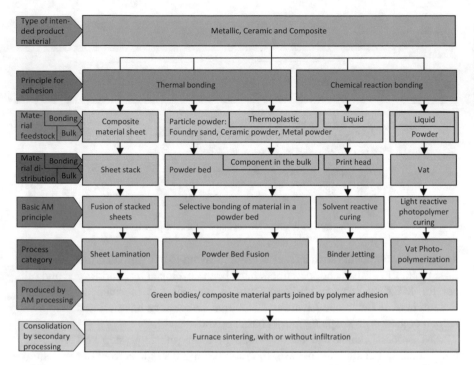

Fig. 14.3 Overview of multi-step AM processing principles for metallic, ceramic, and composite materials [6]

otherwise a liquid mixing system is required. The viscosity of photosensitive thermosets must carefully be taken into account, and additive particulate must not prevent reflow in the processing area [1, 7].

In Powder Bed Fusion (PBF) a heat source fuses the powder feedstock. PBF usually uses a laser or electron beam, but an intensive light source (exposed through physical or chemical masks) can be used to process polymer, ceramic, composite, or metallic material. For polymer PBF, most polymers are semi-crystalline, although some amorphous polymers are used. To obtain a good quality part and prevent material distortion, the difference between melt temperature (during heating) and crystallization temperature (during cooling) should be large, effectively providing a wide processing window (see Fig. 5.12). Preheating of the powder bed and substrate can be helpful to improve polymer part quality. For metal PBF, materials with high heat conductivity, better absorptivity, and less emissivity in the temperature range between melting and boiling are favorable, although most metals that can be cast or welded are suitable. Materials with smoothly varying density and specific heat versus temperature characteristics are also more desirable for PBF systems [8, 9].

MEX uses two principal methods to process the materials, (I) by heating and melting of feedstock prior to deposition and (II) depositing the feedstock at room temperature where a combination of solvent drying or curing solidifies the parts. Temperature-based MEX uses amorphous polymers, which have highly viscous softening over a temperature range that is suitable for deposition. The rheological characteristics of semi-crystalline polymers are sensitive to temperature changes and are difficult to control for such systems. Shrinkage during crystallization can cause part warpage as well.

In MJT photosensitive thermoset polymers are mostly used, which cure after deposition. Wax, suspensions of metal or ceramic particles, and even some metals are also processed using MJT. Material in the form of droplets is selectively deposited on a build plate using inkjet printing technologies. Additives can be included but must be added with care to keep resin viscosity low enough to pass through the nozzles [7].

In Binder Jetting (BJT) the binder has adhesive characteristics and is ink-jetted onto the powder layer surface. The feedstock in this method is therefore both binder material and powders. However, in order to gain high density and improve part strength in metal parts, the binder materials are removed by post-processing such as thermal treatment [1, 10]. BJT has few limitations on the type of material and almost any powder feedstock can be used. For nonmetal powders, an infiltrant is often used to strengthen the part, provide color, or modify other properties.

Sheet Lamination (SHL) uses a sheet form of feedstock to form layers. The sheets can be plastic, metal, paper, or a combination. The sheets are either bonded together followed by cutting or cut followed by bonding. Similar to BJT, the quality of SHL is related to the bonding material or process. Any feedstock that can be formed to sheets can be used for SHL.

Directed energy deposition (DED) uses a thermal source (laser, electric arc, or electron beam) to melt the feedstock material as it is being deposited. DED feedstock is solid and can be either wire, strip, or powder. When processing composite materials, even metal on metal components, the thermophysical properties of the materials must be considered carefully. Mismatches between thermophysical properties of the materials in a composite can result in defects at the interfaces of the materials.

Table 14.1 provides a summary of the common commercial materials for direct processing by AM.

Due to their prevalence in AM, overviews of polymer and metal materials are given here that relate their types, processing principles, and forms to AM processes. Figure 14.4 shows an overview of these relationships for polymers. As can be seen, polymers can be used in all seven AM process classes.

Similarly, an overview of single-step processing and principles for metallic materials is shown in Fig. 14.5.

14.3 Liquid-Based Material

Use of polymeric liquids in additive manufacturing requires a phase change via cooling, drying, or curing [11]. In this section, we introduce polymer and composite liquid materials. Metals can also be processed using MJT, provided that the material phase is changed to liquid before ejection. Other materials like ceramics can be processed this way in powder form using the liquid as a transport mechanism.

Table 14.1 Commercial materials directly processed by AM, by AM process category [1]

Material	Amorphous	Semi-crystalline	Thermoset	MEX	VPP	MJT	PBF	BJT	SHL	DED
ABS [acrylonitrile butadiene styrene]	×			×						
Polycarbonate	×			×			×			
PC/ABS blend	×			×						
PLA [polylactic acid]	×			×						
Polyetherimide [PEI]	×			×						
Acrylics			×		×	×				
Acrylates			×		×	×				
Epoxies			×		×	×				
Polyamide (Nylon) 11 and 12		×					×			
Polyamide neat		×					×			
Polyamide glass filled		×					×			
Polyamide carbon filled		×					×			
Polyamide metal (Al) filled		×					×			
Polyamide polymer bound	×	×		×			×			
Polystyrene	×						×			
Polypropylene		×					×			
Polyester							×			
Polyetheretherketone (PEEK)		×		×			×			
Thermoplastic polyurethane (Elastomer)				×			×			
Chocolate	×	×		×						
Paper									×	
Aluminum alloys							×	×	×	×
Co–Cr alloys							×	×	×	×
Copper alloy							×		×	

Material	Amorphous	Semi-crystalline	Thermoset	MEX	VPP	MJT	PBF	BJT	SHL	DED
Gold							×	×		
Nickel alloys							×	×		×
Silver							×			
Stainless steel				×			×	×	×	×
Titanium, commercial purity							×	×	×	×
Ti–6Al–4V							×	×	×	×
Tool steel							×	×	×	×

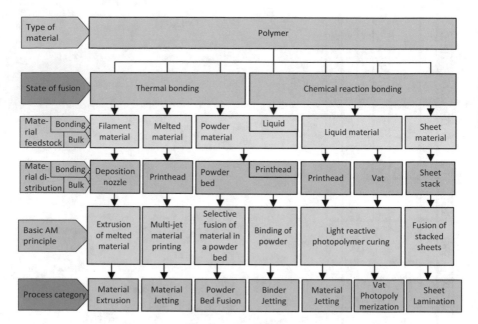

Fig. 14.4 Overview of single-step AM processing principle for polymer materials [6]

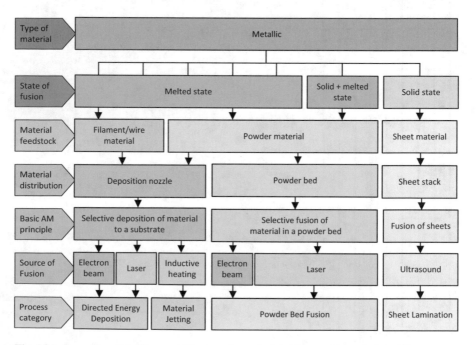

Fig. 14.5 Overview of single-step AM processing principles for metallic materials [6]

14.3.1 Liquids for VPP

Polymers are classified into two main categories comprising thermoplastics and thermosets. Thermosets are permanently set during curing or polymerization and cannot be remelted. Photopolymers are thermosets that are activated upon exposure to radiation, typically in the UV or visible light spectrum, and are widely used as feedstock for VPP.

Common ingredients that are used in VPP are monomers, photoinitiators, and oligomers. The first photopolymer used in the late 1980s were a combination of acrylate monomers and UV photoinitiators [12]. Free-radical photopolymerization produces acrylate-based polymers. This material is well-known for its transparency, resistance to breakage, and elasticity. Acrylate components in resins can be monomers or oligomers (with higher molecular weights than monomers) and can be monofunctional or polyfunctional, which allows them to cross-link. Urethane molecules can be added into acrylate formulations to impart toughness to the solidified resin. Vinyl ester resin is yielded by the esterification of an epoxy resin with methacrylic or acrylic acid. These resins present some issues during processing where atmospheric oxygen prevents reactions and induces high shrinkage and residual stresses accumulate to produce dimensional deviations [1, 13] and are now not as common as acrylates for commercial resins.

Epoxies, which are cationically polymerized photopolymers, were the next generation of VPP feedstock appearing in the early 1990s. They exhibit better mechanical properties than acrylates. They react as a ring-opening polymerization reaction, in which the terminals at the end of a polymer chain act as reactive centers. Due to ring opening the produced epoxy parts present less dimensional deviation compared to acrylates since they shrink less. Cationic photopolymerization utilizes a different reaction mechanism than free-radical polymerization. In this case, cations transfer charge (positively charged ion) during the reaction, while charge transfer occurs through electrons in free-radical polymerization. Many types of epoxy compounds can be used in VPP resins, for example, based on polyglycidyl ester and ether, phenol, or diol chemistries. Further, other cationically polymerized compounds are often used in resins, including oxetanes and polyols with hydroxyl groups [14, 15]. The photoresist material SU-8 is an example of a photopolymerized epoxy, which is used extensively in microelectronics fabrication [16].

Most commercially available stereolithography resins are a mixture of acrylates, epoxies, and other cationically reacting compounds [17]. Each of these primary compounds offers characteristics that contribute to either the manufacturing process, final part properties, or both. For instance, acrylates react quickly which builds structural integrity, while epoxies solidify slowly. Since these components react using different reaction pathways, two different photoinitiators are needed. Note that epoxies do not react with the acrylates; as a result, the mixture forms an interpenetrating polymer network upon solidification [18, 19]. Such mixtures exhibit interesting properties, as described in Sect. 4.2.3.

To improve resin performance and mechanical properties of fabricated parts, dyes, antioxidants, toughening agents, photosensitizers, stabilizing agents, etc. can be added to the feedstock (see Chap. 4). Stabilizing agents can include inhibitors that prevent resins from polymerizing accidentally through heat or ambient light. Photosensitizers act to increase the photopolymerization rate or shift the photoinitiating wavelength. Toughening agents improve the toughness of parts, since acrylates and epoxies can be brittle [1, 20]. Often the toughening agents are elastomer particles, such as polypropylene or rubber, in size ranges of less than 10 μm. Sometimes they are prepared with reactive groups so that they cross-link with the monomers or oligomers, which ensures that they do not settle due to gravity, while the resin is liquid. They may take the form of an elastomeric core and glassy shell that are used as impact modifiers. A very wide variety of resin formulations are available commercially and have been investigated in research.

14.3.2   Liquid Polymer Material for MJT and BJT

Generally, materials for MJT are photopolymers or waxes. Photopolymer formulations for MJT are similar to those used in VPP, while waxes are low-melting-temperature thermoplastics, such as casting wax, that have low viscosities upon melting. Liquid materials for BJT are binders that are jetted onto the powder bed and infiltrants that strengthen the part during a post-processing stage.

Whether for MJT or BJT, the primary constraint for liquid materials is they must be printable by inkjet print heads. This typically means that their viscosity must be less than about 20cP at the printing temperature, which greatly constrains the material formulations. Since most liquid viscosities exhibit considerable temperature sensitivity, a common strategy is to develop a printable liquid polymer material that has a viscosity above 50cP at room temperature, while at printing temperatures (typically 40–80 °C), their viscosity drops to below 20cP. This ensures that upon printing the droplets cool and become viscous enough that they do not flow appreciably, which could cause poor feature definition. See Chap. 7 for more considerations on printability.

The photopolymer MJT materials are typically acrylate-based. A range of stiffnesses can be achieved by using different amounts of polyurethane in the acrylate formulations. For example, the range of elastic modulus found in resins from Stratasys range from 2 to 3 GPa for Vero and about 1 MPa for TangoPlus. More recently, mixtures of acrylates and epoxies have been introduced. Similar to the VPP materials, these mixtures form interpenetrating polymer networks that result from the separate reactions of acrylates, via free-radical polymerization, and epoxies, via cationic polymerization [21]. It is noteworthy that MJT materials are often dyed, which enables multi-color parts.

Wax materials for MJT are typically compounds containing mostly paraffin wax with small amounts of toughening agents, such as butyl rubber, stabilizers, and dyes for color. Typical melting temperatures for MJT waxes are approximately

60 °C. Applications for waxes include inexpensive concept models and patterns for investment casting (lost wax casting). Jewelry; dental restorations, such as copings for crowns and bridges; and industrial parts are significant application areas for these wax prints.

Binder materials used in BJT are typically polymer solutions that bind powder particles together to form layers and parts. These polymer solutions have binding agents, typically a thermoplastic, and one or more solvents. Different formulations are used for plastic, metal, and sand powders since the requirements for each powder feedstock are quite different. Note that machine vendors rarely provide any chemical compositions for their binders. For polymer powders, polyvinyl alcohols may serve as the binding agent, while propanols, pyrrolidone, or butanediol are used as solvents. For metal powders, higher strength binders may be used, but another important consideration is that the binder is burned out in a furnace cycle, and it is desirable for little binder residue to remain. For sand, binders that are consistent with foundry sand formulations are used. Sometimes, a low temperature "bake" is used to cure the binder.

Because MJT and BJT utilize inkjet printing heads, MJT and BJT machines typically contain hundreds or thousands of nozzles. This enables multiple materials to be printed in a single build since subsets of these nozzles can be dedicated to a single material. As a result, parts can be fabricated with materials that vary in mechanical properties, vary in color, or both. Both MJT and BJT machines can utilize the CMYK (cyan–magenta–yellow–black) color model to create parts with thousands of colors. Since droplets in MJT are 30–50 μm in size, dithering patterns used to mix materials result in the perception of solid colors. MJT machines can fabricate functionally graded materials; that is, they print different combinations of materials with different mechanical properties that result in continuous changes in properties [22]. Figure 14.6 shows some example color prints using MJT from Stratasys.

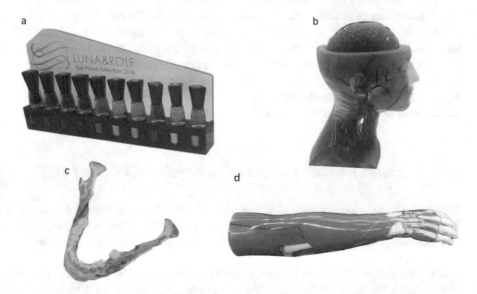

Fig. 14.6 Color printing using MJT from Stratasys

Researchers have also investigated MJT and BJT material systems that react upon deposition. Some groups have attempted to jet reactants to produce polyamide parts; that is, caprolactam and other reactants are co-deposited, and they react locally to produce polyamide in a voxel-by-voxel manner [23]. This may be a strategy for the fabrication of ceramic or ceramic-composite parts.

14.3.3   Liquid Metal Material for MJT

Different metal alloys from such as tin, mercury, aluminum, copper, bismuth, etc. have been printed by MJT printers. When processing by MJT, the feedstock needs to be heated to its melting temperature and then ejected from the nozzle. However, due to high melting temperature of metals compared to polymers, they may more easily damage the components of the printer during melting and ejection. Another difficulty of metal droplet deposition in MJT systems is surface tension instability which is difficult to predict and control [24]. Therefore, MJT has been used rarely for metal printing, except for depositing soldering pastes or conductive metal tracks (see Chap. 7).

Solder droplet printing can be delivered in two methods: continuous and drop on demand, as presented in Chap. 7. In continuous mode, a molten material with 50–300 µm is deposited onto the substrate. A piezoelectric transducer is excited, and, along with the pressure of molten liquid, this induces a sinusoidal disturbance which produces a stable stream of molten material. A flow of inert gas can be used to prevent the negative effect of oxidation. In drop-on-demand mode, a droplet is created only when desired. The mechanism for droplet ejection is either heating the fluid to form a vapor bubble in the liquid or displacement of a piezoelectric plate which is coupled to the feedstock.

A recent development is the application of electromagnetics as a mechanism for ejecting droplets of molten metal. Canon Océ utilizes the Lorentz force to eject a droplet that is formed by an electrical current in the molten metal that is induced by an applied magnetic field. Much of the development work has focused on silver and tin metals. In parallel, a team that formed the Vader Systems Company (purchased by Xerox Corp. in 2019) developed a variant that utilizes a different print head configuration to induce a radially inward Lorentz force to eject droplets. They demonstrated their print head on aluminum.

14.3.4   Liquid Ceramic Composite Materials for VPP and MJT

Ceramic materials are not easy to process by AM due to challenges in delivering feedstock, processing the materials, and/or post-processing. Two broad classes of materials are presented here: photopolymer–ceramic matrix composites and ceramic materials processed indirectly using VPP and MJT. In the former case, the resulting part is a polymer–ceramic composite, while in the latter case the polymer acts as a binder that is burned off before sintering the ceramic material to full density.

Liquid ceramic feedstock materials used in VPP systems are a mixture of powder, liquid resin, and particulates such as toughening agents. The amount of solids and the feedstock temperature have to be precisely maintained to satisfy the necessary viscosity for successful processing [25]. Ceramic-filled photopolymer resins have been marketed for many years by 3D Systems and EnvisionTEC and more recently by Formlabs. The ceramic additives are meant to improve mechanical properties, including at elevated temperatures. Typically, the particles are very small (<1 mm in size) and are often chemically bonded to precursor resin ingredients to prevent settling. For example, the Accura CeraMAX material from 3D Systems has a tensile modulus of about 9.5 GPa and tensile strength of 80–85 MPa. The Ceramic Resin from Formlabs is meant to be fired to produce ceramic parts, but can be left unfired if a stiffened photopolymer part is desired.

An interesting application of photopolymer–ceramic composites is for bone tissue scaffolds. For example, one group reported on scaffolds fabricated using a bioglass and polycaprolactone composite using a commercial VPP system from EnvisionTEC. A methacrylated PCL was synthesized to enable photopolymerization; bioglass particles of up to 20% volume loading were added to the photopolymer, and scaffolds were fabricated that had fully connected pores of about 0.5 mm in size. Results demonstrated the benefits of having bioglass particles on the scaffold surfaces to promote cell growth.

The second broad class of materials is pure ceramics that are realized through VPP and MJT fabrication of photopolymer–ceramic composite feedstock materials. VPP/MJT-fabricated parts are fired at high temperature to burn off the polymer binder and sinter the ceramic to produce a solid ceramic part.

As mentioned earlier in this subsection, Formlabs markets their Ceramic Resin that can be fired to produce ceramic parts. The company Lithoz markets a line of VPP machines for ceramic part production in a wide range of ceramics, including aluminum oxide, zirconium oxide, silicon nitride, tricalcium phosphate, hydroxyapatite, and silica-based materials for casting cores [26]. Tricalcium phosphate and hydroxyapatite materials are intended for bone scaffolds. The VPP feedstock materials are highly loaded slurries of liquid photopolymers and ceramic powders. To facilitate materials development, feedstock powders for structural ceramics, as well as furnaces, typically are sourced from the powder injection molding industry. It should be noted that linear shrinkages of 24.5% were reported in the fabrication of alumina parts with final densities above 99% [27].

Three primary approaches to ceramic printing have been used in MJT. The first uses ceramic powders suspended in wax carrier materials. To print, the print heads melt the wax and eject composite droplets. In a second approach, ceramic powders are suspended in a photopolymer, which is jetted and then cured to fabricate a green part. A third approach is referred to as solution or dispersion-based deposition. In this approach, upon deposition, the carrier fluid evaporates, leaving the ceramic powders to form the part. In all approaches, the solids loading must be low enough to ensure that the material's viscosity is not too high to be jettable. Solids loads of 30–40% are typical. In the first and second approaches, significant shrinkage occurs during binder burn-off and sintering due to the low solids content. In the third approach, the deposition rate of ceramic material is low due, again, to the low solids content and the time required for the carrier fluid to evaporate (often an elevated temperature is used) [1]; however shrinkage upon firing is lower, as the carrier evaporates during the printing process.

The company XJet markets MJT machines for metal and ceramic parts. Their feedstock materials are colloids with nano-scale metal or ceramic powders. After fabrication, parts are sintered to high density in a furnace.

14.3.5 Support Material

Like many AM processes, MJT uses a secondary support system so the formulation of support materials must enable easy removal as well as jettability. The liquid materials for MJT supports are typically dissolvable. Over the years, inkjet printable and dissolvable support materials have been based on polyvinyl alcohol (PVA), glycerol, waxes, or other proprietary materials. PVA and glycerol dissolve in water, while other support materials dissolve in solutions of sodium hydroxide. Sometimes water jets are used to remove support structures that contain some PVA or glycerol. In contrast, waxes are removed using heat.

14.3.6 Other Liquid Polymer Feedstock

MEDFLX 625 is a flexible biocompatible liquid material developed by Stratasys and can be mixed with a rigid material to increase rigidity for their MJT systems. EnvisionTEC markets several medical-grade materials for their Bioplotter MEX systems. One is a biocompatible, non-biodegradable liquid silicone. This gel form resin (UV silicone 60A MG) can be translucent or colorful and is used in microfluidic and implant prototyping. Although deposited by extrusion, it must be UV cured. EnvisionTEC also supply biodegradable thermoplastic polymers (HT PCL MG), based on polycaprolactone, that are intended for bone and cartilage regeneration and drug release applications. Flexible rubber-like materials with up to 400% strain and 9 MPa ultimate strength are available from Adaptive3D Technologies for VPP processes. Tethon 3D offers a range of ceramic-filled resins for VPP processes that have applications for structural components, tissue scaffolds, and investment casting.

14.4 Powder-Based Materials

The material feedstock for powder-based AM processes can be polymer, metal, or ceramic powder and, as shown in Fig. 14.2, can be fed into BJT, MJT, PBF, and DED processes.

14.4.1   Polymer Powder Material

14.4.1.1   Polymer Powder Material for PBF

As discussed in Chap. 5, polymer materials for PBF are typically semi-crystalline thermoplastics since they have well-defined melting temperatures, low thermal conductivity, and low tendency to ball when melted. Several types of semi-crystalline thermoplastics are available commercially, the most common being the polyamides (also known as nylons). Polyamide 6, 11, and 12 are offered commercially; polyamide 12 is available in neat, glass-filled, or fiber-filled variants.

Other thermoplastics available commercially include thermoplastic polyurethane (TPU) and thermoplastic elastomers (TPE), for applications where flexible materials are required. Polystyrene is typically used for patterns for vacuum or investment casting. Polypropylene is stiffer than TPU or TPE, but more flexible than the polyamides; it has a lower density than the other polymers as well (Table 14.2).

In PBF of polymers the feedstock is normally pulverized. Cryogenic grinding, dissipation, or coextrusion of non-miscible polymers produces pulverized powder. Care must be taken to avoid powder contamination since that can lead to formation of defects such as cracks.

Recycling of powder is common in polymer PBF. Over several reuse cycles, however, the polymer molecular weight can increase which changes the melt characteristics during melting and solidification. To prevent this, effective powder recycling management is needed. It is recommended [1, 28] to use 30–50% fresh powder with each new build to improve recycling of feedstock. Mixing fresh and recycled powder to achieve a specific melt flow index has been shown to provide more consistent part characteristics than just mixing a specific percentage of fresh to recycled powder.

Dimensional deviation in PBF of polymers is related to laser spot diameter, the powder particle size, percent recycled powder, laser deflection angle, the ratio of surface to volume of the designed components, selection of appropriate process

Table 14.2   Polymer powder feedstock materials and vendors

Material class	Variant	Vendor
Polyamide	PA 6	BASF
	PA 11	3D Systems, Arkema, EOS
	PA 12	3D Systems, Arkema, EOS
Polystyrene	–	3D Systems, EOS
TPU	–	3D Systems, EOS
TPE	–	3D Systems, CRP Technology, EOS
Polypropylene	–	EOS
Polyaryletherketone	–	EOS
Composite	Glass particles + PA 12	3D Systems, CRP Technology, EOS
	Glass fibers + PA 12	3D Systems, CRP Technology, EOS
	Carbon filled PA	CRP Technology
	Aluminum and glass-filled PA	CRP Technology

parameters, and more. The resolution of PBF components is also related to thermo-physical properties of the feedstock such as thermal expansion and conductivity.

An alternative to laser melting can be achieved by selectively printing IR absorbing particles into the polymer powder and then heating the powder using a heat lamp. This can significantly speed up the fusion of each layer since multi-jet printing of IR particles can be much faster than laser scanning and heating can be applied over the whole layer simultaneously [29].

14.4.2   Metal Powder Material for PBF, DED, and BJT

14.4.2.1   Metal Powders

In principle, any metal that is weldable should be a candidate for PBF, DED, and BJT processes. Common commercially available alloys are shown in Table 14.3, with new additions continually expanding the material supply market.

A growing area for AM feedstock is precious metals. These materials have previously been processed through lost wax casting, but many can now be printed using PBF [33] and have been used in jewelry, dental restorations, and other specialty applications

When processing powder-based metal feedstock, several factors contribute to limit the process. Small layer thickness with a correspondingly low volume of material results in formation of a small melt pool which is somewhat unstable. This high-temperature zone cools very quickly, resulting in non-equilibrium microstructure, and significant thermal residual stress can occur, leading to uncertainty and variations in mechanical properties.

14.4.2.2   Production of Metal Powders

Metal powder in AM processes is produced typically by the gas atomization technique. Atomization produces fine particles from bulk material by breaking them up during the liquid phase. A stream of liquid metal is hit by pressurized gas and broken up by kinetic energy, scattering the droplets. The droplets rapidly solidify, and powders are collected in an atomization tank, which is filled with inert gas. Gas atomization produces highly spherical particles. However, particles can attach to one another to form agglomerates. The ratio of gas to metal flow rates is used to calculate median particle size, d_m [34].

$$d_m = \frac{k}{\sqrt{G/M}} \tag{14.1}$$

where G is gas flow rate, M is metal flow rate, and k is a constant for the specific material. Other important factors in the atomization process are free fall distance,

Table 14.3 Selected alloys commercially used in AM processing [3, 30–32]

Material	Alloy	Process
Iron-based	Stainless steel 316 and 316L	PBF, DED, BJT
	Stainless steel 304	PBF
	Stainless steel 17-4PH	PBF, DED
	Maraging steel 18Ni300	PBF, DED
	Stainless steel 15-5PH	PBF, DED
	Stainless steel 420	PBF, DED
	Stainless steel 347	PBF, DED
	AISI 1033 steel	PBF
	4130 steel	PBF
	HY 100 steel	PBF
	Tool steel H13	PBF, DED
	Tool steel 4340	PBF, DED
	Tool steel M2	PBF, DED
	FeCrMoCB	PBF
	Invar 36	PBF
	AISI 1005 carbon steel	PBF
	AISI 1050 carbon steel	PBF
	AISI 1075 carbon steel	PBF
	Pure Fe	PBF, BJT
Titanium-Based	Commercially pure	PBF, DED, BJT
	Ti–6Al–4V	PBF, DED, BJT
	Ti–6Al–4V ELI	PBF, DED, BJT
	$\gamma - $ TiAl	PBF, DED
Aluminum-based	A100 AM	PBF, DED
	A1000 RAM 10	PBF, DED
	2024 RAM 2	PBF, DED
	2024 RAM 10	PBF, DED
	6061 RAM2	PBF, DED
	7050 RAM2	PBF, DED
	Al–Si–10Mg	PBF, DED, BJT
	Al–12Si	PBF, DED
Ni-based	IN625	PBF, DED, BJT
	IN718	PBF, DED, BJT
	Rene142	PBF
	CMSX-4	PBF
Refractory	Ta–W	PBF
	Co–Cr–MO	PBF, DED
	Alumina	PBF, DED
	Stellite	PBF, DED
	Mo–Re	PBF
	W	PBF, BJT
	WC	PBF, BJT

(continued)

Table 14.3 (continued)

Material	Alloy	Process
Copper-based	Cu–Cr–Zr–Ti	PBF
	Pure Copper	PBF, BJT
Precious metals	Gold	PBF
	Silver	PBF, BJT
	Platinum	PBF
	Palladium	PBF
	Tantalum	PBF
Magnesium-based	Pure Mg	PBF
	Mg-9%Al	PBF
	AZ91D	PBF
	ZK60	PBF
Other metals	Bronze	BJT
	Nb	PBF

atomization tank temperature, tapping temperature, and cooling rate. Different companies like Carpenter, Hoganas AB, 3Dom, ColorFabb, etc. produce a wide range of metal and polymer powders.

ATO LAB uses ultrasonic atomization to [35] produce powder feedstock from both reactive and non-reactive materials, with powders ranging 20–120 µm. This system provides the ability to produce powder on the shop floor to reduce feedstock costs.

Table 14.4 shows common layer thicknesses and powder particle sizes for Laser-Based Powder Bed Fusion (LB-PBF), Electron Beam Powder Bed Fusion (EB-PBF), and DED processes. The larger powder sizes and layer thicknesses produce rougher surfaces and larger dimensional deviations. In general, spherical powders flow and pack better than other shapes. Figure 14.7 shows representative powder particles for LB-PBF.

Generally, metal powders for AM are expensive in comparison with feedstocks for conventional manufacturing since the AM market remains small and has specialized requirements for alloy composition and particle size distributions. Metal AM has attracted the attention of third-party materials suppliers, and many material suppliers now provide metal powders for PBF and DED.

For powder-based DED systems, a fully inert chamber is needed when working with reactive metals. When DED is use for less-reactive alloys, only shielding gas is needed, providing the opportunity to build without a closed chamber, thus enabling printing of bigger parts. In DED, the stream of gas, such as argon, and selection of proper process parameters can prevent material being stuck to the nozzle. Not all powder sprayed from the nozzle is deposited on the designated place, and therefore it is collected after the build, and either discarded or recycled, which increases the cost of the process.

In BJT, some vendors utilize feedstock materials from the metal injection molding (MIM) industry, which tends to reduce costs since this is an established industry. In cases where MIM feedstock is not used, powder costs can be expected to be

Table 14.4 Comparison of powder particle size and layer thickness for LB-PBF, EB-PBF, and DED systems

Characteristics	LB-PBF	EB-PBF	DED (both low and high powder feed laser-based systems)
Powder particle size	15–80 μm	45–10 μm	100–200 μm
Layer thickness	20–100 μm	100–200 μm	100–4000 μm
Material deposition (up to)	0.2 kg/h	3 kg/h	3 kg/h

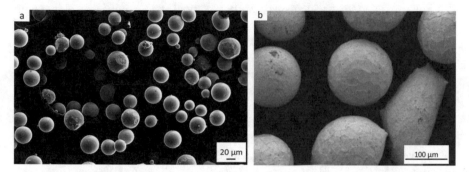

Fig. 14.7 SEM image of (**a**) Ti-6Al-4V PBF powder (**b**) WC/Ni DED powder (Elsevier license number 4661791434964) [36]

higher. With some materials, such as stainless steel, parts can be lightly sintered and infiltrated with a low-melting-temperature metal such as bronze.

Nanoparticle MJT uses inkjet heads to deposit droplets of metallic nanoparticles in a carrier liquid. Similar to BJT this process is performed at low temperature (up to 300 °C). Nanopowders are suspended in a liquid and transferred from a cartridge, so, unlike most PBF and DED systems, the operator is not in direct contact with the metal powder, adding to process safety. The powder particles have stochastic shape and are not fully spherical which increases the packing and bonding of the parts being built. The deposited layers are finer compared to most AM processes, with achievable layer thickness of 2 μm providing superior surface finish for microfabrication. However, this process is limited to specialized material feedstocks from the vendors, reducing the user's freedom to customize the powder or purchase "off the shelf" feedstock [37]. Furnace cycles are necessary to remove binders and sinter the particles to produce full density components. In some cases nanoparticle-filled binders are printed in BJT, which results in green parts with higher overall particle loading, resulting in less overall shrinkage and higher dimensional accuracy during the sintering process.

14.4.2.3   Part Fabrication with Metal Powders

In powder-based metal AM, even though the powder is fully melted, loose powder particles can attach to part surfaces so the surface quality is significantly lower than machined parts. The surface quality of a powder-based AM component is a function of process parameters, powder particle size, and the type of feedstock. Since metal powder particles are small, they tend to oxidize easily, so a protective environment of argon, nitrogen, or vacuum is used typically.

In EBM the temperature of the powder bed is typically kept much higher compared to LB-PBF. Before each layer is built, the electron beam is defocused and scans the loose powder. This leads to light bonding and sintering of the powder and better stability to support the component being built. As a result, powder particle attachment occurs and forms a cake that needs more sieving when reusing the powder. Powder sintering also increases powder electrical conductivity, thus aiding with electron flow and reducing the changes of negative charge buildup and ejection of powder particles due to magnetic repulsion.

One of the chief objectives of most AM metal printers is to create parts with nearly 100% density. Other significant requirements may include good fracture toughness and fatigue properties. Porosity or poor interlayer bonding (light delamination) can act as a crack initiation site that can result in premature failure of components. Porosity and dynamic mechanical properties are very important for medical and aerospace applications. For instance, due to inability to fully eliminate porosity for cast components, the aerospace industry does not accept as-cast Ti-6Al-4V parts for critical applications [2, 3].

Metal AM printers are able to customize microstructure and mechanical properties that are not possible with traditional processes. Changing parameters can affect the cooling and solidification rate. When solidification slows, the grains tend to become larger, reducing some mechanical properties such as tensile strength, but increasing others. In AM of metals, cooling starts from the surface toward the bulk; therefore grains may have different concentrations of alloying elements and different sizes and orientations, causing anisotropy [38]. Due to rapid cooling, segregation is limited, and the chemical composition of metal AM parts appears more uniform in bulk compared with cast and wrought material.

14.4.2.4   Metal Powder Reuse

To achieve a high level of feedstock utilization, unmelted powder can be used for subsequent builds. However, the physical and chemical properties of powder particles may change during the build process, which may cause defects and produce parts out of specifications. In this case, the optimum process parameters may no longer be valid. The loss of small particles changes the flow characteristics of the powder. It is reported for LB-PBF that the morphology of particles stays spherical after 38 builds. For EBM, due to higher bed temperatures and pre-sintering before melting, the powder particles lose shape after 10–20 builds. The oxygen level during processing is also important, affecting the mechanical properties, corrosion resistance, and powder recyclability. Oxygen and oxide levels tend to increase, particularly in reactive metals, as the amount of reused powder increases and as powder is recycled repeatedly. When reusing powder, contamination and loss of elemental alloys are other potential problems. The loss of elemental alloys is related to the lower boiling point of alloy elements. For instance, in Ti-6Al-4V at melting temperature, both Al and V can partially vaporize. As a consequence, alloy feedstocks are often prepared with higher than specification percentages of low-temperature alloying elements such that after melting, and partial vaporization, the chemical composition in fabricated parts falls within acceptable ranges. Additionally, to help prevent significant changes in chemical composition, reused powder should be mixed with fresh powder. The percent fresh to recycled powder depends upon the reactivity of the powder, the oxygen percentage, and bed temperature maintained in the machine during the build and how densely the machine is packed with parts during the build. The higher any of these conditions, the faster the powder degrades and the higher the percentage of fresh powder mixed with recycled powder you need to maintain part consistency.

In PBF, the part may be surrounded by a cake of partially sintered powder. To reuse the material the cake has to be removed and broken down to individual particles. Metal powder identical to the feedstock in the machine can be used as beads to bombard and break down the cake and remove particles attached to part surfaces. The mixture created by this bead blasting can be sieved to remove any agglomerated powder particles and recycled back into the machine.

Recycling powder in DED systems is more difficult compared to PBF. It is nearly impossible to recycle powders when using multiple materials in a machine. To separate different particles, a mixture can be placed in a centrifuge because of the difference in density of each material. The lighter powder particles will be concentrated near the center of the container, and particles with higher density will migrate toward the edge of the centrifuge barrel. However, small contamination can have devastating consequences, so recycling of this type can be risky.

14.4.3   Ceramic Powder Material

Ceramics, due to their combination of low toughness and high melting point, are difficult to process through single-step AM. Direct processing of alumina and its alloy were reported using DED [39] and PBF [40]. However, to obtain fully dense components, multi-step AM processing is needed typically.

14.4.3.1   Ceramic Powder Material for BJT

BJT can produce various complex ceramic components. Alumina powder using a colloidal silica binder has produced parts with high dimensional accuracy. Table 14.5 shows common ceramics processed by BJT.

In BJT of ceramics two methods can be used:

- Dry powder spreading: Sequential deposition of dry powder using a scraper or roller system, followed by the printing of binder material by inkjet printing to produce solid components.

Table 14.5 Application areas of ceramic BJT technology

Application area	Material
Bioceramics – bone tissue	$Ca_5(PO_4)_3(OH)$
	Si/SiC
	$Ca_3(PO_4)_2$
	Bioactive glass
Bioceramics – dental parts	Dental porcelain
Bioceramics – others	$Ca_5(PO_4)_3(OH)$
Structural ceramics	Al_2O_3
	TiC/TiO_2
	Si_3N_4
	SiC
Electric functional parts	Al_2O_3
	$BaTiO_3$

- Slurry-based spreading: A slurry-based material is used for the powder bed, consisting typically of fine powders (less than 20 µm). Typically the liquid carrier quickly evaporates, leaving behind a dry powder bed with a higher powder packing density than dry powder spreading. Binders are printed subsequently onto the powder. If the liquid carrier is not designed to evaporate quickly, binder printed onto the slurry can cause a chemical reaction or dissolve a slurry component to cause binding. Higher final part densities were reported with slurry-based, compared to dry powder based, beds.

The mechanical properties and density of BJT printed components are related to powder material, feedstock size and shape, type and proportion of the binder material, and sintering specifics. In BJT of ceramics, different binders such as carbohydrate (dextrin, maltodextrin, and starch), acid (phosphoric acid and acrylic acid), polymer, and commercial binders are used [41]. The shape and size of the ceramic powder particles in feedstock determine the flowability and sinterability.

The particle shape of the ceramic powder can be irregular or spherical. Spherical powder due to low interparticle friction has higher flowability and tap density compared to irregular powder particles. It is reported [42] that the flow rate of spherical and irregular 55 µm calcium alkaline phosphate ceramic particles are 121 and 166 seconds, respectively. In contrast, higher packing density for irregular milled hydroxyapatite ceramic powder, compared to spherical spray-dried powder, was found by Suwanprateeb et al. [43] that can be related to interparticle friction which restrained the flowing between particles. Higher packing density was related to higher densities in fabricated parts.

The particle size of feedstock powder for ceramics varies between 0.2 and 200 µm. The finer powder has better sinterability, while larger particle size has better flowability. The particle size distribution of ceramic feedstock is a significant factor and determines the density of the parts being built. For powders with multimodal size distributions, fine particles can fill voids between large particles and produce samples with less interparticle voids and higher densities [41].

14.4.3.2   Ceramic Powder Material for PBF

In PBF of ceramics, a laser beam is used to fuse powder particles, since electron beams can only be used with electrically conductive materials. Ceramics are produced by single-step and multi-step PBF. The thermal laser energy irradiating the powder particles can initiate a chemical reaction (including gelation reactions), solid-state sintering, partial melting, and fully melting of the powder particles. Three different powder layer deposition systems including conventional deposition (i.e., roller or scraper), aerosol assisted spray, and slurry-based deposition are known from the literature [44]. As shown in Fig. 14.8, ceramics can be processed by PBF in a single-step. (Multi-step PBF of ceramic was shown in Fig. 14.3.) Table 14.6 shows the common methods, deposition system, and material types for PBF of ceramics.

Although not as common as ceramic PBF, ceramic DED has been demonstrated by several researchers. Any ceramic powder which can be fed through the powder system and which can melt rather than sublimate at elevated temperature can potentially be used. However, the low thermal conductivity of ceramic powders combined with their low thermal shock resistance makes the formation of complex-shaped dense parts almost impossible. As such, ceramic deposition by DED is only advisable when making small ceramic coatings on metal parts or mixing metal and ceramic powders together to deposit metal matrix ceramic components, as described below.

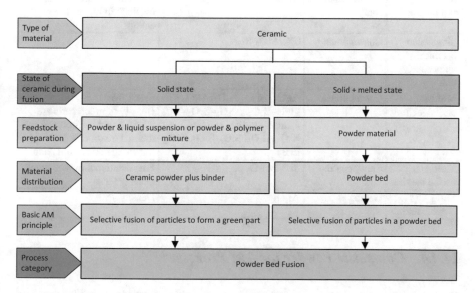

Fig. 14.8 Overview of single-step AM processing principles for ceramic materials [6]

Table 14.6 Reported methods for creating parts from ceramic materials in PBF [44]

Method	Deposition system	Materials	Characteristics
Single-step solid-state sintering	Conventional	SiC, Ta$_2$O$_5$, SiO$_2$, Bi$_4$Ti$_3$O$_{12}$, Bi$_4$Ge$_3$O$_{12}$	Packing density 20–50 vol%
Single-step partial melting	Conventional	AlH$_{12}$N$_3$O$_4$P, WC-10Co/Cu Al$_2$O$_3$–B$_2$O$_3$, ZrO$_2$ Hydroxyapatite phosphate glass	Packing density 20–50 vol%
Single-step partial melting	Layer-wise slurry coating	Hydroxyapatite, SiO$_2$, Al$_2$Si$_2$O$_5$, porcelain	Packing density > 50 vol%
Single-step partial melting	Slurry spraying	ZrO$_2$	24–32% open porosity in final parts
Single-step partial melting	Ring blade	Alumina-feldspar, Al$_2$Si$_2$O$_5$, feldspar, Si–SiC–C	Accuracy is 50–80 μm Poor pressure strength
Single-step partial melting	Electrophoretic	Al$_2$O$_3$	Packing density 50 vol%, bulk density up to 85%
Single-step full melting	Conventional	SiO$_2$–CaP, ZrO$_2$, Al$_2$O$_3$, MgAl$_2$O$_4$	Fine microstructure can be obtained by preheat to 1800 °C and eutectic powder ratio
Single-step full melting	Slurry coating	SiO$_2$–clay parts	Ceramic green substrate is used, and high density parts are obtained
Single-step full melting	Aerosol-assisted spray	Al$_2$O$_3$	Full density parts can be obtained in higher laser energy
Single-step chemically induced binding	Conventional	SiC, SiO$_2$, Si$_3$N$_4$	
Multi-step partial melting	Conventional	Al$_2$O$_3$, Al$_2$O$_3$-B$_2$O$_3$, Al$_2$O$_3$–glass–B$_2$O$_3$, Al$_2$O$_3$–ZrO$_2$–TiC, apatite–mullite, graphite, K$_2$O–Al$_2$O$_3$–SiO$_2$, SiO$_2$, ZrO$_2$, ZrB$_2$	Crack free Density of green parts are 40–80% and is improved by thermal treatments and remelting each multilayers
Multi-step partial melting	Slurry coating	SiO$_2$, Al$_2$O$_3$	Free delamination and crack, Density 98%
Multi-step partial melting	Gelling	Al$_2$O$_3$	Density 75%

14.4.4   Composite Powder for AM Processes

Composite components can be produced by processing of powder particles through different AM processes such as PBF, DED, and BJT. When processing ceramics the interface between the matrix and the dispersed or embedded phases must be designed and engineered to have few defects and good bonding.

14.4.4.1   Composites Produced by PBF

Using PBF, several types of composite materials have been produced, including polymers with ceramic reinforcements, ceramic–metal composites (cermets), ceramic–ceramic composites, and metal–metal composites. Several approaches have been taken to fabricate these materials. Liquid-phase sintering (LPS) can produce a polymer–matrix composite by pre-mixing powders. LPS has been used to produce biomedical-related composites like tricalcium phosphate (TCP)/poly-L-lactic acid (PLLA), polyetheretherketone (PEEK)/hydroxyapatite (HA), polycaprolactone (PCL)/HA, tricalcium phosphate (TCP)/poly-L-lactic acid (PLLA), and PCL microspheres + HA/PCL [1, 45]. LPS has also been used to create structural reinforced polymer composites, including nano-alumina in polystyrene; glass, nano-clay, or silica in nylon; carbon fiber, silicon carbide, or potassium–titanium whiskers in polyamide; and nano-alumina in polystyrene [46]. PBF by LPS is also used to produce ceramic-polymer and ceramic-metal composites such as SiC-PA and SiC-Si [47].

To produce complex shapes of polymer–ceramic composites the powder particles should be spherical, since irregular powder feedstock will lead to excessive shear forces during powder spreading and the formation of wavy or crater-filled surfaces during layering. Temperature-induced phase separation (TIPS) is a promising process for producing powders for PBF. Agglomerates produced by TIPS have regular shape (spherical) and are easily deposited by conventional deposition methods in PBF. The TIPS process is very flexible and can use different binders such as nylon 12, polypropylene, and wax in combination with ceramics like alumina, hydroxyapatite, and zirconia.

Metal–ceramic and metal–metal composite components can be produced using powder precursors to bind a lower-melting-temperature matrix and higher-melting-temperature secondary phase. This improves sinterability and consolidation of composites such as WC-Co/Cu with WC particulates reinforcing the Co matrix, bronze (Cu–Sn), or Cu. Different additives can improve mechanical, physical, chemical, and binding properties. For instance, lanthanum oxide reduces the surface tension and increases the work of adhesion between layers. Cu–Ag composites can be produced using partial melting.

Si–SiC parts can be printed using a single-step chemically induced binding processes. A laser is used to decompose a fraction of the irradiated SiC powder to Si. The decomposed Si becomes the matrix or bridging material for unreacted SiC and forms the ceramic–ceramic composite.

Ceramic composites, such as Al_2O_3–B_2O_3, Al_2O_3–ZrO_2–TiC, apatite–mullite, and K_2O–Al_2O_3–SiO_2, can be produced through multi-step partial melting PBF. To print a ceramic–ceramic matrix, a combination of binders is used. For instance, a mixture of ceramic with a polymer binder that decomposes to produce graphite or another carbide when heated can then react with ceramic powders to produce composite ceramic such as Al_2O_3–ZrO_2–TiC [48].

14.4.4.2   Composites Produced by DED

DED is able to print complex near-net-shape parts for functionally gradient material (FGM) structures. For example, it is reported [49] that a stainless steel alloy (304L) has been graded to a nickel-rich alloy (Invar 36) with no defects. It is even possible to grade across a component through more than one material while controlling the anisotropic behaviors [50]. Use of additives enhances sinterability, adjusts coefficient of thermal expansion, and controls grain growth of functionally graded materials which is vital to produce components with fewer defects. FGMs have been graded from metal–metal or metal–ceramic using DED. The key point to produce functionally graded components is selecting materials with similar thermophysical properties like thermal expansion coefficient [51]. DED processes can be applied to various metal matrix composites with ceramic reinforcing phase such as Ti–SiC, TiC–Ni–Inconel, Ti6ALV4-TiB, W-Co cermets, and TiC–Ni–Inconel [1]. DED can produce high entropy alloys and quaternary metal matrix reinforced with borides which are seamless and suitable for creating a limited stress junction [52]. Figure 14.9 shows different examples of FGM composites made by DED processes.

Laser techniques can be used to produce composites by in situ reactions. For instance, Ti–aluminides, copper matrix composites, and TiC–Al_2O_3 were produced using laser-based PBF and DED techniques. The laser can provide the high thermal energy that is necessary to form new compounds or trigger a chemical reaction to produce new compositions [1, 31, 53].

14.4.4.3   Composites Produced by BJT

Similar to biopolymer composites, bioceramic composites are particulates blended and consolidated via AM processes like BJT. This exhibits the potential for fabrication of components with good dimensional accuracy and complex geometry. Due to the support of the powder, this process can easily produce overhangs, arches, and cellular structures. Si–SiC composites can be produced via BJT followed by sintering to near-net-shape [54].

Porous biocomposites based on a matrix of stainless steel filled with tricalcium phosphate (TCP) have been processed by BJT. By varying the amount of TCP, different biodegradation rates can be achieved that be suitable for implants, while other blends would be better for scaffolds.

Metal–ceramic composites such as iron chromium alloy/17–4 PH and zirconia in a thermoplastic binder can be processed by BJT. Post heat treatment improves the density and removes the pores formed as a secondary phase by precipitation of oxides and carbides within the grain boundaries. This process can produce different composites provided the thermal expansion and shrinkage of the material feedstock are similar.

Fig. 14.9 Gradient alloys printed by DED (**a–e**) Ti6Al4V to Nb rocket nozzle, (**f–j**) 304L to Inconel 626 valve stems, (**k–n**) 304L to Invar 36 gradient mirror, (**o**) 304L to Invar 36 gradient alloy, showing the difference in oxide between the two alloys (**p**). The alloy from (**o**) after surface machining and laser welding of a 304L plate prove weldability of the gradient alloy. (**q**) Ti6Al4V to V, then from V to 420 stainless steel. (**r**) A similar gradient alloy to (**q**) but with a transition from Ti6Al4V to V, then from V to 304L. (**s**) After polishing, the crack at a specific composition is visible. (**t**) SEM micrograph of a brittle Fe–V–Cr phase. (**u**) Ti6Al4V to TiC gradient alloy. (**v**) Tiled micrographs from the gradient in (**u**) where the black phase is the TiC particles (Cambridge license number 4665950028024) [49]

14.5 Solid-Based Materials

14.5.1 Solid Polymer Feedstock for MEX

MEX uses solid feedstock (filaments) to produce components. A wide range of commercial polymer filaments are available. MEX typically uses amorphous thermoplastic polymers to produce parts by thermal layer adhesion. Thermoplastics can be selected for MEX based on different properties such as transparency, rigidity, biocompatibility, moisture resistance, fire retardancy, viscosity, etc. Thermoplastics are generally stable and maintain dimensional accuracy over time. Amorphous thermoplastics are softened up to the glass temperature, producing a high-viscosity material that can be extruded. Filaments for MEX are available with different hardness and toughness properties. A list of common MEX materials is shown in Table 14.7. Note that mechanical properties for several common thermoplastics and fiber-reinforced polymer materials were given in Chap. 6.

In MEX two different support structures are commonly used:

- Supports made of the same material as the build material, but using a lattice structure with low strength connections to the main component
- Using wax or poly-vinyl alcohol (PVA) materials which can be removed by melting or dissolving. For instance, water-soluble PVA is often used as a support structure for PLA parts

Great effort has been made to develop new materials for MEX beyond those commonly used. Many feedstock vendors have developed flexible thermoplastic polyurethane (TPU)-based filaments that exhibit good impact resistance and damping characteristics. For example, Pro FLEX is a TPU-based material from BigRep that has a Shore hardness of 98A, tensile strength of 40 MPa, and elongation at break of 470% with high-temperature resistance and reasonable impact resistance at low temperatures. Pro FLEX exhibits good damping behavior and resistance against dynamic loading that can be used in door handles, athletic shoes, skateboard wheels, etc. Some vendors have developed food-safe materials, and some of these are biodegradable. Many are based on biopolymers, such as the HDglass material from FormFutura, t-glase from Taulman3D, and the HI-TEMP material from BigRep. Note that HDglass is based on PETG (polyethylene terephthalate glycol) and exhibits high strength, high gloss, and transparency. DURABIO is a combination of polycarbonate and polymethacrylate produced by Verbatim that has good transparency, gloss, and impact resistance. Many vendors offer a range of polypropylene (PP) filaments that have good flexibility and toughness. This material is chemically resistant and can be used for food containers. Wood-filled materials are being offered by some vendors; typically, these are composed of PLA and wood or cork fibers. Its wood-like appearance and finishing characteristics have appeal for some applications. A range of other filled materials are being offered that often provide better mechanical properties than neat (unfilled) polymer materials. The MEX materials market is dynamic with new companies and new materials being introduced frequently. Some of the prominent materials vendors, which often target users of low-cost systems, include Airwolf 3D, colorFabb, Gizmo Dorks, MatterHackers, and Taulman3D.

Along with the improvement of polymers for AM, materials for support structure have also been enhanced. Water-soluble support materials have been available for many years, starting with PVA blends. More recently, additional water-soluble materials have been developed for a wide range of thermoplastic part materials. Some of these newer support materials are reportedly based on butenediol–vinyl

Table 14.7 Common commercially available thermoplastic polymers for MEX

Material	Characteristics
ABS (acrylonitrile butadiene styrene)	Good quality engineering thermoplastic Suitable for many applications
ABS blends	Stronger than ABS Better impact strength than ABS Suitable for more demanding applications
Medical grade ABS	Biocompatible Suitable for food packaging, pharmaceutical, and medical applications Sterilizable by gamma radiation or ethylene oxide sterilization methods
PC (polycarbonate)	Durable, stable, and accurate for strong parts High heat resistance and mechanical properties such as tensile strength
PC-ABS	Better heat resistance and mechanical properties compared to PC Good surface quality, similar to ABS High impact strength
Biocompatible PC	Good biocompatibility Suitable for food packaging industries, medical, and pharmaceutical Sterilizable using ethylene oxide sterilization methods or gamma radiation
PLA (polylactic acid)	Inexpensive, prints quickly Good strength and stiffness, compared to ABS Suitable for fast prototypes
Nylon 6 and 12	Good impact strength and toughness Good fatigue properties Suitable for snap-fits, repetitive closures, vibration-resistance
PPSF/PPSU (polyphenylsulfone)	High strength and chemical resistance Suitable for applications in high heat or caustic environments Sterilizable via steam autoclave, chemical, plasma and radiation
PEI (polyetherimide) – ULTEM 9085 and 1010 from Stratasys	High strength to weight ratio V-Zero rating for flame, smoke, and toxicity High heat deflection
PEKK (polyetherketoneketone)	Best strength characteristics of all MEX materials Good heat and wear resistance Good chemical resistance Flame, smoke, and toxicity resistance for aerospace applications

alcohol (BVOH), including HydroFill from Airwolf 3D, eSoluble from eSun, and BVOH from BigRep. EVOLV3D USM from Dow Chemical is another material for support structures that is water soluble and suitable for a wide range of popular build materials. It is based on hydroxypropyl methyl cellulose, not BVOH. Some new mechanically removable support materials have been developed for nylon materials, including carbon and glass filled nylons, including SAC1060 from Taulman3D.

14.5.2 Solid Metal Feedstock for DED and MEX SHL

Metals that are weldable are typically suitable for DED. One interpretation of that characteristic is that they are stable during melting. However, other factors can affect a metal's processability. For example, metals with high reflectivity and thermal conductivities such as gold, copper, and some grades of aluminum are hard to process. Other factors such as thermal expansion, thermal shock resistance, and phase transformation are also important for processing by DED. Figure 14.10 shows different solid material types in AM processes.

14.5.2.1 Wire Metal Feedstock

In wire feed AM systems, wire is fed through a nozzle and melted by an energy source. Almost all wire feedstocks in AM are used in DED processes, but a few researchers have demonstrated limited success with wire filaments utilized in MEX systems, though none has yet been commercialized. The metal DED process is carried out in either a sealed gas enclosure or an open environment with shield gas surrounding the processing location. Using solid feedstock provides higher deposition rates compared to powder feed. Wire feed DED is scalable for components from millimeters to meters in size. The Grainger Catalogue lists 380 different nickel and steel alloy welding wires with many different diameters [55], but there are many others available. A partial list includes:

- Titanium and titanium alloys
- Inconel 600, 625, 718

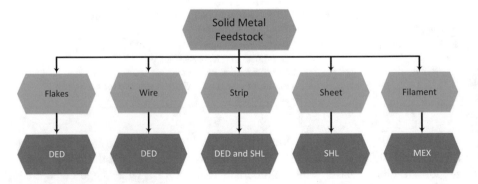

Fig. 14.10 Solid feedstock in AM

- Nickel and copper–nickel alloys
- Stainless steels
- Aluminum alloys 1100, 2318, 2319, 3000 Series, 4043, 4047, 5183, 5356, 5554, 5556
- Alloy steels
- Cobalt alloys
- 4340 steel
- Zircalloy
- Tantalum
- Tungsten
- Niobium
- Molybdenum

Powder feedstock is more expensive compared to wire feedstock across the board. For instance, on average, AM-grade tantalum wire is 7% cheaper than the average cost of tantalum powder. AM-grade Ti-6Al-4V powder is about 140% more expensive than the average cost of wire. Wire Ti-6Al-4V usually has a faster turnaround time compared to powder feedstock, as well. AM-grade Inconel 625 wire and Stainless Steel 316 wire are half of the cost of their powder counterparts.

Breathing in fine particles of any metal can be harmful, and powders such as aluminum and titanium are highly flammable when fluidized in the presence of oxygen. Wire feedstock for DED systems shows none of these problems and therefore is generally safer.

Wire feedstock can be used to create a larger melt pool compared to powder-based DED, providing a higher production rate [56]. Deposition rates vary from 3.18 to 18 kg of material per hour, depending on the type of the material, part complexity, and features, making it one of the fastest metal AM processes in the market. Figure 14.11 shows the build rate for laser and EB-PBF, laser and electron beam

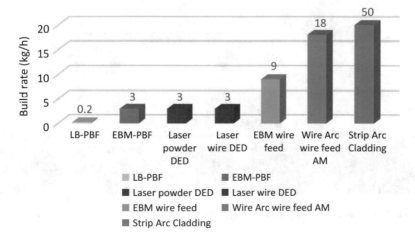

Fig. 14.11 Build rate for different AM processes

DED, and wire-arc DED. With high build rates, large parts become feasible to fabricate, with wire-arc being used to fabricate structures up to 6 m in length.

Wires come in different types including solid wire ($\emptyset$ = 0.8–3.2 mm) and tubular wires ($\emptyset$ > 1.2 mm). Figure 14.12 shows the wire classification for DED. Solid wires are soft and ductile made by the wire drawing process, which is cheaper compared to tubular wires. Tubular wires have two classifications, metal sheath (containing graphite and MoS_2) to improve the melting pool stability and powder encapsulated wire to improve the build quality in different ways. Powder in encapsulated wire can be used to make a composite and for other purposes, as explained below:

- Arc stabilizers: This approach helps to stabilize the arc and prevent the electrode from adhering to the component. For DED arc stabilizer powder can be added in tubular wire to stabilize and prevent deviation of the arc. This reduces the effective area of the arc and improves process quality.
- Fluxing agent: Flux is a chemical cleaning agent that helps to improve flow or purify the melt pool. Fluxes may have more than one function, for instance, preventing oxidation and decreasing viscosity to produce a stable melt pool.
- Slag former: This helps remove contamination waste, control the temperature, and minimize any re-oxidation.
- Shielding gas producer: Carbon dioxide can be injected through the center of the wire, which can reduce weld spatter. Gases such as helium or argon can also be used to prevent oxidation and limit chemical reactions in melt pool.

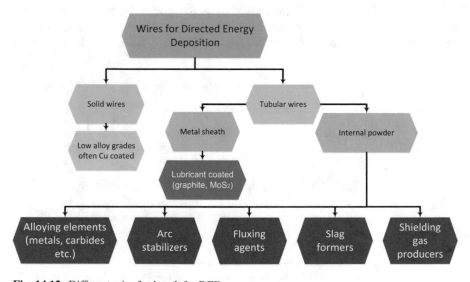

Fig. 14.12 Different wire feedstock for DED process

14.5.2.2   Strip Metal Feedstock

Strip metal feedstock is primarily used for electro-slag welding (ESW) and submerged-arc welding (SAW) and was developed to improve deposition rate and reduce dilution.

Strip can be used as solid feedstock for DED, which can be processed by plasma or electric arc. This is called strip cladding, which is suitable for larger components when lower dimensional accuracy is needed. SAW has been widely used with strip electrodes since the mid-1960s. A strip electrode, usually measuring 15–120 × 0.5 mm, is used as the (normally positive) electrode and an electric arc is formed between the workpiece and strip feedstock. To protect the weld pool from the atmosphere and improve the surface quality, flux is used. Figure 14.13 shows material and the strip cladding process for a big prototype.

Strip feedstock can be used for electro-slag cladding (ESC) which is a development of submerged arc strip cladding. Strip ESC has the highest deposition rate for metal among all AM processes. ESC is a resistance-welding process, and there is no arc between the parent material and strip electrode. The generated heat from molten slag melts the surface of the base material, and then the edge of the strip is submerged in the slag and flux and forms a new shape. The penetration depth in ESW is less than SAW because the molten slag pool is utilized to melt both parent material and feedstock strip and some of the parent material [57].

The deposition rate in strip cladding is a function of amperage which is shown in Fig. 14.14. As can be seen for ESW strip cladding the deposition rate can go up to 50 kg/h. Table 14.8 shows the list of common strip material for arc cladding.

14.5.2.3   Sheet Metal Feedstock

SHL can produce large parts with faster production rates than most other AM processes. The cost of the process can also be lower due to the cheaper feedstock. Ultrasonic Additive Manufacturing (UAM) is a solid-state SHL process to join layers using ultrasonic bonding followed by sequential post-processing such as machining to form geometry. Fabrisonic's line of SonicLayer machines can print up to

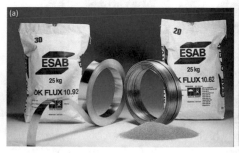

Fig. 14.13 Strips and printed parts by ESC from. (Photo courtesy of ESAB)

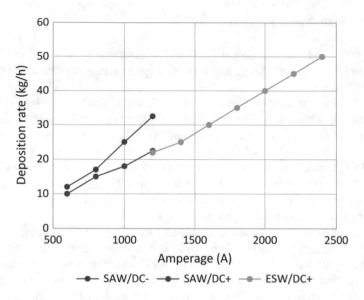

Fig. 14.14 Deposition rate for different strip-based DED processes [57]

Table 14.8 Common steels and nickel-based alloys for strip cladding by EABS and Lincoln Electrode [57, 58]

Material	Alloy	Process	Material	Alloy	Process
Nickel-based	Alloy 82	SAC, ESC	*Stainless steel*	308L SS	SAC, ESC
	NiCr3	SAC, ESC		347 SS	SAC, ESC
	NiCrMo3	SAC, ESC		316L SS	SAC, ESC
	NiCrMo4	SAC, ESC		2209 SS	SAC, ESC
	NiCrMo13	SAC, ESC		2594	SAC, ESC
	NiCu7	SAC		309L SS	SAC, ESC
	Ni 400	SAC, ESC		309 LNB SS	SAC, ESC
	Ni 600	SAC, ESC		317L SS	SAC, ESC
	Ni 625	SAC, ESC		310Mol SS	SAC, ESC
	Ni 825	SAC, ESC		385 SS	SAC, ESC
	NiFeCr1	SAC, ESC		309L SS	ESC
Mild steel	7018 mild steel	SAC, ESC		309LNB SS	ESC
	EH12K	SAC		430 SS	SAC, ESC
	EL 12	SAC		904L	SAC, ESC
	EM12K	SAC		318	SAC
	EM12	SAC			

4 kg/h. Aluminum, titanium, copper, and stainless steel have been processed by SHL [1]. This process protects the sheet microstructure and material properties because the process temperature is not very high and deformation of the microstructure occurs in a very small band at the sheet interfaces. In addition to UAM, joining

of many types of metal sheet using, primarily, form-then-bond (cut then stack) has been demonstrated for customer purposes by many research and commercial organizations.

14.5.2.4   Flakes for Friction Stir Additive Manufacturing

Friction Stir Additive Manufacturing (FSAM) can process different materials ranging from submicron–millimeter powders, wire, and flake. In FSAM the rougher the feedstock, the higher the production rate and lower dimensional accuracy. For instance, when using micrometer powders, near-net shapes can be achieved. However, using cheaper flakes and wires increases the deposition rate, but the chance of defects is higher. Generally, FSAM is more suitable for alloying rather than producing final shapes. Table 14.9 shows materials processed through FSAM.

14.5.3   Solid Ceramic Feedstock for SHL and MEX

14.5.3.1   Sheets

Ceramic tape (sheets/laminates of ceramic powders joined together by polymer binders) can be used as a feedstock for SHL. It is reported [60] that the feedstock for ceramic SHL such as Al_2O_3 green tape of 0.7 mm thickness can be produced by a roll-forming technique. As-built SHL parts with 65.4% relative density were obtained, and after pressure-less sintering, the density increased to 97.1%. An AM method to produce ceramic parts is Cold Low-Pressure Lamination (CLPL). This is based on gluing the tape by means of adhesive film performed at room temperature. No mass flow occurs in CLPL, and therefore fine detail as shown in Fig. 14.15a, b

Table 14.9 Reported materials processed by FSAM [59]

Material	Alloy	Type
Aluminum composite	Al–SiC, Al–Fe, Al–W, Al–Mo	Powder
Copper composite	Cu–tungsten, Cu–tantalum	Powder
Steel	HY 80, stainless 316	Solid
Aluminum	1, 2, 5, 6, 7 series	Solid
Magnesium	AZ31, WE43, E675, AMX602, E21	Solid
Titanium	Ti–6Al–4V	Solid
Nickel	Nickel aluminum bronze	Solid
Steel composite	ODS 14YWT–F82H, 300M–4140, Aermet 100–4140	Solid
Nickel composite	In625–HY80, In600–SS304, Cu–Ni200-Mo	Solid
Aluminum composite	Al–steel	Solid
Copper composite	Cu–Ta, Cu–Nb, Cu–Mo	Solid

Fig. 14.15 (**a** and **b**) Low temperature co-fired ceramic gear made by SHL [62] (Elsevier license number 4672200242505) (**c, d**) SiC green ceramics by robocasting [63] (Wiley license number 4672301356576)

can be obtained with low temperature firing. Layer thickness of ceramic tapes can be in the range of 200–2000 μm. Other ceramic materials such as alumina and Si_3N_4 have been produced for SHL [61].

14.5.3.2   Other Solid Feedstock for Ceramics

Freeze form Extrusion Fabrication (FEF) is an environmentally friendly AM process that can build 3D ceramic components. It produces green ceramic parts by depositing aqueous pastes in controlled freezing conditions and can produce relatively large components. Sizable ice crystals may form during freezing that can produce pores and reduce green part density. To solve this problem, room temperature MEX may also be used with radiation heating for uniform drying of feedstock between subsequent layers [64].

In Robocasting, a highly solid-loaded aqueous ceramic slurry is processed via MEX onto a hot plate to make a transition from a flowable feedstock to a dilatant mass by drying to maintain the shape [65]. Different ceramic feedstocks such as Al_2O_3, $Pb(Zr0.95Ti0.05)O_3$, hydroxyapatite, and SiC have been successfully produced by Robocasting.

14.5.4 Solid-Based Composite Materials for SHL, MEX, and DED

14.5.4.1 Composite Laminate

During UAM, fibers can be embedded in a metal matrix to improve the tensile strength. Because UAM is operated at low or room temperatures, no melting, shrinkage, residual stress, and dimensional deviation occur and complex geometries are produced with good accuracy. Figure 14.16 shows metal–metal and metal–ceramic composites produced by Fabrisonic through UAM.

SHL of composites improves different properties of materials such as conductivity, fatigue properties, strength, etc. Comprehensive research on different composites has been done by Stucker [68] and his team to show the performance of UAM over a wide range of materials as shown in Table 14.10. Despite the advantages of UAM, consideration for design for AM is still crucial to prevent defects that can reduce the advantage of the embedded phase [1]. Another composite material for SHL systems is paper that can be cut into appropriately shaped layers, often by lasers or a very sharp blade. For paper, the binding is achieved using adhesive. Figure 14.17 shows a model of an orange made by SHL using paper, with inkjet printing used for colors.

14.5.4.2 Composite Filament

Due to the naturally limited mechanical properties and functionalities of pure polymer parts, it is useful to develop printable polymer composites with better performance. AM offers many advantages in the production of composites, comprising

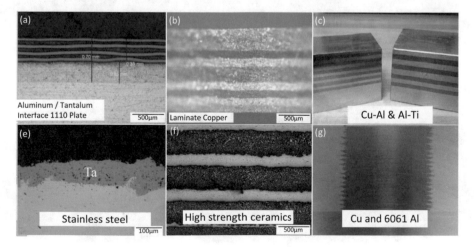

Fig. 14.16 Different composite materials by SHL. (Courtesy Fabrisonic) [66, 67]

Table 14.10 Laminate composites and combinations for UAM [68]

Composite	Combination	Matrix	Composition
Metal-Metal	SS304–Al5052	Metal–metal	Al–Fe$_2$O$_3$
	Al6061–Cu		Al–Cu$_2$O
	Ni–Ti		Steel–steel
	Ti–Al		FeGa
	Al1050–Al3003		Al 3003–H18
	Al–1.2Mn–0.12Cu	Metal–ceramic	Al–SiC
	Al-4.5Cu–1.5Mg	Ceramic–polymer	Al$_2$0$_3$ short fiber reinforced
	Al–Mn		Al–Sic fiber
	Ni–15Cr–8Fe–0.15C		Si–SiC
	Cu–30Zn		Zirconia–Al
	Fe–18Cr–11Ni–1Nb–0.08C	Ceramic–ceramic	Alumina–zirconia
	Fe–18Cr–8Ni–0.08C		Monolithic SiC-continuous fiber SiC/SiC
	Ni–Ti–Al		Si$_3$n$_4$
	MetPreg/Al		

Fig. 14.17 A full-color print from SHL. (Photo courtesy of Mcor) [69]

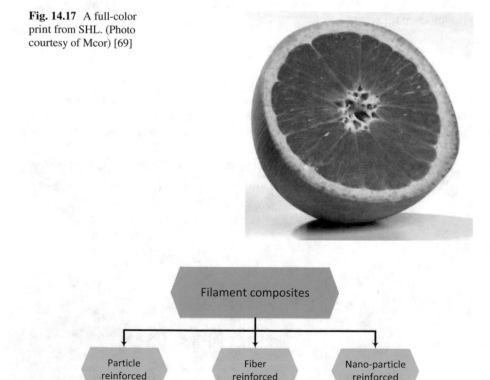

Fig. 14.18 Classification of filament composites

customized mechanical properties and geometry, cost-effectiveness, and high precision. Thus, in recent years, development of composite materials has attracted tremendous attention. Figure 14.18 shows the classification of filament composites.

14.5.4.3   Particle-Reinforced Filament

Particles are widely used as polymer matrix reinforcement due to their low cost. Reinforcement by glass beads or metal particles can improve tensile and flexural modulus (Table 14.11 shows additives for MEX filaments). Aluminum and alumina particles improve wear resistance, while permittivity of the printed parts can be improved by adding ceramic particles. Diamond photonic crystal structures using barium titanate $BaTiO_3$/ABS can improve permittivity by 240% compared to pure ABS filaments [70]. One of the issues of printed components by MEX is distortion related to thermal expansion/contraction and crystallization of polymers. Adding metal or ceramic particles to polymers can reduce its coefficient of thermal expansion and reduce crystallization and thus control distortion.

Table 14.11 A summary of composite materials and improved properties for filaments [71]

Additive type	Material	Enhancement in properties
Particles	Iron/ABS and copper/ABS	Storage modulus and thermal conductivity enhancement Reduction in the coefficient of thermal expansion
	Al and Al_2O_3/nylon-6	Reduction in frictional coefficient
	$BaTiO_3$/ABS and $CaTiO_3$/ polypropylene	Dielectric permittivity and controllable resonance frequency improvement
	Tungsten/PC	Dielectric permittivity, X-ray attenuation factor, and impact resistance improvements
	$BaTiO_3$/ABS	Tuning and improving dielectric permittivity
	Thermoplastic elastomer/ ABS	Reduction in anisotropy
Fibers	Short glass fiber/ABS	Tensile strength improvement
	Short carbon fiber/ABS	Tensile strength improvement
	Continuous carbon fiber/ nylon	Tensile strength improvement
	Continuous carbon fiber/ PLA	Tensile strength improvement
Nanoparticles	TiO_2/ABS	Enhances in tensile modulus and strength, reduction in elongation
	Carbon nanofiber/ABS	Improved tensile modulus and strength, reduction in elongation
	Montmorillonite/ABS	Improved flexural strength and modulus, tensile strength and modulus, and thermal stability Reduction in the thermal expansion coefficient
	Graphene/ABS	Electrical conductivity and thermal stability improvement
	Carbon nanofiber/graphite/ polystyrene	Better voltammetric characteristics, less capacitive background current

Fig. 14.19 (a) Cannon (b) hammer made by stainless steel-polymer filament through MEX. (Courtesy ColorFabb) [72]

ABS/graphene oxide, nylon/carbon, or conductive graphene and ferromagnetic PLA are other types of available filaments. Tensile strength and elastic modulus of ABS are enhanced by the effect of graphene oxide, for example, although elongation of break reduces.

Most AM for metals requires expensive equipment (around $200 K–$2000 K). Moreover, the cost of feedstock and post-process facilities add to the cost of establishing a workable process. An alternative is to use a solid filament that is highly loaded with metal powder for MEX printers. Filaments filled with metal powders such as bronze, aluminum, brass, and steel are available as feedstock. High density parts can be produced after post-treatment such as debinding and sintering. Figure 14.19 shows a component made of stainless steel using MEX. Commercial vendors of machines and materials for metal MEX include Desktop Metal and Markforged.

There are some significant considerations for printing metal filaments through MEX systems:

- Brass print nozzles can wear quickly because of abrasion.
- The filament may be slightly brittle and should be handled with more care.
- To reduce degradation the feedstock should be stored in a dry and cool place.
- The more metal in the filament, the more difficult to adjust printing parameters.

14.5.4.4 Nanomaterial-Reinforced Filament

Nanomaterials often exhibit unique mechanical and chemical properties and are used in high-performance composite filaments for MEX. Composites of polymers and conductive nanoparticles can improve the electrical properties of parts. Adding nano-TiO_2 particles enhances the thermal stability of printed components. Uniform dispersion of nanoparticles into the polymer matrix is key for homogenous properties. In addition, functionally graded polymer nano-composites are possible by delivering different volume fraction of nano-particles to specific building areas.

14.5.4.5 Fiber-Reinforced Filament

Fiber-reinforced polymer composites have unique properties. Filament composites are produced by two main methods:

- Polymer pellets and fibers are blended and then extruded.
- Polymer and fibers are co-extruded from two different spools.

Usually, short carbon or glass fibers are used in composite filaments. The void and orientation of composites are key factors to govern the mechanical properties of composite components. When printing composite parts using MEX, significant voids and pores can exist. The negative effect of porosity can be compensated by fibers aligned in the load-bearing direction. In general short fibers become aligned with the filament extrusion direction, enhancing properties such as strength and stiffness in the direction of the fiber. By carefully designing the deposition path per layer, tailored anisotropic part properties can be produced.

Expandable microspheres can be added into MEX filaments to reduce porosity. It is reported [73] that adding 11 wt% of microspheres improves density from 83% to 93% and subsequently further enhances mechanical properties. The hygroscopic behavior caused by voids can be used to an advantage for moisture-actuated biocomposites. For instance, a wood fiber-reinforced biocomposite by MEX can exhibit potential for increasing swelling ability [71]. The content of fibers affects mechanical properties, but if the volume of the fibers is too high, nozzle clogging can be an issue. Moreover, composites with higher fiber content are difficult to make into filaments due to reduction in toughness. Use of long fibers drastically improves mechanical properties. Shape memory composite polymer is also an area worthy of consideration in AM and can enable 4D printing. In shape memory composites, the alignment of cellulose fiber can be manipulated to determine and control the swelling behavior over time and thus result in significant, designed shape changes after production.

14.5.4.6 Solid-Based Composite Wires

To be in wire form, the material feedstock has to be ductile enough to withstand the forces necessary to reduce the diameter during the wire-drawing process. This is a limitation of advanced materials, but wide categories of materials like titanium, aluminum, nickel, and cobalt can be used as wire feedstock.

Most of the composite components in wire-feed DED systems are made using different wires simultaneously. Using multiple wire-feed nozzles utilized with a single energy source provides the ability to simultaneously feed with independent program control different materials into a single molten pool. This can produce functionally graded and "superalloy" materials. Material compositions can be changed using this system and can provide a new generation of high-performance alloys currently unavailable. Another advantage of this build strategy is making different alloy mixtures in different parts of the components. Using different wire gauges provides the ability for large deposition for one feature while also allowing fine deposition for features requiring higher resolution [55]. Composite materials also can be deposited by using tubular wire that has a metal powder inside, which fuse due to the heat source [74]. This method is therefore a combination of wire and powder DED.

14.6 Material Issues in AM

AM defects and their influence on the quality and performance of produced components are currently a major concern. It is essential to understand the phenomena that drive defect generation and mechanisms and the inspection methodologies for quality control. Issues include those related to fabrication processes, final part properties, and feedstock characteristics.

14.6.1 Build Orientation

Differences in thermal history across a part are associated with anisotropic heat conduction along the build direction and the unique interaction of specific build styles interacting with a specific geometry cross-section in a layer. These affects change significantly if build orientation is changed. This typically leads to inhomogeneity in microstructure and produces anisotropic mechanical properties. Postprocessing such as heat treatment can reduce or eliminate anisotropy but adds to the cost of production. Hot isostatic pressing (HIP) can significantly reduce anisotropy by removing porosity and normalizing microstructure, but it can also result in part deformation and troubles controlling dimensional accuracy.

14.6.1.1 Balling Phenomenon

Balling can occur anytime a molten material is present. It is of most concern in metal AM when surface tension effects are greater than wetting between the molten metal and the underlying substrate. The mechanism is based on Rayleigh Taylor instability and forms ball/bead-shapes instead of continuous melting tracks. This leads to rough surfaces, increase in porosity, and microcracks. Different factors affect balling such as contamination, oxidation, and using poorly adjusted process parameters such as laser power, scan speed, layer thickness, etc. [1, 75]. The residence time of the meltpool also affects balling. When a meltpool solidifies quickly, the material may have no time to form a ball. When a meltpool solidifies slowly, there may be sufficient time for surface tension forces to pull the molten material into a ball.

14.6.2   Keyholes

In thermal-based metal AM processes, keyholes are defects appearing in the bulk and surfaces of parts. When the heat source creates a plasma, it results in high penetration depth and vapor creation. When vapor is generated, it can become trapped at the bottom of the meltpool, forming porosity in the final part. Appropriate control of process parameters can help a plasma vapor bubble collapse from bottom to top and thus not trap vapor nor form porosity. The physics behind plasma formation and collapse and methods for avoiding keyhole porosity is addressed in numerous publications, including by Khorasani et al. [76].

14.6.3   Chemical Degradation and Oxidation

In high-temperature AM processes, the environment must be controlled to reduce contamination, humidity, and oxygen to minimize degradation and oxidation. Degradation may cause depolymerization in polymers or formation of oxide inclusions in metals, reducing part performance. For example, in powder-based metal AM when the meltpool temperature is high, the combination of high temperature and atmosphere may affect the chemical composition of the feedstock and part. Losses of Al in Ti-6Al-4V were reported in EBM due to high process temperature and evaporation of Al [37]. High energy and temperature in polymer PBF can produce smoke from degradation and depolymerization, reducing the mechanical properties.

14.6.4   Reactive Processes

In high-temperature AM systems reaction with the environment coupled with high cooling and rapid solidification produces high-temperature chemical or metallurgical phases for as-built parts that are not observed in conventional manufacturing. Phase types and concentrations are related to changes in energy input applied in building or by changing the composition of the feedstock powder. It is widely reported that changing metallurgical phases can produce different mechanical properties such as hardness, ductility, strength, corrosion resistance, etc. [77]. Chemical reactions can also happen for low-temperate AM processes when using liquid feedstock such as VPP when changing the curing conditions of the photopolymer.

14.6.5   Assistive Gas and Residual Particles

In powder based systems loose powders may fall into the build area and cause problems such as pores, cracks, and increased roughness. The effect of residual particles on part quality intensifies when the system uses the flow of inert gas in conjunction with the laser movement. If the laser moves with the direction of airflow flow, sparks and sputter may impact on the build area to produce rough surfaces and increase the risk of cracks and other defects. Furthermore, flying particles can temporarily block the laser, which can cause defects. This problem is solvable, in part by selecting a proper scan strategy with respect to flow direction and maintaining a clean build chamber environment.

14.6.6   Cracks

Cracking is a problem observed in most AM processes. AM inherently involves bonding material side-by-side and layer-by-layer. Any lack of bonding can result in a crack. Additionally, in AM systems that use temperature, large amounts of thermal stresses occur due to rapid cooling and shrinkage of materials, generating forces that can result in cracks. Impurities and contamination in feedstock lead to non-uniform solidification due to differences in the solidification temperature and can also lead to formation of cracks. Brittle material or material with lower resistance to thermal shock and lower heat conductivity such as ceramic and Ti alloys are more susceptible to crack formation [78]. Shrinkage, drying, and compositional segregation of binder material are other factors that may cause cracking [79].

14.6.7   Delamination

Delamination is basically the separation of two or more layers caused by various factors such as residual stress or insufficient layer bonding (e.g., lack of energy, low adhesive condition, low coverage of binder, etc.). Delamination is related to material characteristics and processing parameters and can also be related to mass transfer in temperature-based AM processes. Interlaminar regions in composites are common matrix regions that are weak that can suffer from delamination. Significantly higher interlaminar stresses appear at the free edges of multi-directional laminated composites because of property mismatch between layers [80].

14.6.8   Distortion

Deflection, distortion, and warpage are related problems that are created from the stresses caused by changes in material volume such as contraction of heated feedstock in PBF, DED, and MEX or shrinkage during polymerization in VPP. In the case of higher distortion, the part can break into two or more sections. Most AM systems use a substrate to which the first layer is bonded. The interaction of contraction forces versus the bonding of the first layer to the substrate can cause distortion which may produce some problems after removing the part from the substrate [78].

14.6.9   Inclusions

The material in AM can chemically react with the assisted gas, forming exogenous intermetallic particles like sulfides and oxides. These can become trapped in the part as inclusions. Inclusions have different sizes ranging from 10 nm to 1 mm, affecting the mechanical properties such as ductility, fracture toughness, and fatigue. Contamination in powders can increase the number and size of inclusions.

14.6.10   Poor Surface Finish

Surface quality is often inadequate in AM components. It is driven by different factors such as the deposition head dimensions, layer thickness, and build orientation. In addition, poor surface quality can originate from the energy source causing recoil pressure, too high or low surface tension, balling, and residual particles. In BJT and MJT non-adjusted pressure of jet flow can produce rough surface quality. Low surface quality can be produced by using recycled material; for instance, used polyamide powders in PBF can result in low surface quality appearing as an "orange peel"-like surface. To improve the surface quality, the use of smaller deposition heads or more finely focused laser beams can be helpful, but will affect build speed and may have other consequences. Another solution to improve the part quality is to use hybrid AM with integrated post-processing [81]. A good rule of thumb is that flat surfaces will have the best surface finish if they are aligned with the z-axis, and upward-facing features will, in general, have better surface finish than downward-facing features.

14.6.11   Porosity

Porosity is a common problem in AM components, especially when the binding mechanisms are driven based on temperature changes, capillary forces, and gravity without applying external pressure. Various factors such as shrinkage, material feed shortage, and lack of deposition/melting/fusion/binding that often occur at the borders of parts can lead to porosity. Also, spherical pores can occur in molten tracks due to trapping of gas.

14.6.12   Shelf Life or Lifetime of the Feedstock

Some companies recommended use time for resins instead of strict shelf time. Different factors such as suspended solids, specific flow rate, frequency of recycling, oxidants, and temperature affect the lifetime of resins.

For powder feedstock, oxidation and moisture are the most important factors which affect performance. It is recommended to use 30–50% of fresh powder with overflow and cake powder to reduce the effect of oxidation and possible contaminations.

Solid feedstock such as laminate, wire, strip, and filament has the lowest contamination and highest life and shelf time. The solid structure prevents chemical reactions and makes the feedstock more stable.

14.6.13   Support Structures

Anytime material is deposited in AM, it needs to be attached to something else by a physical connection. In addition, in a process which deposits molten material, that material must be supported by something underneath it. These reasons, plus the need to mitigate the effects of residual-stress-induced and shrinkage-induced warpages means that support structures are needed in many AM processes. Most AM processes need support structure if the printed surface is less than 45° from the horizontal. This is particularly true for MEX and for PBF of metals, which has been discussed by Gibson and his team [78] with respect to the interaction of dynamic shear force in the liquid phase and gravity forces in the solid phase. When using supports in AM processes, differences in cooling rates between the bulk of the material and contact points with supports increases the risk of the cracks. In addition, supports must be removed after the build process by post-process operations such as machining, which will be discussed in Chap. 16.

14.7   Questions

1. Specify the physical type of feedstock needed for each of the seven AM process.
2. What are the differences between polymers used in VPP and MEX?
3. Name the common photopolymer feedstocks for VPP.
4. Explain the most significant difference in feedstock requirements when using MJT vs VPP.
5. What are the main differences between multi-materials created within a layer versus between layers during an AM process?
6. How is it possible to print multi-material and multi-color components in MJT?
7. What are solder droplet printing methods and how do they operate?
8. What are the possible problems when using recycled powder? What are the best methods for recycling powder in PBF?
9. What are the common single-step processes to produce ceramics by PBF?
10. Explain the different advantages and disadvantages of solid versus powder feedstock for DED systems.

11. What are the most important properties for determining material compatibility when creating a metal composite in DED?
12. How we can make filament composites?
13. What are five common defects in AM components?
14. What is the difference between the polymers that are used in MEX and PBF?

References

1. Bourell, D., et al. (2017). Materials for additive manufacturing. *CIRP Annals, 66*(2), 659–681.
2. Wohlers, T. (2018). *Wohlers report. 3D Printing and Additive Manufacturing State of the Industry, Annual Worldwide Progress Report*. Wohlers Associates.
3. Wohlers, T. (2019). *Wohlers report. 3D Printing and Additive Manufacturing State of the Industry*. Wohlers Associates.
4. Wohlers, T. (2020). *Wohlers report. 3D Printing and Additive Manufacturing Global State of the Industry*. Wohlers Associates.
5. GE Additive. (2020). https://www.ge.com/additive/additive-manufacturing/information/metal-additive-manufacturing-materials
6. ASTM International. (2015). *SO/ASTM52900–15, standard terminology for additive manufacturing – general principles – terminology*. West Conshohocken: ASTM International.
7. Ledesma-Fernandez, J., Tuck, C., & Hague, R. (2015). High viscosity jetting of conductive and dielectric pastes for printed electronics. In *Proceedings of the International Solid Freeform Fabrication Symposium*.
8. Valencia, J. J., & Quested, P. N. (2013). Thermophysical properties.
9. Mills, K. C. (2002). *Recommended values of thermophysical properties for selected commercial alloys*. Woodhead Publishing, Ohio USA.
10. Ngo, T. D., et al. (2018). Additive manufacturing (3D printing): A review of materials, methods, applications and challenges. *Composites Part B: Engineering, 143*, 172–196.
11. RepRap. (2020). https://www.3ders.org/articles/20181105-german-reprap-introduces-l280-first-liquid-additive-manufacturing-lam-production-ready-3d-printer.html
12. Jacobs, P. F. (1995). *Stereolithography and other RP&M technologies: From rapid prototyping to rapid tooling*. New York: Society of Manufacturing Engineers.
13. Crivello, J., & Dietliker, K. (2013). *Volume III: Photoinitiators for free radical cationic & anionic photopolymerisation*. Chichester: Wiley.
14. Southwell, J., et al. (2013). *Radiation curable resin composition and rapid three-dimensional imaging process using the same*. US Patents.
15. Das, S., et al. (2017). *High temperature three dimensional printing compositions*. US Patents.
16. Wilson, J. E., & Burton, M. (1974). Radiation chemistry of monomers, polymers, and plastics. *Physics Today, 27*, 50.
17. Sager, B., & Rosen, D. W. (2008). Use of parameter estimation for stereolithography surface finish improvement. *Rapid Prototyping Journal, 14*(4), 213–220.
18. Sperling, L. H. (2012). *Interpenetrating polymer networks and related materials*. Springer Science & Business Media, New York, USA.
19. Crivello, J. V., Lee, J. L., & Conlon, D. A. (1983). Photoinitiated cationic polymerization with multifunctional vinyl ether monomers. *Journal of Radiation Curing, 10*(1), 6–13.
20. Chaudhary, S., et al. (2014). Poly (ethyleneterephthalate) glycolysates as effective toughening agents for epoxy resin. *Journal of Applied Polymer Science, 131*(4).
21. Napadensky, E., Kritchman, E. M., & Cohen, A. (2007). *Compositions and methods for use in three dimensional model printing*. US Patents.

22. Vaezi, M., et al. (2013). Multiple material additive manufacturing–Part 1: A review: This review paper covers a decade of research on multiple material additive manufacturing technologies which can produce complex geometry parts with different materials. *Virtual and Physical Prototyping, 8*(1), 19–50.
23. Khodabakhshi, K., et al. (2014). Anionic polymerisation of caprolactam at the small-scale via DSC investigations. *Journal of Thermal Analysis and Calorimetry, 115*(1), 383–391.
24. Chaudhary, K. C., & Redekopp, L. G. (1980). The nonlinear capillary instability of a liquid jet. Part 1. Theory. *Journal of Fluid Mechanics, 96*(2), 257–274.
25. Zocca, A., et al. (2015). Additive manufacturing of ceramics: Issues, potentialities, and opportunities. *Journal of the American Ceramic Society, 98*(7), 1983–2001.
26. Lithoz GmbH. (2020). Available Materials for the LCM process. https://www.lithoz.com/en/our-products/materials
27. Schwentenwein, M., & Homa, J. (2015). Additive manufacturing of dense alumina ceramics. *International Journal of Applied Ceramic Technology, 12*(1), 1–7.
28. Amado, A., et al. (2011). Advances in SLS powder characterization. In *22nd Annual International Solid Freeform Fabrication Symposium - An Additive Manufacturing Conference, SFF 2011*.
29. Hopkinson, N., & Erasenthiran, P. (2004). High speed sintering - Early research into a new rapid manufacturing process. *Solid Freeform Fabrication Symposium*, 312–320.
30. Frazier, W. E. (2014). Metal additive manufacturing: A review. *Journal of Materials Engineering and Performance, 23*(6), 1917–1928.
31. Bourell, D., et al. (2011). Modeling effects of oxygen inhibition in mask-based stereolithography. *Rapid Prototyping Journal, 17*(3).
32. DebRoy, T., et al. (2018). Additive manufacturing of metallic components–process, structure and properties. *Progress in Materials Science, 92*, 112–224.
33. Zito, D., et al. (2014). Optimization of SLM technology main parameters in the production of gold and platinum jewelry. In *The Santa Fe Symposium on Jewelry Manufacturing Technology 2014*.
34. Pettersson, T. (2015). Characterization of metal powders produced by two gas atomizing methods for thermal spraying applications.
35. ATO LAB. (2020). http://metalatomizer.com/
36. Van Acker, K., et al. (2005). Influence of tungsten carbide particle size and distribution on the wear resistance of laser clad WC/Ni coatings. *Wear, 258*(1), 194–202.
37. Zhang, J., & Jung, Y.-G. (2018). *Additive manufacturing: Materials, processes, quantifications and applications*. Cambridge, MA: Butterworth-Heinemann.
38. Mahbooba, Z., et al. (2018). Additive manufacturing of an iron-based bulk metallic glass larger than the critical casting thickness. *Applied Materials Today, 11*, 264–269.
39. Niu, F. Y., et al. (2017). Process optimization for suppressing cracks in laser engineered net shaping of Al2O3 ceramics. *JOM, 69*(3), 557–562.
40. Wilkes, J., et al. (2013). Additive manufacturing of ZrO2-Al2O3 ceramic components by selective laser melting. *Rapid Prototyping Journal, 19*(1), 51–57.
41. Du, W., et al. (2017). Binder jetting additive manufacturing of ceramics: A literature review. In *ASME 2017 International Mechanical Engineering Congress and Exposition*. American Society of Mechanical Engineers Digital Collection.
42. Gildenhaar, R., et al. (2012). Calcium alkaline phosphate scaffolds for bone regeneration 3D-fabricated by additive manufacturing. *Key Engineering Materials, 493-494*, 849–854. Trans Tech Publ.
43. Suwanprateeb, J., Sanngam, R., & Panyathanmaporn, T. (2010). Influence of raw powder preparation routes on properties of hydroxyapatite fabricated by 3D printing technique. *Materials Science and Engineering: C, 30*(4), 610–617.
44. Deckers, J., Vleugels, J., & Kruth, J. P. (2014). Additive manufacturing of ceramics: A review. *Journal of Ceramic Science and Technology, 5*(4), 245–260.

45. Liu, D., et al. (2013). Mechanical properties' improvement of a tricalcium phosphate scaffold with poly-l-lactic acid in selective laser sintering. *Biofabrication, 5*(2), 025005.
46. Chivel, Y. (2014). Ablation phenomena and instabilities under laser melting of powder layers. In *8th International Conference on Photonic Technologies LANE*.
47. Evans, R. S., et al. (2005). Rapid manufacturing of silicon carbide composites. *Rapid Prototyping Journal, 11*(1), 37–40.
48. Bai, P.-K., Cheng, J., & Liu, B. (2005). Selective laser sintering of polymer-coated Al_2O_3/ ZrO_2/TiC ceramic powder. *China Nonferrous Metals Society Journal (English version), 15*(2), 02.
49. Hofmann, D. C., et al. (2014). Compositionally graded metals: A new frontier of additive manufacturing. *Journal of Materials Research, 29*(17), 1899–1910.
50. Kieback, B., Neubrand, A., & Riedel, H. (2003). Processing techniques for functionally graded materials. *Materials Science and Engineering A, 362*(1–2), 81–106.
51. Carroll, B. E., et al. (2016). Functionally graded material of 304L stainless steel and inconel 625 fabricated by directed energy deposition: Characterization and thermodynamic modeling. *Acta Materialia, 108*, 46–54.
52. Nag, S., et al. (2009). Characterization of novel borides in Ti–Nb–Zr–Ta+2B metal-matrix composites. *Materials Characterization, 60*(2), 106–113.
53. Gasper, A. N. D., Catchpole-Smith, S., & Clare, A. T. (2017). In-situ synthesis of titanium aluminides by direct metal deposition. *Journal of Materials Processing Technology, 239*, 230–239.
54. Fu, Z., et al. (2013). Three-dimensional printing of SiSiC lattice truss structures. *Materials Science and Engineering: A, 560*, 851–856.
55. SciakyINC. (2020). https://www.sciaky.com/additive-manufacturing/wire-vs-powder
56. Ding, D., et al. (2015). Wire-feed additive manufacturing of metal components: Technologies, developments and future interests. *The International Journal of Advanced Manufacturing Technology, 81*(1), 465–481.
57. Elektriska Svetsnings-Aktiebolaget (ESAB), Editor. Technical handbook strip cladding, Sweden, 2020.
58. The new dimension in stip cladding, L. Electric, Publisher Lincoln Electric, Ohio, USA, 2020.
59. MELD Manufacturing. (2020). http://meldmanufacturing.com/
60. Zhang, Y., et al. (2001). Al2O3 ceramics preparation by LOM (Laminated Object Manufacturing). *The International Journal of Advanced Manufacturing Technology, 17*(7), 531–534.
61. Das, A., et al. (2003). Binder removal studies in ceramic thick shapes made by laminated object manufacturing. *Journal of the European Ceramic Society, 23*(7), 1013–1017.
62. Schindler, K., & Roosen, A. (2009). Manufacture of 3D structures by cold low pressure lamination of ceramic green tapes. *Journal of the European Ceramic Society, 29*(5), 899–904.
63. Cai, K., et al. (2012). Geometrically complex silicon carbide structures fabricated by robocasting. *Journal of the American Ceramic Society, 95*(8), 2660–2666.
64. McMillen, D., et al. (2016). Designed extrudate for additive manufacturing of zirconium diboride by ceramic on-demand extrusion. In *Proceedings of the 26th Annual International Solid Freeform Fabrication Symposium University of Texas*, Austin.
65. Travitzky, N., et al. (2014). Additive manufacturing of ceramic-based materials. *Advanced Engineering Materials, 16*(6), 729–754.
66. Norfolk, M. (2019). Ultrasonic additive manufacturing overview.
67. Fabrisonic. (2020). https://fabrisonic.com/gradient-material-solutions/
68. Janaki Ram, G., et al. (2007). Use of ultrasonic consolidation for fabrication of multi-material structures. *Rapid Prototyping Journal, 13*(4), 226–235.
69. Sculpteo. (2020). *LOM (Laminated Object Manufacturing): 3D Printing with Layers of Paper.* https://www.sculpteo.com/en/glossary/lom-definition/
70. Castles, F., et al. (2016). Microwave dielectric characterisation of 3D-printed BaTiO 3/ABS polymer composites. *Scientific Reports, 6*, 22714.

71. Wang, X., et al. (2017). 3D printing of polymer matrix composites: A review and prospective. *Composites Part B: Engineering, 110*, 442–458.
72. ColorFabb. (2020). https://colorfabb.com/steelfill
73. Wang, J., et al. (2016). A novel approach to improve mechanical properties of parts fabricated by fused deposition modeling. *Materials & Design, 105*, 152–159.
74. Tuominen, J. (2017). Directed energy deposition - Review of materials, properties and applications.
75. Kruth, J. P. et al. (2015). Additive manufacturing of metals via Selective Laser Melting: Process aspects and material developments.
76. Khorasani, A. M., Gibson, I., & Ghaderi, A. R. (2018). Rheological characterization of process parameters influence on surface quality of Ti-6Al-4V parts manufactured by selective laser melting. *The International Journal of Advanced Manufacturing Technology, 97*(9–12), 3761–3775.
77. Bandyopadhyay, A., & Heer, B. (2018). Additive manufacturing of multi-material structures. *Materials Science and Engineering: R: Reports, 129*, 1–16.
78. Khorasani, A., et al. (2017). Production of Ti-6Al-4V acetabular shell using selective laser melting: Possible limitations in fabrication. *Rapid Prototyping Journal, 23*(1), 110–121.
79. Olakanmi, E. O., Cochrane, R. F., & Dalgarno, K. W. (2015). A review on selective laser sintering/melting (SLS/SLM) of aluminium alloy powders: Processing, microstructure, and properties. *Progress in Materials Science, 74*, 401–477.
80. Islam, M. S., & Prabhakar, P. (2017). Interlaminar strengthening of multidirectional laminates using polymer additive manufacturing. *Materials & Design, 133*, 332–339.
81. Khorasani, A. M., et al. (2018). A comprehensive study on surface quality in 5-axis milling of SLM Ti-6Al-4V spherical components. *The International Journal of Advanced Manufacturing Technology, 94*(9–12), 3765–3784.

Chapter 15
Guidelines for Process Selection

Abstract AM processes, like all material processing, are constrained by material properties, speed, cost, and accuracy. The performance capabilities of AM materials and machines may lag behind conventional manufacturing technology (e.g., injection molding) for mass production of identical geometries, but it can outperform traditional manufacturing for small- and medium-lot production. Speed and cost, in terms of time to market, are where AM technology contributes, particularly for complex or customized geometries. The variation between different AM processes and materials is immense. As a result, it can be overwhelming to decide which AM process and material to use. In this chapter we introduce methods for applying decision theory to AM to aid in process–material combination selection for a desired component.

15.1 Introduction

The initial purpose of rapid prototyping technology was to create parts as a means of visual and tactile communication and for initial testing of a concept for fit and function. Since those early days of rapid prototyping, the applications of additive manufacturing processes have expanded considerably. According to Wohlers and associates [1], parts from AM machines are used for a number of purposes, including:

- Visual aids
- Presentation models
- Functional models
- Fit and assembly
- Patterns for prototype tooling
- Patterns for metal castings
- Tooling components
- Direct digital/rapid manufacturing

AM processes are rapidly increasing in capabilities, including production speed, material properties, and cost competitiveness. AM is taking its place beside

I. Gibson et al., *Additive Manufacturing Technologies*,
https://doi.org/10.1007/978-3-030-56127-7_15

traditional manufacturing operations, like injection molding machinery, as a process which has specific benefits and drawbacks. Although AM is slower at mass production of identical components, AM technology can have considerable speed and cost savings, especially in terms of time to market, for complex or customized geometries.

With the growth of AM, there is going to be increasing demand for software that supports making decisions regarding which machines to use and their capabilities and limitations for a specific part design. In particular, software systems can help in the decision-making process for capital investment of new technology, providing accurate estimates of cost and time for quoting purposes, and assistance in process planning.

This chapter deals with three typical problems involving AM that may benefit from decision support:

- Quotation support. Given a part, which machine and material should I use to build?
- Capital investment support. Given a design and industrial profile, what is the best machine that I can buy to fulfill my requirements?
- Process planning support. Given a part and a machine, how do I set it up to work in the most efficient manner alongside my other operations and existing tasks?

Examples of systems designed to fulfill the first two problems are described in detail. The third problem is much more difficult and is discussed briefly.

15.2   Selection Methods for a Part

15.2.1   Decision Theory

Decision theory has a rich history, evolving in the 1940s and 1950s from the field of economics [2]. Although there are many approaches taken in the decision theory field, the focus in this chapter will be only on the utility theory approach. Broadly speaking, there are three elements of any decision [3]:

- Options – the items from which the decision-maker is selecting
- Expectations – of possible outcomes for each option
- Preferences – how the decision-maker values each outcome

Assume that the set of decision options is denoted as $A = \{A_1, A_2, ..., A_n\}$. In engineering applications, one can think of outcomes as the performance of the options as measured by a set of evaluation criteria. More specifically, in AM selection, an outcome might consist of the time, cost, and surface finish of a part built using a certain AM process, while the AM process itself is the option. Expectations of outcomes are modeled as functions of the options, $X = g(A)$, and may be modeled with associated uncertainties.

Preferences model the importance assigned to outcomes by the decision-maker. For example, a designer may prefer low cost and short turnaround times for a concept model while being willing to accept poor surface finish. In many ad hoc decision support methods, preferences are modeled as weights or importances, which are represented as scalars. Typically, weights are specified so that they sum to 1 (normalized). For simple problems, the decision-maker may just choose weights, while for more complex decisions, more sophisticated methods are used, such as pair-wise comparison [4]. In utility theory, preferences are modeled as utility functions on the expectations. Expectations are then modeled as expected utility. The best alternative is the one with the greatest expected utility.

F. Mistree, J.K. Allen, and their coworkers have been developing the Decision Support Problem (DSP) Technique over several decades. The advantages of DSPs, compared with other decision formulations, are that they provide a means for mathematically modeling design decisions involving multiple objectives and supporting human judgment in designing systems [5, 6]. The formulation and solution of DSPs facilitate several types of decisions, including:

Selection – the indication of preference, based on multiple attributes, for one among several alternatives [7]

Compromise – the improvement of an alternative through modification [4, 6]

Coupled and hierarchical – decisions that are linked together, such as selection–selection, compromise–compromise, or selection–compromise [6].

The selection problems being addressed in this chapter will be divided into two related selection subproblems. First, it is necessary to generate feasible alternatives, which, in this case, include materials and processes. Second, given those feasible alternatives, a quantification process is applied that results in a rank-ordered list of alternatives. The first subproblem is referred to as "determining feasibility," while the second is simply called "selection." Additional feasibility determination and selection methods will be discussed in this section as well.

15.2.2   Approaches to Determining Feasibility

The problem of identifying suitable materials and AM machines with which to fabricate a part is surprisingly complex. As noted previously, there are many possible applications for an AM part. For each application, one should consider the suitability of available materials, fabrication cost and time, surface finish and accuracy requirements, part size, feature sizes, mechanical properties, resistance to chemicals, and other application-specific considerations. To complicate matters, the number and capability of commercial materials and machines continues to increase. So, in order to solve AM machine and process chain selection problems, one must navigate the wide variety of materials and machines, comparing one's needs to their capabilities while ensuring that the most up-to-date information is available.

To date, most approaches to determining feasibility have taken a knowledge-based approach in order to deal with the qualitative information related to AM process capability. One of the better developed approaches was presented by Deglin and Bernard [8]. They presented a knowledge-based system for the generation, selection, and process planning of production AM processes. The problem as they defined it was: "To propose, from a detailed functional specification, different alternatives of rapid manufacturing processes, which can be ordered and optimized when considering a combination of different specification criteria (cost, quality, delay, aspect, material, etc.)." Their approach utilized two reasoning methods, case-based and the bottom-up generation of processes; the strengths of each compensated for the other's weaknesses. Their system was developed on the KADVISER platform and utilized a relational database system with extensive material, machine, and application information.

A group at the National University of Singapore (NUS) developed an AM decision system that was integrated with a database system [9, 10]. Their selection system was capable of identifying feasible material/machine combinations, estimating manufacturing cost and time, and determining optimal part orientations. From the feasible material/machines, the user can then select the most suitable combination. Their approach to determining feasible materials and processes is broadly similar to the work of Deglin and Bernard. The NUS group utilized five databases, each organized in a hierarchical, object-oriented manner: three general databases (materials, machines, and applications) and two part-specific databases (geometric information and model specifications).

Several web-based AM selection systems have been developed over the years. The most comprehensive were University-led initiatives, whereas AM selection systems by specific manufacturers have focused more on the technologies they sell. Comprehensive, quantitative web-based AM selection systems require an overwhelming amount of work to keep them current with rapid technology innovation. As a result, most attempts quickly become out of date. One early example of a web-based selection system was developed at the Helsinki University of Technology. Through a series of questions, the selector acquired information about the part accuracy, layer thickness, geometric features, material, and application requirements. The user chooses one of 4–5 options for each question. Additionally, the user specifies preferences for each requirement using a 5-element scale from insignificant to average to important. When all 10–12 questions are answered, the user receives a set of recommended AM machines that best satisfy their requirements. Today, most web-based selection guides are more qualitative than quantitative, such as the guide by 3D Hubs, an online shop that connects AM service providers with customers for those services (*see* [11]).

The problem of determining process and material feasibility can be represented by the Preliminary Selection Decision Support Problem (ps-DSP) [12]. The word formulation of the ps-DSP is given in Fig. 15.1. This is a structured decision formulation and corresponds to a formal decision method based on decision theory. Qualitative comparisons among processes and materials, with respect to decision criteria, are sufficient to identify feasible alternatives and eliminate infeasible ones. After more quantitative information is known, more detailed evaluations of alternatives can be made, as described in the next subsection.

The key step in the ps-DSP is how to capture and apply experience-based knowledge. One chooses a datum concept against which all other concepts are compared. Qualitative comparisons are performed, where a concept is judged as better, worse, or about the same (+1, −1, 0, respectively) as the datum with respect to the principal

Given:	a set of concepts
Identify:	The principal criteria influencing selection. The relative importance of the criteria
Capture:	Experience-based knowledge about the concepts with respect to a datum and the established criteria.
Rank:	The concepts in order of preference based on multiple criteria and their relative importance.

Fig. 15.1 Preliminary Selection Decision Support Problem word formulation

criteria for the selection problem. Then, a weighted sum of comparisons with the datum is computed. Typically, this procedure is repeated for several additional choices of datums. In this manner, one gets a good understanding of the relative merits and deficiencies of each concept.

The ps-DSP has been applied to various engineering problems, most recently for a problem to design an AM process to fabricate metal lattice structures [13].

15.2.3   Approaches to Selection

As stated earlier, there have been a number of approaches taken to support the selection of AM processes for a part. Most aid selection, but only in a qualitative manner, as described earlier. Several methods have been developed in academia that are based on the large literature on decision theory. For an excellent introduction to this topic, see the book by Keeney and Raiffa [2]. In this section, the selection DSP is covered in some detail, and selection using utility theory is summarized.

While the basic advantages of using DSPs of any type lie in providing context and structure for engineering problems, regardless of complexity, they also facilitate the recording of viewpoints associated with these decisions, for completeness and future reference, and evaluation of results through post solution sensitivity analysis. The standard Selection Decision Support Problem (s-DSP) has been applied to many engineering problems, including AM selection [14]. The word formulation of the standard s-DSP is given in Fig. 15.2. Note that the decision options for AM selection are feasible material–process combinations. Expectations are determined by rating the options against the attributes. Preferences are modeled using simple importance values. Rank ordering of options is determined using a weighted-sum expression of importance and attribute ratings. An extension to include utility theory has been accomplished, as described next.

For the Identify step, evaluation attributes are to be specified. For example, accuracy, cost, build time, tensile strength, and feature detail (how small of a feature can

Given:	Set of AM processes/machines and materials (alternatives)
Identify:	Set of evaluation attributes. Create scales and determine importances.
Rate:	Each alternative relative to each attribute.
Rank:	AM methods from most to least promising

Fig. 15.2 Word formulation of the Selection Decision Support Problem

be created) are typical attributes. Scales denote how the attribute is to be measured. For example, the cost scale is typically measured in dollars and is to be minimized. Tensile strength is measured in MPa and is to be maximized. These are examples of ratio scales, since they are measured using real numbers. Interval scales, on the other hand, are measured using integers. Complexity capability is an example attribute that could be measured using an interval scale from 1 to 10, where 10 represents the highest complexity. The decision-maker should formulate interval scales carefully so that many of the integers in the scale have clear definitions. In addition to specifying scales, the decision-maker should also specify minimum and maximum values for each attribute. Finally, the decision-maker is to specify preferences using importance values or weights for each attribute.

For the Rate step of the s-DSP, each alternative AM process or machine should be evaluated against each attribute. From the Identify step, each attribute, a_i, has minimum and maximum values specified, $a_{i,\min}$ and $a_{i,\max}$, respectively. The decision-maker specifies a rating value for attribute a_{ij} for each alternative, j, that lies between $a_{i,\min}$ and $a_{i,\max}$. The final step is to normalize the ratings so that they always take on values between 0 and 1. For cases where the attribute is to be maximized, (15.1) is used to normalize each attribute rating, where r_{ij} is the normalized rating for attribute i and alternative j. (15.2) is used to normalize attribute ratings when the attribute is to be minimized:

$$r_{ij} = \frac{a_{ij} - a_{ij,\min}}{a_{ij,\max} - a_{ij,\min}} \tag{15.1}$$

$$r_{ij} = \frac{a_{ij,\max} - a_{ij}}{a_{ij,\max} - a_{ij,\min}} \tag{15.2}$$

After all attributes are rated, the total merit for each alternative is computed using a weighted sum formulation, as shown in (15.3). The I_i are the importances, or weights. Note that the merit value M_i is always normalized between 0 and 1:

$$M_j = \sum_{j=1} I_i r_{ij} \tag{15.3}$$

After computing the merit of each alternative, the alternatives can be rank ordered from the most favorable to the least. If two or more alternatives are close to the highest rank, additional investigation should be undertaken to understand under which conditions each alternative may be favored over the others. Additionally, the alternatives could be developed further so that more information about them is known. It is also helpful to run multiple sets of preferences (called scenarios) to understand how emphasis on certain attributes can lead to alternatives becoming favored.

Decision theory has a rich history, evolving in the 1940s and 1950s from the field of economics [2]. In order to provide a rigorous, preference-consistent alternative to the traditional merit function for considering alternatives with uncertain attribute values, the area of utility theory is often applied. This requires the satisfaction of a set of axioms such as those proposed by von Neumann and Morgenstern [15], Luce and Raiffa [16], or Savage [17], describing the preferences of rational individuals. Once satisfied, there then exists a utility function with the desirable property of assigning numerical utilities to all possible consequences.

In utility theory, preferences are modeled as utility functions on the expectations. Mathematically, let alternative A_i result in outcome $x_i \in X$ with probability p_i, if outcomes are discrete. Otherwise, expectations on outcomes are modeled using probability density functions, $f_i = f_i(x_i)$. Utility is denoted $u(x)$. Expectations are then modeled as expected utility as shown in (15.4):

$$E\big[u(x)\big] = \Sigma\, p_i u(x_i) \quad \text{for discrete outcomes} \tag{15.4}$$

$$E\big[u(x)\big] = u(x)g(x) \quad \text{for continuous outcomes}$$

This leads to the primary decision rule of utility theory:

Select the alternative whose outcome has the largest expected utility.

Note that expected utility is a probabilistic quantity, not a certain quantity, so there is always risk inherent in these decisions.

Utility functions are constructed by determining points that represent the decision-maker's preferences and then fitting a utility curve to these points. The extreme points indicate ideal and unacceptable values. These points are labeled as x_* and x_0, respectively, and are assigned utilities of 1 and 0, respectively, in Fig. 15.3. The remaining points are usually obtained by asking the decision-maker a series of questions (for more information, please consult a standard reference on utility theory, e.g., [2]). Specifically, a decision-maker is asked to identify his/her certainty equivalent for a few 50–50 lotteries. A lottery is a hypothetical situation in which the outcome of a decision is uncertain; it is used to assess a decision-maker's

Fig. 15.3 Utility curve from five data points

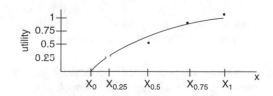

preferences. A certainty equivalent is the level of an attribute for which the decision-maker would be indifferent between receiving that attribute level for *certain* and receiving the results of a specified 50–50 lottery. For example, to obtain the value of $x_{0.5}$ in Fig. 15.3, the decision-maker is asked to identify his/her certainty equivalent to the lottery. Generally, at least five points are identified along the decision-maker's utility curve. This preference assessment procedure must be repeated for each of the attributes of interest. In the 5-point form, typical utility functions have the form:

$$u(x) = c + be^{-ax}. \tag{15.5}$$

By complementing the standard selection DSP with utility theory, an axiomatic basis is provided for accurately reflecting the preferences of a designer for trade-offs and uncertainty associated with multiple attributes. The utility selection DSP has been formulated and applied to several engineering problems, including AM selection [5, 18]. Quite a few other researchers have applied utility theory to engineering selection problems; one of the original works in this area is [19].

15.2.4   Selection Example

In this section, we present an example of a capital investment decision related to the application of metal AM processes to the production manufacture of steel caster wheels. This selection problem is very similar to a quotation problem, but includes a range of part dimensions, not single dimension values for one part. In this scenario, the caster wheel manufacturer is attempting to select an AM machine that can be used for production of its small custom orders. It is infeasible to stock all the combinations of wheels that they want to offer; thus they need to be able to produce these quickly while also keeping the price down for the customer. The technologies under consideration are two different Directed Energy Deposition (DED) systems, two laser and one Electron Beam Powder Bed Fusion (EB-PBF) machines, and one polymer Powder Bed Fusion (PBF) machine. A readily available stainless steel material (whatever was commercially available for the process, which in the case of the polymer PBF machine was a two-step process involving furnace post-processing) was used for this example. The processes will be numbered randomly (Processes 1–6) for the purposes of presentation, since this example was developed in the mid-2000s [20], and, as a result, the data are obsolete.

Before beginning the selection process, the uncertainty involved in the customization process was considered. Since these caster wheels will be customized, there is a degree of geometric uncertainty involved.

A model of a caster wheel is displayed in the Fig. 15.4a, while its main dimensions are shown in Fig. 15.4b. In this example, we have decided to only allow customization of certain features. Only standard 12 mm diameter × 100 mm length bolts will be used for the inner bore; therefore, these dimensions will be constrained. Customers will be allowed to customize all other features of the caster wheel within allowable ranges for this model wheel, as displayed in Table 15.1.

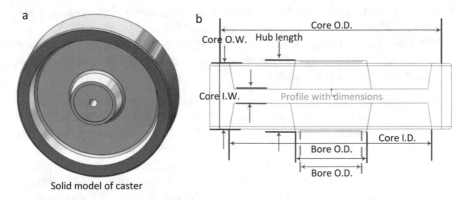

Fig. 15.4 Model of steel caster wheel

Table 15.1 Caster wheel dimensions

Measurement point	Dimensions (mm)	
	Min	Max
Core outer diameter	100	150
Core inner diameter	90	140
Bore outer diameter	38	58
Bore inner diameter	32	32
Hub length	64	64
Core outer width	38	125
Core inner width	12	32

The alternative AM technologies will be evaluated based on seven attributes that span a typical range of requirements, as shown in the following section. Scale type refers to the method used to quantify the attribute. For example, ultimate tensile strength is a ratio scale, meaning that it is represented by a real number, in this case with units of MPa. Geometric complexity is an example of an interval scale, in this case with ratings between 1 and 10, with 1 meaning the lowest complexity and 10 meaning the greatest amount of complexity.

1. Ultimate Tensile Strength (UTS): UTS is the maximum stress reached before a material fractures. Ratio scale [MPa].
2. Rockwell Hardness C (Hard): Hardness is commonly defined as the resistance of a material to indentation. Ratio scale [HRc].
3. Density (Dens.): The density refers to the final density of the part after all processing steps. This density is proportional to the amount of voids found at the surface. These voids cause a rough surface finish. Ratio scale [%].
4. Detail Capability (DC): The detail capability is the smallest feature size the technology can make. Ratio scale [mm].
5. Geometric Complexity (GC): The geometric complexity is the ability of the technology to build complex parts. More specifically, in this case, it is used to refer to the ability to produce overhangs. Interval scale (1–10).

6. Build Time (Time): The build time refers to the time required to fabricate a part, not including post-processing steps. Ratio scale [h].
7. Part Cost (Cost): The part cost is the cost it takes to build one part with all costs included. These costs include manufacturing cost, material cost, machine cost, operation cost, etc. Ratio scale [$].

In this example, we examine two weighting scenarios (relative importance ratings). In Scenario 1, geometric complexity was most heavily weighted because of the significant overhangs present in the build orientation of the casters. Build time and part cost were also heavily weighted because of their importance to the business structure surrounding customization of caster wheels. Because of the environment of use of the caster wheels, UTS was also given a high weighting. Detail capability was weighted least because of the lack of small, detailed features in the geometry of the caster wheels. In Scenario 2, all selection attributes were equally weighted.

Table 15.2 shows the results of the evaluation of the alternatives with respect to the attributes. Weights for the two scenarios, called Relative Importances, are included under the attribute names.

On the basis of these ratings, the overall merit for each alternative can be computed. Merit values for each scenario are given in Table 15.3, along with their rankings. Note that slightly different rankings are evident from the different scenarios. This indicates the importance of accurately capturing decision-maker preferences. Process 4 is the top ranking process in both scenarios. However, the second choice could be Process 2, 3, or 6, depending upon preferences. In cases like this, it is a good idea to run additional scenarios in order to understand the trade-offs that are relevant.

This capital investment example illustrated the application of selection decision support methods. As mentioned, it is very important to explore several scenarios (sets of preferences) to understand the sensitivities of ratings and rankings to changes in preferences. Modifications to the method are straightforward to achieve target values, instead of minimizing or maximizing an attribute, and to incorporate other types of uncertainty.

15.3 Challenges of Selection

The example from the previous section illustrates some of the difficulties and limitations of straightforward application of decision methods to real decision-making situations. The complex relationships among attributes, and the variations that can arise when building a wide range of parts, make it difficult to decouple decision attributes and develop structured decision problems. Nonetheless, with a proper understanding of technologies and attributes, and how to relate them together, meaningful information can be gained. This section takes a brief look at these issues.

Different AM systems are focused on slightly different markets. For example, there are large, expensive machines that can fabricate parts using a variety of materials with relatively good accuracy and/or material properties and with the ability to fine-tune the systems to meet specific needs. In contrast, there are cheaper systems,

Table 15.2 Rating alternatives with respect to attributes

Atributes		UTS	Hardness	Density	Detail cap.	Geom. compl.		Build time		Part cost	
						mm	Max	mm	Max	mm	Max
Rel. Imp.	Scen 1	0.167	0.143	0.071	0.024	0.214	0.214	0.19	0.19	0.19	0.19
	Scen 2	0.143	0.143	0.143	0.143	0.143	0.143	0.143	0.143	0.143	0.143
Alternatives	Proc 1	600	21	95	0.3	7	10	2.8	59	390	2050
	Proc 2	1430	50	100	1.2	7	10	1.42	30	134	510
	Proc 3	1700	53	100	0.762	4	6	0.26	5	64	310
	Proc 4	2000	60	99.5	0.15	7	10	1.9	39	240	1340
	Proc 5	600	15	100	0.6	7	10	1.4	24	180	890
	Proc 6	1800	53	100	1	4	6	0.12	2	30	170
Scales	Type	Ratio	Ratio	Ratio	Ratio	Interval		Ratio		Ratio	
	Low	500	10	95	0.1	1		2		25	
	High	2500	70	100	2	10		120		1000	
	Pref.	High	High	High	Low	High		Low		Low	
	Units	MPa	HRc	Percent	mm	nmu		h		$	

Table 15.3 Merit values and rankings

	Scenario 1		Scenario 2	
	Merit	Rank	Merit	Rank
Proc 1	0.254	6	0.284	6
Proc 2	0.743	2	0.667	4
Proc 3	0.689	4	0.703	2
Proc 4	0.753	1	0.808	1
Proc 5	0.528	5	0.539	5
Proc 6	0.72	3	0.697	3

which are designed to have minimal setup and to produce parts of acceptable quality in a predictable and reliable manner. In this latter case, parts may not have high accuracy, material strength, or flexibility of use.

Different users will require different things from an AM machine. Machines vary in terms of cost, size, range of materials, accuracy of part, time of build, etc. It is not surprising to know that the more expensive machines typically provide the wider range of options, and, therefore, it is important for someone looking to buy a new machine to be able to understand the costs vs. the benefits so that it is possible to choose the best machine to suit their needs.

Approaching a manufacturer or distributor of AM equipment is one way to get information concerning the specification of their machine. Such companies are obviously biased toward their own product, and, therefore, it is going to be difficult to obtain truly objective opinions. Conventions and exhibitions are a good way to make comparisons, but it is not necessarily easy to identify the usability of machines. Contacting existing users is sometimes difficult and time-consuming, but they can give very honest opinions. Attending a Users Group, such as the Additive Manufacturing Users Group, is a great way to meet and interact with manufacturers and users simultaneously. To gather information effectively, it works best if you are already equipped with background information concerning your proposed use of the technology.

When looking for advice about suitable selection methods or systems, it is useful to consider the following points. One web-based system was developed to meet these considerations [21]. An alternative approach will be presented in the next section.

- The information in the system should be unbiased wherever possible.
- The method/system should provide support and advice rather than just a quantified result.
- The method/system should provide an introduction to AM to equip the user with background knowledge as well as advice on different AM technologies.
- A range of options should be given to the user in order to adjust requirements and show how changes in requirements may affect the decision.
- The system should be linked to a comprehensive and up-to-date database of AM machines.
- After the search process has completed, the system should give guidance on where to look next for additional information.

The process of accessing the system should be as beneficial to the user as the answers it gives. However, this is not as easy a task as one might first envisage. If it were possible to decouple the attributes of the system from the user specification, then it would be a relatively simple task to select one machine against another. To illustrate that this is not always possible, consider the following scenarios:

1. In a PBF machine, warm-up and cool down are important stages during the build cycle that do not directly involve parts being fabricated. This means that large parts do not take proportionally longer times to build compared with smaller ones. Large builds are more efficient than small ones. In Vat Photopolymerization (VPP) and Material Extrusion (MEX) machines, there is a much stronger correlation between part size and build time. Small parts would therefore take less time on a VPP or MEX machine than when using PBF, if considered in isolation. Many users, however, batch process their builds, and the ability to vertically stack parts in a PBF machine makes it generally possible to utilize the available space more efficiently. The warm-up and cool-down overheads are less important for larger builds, and the time per layer is generally quicker than most VPP and MEX machines. As a result of this discussion, it is not easy to see which machine would be quicker without carefully analyzing the entire process plan for using a new machine.

2. Generally, it costs less to buy a MEX machine compared to a Binder Jetting (BJT) machine. There are technical differences between these machines that make them suitable for different potential applications. Even if a specific MEX and BJT machine are similar in price, comparison may not be straightforward. A MEX machine that uses a cartridge-based material delivery system requires a complete replacement of the cartridge when empty. This makes material use more expensive when compared with a 3D Systems BJT (aka "Z Corp") machine. For occasional use it is therefore perhaps better to use a MEX cartridge machine when all factors are equal. On the other hand, the more parts you build, the more cost-effective the BJT machine becomes.

3. Identifying a new application or market can completely change the economics of a machine. For example, in the metals area, DED machines tend to be slower and have worse feature detail capability than PBF metal machines. This has led to many more machine sales for Arcam, Renishaw, EOS, etc. than for Optomec. However, some companies need an ability to use AM to repair or modify existing molds and metal parts, which is very difficult, if not impossible, with a powder bed machine.

These examples indicate that selection results depend to a large extent on the user's knowledge of AM capabilities and applications. Selection tools that include expert systems may have an advantage over tools based on straightforward decision methods alone. Expert systems attempt to embody the expertise resulting from extensive use of AM technology into a software package that can assist the user in overcoming at least some of the learning curve quickly and in a single stage. See [21] for a more complete coverage of this idea

15.4   Example System for Preliminary Selection

A preliminary selection tool was developed for AM, called AMSelect,[1] that walks the user through a series of questions to identify feasible processes and machines [22]. Build times and costs are computed, but quantitative rating and rank-ordering are not performed. More specifically, the software enables designers, managers, and service bureau personnel to:

- Explore AM technologies for their application in a possible Direct Digital Manufacturing (DDM) project
- Identify candidate materials and processes
- Explore build times, build options, and costs
- Explore manufacturing and life-cycle benefits of AM
- Select appropriate AM technologies for DDM applications
- Explore case studies and anticipate benefits
- Support quotation and capital investment decisions

Figure 15.5 illustrates the logic underlying AMSelect. A database of machine types and capabilities is read, which represents the set of machines that the software will consider. The software supports a qualitative assessment of the suitability of DDM for the application and then enables the user to explore the performance of various AM machines. Build time and cost estimates are provided, which enable the user to make a selection decision.

To use AMSelect, the user first enters information about the production project, including production rate (parts per week), target part cost, how long the part is expected to be in production, and the useful life of the part. After the user enters information about the part to be produced and its desired characteristics (Fig. 15.6), the user answers questions about how the application may take advantage of the unique capabilities of AM processes, as shown in Figs. 15.7 and 15.8. In this version of the software, the questions ask about part shape similarity across the production volume, part geometric complexity, the extent of part consolidation compared to a design for conventional manufacturing processes, and the part delivery time. Based on the responses, the software responds with general statements about the likelihood of AM processes being suitable for the user's application; for example, see the responses for the fictitious problem from Fig. 15.6. If the user is satisfied that his/her application is suitable for DDM, then they can proceed with a more quantitative exploration of AM machines.

The AMSelect software enables the user to explore the capabilities of various AM machines for their application. As shown in Fig. 15.9, AMSelect first segregates machines that appear to be feasible from those that are infeasible, based on material, surface finish, accuracy, and minimum feature size requirements. The user

---

[1] The content of AMSelect is available at the link that will appear at the bottom of the chapter opening page.

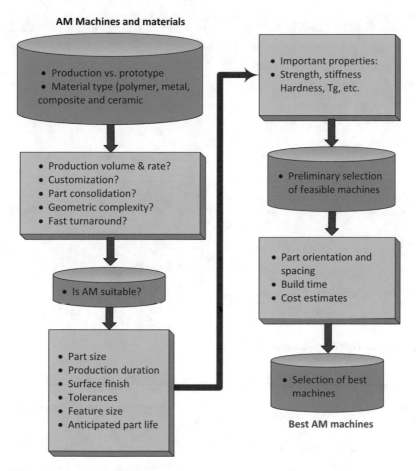

Fig. 15.5 Flowchart of AMSelect operation

can select from both the sets of feasible and infeasible machines, which can be useful for comparison purposes.

If the user wants to see the layout of parts in a machine's build chamber, they can hit the Display button, while the machine of interest is selected. For example, the build chamber of a SLA ProX 950 machine is shown in Fig. 15.10 for parts with bounding box dimensions of $100 \times 100 \times 100$ mm (part size from Fig. 15.6). Since the ProX 950 has platform dimensions of 1500×550 mm, a total of 78 parts can fit on the platform with 10 mm spacing, as shown. The user has control over the spacing between parts. Entering negative spacing values effectively "nests" parts within one another, which may be useful if parts are shaped like drinking cups, for example. Note that AMSelect will stack parts vertically if that is a typical build mode for the technology; parts are often fabricated in stacked layers in PBF as one example. Also note that serial manufacturing of end-use products is assumed for the AMSelect software. As such, the software assumes that a large quantity of parts must be

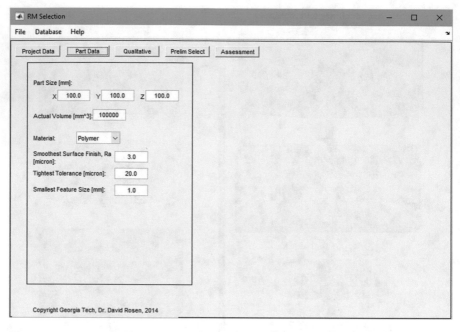

Fig. 15.6 Part Data entry screen for AMSelect

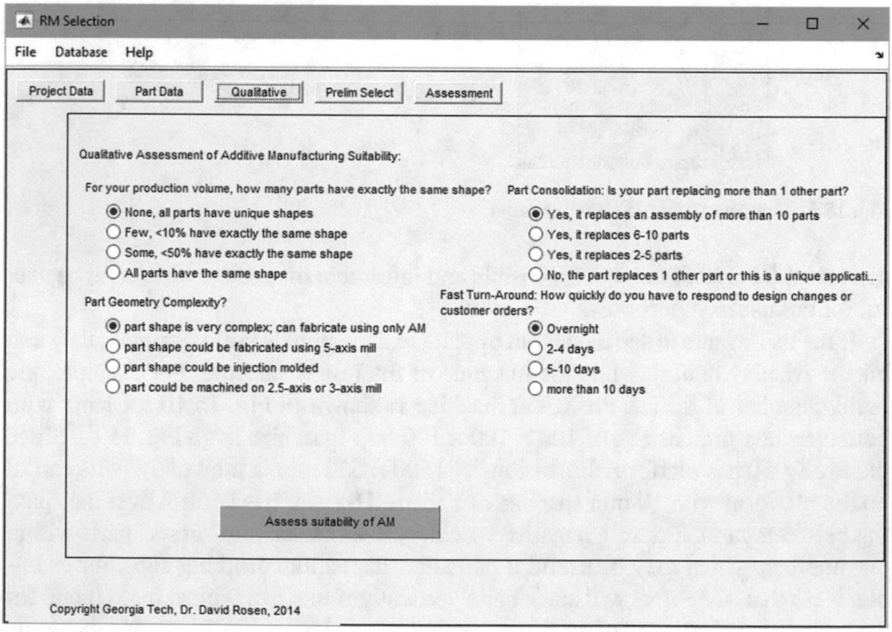

Fig. 15.7 Qualitative assessment question screen

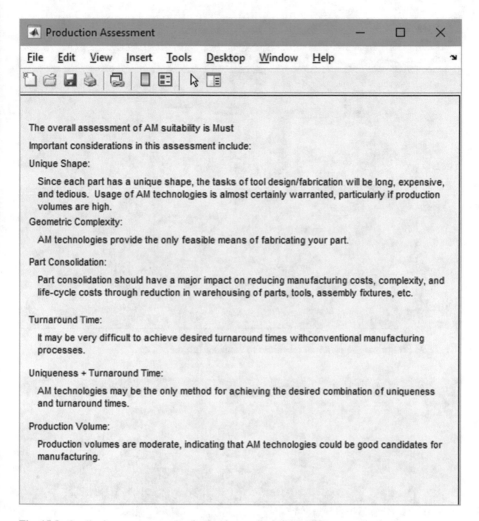

The overall assessment of AM suitability is Must

Important considerations in this assessment include:

Unique Shape:

Since each part has a unique shape, the tasks of tool design/fabrication will be long, expensive, and tedious. Usage of AM technologies is almost certainly warranted, particularly if production volumes are high.

Geometric Complexity:

AM technologies provide the only feasible means of fabricating your part.

Part Consolidation:

Part consolidation should have a major impact on reducing manufacturing costs, complexity, and life-cycle costs through reduction in warehousing of parts, tools, assembly fixtures, etc.

Turnaround Time:

It may be very difficult to achieve desired turnaround times withconventional manufacturing processes.

Uniqueness + Turnaround Time:

AM technologies may be the only method for achieving the desired combination of uniqueness and turnaround times.

Production Volume:

Production volumes are moderate, indicating that AM technologies could be good candidates for manufacturing.

Fig. 15.8 Qualitative assessment results for the entries in Fig. 15.7

produced and fills the platform or build chamber with only one type of part (part described in the Part Data screen, Fig. 15.6). The user can also change the part orientation in an attempt to fit additional parts into a build.

In the last major step in AMSelect, the Assessment button (*see* Figs. 15.6, 15.7, and 15.9) can be selected to estimate the build time for the platform of parts, as well as the cost per part. These assessments can be particularly useful in comparing technologies and machines for an application. Considerable uncertainty exists regarding build speeds, so ranges of build times are calculated based on typical ranges of scanning speeds, delay times, recoating speeds, etc. Part costs are broken down into machine, material, operation, and maintenance costs, similar to the cost model to be presented in Chap. 18.

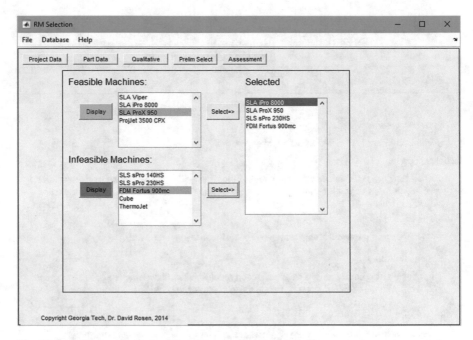

Fig. 15.9 Preliminary selection of machines to consider further

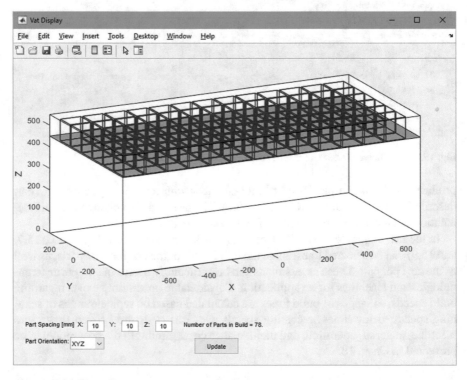

Fig. 15.10 Layout of parts on the machine platform

As shown in Fig. 15.11, long build times do not necessarily translate to high part costs, particularly if many parts can be built at once. The SLA iPro 8000, SLS sPro 230, and the FDM Fortus 900mc have similar build times for a platform full of parts, but part costs are several times smaller for the sPro 230, compared to the SLA iPro 8000, since many more parts can be built in about the same amount of time. On the other hand, the SLA Viper Si2 takes almost as long for its platform, but parts are very expensive since the Viper is building in high-resolution mode (built-in assumption) and few parts can fit on its relatively small platform.

Maintenance of the database for AMSelect can be problematic, since machine capabilities may be upgraded, costs may be reduced, and new machines may be developed. AMSelect allows users to edit its database, either by modifying existing machines or by creating new ones. The screen that shows this capability is shown in Fig. 15.12.

Armed with these results, the user can make a selection of AM machines to explore further. The decision may be based on part cost. But, the user needs to take all relevant information into account. Recall that the Sinterstation machines were not feasible for this application (due to feature size requirements, although this was not shown).This was why the Sinterstation appeared in the infeasible column in Fig. 15.9. With these results, the user can determine whether or not she/he wants to relax the feature size requirement to reduce costs, or maintain requirements with a potential cost penalty. Hence, trade-off scenarios can be explored with AMSelect.

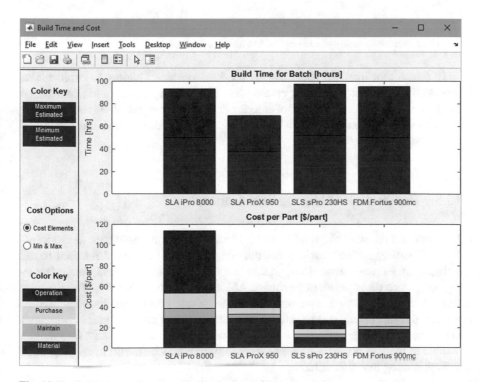

Fig. 15.11 Build time and cost results for the example

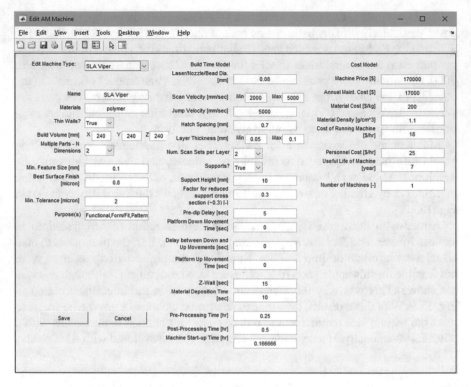

Fig. 15.12 Screen for adding or editing an AM machine definition

In fact, a tool like AMSelect can be used by machine vendors to explore new product development. They can "create" new machines by adding a machine to the database with the characteristics of interest. Then, they can test their new machine by quantitatively comparing it with existing machines based on build times and part costs.

15.5   Production Planning and Control

This section addresses the third type of selection decision introduced in the introduction, namely, support for process planning. It is probably most relevant to the activities of service bureaus (SBs), including internal organizations in manufacturing companies that operate one or more AM machines and processes. The SB may know which machine and material a part is to be made from, but in most circumstances, the part cannot be considered in isolation. When any new part is presented to the process planner at the SB, it is likely that he has already committed to build a number of parts. A decision support software system may be useful in keeping track and optimizing machine utilization.

Consider the process when a new part is presented to the SB for building. In general, the information presented to the process planner will include the following:

1. Part geometry
2. Number of parts
3. Delivery date or schedule for batches of parts
4. Processes other than AM to be carried out (pre-processing and post-processing)
5. Expectations of the user (accuracy, degree of finish, etc.)

Furnished with this small amount of data, it is possible to start integrating the new job with the existing jobs and available resources. Four topics will be explored further, namely, production planning, pre-processing, part build, and post-processing.

15.5.1   Production Planning

Several related decision are needed early in the process. A suitable AM process and machine must be identified from among those in the facility. This was probably done during the quoting stage before the customer selected the SB. After that is settled, AM machine availability must be considered. If the SB has more than one suitable machine, a choice must be made as to which machine to use. If the job is for a series of part batches, the SB may choose to run all batches on the same machine, or on multiple machines. If multiple machines, the SB must ensure that all selected machines can provide repeatable results, which is not always the case. Otherwise, potentially lengthy calibration builds may be needed to ensure consistent part quality from all machines used.

A job scheduling system should be used, particularly for production manufacturing applications, so that part batches can be produced to meet deadlines. A number of software companies provide AM scheduling software that can aid the user in these scheduling and monitoring tasks. If the SB has insufficient machine capacity, it may need to invest in further capacity, necessitating a machine selection scenario. Alternatively, the SB could retain the services of other SBs if the economics of further machine investments is questionable.

15.5.2   Pre-Processing

Pre-processing means software-based manipulation. This will be carried out on the file that describes the geometry of the part. Such manipulation can generally be divided into four areas, (I) modification of the design, (II) determination of build parameters (III) part orientation and placement on the build plate, and (IV) support generation.

Modification of the design may be required for two reasons. First, part details may need adjustment to accommodate process characteristics. For example, shaft or pin diameters may need to be reduced, to increase clearance for assembly, when building in many processes since most processes are material-safe (i.e., features become oversized). Second, models may require repair if the STEP, IGES, AMF, or STL file has problems such as missing triangles, incorrectly oriented surfaces, or the like. An increasing number of companies are adopting distortion compensation software to help with design modification. In distortion compensation a finite element simulation is performed to predict the shape changes during the part build. The software then modifies the original geometry in the opposite direction to the predicted shape changes. When built, the part then deforms back toward the intended shape. This type of tool can reduce total dimensional deviation after the build by 80–90% or more if done iteratively.

Determination of the build parameters is very specific to the AM process to be used. This includes selecting a part orientation, support generation, setting of build styles, layer thickness selection, temperature setting, etc. In general, this is either a very quick process, or it takes a predictable length of time to set up. On occasion, and for some particular types of machine, this process can be very time-consuming. This usually corresponds to instances when the user expectations closely meet the upper limits of the machine specification (high accuracy, build strength, early delivery date, etc.). Under such conditions, the user must devote more time and attention to parameter setting. The decision support software should make the process planner aware under which circumstances this may occur and allocate resources appropriately. This process also is time-consuming when samples with different process parameters have to be printed in a single build job. In this case, operators should manually define the process parameters for each sample. It should be noted that in the case of printing samples with different process parameters, the value of layer thickness has to be constant.

In most cases machine users simply select build parameters from a set of predefined parameters. In some instances, a user may want to develop a completely new set of build parameters to explore new materials and/or material property versus build time versus accuracy trade-offs. To comprehensively explore trade-offs when developing new build parameters, particularly for metals, is a monthslong process involving the manufacture and testing of hundreds or thousands of coupons. Newly introduced simulation tools, such as Ansys Additive Science, which is focused on metal PBF, were designed to aid users in process parameter development by predicting the effects of different process parameter variations on melt pool dimensions, porosity, and microstructure. As these types of tools become more prevalent and as they are expanded to more machine and material types, it will become possible to develop build-specific process parameter combinations to optimize outcomes for a specific component's characteristics.

The placing of samples within the build volume is an important decision when setting up a build, particularly when using an AM process that uses heat for bonding. In heat-based processes, high thermal gradients lead to different temperatures and cooling rates for the samples that are printed in different places on the substrate. In addition, parts that have thicker cross-sections generate more heat than thin parts.

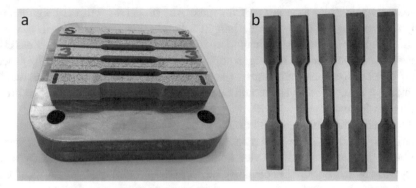

Fig. 15.13 Building in (**a**) bulk and (**b**) cutting by wire EDM

When parts are placed too near to another part, the heat in one part will influence the heating and cooling behavior in nearby parts and thus accuracy and properties. Therefore, part placement is also an important design consideration. Finite element analysis is also proving to be a helpful tool for part placement, as it can guide the user to understand when parts start to thermally impact each other.

Build orientation is an important decision for any part. Orientation affects the final properties of the part, since all AM processes result in anisotropic properties, such as for density, strength, and elongation. Orientation affects surface finish and which surfaces might have noticeable stair-step effects. Orientation affects build time, since for a given volume, most AM processes take longer to build a part the taller it is in the z-axis. Orientation also affects distortion. Large and flat surfaces tend to distort more when aligned in the x–y plane than when tilted away from a horizontal position.

Also part-building strategy is to consider whether merging geometry together and then separating it later might be more efficient than creating every specimen separately. For instance, to produce a large number of dog bone samples in a single build job, they can be designed in bulk and cut by water jet or wire EDM faster than they can be built individually (Fig. 15.13).

In many AM processes, such as metal PBF and VPP, supports are always required. While it is possible to design supports in a CAD system, it is typically more time and cost-effective to use a dedicated software tool to aid in support design. Many tools exist on the market today that quickly guide users through the process of support design, helping them select the best supports and support strategy for a part. For some processes, such as VPP when using a well-characterized material, supports can be generated completely automatically with very low probability of support failure.

15.5.3   Part Build Time

For some processes, like MEX or DED process, it does not really matter in terms of time whether parts are built one after another (batches of 1) or parts are grouped together in batches since the layering time is insignificant compared to the deposition time. However, build time per part for most AM processes will vary significantly when comparing building a single part in a build versus many parts in the same build. This may be due to significant preparation time before the build process takes place (such as powder bed heating in polymer PBF), or because the layering process itself takes a significant amount of time. In the latter case, it is obvious that the cumulative number of layers should be as low as possible to minimize the overall build time for many parts.

Orientation to minimize build time is often at odds with the preferred build orientation to maximize part properties. This can cause difficulties when organizing the batch production of parts. Orientation of parts so that they fit efficiently within the work volume does not necessarily mean optimal build quality and vice versa. For those machines that need to use support structures during the build process, this represents an additional problem, both in terms of build time (allocation of time to build the support structures for different orientations) and post-processing time (removing the supports). Many researchers have discussed these dilemmas [23].

What this generally means to a process planner is compromise. Compromise is not unusual to a process planner; in fact, it is a typical characteristic, but the degree of flexibility provided by many AM machines makes this a particularly interesting problem. Just because an AM machine is being used constantly does not mean it is being used efficiently.

15.5.4 Post-Processing

All AM parts require a degree of post-processing. This may only require simple removal of support structures or excess powder for certain parts and applications. Or it may require a very significant time overhead and the use of expensive tools such as CNC machining or wire EDM. Parts may require a large amount of skilled manual work in terms of surface preparation and coating. Alternatively, the AM part may be one stage in a complex rapid tooling process that requires numerous manual and automated stages. All of these post-processing requirements may result from the AM machine depending upon part geometry and specifications for the completed component. It can even be an iterative process involving all of the above steps at different stages in the development cycle based on the same part CAD data.

15.5.5 Summary

AM process and material selection can be a complex process, requiring substantial trade-offs and expert judgment. It is clear that a detailed understanding of AM processes, materials, and applications is needed to utilize AM resources effectively and efficiently. Decision theory and software tools have been shown to be effective means for process selection, but these tools can be difficult to keep current due to the large number of variables involved. Today, there are no comprehensive AM process selection tools available in the marketplace, so designers and machine users must carefully consider their specific problem and available resources to manage this complexity.

15.6   Future Work

Some summary statements and open problems to motivate continued research are presented here.

(a) Selection methods and systems are only as good as the information that is utilized to make suggestions. Maintaining up-to-date and accurate machine and material databases will likely be an ongoing problem. Centralized databases and standard database and benchmarking practices will help to mitigate this issue.

(b) Customers (people wanting parts made) have a wide range of intended applications and needs. A better representation of those needs is required in order to facilitate better selection decisions. Improved methods for capturing and modeling user preferences are also needed.

(c) Related to the wide range of applications is the wide range of manufacturing process chains that could be used to construct parts. Better, more complete methods of generating, evaluating, and selecting process chains are needed for cases where multiple parts or products are needed or when complex prototypes need to be constructed. An example of the latter case is a functional prototype of a new product that consists of electronic and mechanical subsystems. Many options likely exist for fabricating individual parts or modules.

(d) More generally, integration of selection methods, with databases and process chain exploration methods, would be very beneficial.

(e) Methods are needed that hide the complexity associated with the wide variety of process variables and nuances of AM technologies. This is particularly important for novice users of AM machines or even for knowledgeable users who work in production environments. Alternatively, knowledgeable users must have access to all process variables if necessary to deal with difficult builds.

(f) Better methods are needed that enable users to explore trade-offs (compromises) among build goals and to find machine settings that enable them to best meet their goals. These methods should work across the many different types of AM machines and materials.

(g) It is not uncommon for AM customers to want parts that are at the boundaries of AM machine capabilities. Tools that recognize when capability limits are reached or exceeded would be very helpful. Furthermore, these tools should provide guidance that assists users in identifying process settings that are likely to yield the best results. Providing estimates of part qualities (e.g., part detail actual sizes vs. desired sizes, actual surface finish vs. desired surface finish, etc.) would also be helpful.

15.7 Questions

1. You have been assigned to fabricate several prototypes of a cell phone housing for assembly and functional testing purposes. Discuss the advantages and disadvantages of commercial AM processes. Identify the most likely to succeed processes.
2. Repeat question 1 for a laptop housing.
3. Repeat question 1 for metal copings for dental restorations (e.g., crowns and bridges). Realize that accuracy requirements are approximately 10 μm. Titanium or cobalt–chrome materials are used typically.
4. For the selection example in Sect. 15.2.4:

   (a) Update the information used using information sources at your disposal (websites, etc.).
   (b) Repeat the selection process using your updated information. Develop your new versions of Tables 15.2 and 15.3.

5. Repeat question 3 using the selection DSP method and the updated information that you found for question 4.
6. Explain the difference between ratio and interval scales in s-DSP method, and give an example of each.
7. Assume that you are assigned to a new company project with the goal of AM production of a new metal carbide drill bit design for oil drilling in deep wells. What AM process would you recommend to your R&D team to fulfill this requirement? Give your response in the form of a short proposal that could include an overview of the process, the benefits and drawbacks, a rough idea of the cost, or other information that would provide sufficient justification of your decision. Assume that the drill bit design can fit within a 25 mm × 25 mm × 25 mm build envelope, has internal cooling channels, and is produced from a single, homogeneous material.
8. Find a website that helps a user choose an AM process. Play around with different assumptions and answers. What type of selection methodology do you think they use for their recommendation? Why do you think this?

References

1. Wohlers, T. (2013). *Wohler report. 3D Printing and Additive Manufacturing State of the Industry, Annual Worldwide Progress Report*. Wohler Association.
2. Keeney, R. L. (1976). *Decisions with multiple objectives: Preferences and value tradeoffs* (pp. 231–232). Cambridge University Press, Cambridge UK.
3. Hazelrigg, G. A. (1996). *Systems engineering: An approach to information-based design*. Upper Saddle River: Prentice Hall.
4. Mistree, F., Smith, W., & Bras, B. (1993). A decision-based approach to concurrent design. In *Concurrent engineering* (pp. 127–158). Springer, Boston, US.
5. Marston, M., Allen, J. K., & Mistree, F. (2000). The decision support problem technique: Integrating descriptive and normative approaches in decision based design. *Engineering Valuation and Cost Analysis, 3*(2), 107–129.
6. Mistree, F., et al. (1990). Decision-based design: A contemporary paradigm for ship design. *Transactions, Society of Naval Architects and Marine Engineers, 98*, 565–597.
7. Bascaran, E., Bannerot, R. B., & Mistree, F. (1989). Hierarchical selection decision support problems in conceptual design. *Engineering Optimization, 14*(3), 207–238.
8. Deglin, A., & Bernard, A. (2002). A knowledge-based environment for modelling and computer-aided process planning of rapid manufacturing processes. In *Integrated design and manufacturing in mechanical engineering* (pp. 85–92). Dordrecht: Springer.
9. Fuh, J., et al. (2002). A Web-based database system for RP machines, processes and materials selection. In *Software Solutions for Rapid Prototyping* (pp. 27–55). London, UK.
10. Xu, F., Wong, Y., & Loh, H. (1999, August). A knowledge-based decision support system for RP&M process selection. In *Proceedings solid freeform fabrication symposium,* Austin.
11. Selecting the right 3D printing process. https://www.3dhubs.com/knowledge-base/selecting-right-3d-printing-process/. 2020.
12. Allen, J. (1996). The decision to introduce new technology: The fuzzy preliminary selection decision support problem. *Engineering Optimization, 26*(1), 61–77.
13. Williams, C. B. (2008). *Design and development of a layer-based additive manufacturing process for the realization of metal parts of designed mesostructure*. Georgia Institute of Technology.
14. Herrmann, A., & Allen, J. K.. (1999). Selection of rapid tooling materials and processes in a distributed design environment. In *ASME Design For Manufacture Conference*.
15. Morgenstern, O., & Von Neumann, J. (1953). *Theory of games and economic behavior*. Princeton University Press, USA.
16. Luce, R. D., & Raiffa, H. (1965). *Games and decisions: introduction and critical survey*. New York: Wiley.
17. Savage, L. (1954). *The foundations of statistics*. Wiley, New York, USA.
18. Fernández, M., et al. (2001). Utility-based decision support for selection in engineering design. In *ASME Design Automation Conference*.
19. Thurston, D. L. (1991). A formal method for subjective design evaluation with multiple attributes. *Research in Engineering Design, 3*(2), 105–122.
20. Wilson, J. O., & Rosen, D. (2005). Selection for rapid manufacturing under epistemic uncertainty. In *ASME 2005 International Design Engineering Technical Conferences and Computers and Information in Engineering Conference*. American Society of Mechanical Engineers.
21. Rosen, D., & Gibson, I. (2002). Decision support and system selection for RP. In *Book chapter in Software solutions for RP*. Rapid Prototyping, London, UK.
22. Rosen, D. (2005, May). Direct digital manufacturing: issues and tools for making key decisions. In *Proceedings SME rapid prototyping and manufacturing conference*, Dearborn.
23. Dutta, D., et al. (2001). Layered manufacturing: Current status and future trends. *Journal of Computing and Information Science in Engineering, 1*(1), 60–71.

Chapter 16
Post-Processing

Abstract Throughout this book we have discussed geometric accuracy, surface finish, and property limitations of AM. When a part comes out of an AM machine, there may still be a number of processes to be carried out before it can be considered ready for use. This may include subtractive manufacturing, material treatments, or coatings. This chapter discussed processes applied to parts once the AM stage has been completed, generally referred to as "post-processing."

16.1 Introduction

Most AM processes require post-processing after part building to prepare the part for its intended form, fit, and/or function. Depending upon the AM technique, the reason for post-processing varies. For purposes of simplicity, this chapter will focus on post-processing techniques which are used to enhance components or overcome AM limitations. These include:

- Post-processing to improve surface quality
- Support Material Removal
- Surface Texture Improvement
- Aesthetic Improvement
- Post-processing to improve dimensional deviations
- Accuracy Improvement
- Post-processing to improve mechanical properties
- Property Enhancement Using Nonthermal Techniques
- Property Enhancement Using Thermal Techniques
- Preparation for Use as a Pattern

The skill with which various AM practitioners perform post-processing is one of the most distinguishing characteristics between competing service providers. Companies which can efficiently and accurately post-process parts to a customer's

© Springer Nature Switzerland AG 2021
I. Gibson et al., *Additive Manufacturing Technologies*,
https://doi.org/10.1007/978-3-030-56127-7_16

expectations can often charge a premium for their services, whereas companies which compete primarily on price may sacrifice post-processing quality in order to reduce costs.

16.2   Post-Processing to Improve Surface Quality

16.2.1   Support Material Removal

The most common type of post-processing in AM is support removal. Support material can be broadly classified into two categories: (a) material which surrounds the part as a naturally occurring by-product of the build process (natural supports) and (b) rigid structures which are designed and built to support, restrain, or attach the part being built to a build platform (synthetic supports). In cases where synthetic supports are used, and the part build process utilizes powder, the powder should be removed from surface, holes, cavities, lattices, and supports before removing the synthetic supports.

16.2.1.1   Natural Support Post-Processing

In processes where the part being built is fully encapsulated in the build material, the part must be removed from the surrounding material prior to its use. Processes which provide natural supports are primarily powder-based and sheet-based processes. Specifically, all Powder Bed Fusion (PBF) and Binder Jetting (BJT) processes require removal of the part from the loose powder surrounding the part; and bond-then-form sheet metal lamination processes require removal of the encapsulating sheet material.

In polymer PBF processes, after the part is built, it is typically necessary to allow the part to go through a cool-down stage. The part should remain embedded inside the powder to minimize part distortion due to non-uniform cooling. The cool-down time is dependent on the build material and the size of the part(s). Once cool down is complete, there are several methods used to remove the part(s) from the surrounding loose powder. Typically, the entire build (made up of loose powder and fused parts) is removed from the machine as a block and transported to a "breakout" station where the parts are removed manually from the surrounding powdered material. Brushes, compressed air, and light bead blasting are commonly used to remove loosely adhered powder, whereas woodworking tools and dental cleaning tools are commonly used to remove powders which have sintered to the surface or powder entrapped in small channels or features. Internal cavities and hollow spaces can be difficult to clean and may require significant post-processing time.

With the exception of an extended cool-down time, natural support removal techniques for BJT processes are identical to those used for PBF. In most cases, parts made using BJT are brittle out of the machine. Thus, until the parts have been strengthened by infiltration, the parts must be handled with care. This is also true for

PBF materials that require post-infiltration, such as some elastomeric materials, polystyrene materials for investment casting, and metal and ceramic green parts.

Automated loose powder removal processes have been developed. These can be stand-alone apparatuses or integrated into the build chamber. One of the first Z Corp (now 3D Systems) BJT machines with this capability is illustrated in Fig. 16.1. Several metal PBF machine manufacturers have started to integrate semiautomated powder removal techniques into their machines as well. One Click Metal [1] has introduced an automated technology for bulk powder removal that includes an auto-mated turntable for undercut and internal powder removal. This system uses an ultrasonically actuated sieve and transfers the oversize and agglomerated powders to a waste bin. Current trends suggest that many future PBF and BJT machines will incorporate some form of automated powder removal after part completion.

Bond-then-form Sheet Lamination (SHL) processes require natural support material removal prior to use. If complex geometries with overhanging features, internal cavities, channels, or fine features are used, the support removal may be tedious and time-consuming. If enclosed cavities or channels are created, it is often necessary to delaminate the model at a specific z-height in order to gain access to de-cube the internal feature and then reglue it after removing excess support materi-als. An example de-cubing operation for LOM is shown in Fig. 16.2.

16.2.1.2   Synthetic Support Removal

Processes which do not naturally support parts require synthetic supports for over-hanging features. In some cases, such as when using PBF techniques for metals, synthetic supports are also required to resist distortion. Synthetic supports can be made from the build material or from a secondary material. The development of

Fig. 16.1 Automated powder removal using vibratory and vacuum assist in a Z Corp 450 machine. (Photo courtesy of Z Corporation)

Solid Sheet Systems
Cubic Technologies' Laminated Object Manufacturing (LOM)
Post Processing

(a) The laminated stack is removed from the machine's elevator plate.
(b) The surrounding wall is lifted off the object to expose cubes of excess material.
(c) Cubes are easily separated from the object's surface.
(d) The object's surface can then be sanded, polished or painted, as desired.

Fig. 16.2 LOM support removal process (de-cubing) process, showing (**a**) the finished block of material; (**b**) removal of cubes far from the part; (**c**) removal of cubes directly adjacent to the part; (**d**) the finished product. (Photo courtesy of Worldwide Guide to Rapid Prototyping website (C) Copyright Castle Island Co., All rights reserved. Photo provided by Cubic Technologies)

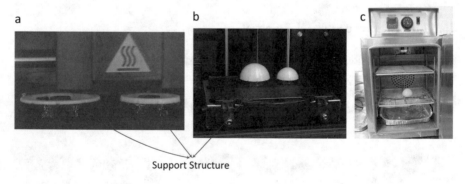

Support Structure

Fig. 16.3 (**a**) and (**b**) secondary support structures are used to support the part (**c**) an oven provides thermal support removal

secondary support materials was a key step in simplifying the removal of synthetic supports as these materials are designed to be either weaker, soluble in a liquid solution, or to melt at a lower temperature than the build material.

The orientation of a part with respect to the primary build axis significantly affects support generation and removal. If a thin part is laid flat, for instance, the amount of support material consumed may significantly exceed the amount of build material (*see* Fig. 16.3a where a thin disk is supported by a thick amount of secondary support material). The orientation of supports also affects the surface finish of the part, as support removal typically leaves "witness marks" (small bumps or divots) where the supports were attached. Additionally, the use of supports in regions of small features may lead to these features being broken when the supports are removed. Thus, orientation and location of supports is a key factor for many processes to achieve desirable finished part characteristics.

16.2.1.3   Supports Made from the Build Material

All Material Extrusion (MEX), Material Jetting (MJT), and Vat Photopolymerization (VPP) processes require supports for overhanging structures and to connect the part to the build platform. Since these processes are used primarily for polymer parts, the low strength of the supports allows them to be removed manually. These types of supports are also commonly referred to as breakaway supports. The removal of supports from downward-facing features leaves witness marks where the supports were attached. As a result, these surfaces may require subsequent sanding and polishing. Figure 16.4a, b shows breakaway support removal techniques for parts made using MEX and Vat Photopolymerization.

PBF and Directed Energy Deposition (DED) processes for metals and ceramics also typically require support materials. An example acetabular cup, oriented so that support removal does not mar the bearing surfaces, is shown in Fig. 16.5. For these processes the metal supports are often too strong to be removed by hand; thus, the use of milling, band saws, cutoff blades, wire EDM, or pneumatically driven chisels can be used for removing support structures. Some types of support removal produce rough surfaces which may need to be polished or further machined. As discussed in Chap. 5, parts made using electron beam melting have fewer supports than those made using metal laser sintering, since EBM holds the part at elevated temperature throughout the build process and less residual stresses are induced. These metal supports are often weak enough to be removed by hand.

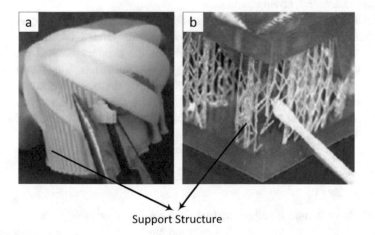

Support Structure

Fig. 16.4 Breakaway support removal for (**a**) an MEX part. (Photo courtesy of Jim Flowers) and (**b**) an SLA part. (Photo courtesy of Worldwide Guide to Rapid Prototyping website. Copyright Castle Island Co., All rights reserved. Photo provided by Cadem A.S., Turkey)

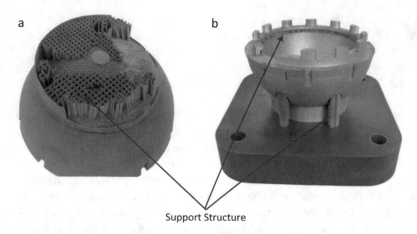

Support Structure

Fig. 16.5 (**a, b**) Laser-Based Powder Bed Fusion (LB-PBF) acetabular cup for a hip replacement surgery, made from Ti-6Al-4 V

16.2.1.4   Supports Made from Secondary Materials

A number of secondary support materials have been developed over the years in order to alleviate the labor-intensive manual removal of support materials. Two of the first technologies to use secondary support materials were the Cubital VPP hybrid process and the Solidscape MJT process. Their use of wax support materials enabled the block of support/build to be placed in a warm water bath; thus, melting or dissolving the wax yields the final parts. Since that time, secondary supports have become common commercially in MEX and MJT processes. Secondary supports have also been demonstrated for form-then-bond sheet metal lamination and DED processes in research environments.

For polymers, the most common secondary support materials are polymer materials which can be melted and/or dissolved in a water-based solvent. The water can be jetted or ultrasonically vibrated to accelerate the support removal process. For metals, the most common secondary support materials are lower-melting-temperature alloys or alloys which can be chemically dissolved in a solvent (in this case the solvent must not affect the build material). An example where secondary support material is melted and collected in a collection tray is shown in Fig. 16.3c.

16.2.2   Surface Texture Improvements

AM parts have common surface texture features that may need to be modified for aesthetic or performance reasons. Common undesirable surface texture features include stair-steps, powder adhesion, fill patterns from MEX or DED systems, and witness marks from support material removal. Stair-stepping is a fundamental issue in layered manufacturing, although one can choose a thin layer thickness to minimize error at the expense of build time. Powder adhesion is a fundamental characteristic of BJT, PBF, and powder-based DED processes. The amount of powder adhesion can be controlled, to some degree, by changing part orientation, powder morphology, and thermal control technique (such as modifying the scan pattern).

The type of post-processing utilized for surface texture improvements is dependent upon the desired surface finish outcome. If a matte surface finish is desired, a simple bead blasting of the surface can help even the surface texture, remove sharp corners from stair-stepping, and give an overall matte appearance. If a smooth or polished finish is desired, then wet or dry sanding and hand polishing are performed. In many cases, it is desirable to paint the surface (e.g., with cyanoacrylate, or a sealant) prior to sanding or polishing. Painting the surface has the dual benefit of sealing porosity and, by viscous forces, smoothing the stair-step effect, thus making sanding and polishing easier and more effective.

Several automated techniques have been explored for surface texture improvements. Two of the most commonly utilized include tumbling for external features and abrasive flow machining for, primarily, internal features. These processes have been shown to smooth surface features nicely, but at the cost of small feature resolution, sharp corner retention, and accuracy.

16.2.3   Aesthetic Improvements

Many times AM is used to make parts which will be displayed for aesthetic or artistic reasons or used as marketing samples. In these and similar instances, the aesthetics of the part is of critical importance for its end application.

Often the desired aesthetic improvement is solely related to surface finish. In this case, the post-processing options discussed elsewhere in Sect. 16.2 can be used. In some cases, a difference in surface texture between one region and another may be desired (this is often the case in jewelry). In this case, finishing of selected surfaces only is required (such as for the cover art for this book).

In cases where the color of the AM part is not of sufficient quality, several methods can be used to improve the part aesthetics. Some types of AM parts can be effectively colored by simply dipping the part into a dye of the appropriate color. This method is particularly effective for parts created from powder beds, as the inherent porosity in these parts leads to effective absorption. If painting is required, the part may need to be sealed prior to painting. Common automotive paints are quite effective in these instances.

Another aesthetic enhancement (which also strengthens the part and improves wear resistance) is chrome plating. Figure 16.6 shows a stereolithography part before and after chrome plating. Several materials have been electroless coated to AM parts, including Ni, Cu, and other coatings. In some cases, these coatings are thick enough that, in addition to aesthetic improvements, the parts are robust enough to use as functional prototypes or as tools for injection molding or as EDM electrodes.

Fig. 16.6 Stereolithography part (**a**) before and (**b**) after chrome plating. (photo courtesy of Artcraft Plating)

16.3   Post-Processing to Improve Dimensional Deviations

16.3.1   Accuracy Improvements

There is a wide range of accuracy capabilities between AM processes. Some processes are capable of submicron tolerances, whereas others have deviations above 1 mm. Typically, the larger the build volume and the faster the build speed, the worse the accuracy. This is particularly noticeable, for instance, in DED processes where the slowest and most accurate DED processes have accuracies approaching a few microns, whereas the larger bulk deposition machines have accuracies of several millimeters.

16.3.2   Sources of Inaccuracy

Process-dependent errors affect the accuracy of the X–Y plane differently from the Z axis accuracy. These errors come from positioning and indexing limitations of specific machine architectures, lack of closed-loop process monitoring and control strategies, and/or issues fundamental to the volumetric rate of material addition (such as melt pool or droplet size). In addition, for many processes, accuracy is highly dependent upon operator skill. Further accuracy improvements in AM may require fully automatic real-time control strategies to monitor and control the process, rather than the need to rely on expert operators as a feedback mechanism.

Integration of additive plus subtractive processing, as discussed in the hybrid AM chapter, is another method for process accuracy improvement.

Material-dependent phenomena also play a role in accuracy, including shrinkage and residual stress-induced distortion. Repeatable shrinkage and distortion can be predicted by finite element analysis and compensated by scaling the CAD model. Predictive capabilities are now enough to compensate for variations in shrinkage and residual stresses that are scan pattern or geometry dependent. Quantitative understanding of the effects of process parameters, build style, part orientation, support structures, and other factors on the magnitude of shrinkage, residual stress, and distortion is needed for accurate predictions.

Another source of inaccuracy is related to the position of parts on the build plate and differential cooling in the build chamber, such as cooling from inert gas flow in PBF processes. For thermal-based AM processes, it is impossible to completely eliminate temperature gradients in the build chamber. The number and shape of the products affect thermal gradients and dimensional deviations. For systems that use inert gas, higher dimensional deviations have been observed in the blowing direction of the gas.

For parts which require a high degree of accuracy, extra material must be added to critical features, which is then removed via milling or other subtractive means to achieve the desired accuracy. In order to marry the benefits of AM and the accuracy of a CNC machined component, a comprehensive strategy for achieving high accuracy should be adopted. One such strategy involves pre-processing of the STL file to compensate for inaccuracies followed by finish machining of the final part. The following sections describe steps to consider when seeking to establish a comprehensive finish machining strategy.

16.3.3   Model Pre-Processing to Compensate for Inaccuracy

For many AM processes, the position of the part within the build chamber and the orientation will influence part accuracy, surface finish, and build time. Thus, translation and rotation operations are applied to the original model to optimize the part position and orientation.

Shrinkage often occurs during AM. Shrinkage also occurs during the post-process furnace operations needed for indirect processing of metal or ceramic green parts. Pre-process manipulation of the STL model will allow a scale factor to be used to compensate for the average shrinkage of the process chain. However, when compensating for average shrinkage, there will always be some features which shrink slightly more or less than the average (shrinkage variation). Even when using newer finite element tools to predict shrinkage and distortion, there is always some error.

In order to compensate for shrinkage variation, if the highest shrinkage value is used, then ribs and similar features will always be at least as big as the desired geometry. However, channels and holes will be too large. Thus, simply using the largest shrinkage value is not an acceptable solution.

For surfaces which require CNC machining to achieve the required surface finish or accuracy, the user must make sure there is enough material left on the surface to be machined. Adding "skin" to the original model may be necessary. This skin addition, such that there is material left to machine on critical features, can be referred to as making the part "steel-safe." Many studies have shown that shrinkage variations are geometry-dependent and not 100% repeatable, even when using the same AM or furnace post-processing parameters. Thus, compensating for dimensional variation for specific geometric features of interest may be required [2]. See Chap. 17 for more information on STL files and software systems to manipulate them.

16.3.4   Machining Strategy

Material removal is also called machining, and the selection of a proper strategy is very important for finishing AM parts and tools. A large number of machining processes have been shown to be useful for AM post-processing, depending upon the complexity of the component and the material being removed. When the cause of material removal is mechanical force, the process is called conventional machining. If the metal cutting process is carried out without mechanical forces, the process is stated as advanced machining. Figure 16.7 shows various machining operations that can be used for AM components.

Selection of proper conventional machining processes is highly related to the final requirements and shape of the components. These requirements comprise the quality of the surface, dimensional accuracy, and clearances as specified.

16.3.4.1   Grinding

Grinding is the most common abrasive machining method and provides high surface quality and dimensional accuracy. AM can produce very hard materials, such as those used in drill and milling bits [3], and to post-process these hard materials, CNC form-grinding is useful. Grinding is highly used for achieving a very smooth surface on an AM-produced part, such as when sample preparation of AM parts for various material characterizations such as scanning electron microscopy (SEM) and electron backscatter diffraction (EBSD) is needed. Grinding has limitations for complex parts, which limits the usage in the AM field.

Lapping and cleaning/polishing pads using rotary devices are also commonly used to remove rough surfaces of AM parts. Due to high speed and soft abrasive materials, these grinding processes are useful for finishing. Lapping and cleaning/polishing are normally performed with simple rotary movement (single axis), so they are not suitable to improve dimensional accuracy in complex geometries.

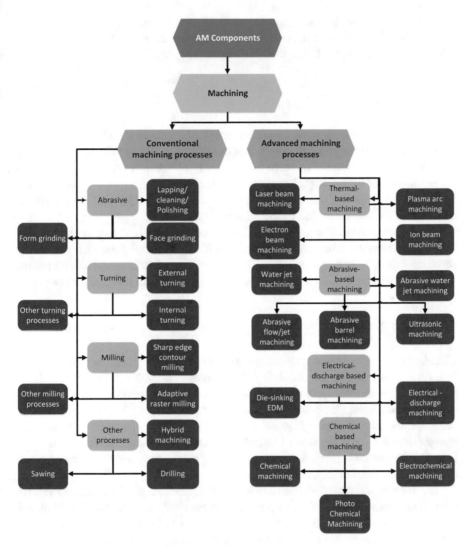

Fig. 16.7 Different machining strategies for AM parts

16.3.4.2   Turning

If the printed part has a simple cylindrical shape or features, turning is an ideal choice for post-processing. Turning is also useful when material removal has to be carried out for internal surfaces. In turning the single cutting point leads to less impact during the material removal process and has better surface quality compared to milling. For certain non-cylindrical shapes, multi-axis CNC turning can be used. In the case of high complexity, the running chuck and tool holder, with rotation around different axes, are used to generate multi-axis passes to cover the curves and patterns [4–6]. However, if the designed part has very intricate shape and complexity, a milling operation is better than turning or grinding.

16.3.4.3 Milling

In milling, a workpiece is fixed, and the cutting tool rotates and translates around the part. Multi-axis milling enables rotation of the milling bed (and thus the part) around X, Y, and Z axes to further increase the flexibility of the process and its usefulness for machining of highly complex parts. 5-axis and 6-axis machines are particularly helpful for situations where the milling tool must approach the part from different orientations to finish machine all the relevant features.

CNC milling is a ubiquitous process in manufacturing and in technical education, and it is assumed the reader is familiar with these techniques. To better understand how the challenges involved with CNC milling of near-net-shape AM parts, particularly those intended as molds, the reader is pointed to this early work [7].

16.3.4.4 Hole Drilling

Circular holes are common features in parts and tools. Using milling tools to create holes is inefficient, and the circularity of the holes is poor. Therefore, a machining strategy which identifies and drills holes is preferable. Although it is easy for a person to identify a circular hole in a design and manually program a drill, it is challenging for software to recognize holes in an STL or AMF file, as the 3D geometry is represented by a collection of unordered triangular planar facets (and thus all feature information is lost). A simple method for automating hole identification and drilling is discussed below.

The intersection curve between a hole and a surface is typically a closed loop. By using this information, a hole recognition algorithm begins by identifying all closed loops made up of sharp edges from the model. These closed loops may not necessarily be the intersection curves between holes and a surface, so a series of hole-checking rules are used to remove the loops that do not correspond to drilled holes. The remaining loops and their surface normal vectors are used to determine the diameter, axis orientation, and depth for drilling. From this information, tool paths can be automatically generated [7].

16.3.4.5 Other Conventional Machining Processes

Hybrid conventional machining such as milling–turning can provide both the surface quality of turning and flexibility of milling and can be used to post-process AM parts. Figure 16.8 shows a hybrid milling–turning machine (OKUMA MULTUS U3000) during machining of complex shapes.

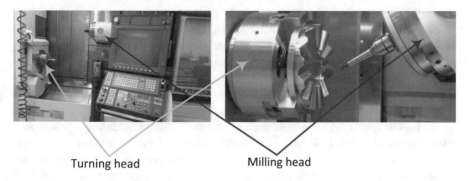

Turning head   Milling head

Fig. 16.8 Milling–turning material removal (OKUMA MULTUS U3000). (Photo courtesy of OKUMA)

In this operation each turning and milling head is moved in X, Y and Z directions to provide 6-axis movements. Also, the heads can rotate around X, Y and Z direction to provide another six degrees of movement, thus totally 12 axes to provide excellent flexibility for complex geometries. However, hybrid milling–turning has limitations for internal surfaces.

Mechanical sawing with straight or rotary blades is commonly used to remove parts from a build plate. Using this method has some disadvantages compared to advanced machining processes. Mechanical saws produce high forces that increase the chance of breaking thin walls. Also, the cutting line is in the range of 1–3 mm which is higher than wire EDM (mentioned later). Another problem for this method is clamping which needs high accuracy. The advantage of this method is low-cost equipment and running operation, no need for highly trained staff, and low maintenance costs. When using sawing to separate parts from the base plate, it is recommended to make the minimum support height 5 mm or more so that the part is not damaged when cutting through the supports.

16.3.4.6   Thermal-Based Advanced Machining Processes

Heating, melting, vaporization, and chemical degradation (i.e., breaking of chemical bonds) will result in removing material from the surface of a workpiece at very high temperature.

Laser Machining Laser machining can be used for surface treatment and to improve accuracy of AM parts [8, 9]. The performance and efficiency of laser machining are related to thermal conductivity and absorptivity of the material and the type of the laser. Continuous wave (CW) lasers can provide 10^{10} W/m^2, while ultrashort pulsed lasers produce 10^{18} W/m^2. This method can be used for many types of materials including polymers, laminated gold thin film, ceramics, metals, etc. Laser machining has some advantages over conventional machining like independency from the hardness and strength of the material and higher movement flexibility. However, the high temperature of the laser beam may create defects below the machining surface and may also reduce dynamic mechanical properties such as fatigue properties. Therefore, more consideration has to be taken into account when selecting this machining strategy.

Plasma Arc Machining (PAM) Most gases have molecules consisting of two or more atoms. When increasing the temperature in the range of 2000 °C, molecules divide into atoms, and if the temperature increases to 3000 °C, electrons are removed from atoms, gases are ionized, and plasma is formed. Plasma is a hot ionized gas with approximately equal numbers of negatively charged electrons and positively charged ions. Plasma has high thermal energy and can be used for material removal for almost all metals and ceramics. In this method, the workpiece is heated to melting temperature, and the liquid phase is removed by plasma pressure and gas flow. The removal mechanism can be aided by chemical reaction between plasma and workpiece or evaporation (due to the high temperature of plasma). PAM is used for cutting, grooving, and face turning. In plasma arc turning, instead of the cutting tool, a plasma arc nozzle is used. This method is suitable for hard and brittle materials (such as PBF AM parts with thin layers), has a fast material removal rate, and can be used to process small cavities with good dimensional accuracy. PAM has some disadvantages such as negative effect on residual stress for the machined and underlying surfaces, high cost, and high consumption of inert gas [6]. Residual stress from PAM is greater than most thermal-based AM operations and so is best suited for ceramic and sintered AM parts.

Electron Beam (EB) Machining EB machining typically operates in pulse mode under 10^{-5} bar vacuum and with a focus area around 25 μm, producing very high energy density. This energy is absorbed by the workpiece and leads to evaporation and thus acts as a material removal process [10]. The surface quality of electron beam machined parts is highly related to the material and the quality of the original part (before machining). For instance, for carbon and gold, the best surface quality was reported around 5 μm and for titanium 10 μm. This process can produce holes with 25 μm diameter and 10 mm length. EB machining is suitable for brittle material and thin walls. Moreover, EB machining is independent of material type and has a higher removal rate compared to electrical discharge machining. EB machining produces residual stress and for nonthermal AM processes is a suitable method. However, for thermal-based AM processes, this machining increases the value of residual stresses and the chance of distortion. Also, electron beam machining needs expensive equipment and an expert operator [11].

Ion Beam Machining (IBM) IBM has the same operational concept as electron beam machining. In this method a source of plasma and an accelerator (electromagnetic coils) ionize the gas atoms. Using high potential (1000 V), a heated tungsten cathode emits electrons, which move toward the anode in a vacuum chamber. Electromagnetic coils are used to move the electrons in a spiral pattern to increase the length of movement. These combined activities result in higher ionization and energy, which can evaporate the workpiece and thus remove material. This method is highly recommended for the machining of small components, even in the range of 100 nm (for microprinting). High surface quality ranging from 1 μm was reported using this machining. Also, this method is used for surface texturing. Due to the high resolution of IBM, it can be used for texturing surfaces for biomedical applications to increase cell proliferation and growth rate. IBM can produce atomically clean surfaces which is one of the advantages of this method over electron beam and electrical discharge machining. The atomically clean surface has a high potential for coating with other materials, improving bonding. IBM is highly used in sample preparation for characterization of AM parts. When preparing scanning electron microscopy (SEM) and transmission electron microscopy (TEM) samples, ion beam is a perfect choice. Preparing TEM samples using IBM offers the ability to create site-specific samples, reduces the radiation field interference, and produces useful cross-sections to better understand the fusion gas behavior along grain boundaries [12]. Other advantages of this method are low-temperature processing (compared to other thermal-based machining processes) independent of material type, no chemical reaction, etching rates that are easily controllable, and no under-cutting as compared to chemical etching. In contrast, this method needs expensive equipment, and the material removal rate is slow. Figure 16.9 shows the ion beam machined surface for material characterization of Laser-Based Powder Bed Fusion (LB-PBF) stainless steel 316 L.

16.3.4.7   Abrasive-Based Advanced Machining Processes

Five types of abrasive-based machining are utilized with AM, including water jet machining (WJM), abrasive water jet machining (AWJM), abrasive flow/jet machining (AFM/AJM), abrasive barrel machining (ABM), and ultrasonic machining (USM).

Water Jet Machining (WJM)   Although water jet machining doesn't technically use an abrasive (just water) to cut samples, we address it here due to its similarity

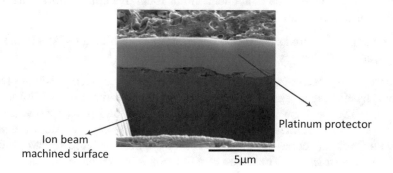

Fig. 16.9 Ion beam machined surface for material characterization of LB-PBF stainless steel 316 L

with the subsequent processes. In water jet machining, a hydraulic pump equipped with an electrical motor increases the pressure of the water from 4 bars to 3800 bars. An accumulator is used to remove the pressure fluctuations and produces uniform water flow. Then water is passed from a hard nozzle (often made of sapphire) at a speed of 900 m/s or more. This method is mainly used for cutting but by increasing the focusing diameter can be also used for surface machining. Water jet machining is a good choice for post-processing of composite AM parts, including reinforced plastics. The cooling effect of water prevents the destruction of fibers; however the stream of water is deviated by impacting different materials which reduces the accuracy. This method is also suitable for the post-processing of AM ceramic parts which are very brittle. Water jet machining can be used for removing sharp edges from contours that are observed in metal powder-based systems [13, 14]. Water jet cutting is an excellent process to use to machine biomedical and food products since only water interacts with the part being processed. Figure 16.10a shows the schematic of water jet machining.

Abrasive Water Jet Machining (AWJM) AWJM has the same concepts and operation as WJM. The only difference between AWJM and WJM is the abrasive particles which are fed into the fluid flow. The abrasive feeder provides a constant flow of abrasive particles into the main water stream. The presence of abrasives in the water jet increases the cutting rate and enables cutting of harder materials compared to water jet machining without abrasives. The advantage of working in low temperature makes AWJM suitable for post-processing of powder-based AM components with small features. AWJM is relatively inexpensive, and the nozzle produces clean surfaces and has good flexibility for dealing with complex shapes. The velocity of water should be within the range of 700–900 m/s. Hard materials such as silicon carbide and silica sand are used as abrasive particles. The advantages of AWJM are operation with low contamination, high cutting speed, and the ability to machine in different axes. In thermal-based AM processes, periodic heating and cooling produce high residual stress. Therefore, AWJM (due to low working temperature) is a good choice for machining when residual stress is a concern. A wide range of materials including metals and nonmetals can be proceeded. Recycling of the abrasives is difficult, and this method is not suitable for rough material removal due to the deviation of the jet stream [10].

Another process similar to AWJM is abrasive jet machining (AJM). In the case of AJM, air or another gas is used to propel the abrasives.

Abrasive Flow Machining (AFM) AFM is an interior surface finishing process operated by flowing an abrasive-laden fluid through a workpiece. In AFM in contrast to AWJM, incompressible fluid is used, and the fluid is pumped at a much lower speed then in AWJM. The fluid may be pushed forward and backward rather than in a single direction. This method can be applied for intricate AM products such as lattice structures. The nature of AFM makes it a proper choice for interior surfaces, holes, slots, cavities, and other areas such as lattices that may be difficult or impossible to reach with other polishing or grinding processes. AFM is

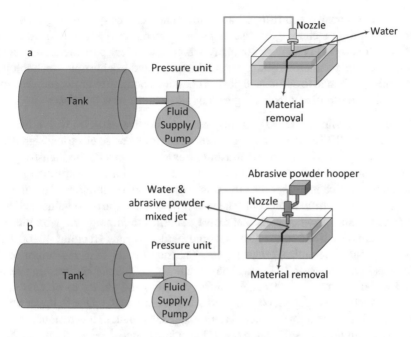

Fig. 16.10 (**a**) Schematic of water jet machining, (**b**) schematic of abrasive water jet machining

successfully used for a broad range of materials, including AM metal and VPP prototypes [15, 16]. AFM needs a sealed operational chamber, and due to low material
removal rate, AFM is not common for large surface area machining [17].

Abrasive Barrel Machining (ABM)   In ABM, water, abrasive particulate media,
and the workpiece which has to be polished are fed in a barrel. The barrel rotates,
and the rotational motion of the abrasive particles interacting with the tumbling
workpiece results in material removal. Many factors, such as size and rotational
speed of the barrel, geometry and quantity of abrasive particles, and different types
of flow, affect the material removal rate (MRR) [18]. This process can also be done
without adding water to the barrel. When using water in the barrel, the operational
temperature is low, and due to uniform surface contact between media and workpiece, no residual stress is generated. Also, AM parts with lattice structures can be
machined by this method using small media. Therefore, this method is a good
choice for complex AM parts. Another advantage of ABM is that batch machining
of several samples at the same time is possible. Drawbacks of ABM are non-uniform
material removal and an inability to process location-specific dimensional deviations [6, 11].

Many companies sell equipment based upon variations of ABM that may use a
tumbling barrel or other mechanisms to create relative motion between a water/
abrasive mixture and workpiece. One example is PostProcess Technologies Inc.
[16] which has introduced a number of machines that utilize a combination of water

abrasive and abrasive centrifugal barrel finishing for post-processing AM parts. Table 16.1 shows technologies, machines, and operation types developed by this company. Iepco [19] is another company which also offers post-processing equipment, including microblasting, nanoblasting, and deburring technologies. Companies dedicated to automated post-processing continue to grow, and we expect this area to see significant technological advances in the next few years.

Ultrasonic Machining (USM) In this method, abrasive particles with a hardness of higher than 40HRc are used. First, a generator produces an electrical signal with high frequency (20 kHz), and an ultrasonic transducer converts the high-frequency electrical energy to linear motion (vibration). This motion is transformed into the concentrator, which increases the vibration amplitude. The vibrations from the concentrator are transferred to the tool. The tool is placed in a slurry container and gives energy to the slurry, which causes abrasive materials to impinge the workpiece for material removal. USM is used to machine hard materials (ranging 40–60 HRc) such as carbides, ceramics, tungsten, etc. The surface finishing resolution of this method is in the range of 7–25 μm. The advantages of this method are high resolution, low temperature in the material removal area, and ability to process brittle materials such as powder-based AM products with thin walls. Due to lower operating temperature and uniform material removal, USM produces low residual stresses. USM is not suitable for soft material and has a low material removal rate [6, 11].

16.3.4.8 Electrical Discharge Machining (EDM)

In EDM, a large voltage is applied between an electrode and a workpiece. The gap between the electrode and workpiece is filled with a dielectric fluid. Sparks are generated between the electrode and workpiece causing material removal.

In die-sinking EDM, the electrode (typically either copper or graphite) is shaped before EDM. The shape of the electrode represents the negative shape of the intended surface of the workpiece. A series of electric pulses are applied, and each pulse results in a spark which causes a small amount of material removal due to vaporization. At the end of the spark, the dielectric fluid washes away the removed

Table 16.1 Different abrasive technologies by PostProcess Company [20]

Technology	Machine	Operation type	AM Process
Volume Velocity Dispersion	BASE™, DECI™	High volume and flow jet streams spraying coupled with perpendicularly linear motion	MEX, MJT, VPP
Submersed Vortex Cavitation	DEMI™, FORTI™, CENTI™	Agitated flow by sink/float process to rotate parts in the chamber	MEX, MJT, VPP
Suspended Rotational Force	NITOR™, LEVO™	A combination of circulating motion with immersion into mix of abrasive and fluid	MEX, MJT, PBF, VPP
Hybrid	DECI Duo™	Thermal atomized fusillade technology uses two perpendicular abrasive jet streams while rotating parts 360^0	MEX, MJT, PBF, VPP, BJT

material. EDM is a noncontact and non-force process, so it is suitable for processing of AM metal parts. It is good for machining thin walls that cannot take the stress of traditional machining. By using EDM, the risk of breaking the parts is less than when using conventional methods. The surface roughness generated by this method is uniform and has the same value independent of the region being machined. EDM leaves no burrs and can process complex shapes without distortion and is not affected by material hardness.

Wire EDM (where a wire is used as the electrode) is commonly used to remove metal AM parts from the build plate. This method doesn't apply any mechanical force, and the accuracy of the cut is similar to the diameter of the wire (0.2 mm or less). Die-sinking EDM and wire EDM are only applicable for conductive materials and are more expensive process than conventional milling or turning. Thus, EDM is suitable for processing AM metal parts [21, 22].

16.3.4.9   Chemical-Based Advanced Machining Processes

Chemical Machining (ChM) ChM uses corrosive solutions or baths of temperature-regulated etching chemicals to remove material. Inert materials known as maskants are used to protect specific areas that shouldn't be removed. ChM is suitable for complex materials that can be damaged by cutting forces in conventional machining. ChM is used to remove the materials from the surface of the workpiece, and material removal rate is uniform, making it suitable for thin walls and lattice structures. ChM can be used to etch samples for microstructural characterization. This method can achieve accurate dimensional tolerances and can be used as a post-process method for jewelry and aerospace parts. The disadvantages of ChM are the generated gas from corrosive substances and low material removal rate. The material range for ChM is limited, and finding a suitable maskant for different materials is problematic [23, 24].

The Austrian company Hirtenberger uses ChM to remove supports and smooth AM-produced metal parts. Using a combination of hydrodynamic flow, electrochemical pulsing, and particle-assisted removal and surface cleaning [25], a multistep automated technique removes supports and performs rough and fine material removal processes. Supports must be printed in unique ways so that they are removed by ChM, without substantially removing material from the part. This system is applicable for most metals and alloys regardless of hardness. Hirtenberger claims a surface quality of $R_a = 0.5$ μm, which is suitable for many applications.

Photochemical Machining (PCM) PCM starts by printing the shape of the part as a picture onto an optically clear photographic film. The "phototool" includes two sheets of this film that illustrates negative images of the object. The two sheets are optically and mechanically placed to generate the top and bottom halves of the tool. The sheets are cut to obtain the desired size followed by cleaning and laminating on both sides with a UV-sensitive photoresist. In the next step, the coated metal is placed between the sheets, and the plate is exposed to UV light and allows the areas of resist (that are in the clear sections of the film) to be hardened. Then the plate is cleaned from unexposed resist, leaving the areas to be etched unprotected (Fig. 16.11). As a result the shape of the photo in negatives is engraved in the material. PCM can economically produce intricate components with very fine detail and can be used to provide surface modification to AM-produced parts. PCM can be applied to any material regardless of the hardness but is limited in thickness. PCM can produce very fine details such as for jewelry components [6, 11].

16.4   Post-Processing to Improve Mechanical Properties

16.4.1   Property Enhancements Using Nonthermal Techniques

Powder-based and extrusion-based processes often create porous structures. In many cases, that porosity can be infiltrated by a higher-strength material, such as cyanoacrylate (Super Glue®). Proprietary methods and materials have also been developed to increase the strength, ductility, heat deflection, flammability resistance, EMI shielding, or other properties of AM parts using infiltrants and various types of nanocomposite reinforcements.

A common post-processing operation for photopolymer materials is curing. During processing, many photopolymers do not achieve complete polymerization. As a result, these parts are put into a Post-Cure Apparatus, a device that floods the

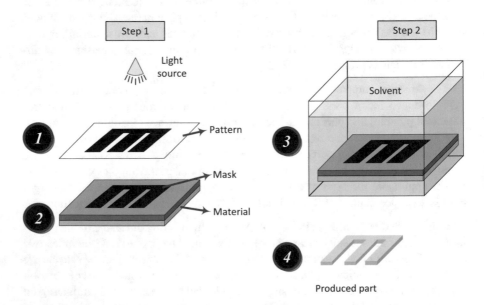

Fig. 16.11 Schematic of photochemical machining

part with UV and visible radiation in order to completely cure the surface and subsurface regions of the part. Additionally, the part can be placed in a low-temperature oven, which helps accelerate and completely cure the photopolymer and in some cases greatly enhance the part's mechanical properties.

16.4.1.1   Shot Peening

Shot peening is a method that induces residual compressive stress by bombarding surface with very hard spherical media in a controlled operation. The media is typically steel, ceramic, or glass, but can be any powder with equal or greater hardness than the workpiece. Shot peening acts like a tiny peening hammer producing a small indentation on the surface. In thermal-based AM techniques such as MEX, PBF, and DED, the periodic heating and cooling creates high residual on the surface of the workpiece. Shot peening can improve surface stress states by inducing beneficial residual compressive stress. The proven results from controlled shot peening show a dramatic beneficial effect on both the life and strength of components, making the material more resistant to fatigue, fretting, and stress corrosion cracking. Shot peening improves the surface characteristics of metals, composites, and polymers [26]. By using small media, the process can be applied for intricate and lattice shapes. By using media that is identical to the powder used to create the AM part, the powder can be recycled and placed into the AM machine for further use. 3D printing of gears and turbine blades is a high demand area in AM which require surface treatment like multi-axis shot peening. Figure 16.12 shows a schematic of shot peening.

16.4.1.2   Cold Isostatic Pressing (CIP)

Cold isostatic pressing (CIP) is post-process method to improve mechanical properties and consolidate material to make a green part that can be further treated using other post-processes such as rolling or machining. The operational pressure for CIP is 15,000 to 60,000 psi with ambient temperature up to 93 °C. There are two types of CIP, wet and dry. In wet condition the process is carried out in a fluid to increase the uniformity of the pressing process. For simple parts, a workpiece is surrounded by a bag or mold and exposed to high pressure, but for complex or near-net-shape parts, the material is in direct contact with air or liquid. CIP is used for consolidation of graphite, polymers, ceramics, and metal AM parts. CIP needs cheaper equipment compared to hot isostatic pressing (HIP) although it produces weaker mechanical properties, so CIP is used before HIP. Both CIP and HIP are ideal for processing complex AM shapes.

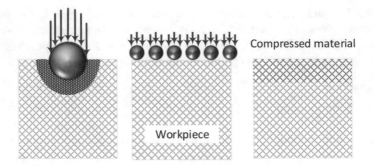

Fig. 16.12 Schematic of shot peening

16.4.2   Property Enhancements Using Thermal Techniques

Thermal techniques for property improvement can be divided into three categories based upon the pressure applied, as seen in Fig. 16.13.

To improve the quality, mechanical properties, and material integrity of AM parts, it is commonly necessary to eliminate porosity and create a homogeneous microstructure. Heat treatment is an excellent way to achieve these goals.

16.4.2.1   Heat Treatment in Ambient Pressure

After AM processing, many parts are thermally processed to enhance their properties. In the case of DED and PBF techniques for metals, this thermal processing is primarily heat treatment to form the desired microstructures and/or to relieve residual stresses. In these instances, traditional recipes for heat treatment developed for the specific metal alloy being employed are often used. In some cases, however, special heat treatment methods have been developed. This may be to retain the fine-grained microstructure within the AM part while still providing stress relief and ductility enhancement. It may also be necessary to create a custom heat treatment recipe when the AM-built part has a nonequilibrium microstructure that doesn't respond as expected to traditional heat treatment recipes.

Before the advent of DED and PBF techniques capable of directly processing metals and ceramics, many techniques were developed for creating metal and ceramic green parts using AM. These were then furnace post-processed to achieve dense, usable metal and ceramic parts. BJT is the main AM process used for these purposes today. The basic approach to furnace processing of green parts was illustrated in Fig. 5.8. In order to prepare a green part for furnace processing, several preparatory steps are typically done. Figure 16.14 shows the steps for preparing a metal green part made from LaserForm ST-100 for furnace infiltration.

Figure 16.15 shows a piece of artwork from Bathsheba Grossman (bathsheba. com). Bathsheba is well-known for her mathematically inspired 3D sculptures. This piece was printed as a green part using BJT and subsequently processed in a furnace,

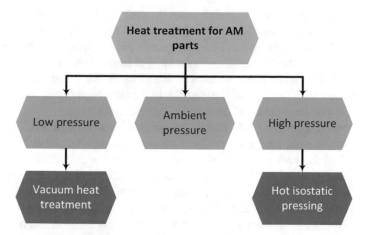

Fig. 16.13 Heat treatment for AM parts based on the operational pressure

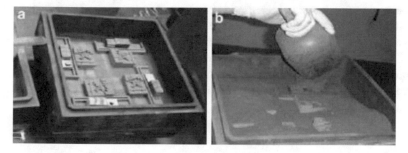

Fig. 16.14 LaserForm ST-100 green parts. (**a**) Parts are placed next to "boats" on which the bronze infiltrant is placed. The bronze infiltrates through the boat into the part. (**b**) The parts are often covered in aluminum oxide powder before placing them in a furnace to help support fragile features during debinding, sintering, and infiltration and to help minimize thermal gradients

as a thermal treatment to improve the mechanical properties and appearance. Bathsheba uses hand polishing and complimentary post-processing techniques to change the surface aesthetics of certain regions of her artwork (e.g., the edges of the sculpture in this image) while leaving other portions of the artwork unfinished. Rough as-built surface finishes contrast nicely with polished surfaces to create aesthetic appeal.

Control of shrinkage and dimensional accuracy during furnace processing is complicated by the number of process parameters that must be optimized and the multiple steps involved. Figure 16.16 illustrates the complicated nature of optimization for this type of furnace processing. The y-axis, $(F1–F2)/F1$, represents the dimensional changes during the final furnace stage of infiltration of stainless steel (RapidSteel 2.0) parts using bronze. $F1$ is the dimension of the brown part before infiltration, and $F2$ is the dimension after infiltration. The data represent thousands of measurements across both internal (channel-like) and external (riblike) features ranging from 0.3 to 3.0 in. Although many factors were studied, only two were found to be statistically significant for the infiltration step, atmospheric pressure in the furnace and infiltrant amount. The atmospheric pressure ranged between 10 torr and 800 torr. The amount of infiltrant used ranged from a low of 85% to a high of 110%, where the percentage amount was based upon the theoretical amount of material needed to fully fill all of the porosity in the part, based upon measurements of the weight and the volume of the part just prior to infiltration.

It can be seen from Fig. 16.16 that the factor combinations with the lowest over-all shrinkage were not the factors with the lowest shrinkage variation. Factor combination A had the lowest average shrinkage, while factor combination E had the lowest shrinkage variation. As average shrinkage can be easily compensated using a scaling factor, the optimum factor combination for highest accuracy and precision would be factor combination E. This study illustrates that furnace processing parameters and strategies for green parts which are subsequently sintered and infiltrated to form a final part need to be carefully considered to achieve high dimensional tolerances.

Laser sintering has also been used to produce complex-shaped ZrB_2/Cu composite EDM electrodes. The approach involved (a) fabrication of a green part from polymer-coated ZrB_2 powder using laser sintering, (b) debinding and sintering of the ZrB_2, and (c) infiltration of the sintered, porous ZrB_2 with liquid copper. This manufacturing route was found to result in a more homogeneous structure compared to a hot pressing route and was shown to be a successful way to produce die-sinking EDM electrodes.

Fig. 16.15 BJT artistic part from Bathsheba Sculpture

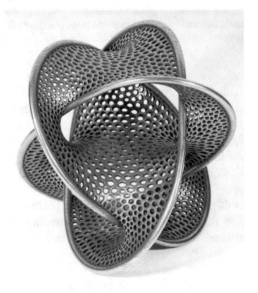

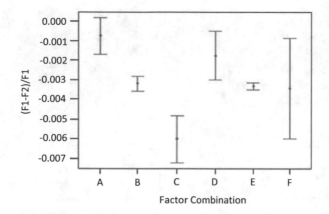

Fig. 16.16 95 percent confidence intervals for variation in shrinkage for stainless steel (RapidSteel 2.0) infiltration by bronze. (Factor combinations are (A) 10 torr, 80%; (B) 10 torr, 95%; (C) 10 torr, 110%; (D) 800 torr, 80%; (E) 800 torr, 95%; (F) 800 torr, 110%)

16.4.2.2   Heat Treatment in Low Pressure

For some AM materials prone to oxidation, such as Ti alloys and ceramics, heat treatment should be performed in a vacuum. The vacuum system for low-pressure heat treatment is more expensive compared to that for ambient pressure. Vacuum furnaces have uniform temperature in the range of 800–3000 °C with low contamination by carbon, oxygen, and other gases. The vacuum pumping system removes low-temperature by-products from the process materials during heating, resulting in a higher purity end product. Moreover, an inert gas, such as argon, is often used to quickly cool the treated metals back to non-metallurgical levels after the desired process in the furnace [27, 28]. Figure 16.17 shows the effect of heat treatment in vacuum on the microstructure of the LB-PBF Ti-6Al-4V part. The annealing increases the grain size and tailors the mechanical properties.

16.4.2.3   Heat Treatment in High Pressure (Hot Isostatic Pressing)

HIP is a forming and densification process using heated gas under very high pressure. Maximum standard operating pressures range from 1500 to 30,000 psi. Unlike mechanical force which compresses workpieces from one or two sides, HIP is applied uniformly (e.g., isostatically) on all sides of an object. HIP can significant reduce porosity with little net shape change. HIPing has the following specific benefits:

- Densification of powdered metal parts to nearly 100 percent theoretical density
- Increased resistance to fatigue and temperature extremes
- Higher resistant to impact wear and abrasion
- Improved ductility
- Little or no secondary machining or manual rework and decreased scrap rate

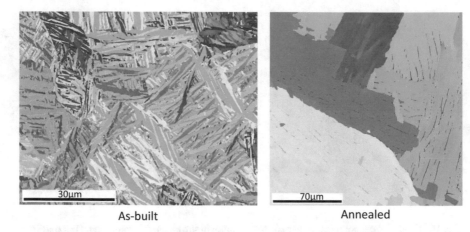

As-built   Annealed

Fig. 16.17 As-built and vacuum annealed Ti-6Al-4V produced by LB-PBF

HIPing can provide diffusion bonding of dissimilar metals, which increases the bonding strength for multi-material AM structures as well as DED and PBF-produced AM parts [29, 30]. HIP can be used to treat metals, ceramics, composites, and polymer parts. The temperature and pressure conditions using in HIP must be related to the physical properties of the materials. For instance, HIP can be successfully used for MEX parts to eliminate the pores. Figure 16.18 shows the effect of temperature and pressure in the densification of such parts.

It is also reported that some pores (entrapped argon gas) can be removed by HIP. The disadvantages of the HIP for AM parts are the cost of the process and solubility of the gas in the printed parts. This problem is more obvious for materials which have a higher tendency to chemical reactions [30].

One unique example of using HIPing with PBF is an approach which utilized laser sintering to produce porous parts with gas-impermeable skins. By scanning only the outside contours of a part during fabrication by SLS, a metal "can" filled with loose powder is made. These parts are then post-processed to full density using hot isostatic pressing (HIP). This in situ encapsulation results in no adverse container–powder interactions (as they are made from the same bed of powder), reduced pre-processing time, and fewer post-processing steps compared to conventional HIP of canned parts. The SLS/HIP approach was successfully used to produce complex 3D parts in Inconel 625 and Ti–6Al–4V for aerospace applications [32].

16.5   Preparation for Use as a Pattern

Often parts made using AM are intended as patterns for investment casting, sand casting, room temperature vulcanization (RTV) molding, spray metal deposition, or other pattern replication processes. In many cases, the use of an AM pattern in a

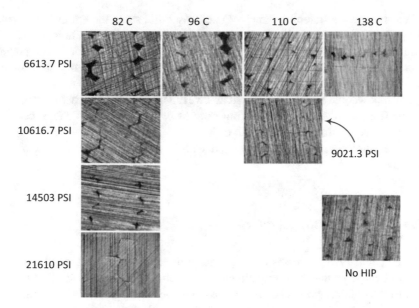

Fig. 16.18   HIP of polycarbonate to remove porosity [31]

casting process is the least expensive way to use AM to produce a metal part, as many of the metal-based AM processes are still expensive to own and operate.

The accuracy and surface finish of an AM pattern will directly influence the final part accuracy and surface finish. As a result, special care must be taken to ensure the pattern has the accuracy and surface finish desired in the final part. In addition, the pattern must be scaled to compensate for any shrinkage that takes place in the pattern replication steps.

16.5.1   Investment Casting Patterns

In the case of investment casting, the AM pattern will be consumed during processing. In this instance, residue left in the mold as the pattern is melted or burned out is undesirable. Any sealants used to smooth the surface during pattern preparation should be carefully chosen so as not to inadvertently create unwanted residue.

AM parts can be printed on a casting tree or manually added to a casting tree after AM. Figure 16.19 shows rings made using a MJT system. In the first picture, a collection of rings is shown on the build platform; each ring is supported by a secondary support material in white. In the second picture, a close-up of the ring pattern is shown. The third picture shows metal rings still attached to a casting tree. In this instance, the rings were added to the tree after AM, but before casting. It is also common to print investment casting patterns and the casting tree simultaneously, rather than attaching patterns to a tree as a secondary step.

When using the stereolithography QuickCast build style, the hollow, truss-filled shell patterns must be drained of liquid prior to investment. The hole(s) used for draining must be covered to avoid investment entering the interior of the pattern. Since photopolymer materials are thermosets, they must be burned out of the investment rather than melted.

When powdered materials are used as investment casting patterns, such as polystyrene from a polymer laser sintering process or starch from a BJT process, the resulting part is porous and brittle. In order to seal the part and strengthen it for the investment process, the part is infiltrated with an investment casting wax prior to investment.

16.5.2   Sand Casting Patterns

Both BJT and PBF processes can be used to directly create sand mold cores and cavities by using a thermosetting binder to bind sand in the desired shape. One benefit of these direct approaches is that complex geometry cores can be made that would be very difficult to fabricate using any other process, as illustrated in Fig. 16.20.

In order to prepare AM sand casting patterns for casting, loose powder is removed, and the pattern is heated to complete cross-linking of the thermoset binder and to remove moisture and gaseous by-products. In some cases, additional binders are added to the pattern before heating, to increase the strength for handling. Once the pattern is thermally treated, it is assembled with its corresponding core(s) and/ or cavity, and hot metal is poured into the mold. After cooling, the sand pattern is removed using tools and bead blasting.

In addition to directly producing sand casting cores and cavities, AM can be used to create parts which are used in place of the typical wooden or metal patterns around which a sand casting mold is created. In this case, the AM part is built as one or more portions of the part to be cast, split along the parting line. The split part is placed in a box, sand mixed with binder is poured around the part, and the sand is compressed (pounded) so that the binder holds the sand together. The box is then

Fig. 16.19 Rings for investment casting, made using a ProJet® CPX 3D Printer. (Photo courtesy of 3D Systems)

disassembled, the sand mold is removed from the box, and the pattern is removed from the mold. The mold is then reassembled with its complementary mold half and core(s); and molten metal is poured into the mold.

16.5.3 Other Pattern Replication Methods

There are many pattern replication methods which have been utilized since the late 1980s to transform the weak "rapid prototypes" of those days into parts with useful material properties. As the number of AM technologies has increased and the durability of the materials that they can produce has improved substantially, these replication processes are finding less use, as people prefer to directly produce a usable part if possible. However, even with the multiplication of AM technologies and materials, pattern replication processes are widely used among service bureaus and companies who need parts from a specific material that is not directly processable in AM.

Probably the most common pattern replication methods are RTV molding and silicone rubber molding. In RTV molding, as shown in Fig. 16.21, the AM pattern is given visual markers (such as by using colored tape) to illustrate the parting line locations for mold disassembly; runners, risers, and gates are added; the model is suspended in a mold box; and a rubberlike material is poured around the model to encapsulate it. After cross-linking, the solid translucent rubber mold is removed from the mold box, a knife is used to cut the rubber mold into pieces according to the parting line markers, and the pattern is removed from the mold. In order to complete the replication process, the mold is reassembled and held together in a box or by placing rubber bands around the mold, and molten material is poured into the mold and allowed to solidify. After solidification, the mold is opened, the part is removed, and the process is repeated until a sufficient number of parts are made. Using this process, a single pattern can be used to make 10's or 100's of identical parts.

Fig. 16.20 Sand casting pattern for a cylinder head of a V6, 24-valve car engine (left) during loose powder removal and (right) pattern prepared for casting alongside a finished casting. (Joint project between CADCAM Becker GmbH and VAW Südalumin GmbH, made on an EOSINT S laser-sintering machine, photo courtesy of EOS)

If the part being made in the RTV mold is a wax pattern, it can subsequently be used in an investment or plaster casting process to produce a metal part. Thus, by combining RTV molding and investment casting, one AM pattern can be replicated into a large number of metal parts for a relatively modest cost.

Metal spray processes have also been used to replicate geometry from an AM part into a metal part. In the case of metal spray, only one side of the pattern is replicated into the metal part. This is most often used for tooling or parts where one side contains all the geometric complexity and the rest of the tool or part is made up of flat edges. Using spray metal or electroless deposition processes, an AM pattern can be replicated to form an injection molding core or cavity, which can then be used to mold other parts.

16.6   Summary

Most AM-produced parts require post-processing prior to implementation in their intended use. Effective utilization of AM processes requires a knowledge not only of AM process benefits and limitations but also of the requisite post-processing operations necessary to finalize the part for use.

When considering the intended form, fit, and function for an AM-produced part, post-processing steps may vary. To achieve the correct form, support material removal, surface texture improvements, and aesthetic improvements are commonly required. To achieve the correct fit, accuracy improvements, typically via milling, are commonly required. To achieve the correct function, the AM part may require preparation for use as a pattern, property enhancements using nonthermal techniques, or property enhancements using thermal techniques. Whether using automated secondary support material removal, labor-intensive de-cubing, high-temperature furnace processing, or secondary machining, choosing and properly implementing the best AM process, material, and post-processing combination for the intended application is critical to success.

Fig. 16.21 RTV molding process steps. (Photo courtesy of MTT Technologies Group)

16.7 Questions

1. What are the key material property considerations when selecting a secondary support material for MJT and MEX? Would these considerations change when considering supporting metals deposited using a DED process?
2. What are the primary benefits and drawbacks when offsetting triangle surfaces versus triangle vertices? (see [7] for reference). Which approach would be better for freeform surfaces, such as the hood of a car or the profile of a face?
3. Assuming that the total shrinkage in an AM process is represented by Fig. 16.16, which process parameter combination would you choose? Why? How would you ensure the most accurate part is built considering your choice?
4. In AM processes often a larger shrinkage value is found in the X–Y plane than in the Z direction before post-processing. Why might this be the case?
5. A common problem with additive manufacturing when compared to traditional manufacturing is the poor quality of the surface finish. Discuss a way to increase surface finish without post-processing.
6. What is the main advantage of using secondary material support structure compared to those made of the build material? What is at least one disadvantage?
7. Would there be a difference in the post-processing for a SLS part that is made of polymer or metal? Explain.
8. Compare and contrast natural supports and synthetic supports. Include a discussion of their uses and common AM processes used for each.
9. What is the "stair-step" effect? How can this be avoided? What is a disadvantage of avoiding it?
10. What process typically follows the creation of a "green" part in indirect powder processing, and why is it necessary?
11. Why are supports necessary when dealing with metal PBF, and how does this impact the overall cost of the process?
12. Additive technologies used to create metal parts can have high part accuracy, but the surface is less than desirable when compared to traditionally machined parts. Do a web search, and give an example of a part that is built in a metal additive process and how traditional machining techniques (turning, milling, and drilling) are used to post-process the part.
13. Research and summarize a real-life example for how additive manufacturing is used to extend the life of traditional manufacturing tools.

References

1. One Click Metal. (2020). https://oneclickmetal.com/
2. Qu, X., & Stucker, B. (2003). A 3D surface offset method for STL-format models. *Rapid Prototyping Journal, 9*(3), 133–141.
3. Desktop Metal. (2020). https://www.businesswire.com/news/home/20181113005159/en/Desktop-Metal-Introduces-New-Model-Production-System

4. Frigieri, E. P., Ynoguti, C. A., & Paiva, A. P. (2019). Correlation analysis among audible sound emissions and machining parameters in hardened steel turning. *Journal of Intelligent Manufacturing, 30*(4), 1753–1764.

5. Bonhin, E. P., et al. (2019). Effect of machining parameters on turning of VAT32® superalloy with ceramic tool. *Materials Manufacturing Processes, 34*, 1–7.

6. Khorasani, A. M., Khavanin Zadeh, M. R., & Vahdat Azad, A. (2012). Advanced Production Processes. Psychology, art and technology. 207.

7. Stucker, B., & Qu, X. (2003). A finish machining strategy for rapid manufactured parts and tools. *Rapid Prototyping Journal, 9*(4), 194–200.

8. Hong, Y., et al. (2019). Fabricating ceramics with embedded microchannels using an integrated additive manufacturing and laser machining method. *Journal of the American Ceramic Society, 102*(3), 1071–1082.

9. Mingareev, I., et al. (2013). Femtosecond laser post-processing of metal parts produced by laser additive manufacturing. In *MATEC Web of Conferences*. EDP Sciences.

10. Bibb, R., Eggbeer, D., & Williams, R. (2006). Rapid manufacture of removable partial denture frameworks. *Rapid Prototyping Journal, 12*(2), 95–99.

11. McGeough, J. A. (1988). *Advanced methods of machining*. Springer Science & Business Media, New York, USA.

12. Miller, B. D., et al. (2012). Advantages and disadvantages of using a focused ion beam to prepare TEM samples from irradiated U–10Mo monolithic nuclear fuel. *Journal of Nuclear Materials, 424*(1), 38–42.

13. Galantucci, L. M., Lavecchia, F., & Percoco, G. (2009). Experimental study aiming to enhance the surface finish of fused deposition modeled parts. *CIRP Annals, 58*(1), 189–192.

14. Gupta, K. (2019). *Near Net Shape Manufacturing Processes*. Springer, Switzerland.

15. Madou, M. J. (2011). *Manufacturing techniques for microfabrication and nanotechnology* (Vol. 2). Boca Raton: USA, CRC Press.

16. Williams, R. E., & Melton, V. L. (1998). Abrasive flow finishing of stereolithography prototypes. *Rapid Prototyping Journal, 4*(2), 56–67.

17. Schrader, G. F., & Elshennawy, A. K. (2000). *Manufacturing processes and materials*. Society of Manufacturing Engineers, Michigan, USA.

18. Boschetto, A., Bottini, L., & Veniali, F. (2013). Microremoval modeling of surface roughness in barrel finishing. *The International Journal of Advanced Manufacturing Technology, 69*(9–12), 2343–2354.

19. iepco. (2020). https://iepco.ch/en/technologies

20. Post Process Company. (2020). https://www.postprocess.com/products/

21. Golinveaux, F., et al. (2019). Fracture fatigue properties of Inconel 718 manufactured with Selective Laser Melting. In *AIAA Scitech* 2019 Forum.

22. Dodun, O., et al. (2009). Using wire electrical discharge machining for improved corner cutting accuracy of thin parts. *The International Journal of Advanced Manufacturing Technology, 41*(9–10), 858.

23. Kasashima, Y., et al. (2019). Investigation of relationship between plasma etching characteristics and microstructures of alumina ceramics for chamber parts. Japanese Journal of Applied Physics.

24. Diaz, A. & Sroka, G. J. (2019). *Chemical Processing of Additive Manufactured Workpieces*. USPatent.

25. Hirtenberger Engineered Surface. (2020). https://hes.hirtenberger.com/en/

26. Madireddy, G., Curtis, M. M. E., Berger, J., Underwood, N., Khayari, Y., Marth, B., Smith, B., Christy, S., Krueger, K., Sealy, M. P., & Rao, P. (2017). Effect of Process Parameters and Shot Peening on the Tensile Strength and Deflection of Polymer Parts Made using Mask Image Projection Stereolithography (MIP-SLA). In *Solid Freeform Fabrication 2017: Proceedings of the 28th Annual International*. Austin.

27. Khorasani, A. M., et al. (2019). Investigation on the effect of heat treatment and process parameters on the tensile behaviour of SLM Ti-6Al-4V parts. *The International Journal of Advanced Manufacturing Technology, 101*(9), 3183–3197.
28. Khorasani, A. M., et al. (2017). On The Role of Different Annealing Heat Treatments on Mechanical Properties and Microstructure of Selective Laser Melted and Conventional Wrought Ti-6Al-4V. *Rapid Prototyping Journal,* **23**(2).
29. Tillmann, W., et al. (2017). Hot isostatic pressing of IN718 components manufactured by selective laser melting. *Additive Manufacturing, 13,* 93–102.
30. Shao, S., et al. (2017). Solubility of argon in laser additive manufactured α-titanium under hot isostatic pressing condition. *Computational Materials Science, 131,* 209–219.
31. Parker, M. E., et al. (2009). *Eliminating voids in FDM processed polyphenylsulfone, polycarbonate, and ULTEM 9085 by hot isostatic pressing.* South Dakota Schools of Mines & Technology, Research Experience for Undergraduate.
32. Das, S., et al. (1999). Processing of titanium net shapes by SLS/HIP. *Materials and Design, 20*(2–3), 115–121.

Chapter 17
Software for Additive Manufacturing

Abstract This chapter deals with software that is commonly used for Additive Manufacturing. In particular, we will discuss the STL file format that is used by most machines for model input data. These files are manipulated in a number of machine-specific ways to create slice data and for support generation. Basic principles of STL files are covered here including some discussion on common errors and other software that can assist with STL files. We consider some of the limitations of the STL format and how it may be replaced by something more suitable like the Additive Manufacturing File format. A proliferation of software tools for AM are emerging and becoming a major productivity aid for AM users. This includes software for build preparation, machine scheduling, process modeling, automated design, and more. As such, the reader is encouraged to search the latest literature, in addition to this chapter, to better understand this dynamic aspect of the AM industry.

17.1 Introduction

It is clear that Additive Manufacturing would not exist without computers and would not have developed so far if it were not for the development of 3D solid modeling CAD. The quality, reliability, and ease of use of 3D CAD have meant that virtually any geometry can be modeled, and it has enhanced our ability to design. Some of the most impressive models made using AM are those that demonstrate the capacity to fabricate complex forms in a single stage without the need to assemble or to use secondary tooling. As mentioned in Chap. 1, the WYSIWYB (What You See Is What You Build) capability allows users to consider the design with fewer concerns over how it can be built.

Virtually, every commercial solid modeling CAD system has the ability to output to an AM machine. This is because, for most cases, the only information that an AM machine requires from the CAD system is the external geometric form. There is no requirement for the machine to know how the part was modeled, any of the features or any functional elements. So long as the external geometry can be defined, the part can be built.

© Springer Nature Switzerland AG 2021
I. Gibson et al., *Additive Manufacturing Technologies*,
https://doi.org/10.1007/978-3-030-56127-7_17

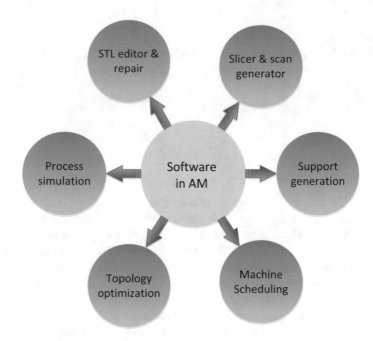

Fig. 17.1 Software classification in AM

This chapter will describe the fundamentals for creating output files for AM. It will discuss the most common technique, which is to create the STL file format, explaining how an STL file works and typical problems associated with it. Some of the numerous software tools for use with AM will be described, and possible effects of new concepts in AM on the development of associated software tools in the future will be discussed. Finally, there will be a discussion concerning the Additive Manufacturing File (AMF) format and 3MF, which have been developed to address the needs of future AM technology where more than just the geometric information may be required.

Software tools have different applications such as for converting and preparing 3D CAD data, support structure optimization, viewing of STL files, topology optimization, process simulation, distortion compensation, data repair, and more. Recently, software companies have become much more active in producing AM software for AM applications. Figure 17.1 shows various AM software and applications.

17.2   AM Software for STL Editing

Editing software for AM takes STL data, repairs errors, performs minor edits on the design, and prepares for the part to be built. Many STL editors are open source and used for different application areas such as medical image informatics, machine learning, and 3D visualization. STL editors can also be used as a geometric

Table 17.1 AM software for STL editing

No	Company name	Product name	Application
1	3D Model to print [2]	3D Model to print	Service for converting and preparing 3D CAD data for 3D printing
2	3D Slicer [3]	Slicer	Conversion of DICOM files to STL
3	3D Tool [4]	3D-Tool Free Viewer	STL viewer and checker
4	Autodesk [5]	Meshmixer	STL editor
		Netfabb	Data preparation and STL editor Data preparation, STL editor, and build packing Data preparation, STL editor, build packing, lattice structure, and topology optimization
5	CADspan [6]	CADspan Pro	Mesh resurfacing of 3D CAD data and STL editing
6	DeskArtes [7]	3Data Expert	Data preparation and STL editor
7	MakePrintable [8]	MakePrintable	STL checker, STL repair
8	Materialise [9]	MAGICS	Data preparation and STL editor with multiple modules available for other functions
		3-matic	STL texturing, remeshing, simplification, and modification
		e-Stage	Automatic support generation for Vat Photopolymerization
9	SourceForge [10]	MeshFix	STL checker and repair
10	Ultimaker [11]	Cura software	Prepare files for 3D printing
11	ViewSTL [12]	ViewSTL	Online STL viewer
12	Ansys	SpaceClaim with Additive Prep	STL creation from all major 3D CAD formats, STL repair and editing, comparison of STL with scan data, support generation, orientation guidance, preparation of files for printing

resurfacer with the ability to repair, reposition, slice, or scale. Table 17.1 shows representative AM software for STL editing. More advanced versions of these editors calculate shrinkage and produce outputs that describe the exterior of the final version of build [1].

17.2.1   Preparation of CAD Models: The STL File

The STL file is derived from the word STereoLithography and is a format created by Charles Hull for the first commercial AM process, produced by the US company 3D Systems in the late 1980s [13]. Some literature has suggested that STL stands for Stereolithography Tessellation Language or Standard Tessellation Language, but Mr. Hull has confirmed that STL was simply intended as an abbreviation for stereolithography. STL files are generated from 3D CAD data within the CAD system. The output is a boundary representation that is approximated by a mesh of triangles.

17.2.1.1   STL File Format: Binary/ASCII

STL files can be output as either binary or ASCII (text) format. The ASCII format is less common (due to the larger file sizes) but easier to understand and is generally used for illustration and teaching. Most AM systems run on PCs using Windows. The STL file is normally labeled with a ".STL" extension and may be case insensitive. A few AM systems require a different or more specific file definition, and it is becoming increasingly common for AM systems to read in native CAD formats.

STL files only show approximations of the surface or solid entities, and so any information concerning the color, material, build layers, or history is ignored during the conversion process. Furthermore, any points, lines, or curves used during the construction of the surface or solid, and not explicitly used in that solid or surface, will also be ignored.

An STL file consists of lists of triangular facets. Each triangular facet is uniquely identified by a unit normal vector and three vertices. The STL file itself holds no dimensions, so the AM machine operator must know whether the dimensions are mm, inches, or some other unit. Since each vertex also has 3 numbers, there are a total of 12 numbers to describe each triangle. The following file shows a simple ASCII STL file that describes a right-angled, triangular pyramid structure, as shown in Fig. 17.2.

Note that the file begins with an object name delimited as a solid. Triangles can be in any order, each delimited as a facet. The facet line also includes the normal vector for that triangle. Note that this normal is calculated from any convenient location on the triangle and may be from one of the vertices or from the center of the triangle. It is defined that the normal is perpendicular to the triangle and is of unit length. The normal is used to define the outside of the surface of the solid, essentially pointing away from the part. The group of three vertices defining the triangle is delimited by the terms "outer loop" and "endloop." The outside of the triangle is best defined using a right-hand rule approach. As we look at a triangle from the outside, the vertices should be listed in a counterclockwise order. Using the right hand with the thumb pointing outside the part, the other fingers curl in the direction of the order of the vertices, the starting vertex being arbitrary. When this approach is used, it avoids having to do any calculations with an additional number (i.e., the facet normal), and therefore STL files may not require the normal to prevent ambiguity.

A binary STL file can be described in the following way:

– An 80 byte ASCII header that can be used to describe the part
– A 4 byte unsigned long integer that indicates the number of facets in the object
– A list of facet records, each 50 bytes long

The facet record will be presented in the following way:

– Three floating values of 4 bytes each to describe the normal vector
– Three floating values of 4 bytes each to describe the first vertex

```
Solid triangular_pyramid
        facet normal 0.0 -1.0 0.0
            outer loop
                vertex 0.0 0.0 0.0
                vertex 1.0 0.0 0.0
                vertex 0.0 0.0 1.0
            end loop
        endfacet
        facet normal 0.0 0.0 -1.0
            outer loop
                vertex 0.0 0.0 0.0
                vertex 0.0 1.0 0.0
                vertex 1.0 0.0 0.0
            end loop
        endfacet
        facet normal 0.0 0.0 -1.0
            outer loop
                vertex 0.0 0.0 0.0
                vertex 0.0 0.0 1.0
                vertex 0.0 1.0 0.0
            end loop
        endfacet
        facet normal 0.577 0.577 0.577
            outer loop
                vertex 1.0 0.0 0.0
                vertex 0.0 1.0 0.0
                vertex 0.0 0.0 1.0
            end loop
        endfacet
        endsolid
```

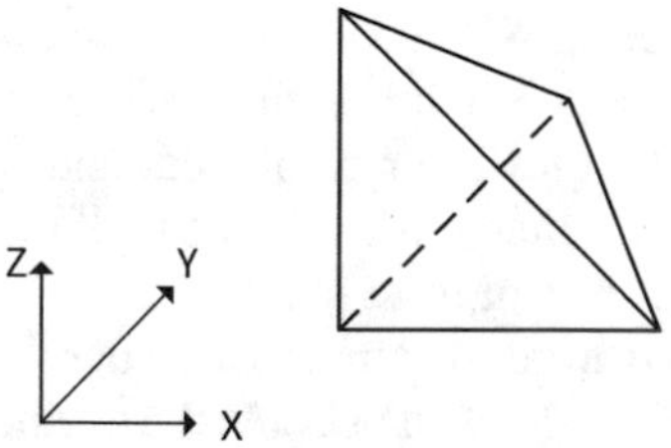

Fig. 17.2 A right-angled triangular pyramid as described by the sample STL file. Note that the bottom left-hand corner coincides with the origin and that every vertex coming out of the origin is of unit length

- Three floating values of 4 bytes each to describe the second vertex
- Three floating values of 4 bytes each to describe the third vertex
- One unsigned integer of 2 bytes, that should be zero, used for checking

17.2.1.2 Creating STL Files from a CAD System

Nearly all geometric solid modeling CAD systems can generate STL files from a valid, fully enclosed solid model. Most CAD systems can quickly tell the user if a model is not a solid. This test is particularly necessary for systems that use surface modeling techniques, where it can be possible to create an object that is not fully closed off. Such systems would be used for graphics applications where there is a need for powerful manipulation of surface detail (like with Autodesk AliasStudio software [14] or Rhino from Robert McNeel & Associates [15]) rather than for engineering detailing. Solid modeling systems, like SolidWorks, may use surface modeling as part of the construction process, but the final result is always a solid that would not require such a test.

Most CAD systems use a "Save as" or "Export" function to convert the native format into an STL file. There is typically some control over the size of the triangles to be used in the model. Since STL uses planar surfaces to approximate curved surfaces, then obviously the larger the triangles, the looser that approximation becomes. Most CAD systems do not directly limit the size of the triangles since it is also obvious that the smaller the triangle, the larger the resulting file for a given object. An effective approach would be to minimize the offset between the triangle and the surface that it is supposed to represent. A perfect cube with perfectly sharp edges and points can be represented by 12 triangles, all with an offset of 0 between the STL file and the original CAD model. However, few designs would be that convenient, and it is important to ensure a good balance between surface approximation and excessively large file. Figure 17.3 shows the effect of changing the triangle offset parameter on an STL file. The exact value of the required offset would largely depend on the resolution or accuracy of the AM process to be used. If the offset is smaller than the basic resolution of the process, then making it smaller will have no effect on the precision of the resulting model. To take advantage of the full resolution of an AM machine, the maximum triangle offset should be set to be smaller than the resolution of the machine. For instance, if an AM process operates around the 0.1 mm layer resolution, and the x/y layer creation accuracy is also around 0.1 mm, then a triangle offset of 0.05 mm or slightly lower will be acceptable for fabrication of most parts.

If distortion compensation or other mesh manipulation software is to be used on the STL file, then not only the maximum triangle offset should be less than the resolution, but all triangle edges should be no longer than the machine resolution. Take,

Fig. 17.3 An original CAD model converted into an STL file using different offset height (cusp) values, showing how the model accuracy will change according to the triangle offset

for example, a straight wall built vertically. Distortion during a build makes the straight wall curved. In order to represent that curve and compensate for that curve accurately, many triangles will be required. However, since a very large triangle can represent a straight wall with no triangle offset, then the original STL may not have enough triangles on that face to enable distortion compensation. As such, a good rule of thumb is that if mesh manipulation is desired for any reason, creation of an STL file with a maximum triangle edge length less than the resolution of the machine will give the most accurate results. The drawback of this approach is that, for large components, the STL file may be hundreds of megabytes or even many gigabytes in size and may take up considerable disk space as well as taking a long time to transfer.

17.3   AM Software for Slicing

Most AM machines have in-built slicing software. However, there are many externally sourced software that convert the digital 3D object into operational toolpath codes understood by many 3D printers. These allow part rotation to any angle, making it easy to change build orientations prior to the printing process.

The proliferation of low-cost Material Extrusion (MEX) systems has resulted in a large number of build preparation software tools for MEX machines. New innovations include features such as adaptive ("Stepover-Controlled") layer height, numerically modeled dynamic plastic deposition ("Preload"), extruder prioritizing (for handling overlapping meshes), and more. Other features of slicing software include "varying build styles per object," "loop-stitching" to optimize bonding of the infill, and "loop-modifying seam hiding." These all aim to generate operational code for more aesthetic and precise results for technically demanding prints [16, 17]. Table 17.2 lists a few of the vendors that provide slicing software for MEX machines. Dozens of software companies provide AM build preparation software that includes slicing, resulting in a rapidly changing, dynamic market. The reader is encouraged to attend a local AM tradeshow or exposition, where they will be able to see numerous packages with newly released features.

Table 17.2 A sample of MEX slicing software

No	Company name	Product name	Application
1	3D Slicer [3]	Slicer	Conversion of DICOM files to STL
2	Craft Unique [17]	CraftWare	Slicer and 3D printer host
3	KISSlicer [16]	KISSlicer	Slicer
4	Printurn [6]	Printurn	3D printer host software, slicer, and print controller
5	Repetier [18]	Repetier	Slicer, 3D printer host
6	ReplicatorG [19]	ReplicatorG	Slicer
7	Slic3r [20]	Slic3r	Slicer

17.3.1   Calculation of Each Slice Profile

Virtually, every AM system will be able to read both binary and ASCII STL files. Since most AM processes work by adding layers of material of a prescribed thickness, starting at the bottom of the part and working upward, the part file description must therefore be processed to extract the profile of each layer. Each layer can be considered a plane in a nominal XY Cartesian frame. Incremental movement for each layer can then be along the orthogonal Z axis.

The XY plane, positioned along the Z axis, can be considered as a cutting plane. Any triangle intersecting this plane can be considered to contribute to the slice profile. An algorithm like the one in Fig. 17.4 can be used to extract all the profile segments for a given STL file.

The resultant of this algorithm is a set of intersecting lines that are ordered according to the set of intersecting planes. A program that is written according to this algorithm would have a number of additional components, including a way of defining the start and end of each file and each plane. Furthermore, there would be no order to each line segment, which would be defined in terms of the XY components and indexed by the plane that corresponds to each Z value. Also, the assumption is that the STL file has an arbitrary set of triangles that are randomly distributed. It may be possible to pre-process each file so that searches can be carried out in a more efficient manner. One way to optimize the search would be to order the triangles according to the minimum Z value. A simple check for intersection of a triangle with a plane would be to check the Z value for each vertex. If the Z value of any vertex in the triangle is less than or equal to the Z value of the plane, then that triangle may intersect the plane. Using the above test, once it has been established that a triangle does not intersect with the cutting plane, then every other triangle is known to be above that triangle and therefore does not require checking. A similar check could be done with the maximum Z value of a triangle.

There are a number of discrete scenarios describing the intersection of each triangle with the cutting plane:

- All the vertices of a triangle lie above or below the intersecting plane. This triangle will not contribute to the profile on this plane.

Fig. 17.5 A vertex taken from an STL triangle projected onto the $y = 0$ plane. Since the height z_i is known, we can derive the intersection point x_i. A similar case can be done for y_i in the $x = 0$ plane

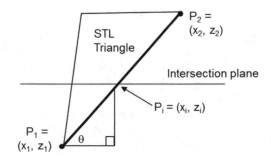

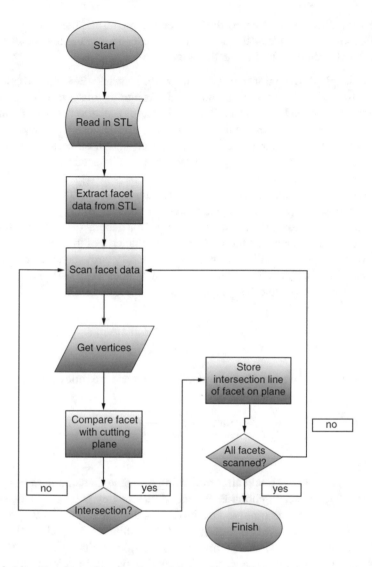

Fig. 17.4 Algorithm for testing triangles and generating line intersections. The result will be an unordered matrix of intersecting lines. (Adapted from [21])

- A single vertex directly lies on the plane. In this case, there is one intersecting point, which can be ignored, but the same vertex will be included in other triangles satisfying another condition below.
- Two vertices lie on the plane. Here one of the edges of the corresponding triangle lies on that plane and that edge contributes fully to the profile.
- Three vertices lie on the plane. In this case, the whole triangle contributes wholly to the profile, unless there are one or more adjacent triangles also lying on the plane, in which case the included edges can be ignored.

- One vertex lies above or below the intersecting plane, and the other two vertices lie on the opposite side of the plane. In this case, an intersecting vector must be calculated from the edges of the triangle.

Most triangles will conform to scenario 1 or 5. Scenarios 2–4 may be considered special cases and require special treatment. Assuming that we have performed appropriate checks and that a triangle corresponds to scenario 5, then we must take action and generate a corresponding intersecting profile vector. In this case, there will be two vectors defined by the triangle vertices, and these vectors will intersect with the cutting plane. The line connecting these two intersection points will form part of the outline for that plane.

The problem to be solved is a classical line intersection with a plane problem. In this case, the line is defined using Cartesian coordinates in (x, y, z). The plane is defined in (x, y) for a specific constant height, z. In a general case, we can therefore project the line and plane onto the $x = 0$ and $y = 0$ planes. For the $y = 0$ plane, we can obtain something similar to Fig. 17.4. Points P_1 and P_2 correspond to two points of the intersecting triangle. P_p is the projected point onto the $y = 0$ plane to form a unique right-angled triangle. The angle θ can be calculated from:

$$\tan \theta = \frac{\left(z_2 - z_1\right)}{\left(x_2 - x_1\right)} \tag{17.1}$$

Since we know the z height of the plane, we can use the following equation:

$$\tan \theta = \frac{\left(z_i - z_1\right)}{\left(x_i - x_1\right)} \tag{17.2}$$

and solve for x_i

A point y_i can also be found after projecting the same line onto the $x = 0$ plane to fully define the intersecting point P_i. A second intersecting point can be determined using another line of the triangle that intersects the plane. These two points will make up a line on the plane that forms part of the outline of the model. It is possible to determine directionality of this line by correct use of the right-hand rule, thus turning this line segment into a vector. This may be useful for determining whether a completed curve forms part of an enclosing outline or corresponds to a hole.

Once all intersecting lines have been determined according to Fig. 17.4, then these lines must be joined together to form complete curves. This would be done using an algorithm based on that described in Fig. 17.6. In this case, each line segment is tested to determine which segment is closest. A "closest point" algorithm is necessary since calculations may not exactly locate points together due to rounding errors, even though the same line would normally be used to determine the start location of one segment and the end of another. Note that this algorithm should really have further nesting to test whether a curve has been completed. If a curve is complete, then any remaining line segments would correspond to additional curves.

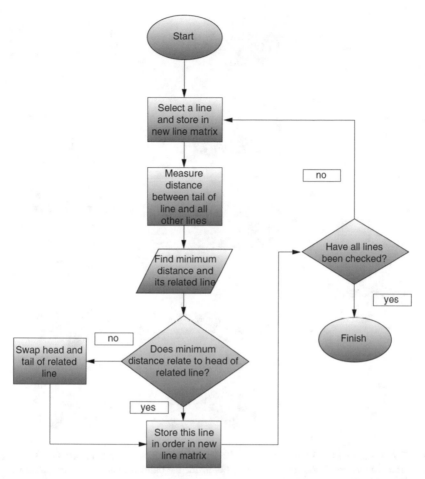

Fig. 17.6 Algorithm for ordering the line intersections into complete outlines. This assumes there is only a single contour in each plane. (Adapted from [21])

These additional curves could form a nest of curves lying inside or outside others, or they could be separate. The two algorithms mentioned here focus on the intersection of triangular facets with the cutting plane. A further development of these algorithms could be to use the normal vectors of each triangle. In this way, it would be possible to establish the external direction of a curve. This would be helpful in determining nested curves. The outermost curve will be pointing outside the part. If a curve set is pointing inward on itself, then it is clear there must be a further curve enveloping this one (*see* Fig. 17.3). Use of the normal vectors may also be helpful in organizing curve sets that are in very close proximity to each other.

Once this stage has been completed, there will be a file containing an ordered set of vectors that will trace complete outlines corresponding to the intersecting plane. How these outlines are used depends somewhat on which AM technology is to be used. Many machines can use the vectors generated in Fig. 17.6 to control a plotting

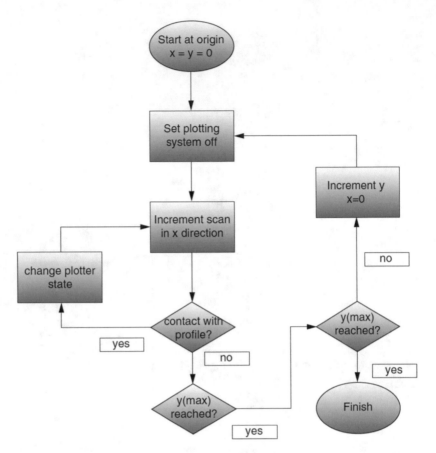

Fig. 17.7 Algorithm for filling in a 2D profile based on vectors generated using Fig. **17.6** and a raster scanning approach. Assume the profile fits inside the build volume, the raster scans in the X direction and lines increment in the Y direction

process to draw the outlines of each layer, such as in Sheet Lamination (SHL). However, most AM machines would also need to fill in these outlines to make a solid. Figure 17.7 uses an inside/outside algorithm to determine when to switch on a filling mechanism to draw scanning lines perpendicular to one of the planar axes. The assumption is that the part is fully enclosed inside the build envelope and therefore the default fill is switched off.

17.3.2 Technology-Specific Elements

Figures 17.4, 17.6, and 17.7 are basic algorithms that are generic in nature. These algorithms need to be refined to prevent errors and to tailor them to suit a particular process. Other refinements may be employed to speed the slicing process up by eliminating redundancy, for example.

As mentioned in previous chapters, many AM systems require parts built using support structures. Supports are normally a lower-density pattern of material placed below the region to be supported. Such a pattern could be a simple square pattern or something more complex like a hexagonal or even a fractal mesh. Furthermore, the pattern could be connected to the part with a tapered region that may be more convenient to remove when compared with thicker connecting edges.

Determination of the regions to be supported can be made by analyzing the angle of the triangle normals. Those normals that are pointing downward at some previously defined minimum angle would require supports. Those triangles that are sloping above that angle would not require supports. Supports are extended until they intersect either with the base platform or another upward-facing surface of the part. Supports connecting to the upward-facing surface may also have a taper that enables easy removal. A technique that is commonly used is to extend supports from the entire build platform and eliminate any supports that do not intersect with the part at the minimum angle or less (*see* Fig. 17.8).

Support structures can be generated directly as STL models and can be incorporated into the slicing algorithms already mentioned. Many times, particularly for MEX processes, they are directly generated by a proprietary algorithm within the slicing process. Some other processing requirements that would be dependent for different AM technologies include:

- *Raster scanning*: While many technologies would use a simple raster scan for each layer, there are alternatives. Some systems use a switchable raster scan, scanning in the X direction of an XY plane for one layer and then moving to the Y direction for alternate layers. As discussed in Chap. 5, some systems subdivide the fill area into smaller regions (e.g., squares or stripes) and use rotating raster scans between regions.
- *Patterned vector scanning*: MEX technology requires a fill pattern to be generated within an enclosed boundary. This is done using vectors generated using a patterning strategy. For a particular layer, a pattern would be determined by choosing a specific angle for the vectors to travel. The fill is then a zigzag pattern along the direction defined by this angle. Once a zigzag has reached an end, there may be a need for further zigzag fills to complete a layer (*see* Fig. 17.9 for an example of a zigzag scan pattern).

Fig. 17.8 Supports
generated for a part build

– *Hatching patterns:* SHL processes like the now discontinued Helisys LOM and
the Solid Centre machine from Kira require material surrounding the part to be
hatched with a pattern that allows it to be de-cubed once the part has been completed (*see* Fig. 17.10).

17.4   AM Software for STL Manipulation

Most CAD packages allow users to develop complex forms and convert them to
STL. There are a dozens, if not hundreds, of STL viewers available, often as a free
download. As an example, consider STL view from Simplify3D [22] (see Fig. 17.11).
This software supports many types of MEX 3D printers. Simplify3D includes pre-
print animations that show the exact actions of a 3D printer before starting the print.
These animations show the extruder as it lays down each individual line of feedstock. Simplify3D helps the user select speed, sequence, and setting for a specific
CAD design. Other features of Simplify3D include support structure options to
obtain a good surface quality and the ability to customize supports to reduce waste
and time during production. These Simplify3D features are a subset of the overall
build preparation functions available for MEX. For other AM processes, as well as
for MEX, dozens of companies sell software for STL manipulation, slicing, support
generation, and other functions specifically tailored for each type of AM process.

Many AM machine manufacturers offer viewers within build preparation software distributed with their machines. These machine-specific tools help users
reduce errors in data transfer, either from incorrect STL conversion or from wrong
interpretation of the design intent.

Fig. 17.9 A scan pattern
using vector scanning in
MEX. Note the outline
drawn first followed by a
small number of zigzag
patterns to fill in the space

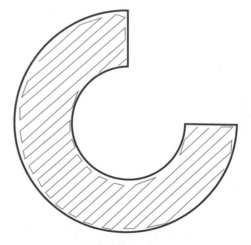

Fig. 17.10 Hatching
pattern for SHL processes.
Note the outside hatch
pattern that will result in
cubes which will be
separated from the solid
part during post-processing

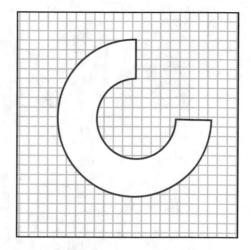

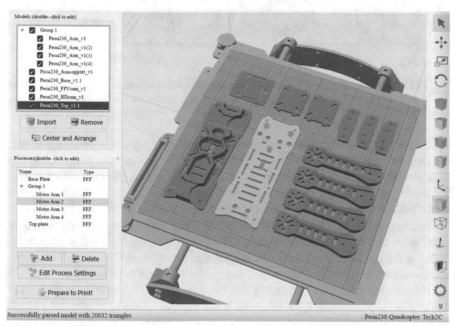

Fig. 17.11 Customize your supports for a better print from Simplify3D. (Photo courtesy of Simplify3D) [22]

17.4.1   STL File Manipulation

Once a part has been converted into STL, all parametric relationships and feature identities are lost. As such, most of the CAD-based functions for model manipulation are no longer available to the user. This is because triangle-based definitions are only collections of points, vertices, and normal vectors. Consider the modeling of a simple geometry, like the cut cylinder in Fig. 17.12a. Making a minor change in one of the measurements may result in a very radical change in distribution of triangles.

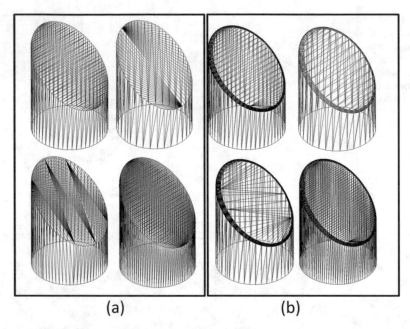

Fig. 17.12 STL files of a cut cylinder. Note that although the two models in (**a**) are very similar, the location of the triangles is very different. Addition of a simple filet in (**b**) shows even greater change in the STL file

While it is possible to simplify the model by reducing the number of triangles, it is quite easy to see that defining boundaries in most models cannot be easily done. The addition of a fillet in Fig. 17.12b shows an even more radical change in the STL file. Furthermore, if one were to attempt to move the oval that represents the cut surface, the triangles representing the filet would no longer show a constant radius curve.

Building models using AM is often done by people working in departments or companies that are separate from the original designers. It may be that whoever is building the model may not have direct access to the original CAD data. There may therefore be a need to modify the STL data before the part is to be built.

The market leader in software for STL file manipulation is MAGICS from Materialise [23]. MAGICS, for example, has a number of modules useful for modifying STL files, including:

- Checking the integrity of STL files based on the problems described below.
- Incorporating support structures including tapered features on the supports that may make them easier to remove.
- Optimizing the use of AM machines, like ensuring the machine build envelope is efficiently packed with parts.
- Automated support structure placement for many machine types.
- Adding in features like serial numbers and identifying marks onto the parts to ensure correct identification, easy assembly, etc.
- Remeshing STL files that may have been created using reverse engineering software or other non-CAD based systems. Such files may be excessively large and can often be reduced in size without compromising the part accuracy.
- Segmenting large models into multiple parts.

- Combining multiple STL files into a single model.
- Performing Boolean tasks like subtracting one model data from another model data and/or adding and subtracting simple features, like cylinders, holes, straight cuts, etc. from a model to create a modified model.

Many commercially available software tools for STL file manipulation have similar modules to those listed above.

One useful tool for STL files is to explore collisions and/or intersections between two models. Mesh analysis is used, for instance, to determine the difference between the original surface information from the original CAD model and scan data from an as-produced part. The surface area, bounding box, and error values can be determined and returned to the user to help them understand, both quantitatively and qualitatively, how closely the scanned part corresponds to the original CAD model.

17.4.2   Mesh Healing

Mesh healing is repairing mesh representations of solids, including filling holes, fixing self-intersections, and deleting redundant shells. An STL file is a surface mesh, and so algorithms for modifying and fixing meshing errors (see Sect. 17.5) are commonly applied to modify and fix STL files. Many software companies, for instance, Polygonica, have algorithms for automatic hole-closing to provide users with finer control over how holes are closed. The curvature of the model close to the boundary of the hole can be used to produce smooth hole-closing. This procedure produces a much cleaner model than simple planar closing and is useful for models that have been generated from scan data.

17.4.2.1   Remeshing

A number of robust algorithms have been developed for the finite element method (FEM) and computational fluid dynamics (CFD) to enable remeshing to improve the quality of a mesh. Remeshing is commonly done to remove long and thin primitives (e.g., triangles, quadrilaterals, etc.) in order to regularize mesh size and shape. Remeshing can also simplify a solid by reducing the number of primitives in the mesh. Using remeshing software provides primitives with better aspect ratio while maintaining the overall shape and features of the original model. Uniform aspect ratios are important when performing certain types of surface offsets, such as distortion compensation. In addition, the file size can be decreased by reducing polygon count to a requested number while maintaining the shape of a model within a specified dimensional tolerance. Robust remeshing capabilities are available in leading STL file repair and manipulation software.

17.4.3   Surface Offsetting

Offsetting is a mechanism where a solid, surface, or cross-section is expanded or contracted. In AM of complex shapes, the thickness of a model can be variable, which affects the weight and the amount of material used in a region, resulting in differential distortion and shrinkage. Offsetting is useful in AM to move lines or surfaces to account for distortion or to account for process-specific features such as the thickness of a deposition road, laser diameter, melt pool size, or other machine-specific characteristic. Offsetting surfaces to account for machine deposition and bonding characteristics is critically important to achieving high accuracy components. For 3D scan data, offsetting can be used to fill in missing scan data or to add thickness to a surface in order to convert it to a solid model.

Many types of offsets are possible. If the offset has a constant distance in all directions, it is called a spherical offset. Alternatively, a solid may be offset by another solid, and in this case, the offset distance may vary according to the normal direction of the surface. Distortion compensation applies an offset to each area of a surface by a magnitude and direction determined by simulation of the distortion that occurs in the build process. Depending on the algorithm used for surface offsetting, the results can be quite error-prone, particularly if the surface offset magnitude approaches the size of the smallest part feature. As such, surface offsetting may have unintended consequences, and users are wise to carefully look at visualizations of their part after surface offsetting to ensure that the intended behavior has occurred.

17.4.4   STL Manipulation on the AM Machine

STL data for a part consist of a set of points defined in space, based on an arbitrarily selected point of origin. This origin point may not be appropriate to the machine the part is to be built on. Furthermore, even if the part is correctly defined within the machine space, the user may wish to move the part to some other location or to make a duplicate to be built beside the original part. Other tasks, like scaling, changing orientation, and merging with other STL files, are all things that are routinely done using the STL manipulation tools on the AM machine.

Creation of support structures may also be done on the AM machine. With some AM operating systems, there is little or no control over placement of supports or manipulation of the model STL data. Considering Fig. 17.8 again, it may be possible to build the handle feature without so many supports, or even with no supports at all. Distortion around the upper part of the handle would likely occur without supports in many AM machines, but the user may prefer this to having to clean up the model to remove the support material inside the handle.

17.5   Problems with STL Files

Although the STL format is quite simple, there can be errors in files resulting from CAD conversion. The following are typical problems that occur with STL files:

Unit changing: This is a common problem since STL files do not store any unit information. Since US machines commonly use imperial measurements and most of the rest of the world uses metric, some files can appear scaled because there is no explicit mention of the units used in the STL format. If the part STL file was created assuming millimeters and the part was read into a machine assuming inches, the part will be approximately 25 times too large. Furthermore, units are also important when deciding the location of the part with respect to the origin within the machine to be used. For most machines, the assumed origin of the machine lies in the bottom left-hand corner, and so all triangle coordinates within an STL file must be positive. However, this may not be the case for a particular part made in the CAD system, and so some linear offset of the STL file may be required. For machines which define the origin as the center of the machine, the STL file vertex location can be both positive and negative, complicating things even further.

Vertex-to-vertex rule: Many slicing algorithms assume each triangle must share two of its vertices with each of the triangles adjacent to it. This means that a vertex should not intersect the side of another triangle, like that shown in Fig. 17.13. This is not explicitly stated in the STL file description, and therefore STL file generation may not adhere to this rule. However, a number of checks can be made on the file to determine whether this rule has been violated. For example, the number of faces of a proper solid defined using STL must be an even number. Furthermore, the number of edges must be divisible by 3 and follow the equation:

$$\frac{\text{No. of faces}}{\text{No. of edges}} = \frac{3}{2} \qquad (17.3)$$

Leaking STL files: As mentioned earlier, STL files should describe fully enclosed surfaces that represent the solids generated within the originating CAD system. In other words, STL data files should construct one or more manifold entities according to Euler's rule for solids:

$$\text{No. of faces} - \text{No. of edges} + \text{No. of vertices} = 2 \times \text{No. of bodies} \qquad (17.4)$$

If this rule does not hold, then the STL file is said to be leaking, and the file slices will not represent the actual model. There may be too few or too many vectors for a particular slice. Slicing software may add in extra vectors to close the outline or it may just ignore the extra vectors. Small defects can possibly be ignored in this way. Large leaks may result in unacceptable final models.

Fig. 17.13 A case that violates the vertex-to-vertex rule

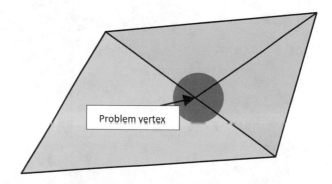

Problem vertex

Leaks can be generated by facets crossing each other in 3D space as shown in Fig. 17.14. This can result from poorly generated CAD models, particularly those that do not use Boolean operations when generating solids.

A CAD model may also be generated using a method which stitches together surface patches. If the triangulated edges of two surface patches do not match up with each other, then holes, like in Fig. 17.15, may occur.

Degenerated facets: A triangle may be so small that all three points virtually coincide with each other. After truncation, these points lay on top of each other causing a triangle with no area, called a degenerated facet. This can also occur when a truncated triangle returns no height and all three vertices of the triangle lie on a single straight line. While the resulting slicing algorithm will not cause incorrect slices, there may be some difficulties with any checking algorithms, and so such triangles should really be removed from the STL file.

It is worth mentioning that, while a few errors may creep into some STL files, most professional 3D CAD systems today produce high-quality and error-free results. In the past, problems more commonly occurred from surface modeling systems, which are now becoming scarcer, even in fields outside of engineering CAD like computer graphics and 3D gaming software. Also, in earlier systems, STL generation was not properly checked, and faults were not detected within the CAD system. Today, potential problems are better understood, and there are well-known algorithms for detecting and correcting such problems. However, the recent surge in home-use 3D printers has resulted in a large selection of software routines that are freely available but not thoroughly tested and could suffer from the problems described in this chapter.

Fig. 17.14 Two triangles intersecting each other in 3D space

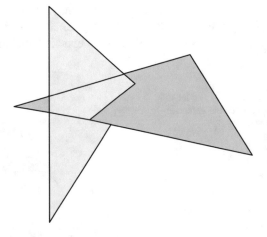

Fig. 17.15 Two surface patches that do not match up with each other, resulting in holes

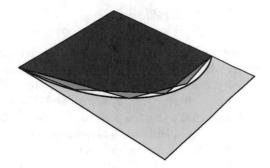

17.6 Beyond the STL File

The STL definition was created in the 1980s and has served the AM industry well. However, there are other ways in which files can be used for build preparation. Furthermore, the fact that the STL file only represents the surface geometry may cause problems for parts that require some heterogeneous content. This section discusses some of the issues surrounding the limitations of STL and methods for overcoming these limitations.

17.6.1 Direct Slicing of the CAD Model

Since generation of STL files may be tedious and introduce errors and approximations, there is increasing interest in using CAD tools to directly generate slice data for AM machines. It is a trivial task for most 3D solid modeling CAD systems to calculate the intersection of a plane with a model, thus extracting a slice. This slice data would ordinarily need to be processed to suit the drive system of the AM technology, but may be handled in most CAD systems with the use of macros. Support structures can be generated using standard geometry specifications and projected onto the part from a virtual representation of the AM machine build platform.

For the first 30 years of the AM industry, this direct slicing in CAD approach was primarily investigated as a research topic [24]. The major barrier to using this approach is that every CAD system must include a suite of algorithms for direct slicing relevant to a variety of machines or technologies. This is cumbersome and may require periodic updates as new machines become available. However, the rapid growth of the AM industry over the past decade has caused many CAD companies to rethink their strategy, and a growing number of CAD vendors now include options for slicing and generating output data for AM machines directly in the CAD systems. Nevertheless, many companies still separate designers from manufacturing engineers, or they outsource production to third-party service providers. As a result, a method for transferring data between CAD and AM machines in a neutral format is still helpful.

17.6.2 Color Models

Currently, several AM technologies are available on the market that can produce parts with color variations, including full-color output. Color machines include color Binder Jetting (BJT) technology from 3D Systems, Material Jetting (MJT) machines from 3D Systems and Stratasys, and SHL systems from CleanGreen3D. Colored parts have proven to be very popular, and it is likely that other color AM machines will make their way to the market. The conventional STL file contains no information pertaining to the color of the part or any features thereon. Coloring of STL files is possible, and there are in fact color STL file definitions available [25] but are limited by the fact that a single triangle can only be one specific color. It is therefore much better to use something like VRML painting options that allow you to assign bitmap images to individual facets [25]. In such a way, it is possible to take advantage of the color possibilities that the AM machine can give you.

17.6.3 Multiple Materials

Color is one of the simplest examples of multiple material products that AM is capable of producing. Multi-material parts can be made in many types of AM processes, from composite materials with varying levels of porosity to regions containing discretely different materials. For many of these AM technologies, STL is an impediment. Since the STL definition is for surface data only, the assumption is that the solid material between these surfaces is homogeneous. While there has been significant thought applied to the problem of representations for heterogeneous solid modeling [26], there is still much to be considered before we can arrive at a standard to supersede STL for multi-material AM technology.

17.6.4 Use of STL for Machining

STL is used for applications beyond just converting CAD to Additive Manufacturing input. Reverse engineering packages commonly convert point cloud data directly into STL files without the need for CAD. Such technology connected directly to AM forms the basis for a 3D fax machine. Another technology that can easily make use of STL files is subtractive manufacturing.

Subtractive manufacturing systems can readily make use of the surface data represented in STL to determine the boundaries for machining. With some additional knowledge concerning the dimensions of the starting block of material, tool, machining center, etc., it is possible to calculate machining strategies. Delft Spline [27] has been using STL files to create machine toolpath profiles for a number of years. Figure 17.16 shows the progression of a model through to a tool to manufacture a final product using their DeskProto software. Another technology that uses a hybrid of subtractive and additive processes to fabricate parts is the SRP (Subtractive Rapid Prototyping) technique developed by Roland [28] for their desktop milling machines.

17.7   AM Software for Process Visualization and Collision Detection

To better understand the effects of scan/deposition strategy on part formation, many slicing software tools give users the ability to visualize and modify specific scan/deposition paths. This is useful to understand and control build errors and to ensure that critical features, particularly thin walls and tiny features, will be built with an appropriate strategy. In some cases, the build process can be animated so that a user can "watch" the build occur prior to starting a machine. In some tools, rendered images can give an understanding of the surface roughness for a given layer thickness, for example [29].

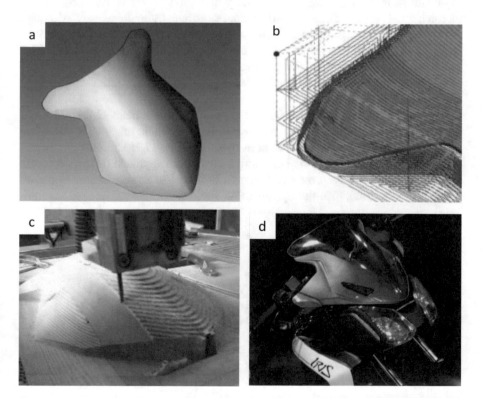

Fig. 17.16 DeskProto software being used to derive machine toolpaths from STL file data to form a mold for creating the windscreen of a motorcycle

Collision and interference avoidance and detection is becoming increasingly important for AM, particularly as Hybrid and multi-step AM processes become broadly adopted. Both static and dynamic analyses are useful to predict interference and detect collisions between various additive functions (extrusion, printing, etc.)_ and subtractive functions (milling, planning, etc.). To increase productivity, multiple additive and/or subtractive actions should occur simultaneously, thus increasing chance for collision and the need for collision detection.

17.8   AM Software for Topology Optimization

Topology optimization (TO) aims at searching for the optimal layout of material within a design domain for a given set of boundary conditions. This results in material distribution which satisfies a set of performance requirements. Topological optimization is an iterative process that helps to improve the design each iteration. The most common type of TO is structural optimization, which aims to minimize part mass while meeting part performance characteristics. An area of TO that is emerging from research into commercial products is multi-physics TO, which enables minimum weight components which satisfy multiple types of constraints, such as thermal, fluids, structural, and manufacturing constraints, simultaneously. Although TO can be used to design optimum structures in general, the complex designs resulting from TO may only be producible using AM, due to its ability to produce lattice and complex structures in a single production step.

Some commercial software that implement topology optimization are listed in Table 17.3.

Topology optimization techniques are easy to perform in many CAD and CAM tools. Topology optimization can be divided into two main categories: truss-like structures (see Fig. 17.17) and continuum structures (Fig. 17.18). Figure 17.18b and c shows topology-optimized frames using a continuum approach for a motorcycle and bicycle, respectively. A method that marries truss-like and continuum optimum is lattice optimization. In lattice optimization, for instance, TO can be used to return

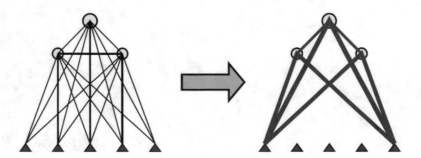

Fig. 17.17 Topology optimization of truss-like structure

Table 17.3 AM software for topology optimization

No	Company name	Product name	Application
1	Abaqus [30]	Abaqus	Topology optimization
2	Ameba [31]	Ameba	Topology optimization
3	Altair Engineering [32]	solidThinking Inspire	Topology optimization
		HyperWorks	HyperWorks
4	Autodesk [33]	Within Medical Autodesk	Lattice structure for orthopedic industry Porous coatings for implants
		Fusion 360	Topology optimization
5	Dassault Systèmes [34]	Tosca Structure	Topology optimization for FEA packages including Abaqus, Ansys, and MSC Nastran
6	Desktop Metals [35]	Live Parts	Generative design and topology optimization
7	DTU [36]	TopOpt	Topology optimization
8	Siemens [37]	NX	High-end CAD that integrates topology optimization, lattice structures, and support generation
9	Topostruct [38]	Topostruct	2D and 3D topology optimization
10	Ansys	Discovery Live	Real-time multi-physics topological optimization using SIMP and level set techniques that include the ability to use AM-specific overhang and orientation constraints. Also includes options for lattice optimization

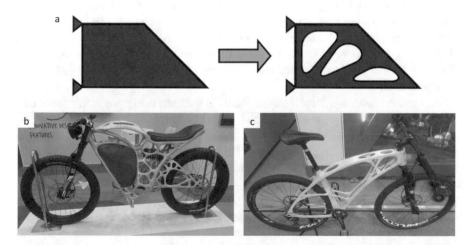

Fig. 17.18 (**a**) Topology optimization of continuum structure, (**b**) topology optimized motorcycle frame by APWORKS, (**c**) 3D printed bicycle frame with integrated endless fiber reinforcement by Fraunhofer

an optimum part density in each location. This local density result is then converted into a local thickness of a lattice truss for the user-selected lattice type.

A range of numerical approaches can be used to implement optimization, such as the homogenization method, Solid Isotropic Microstructure with Penalization (SIMP) method, level set methods, and evolutionary structural optimization algorithms. These algorithms have been successfully used in a wide range of engineering applications ranging from nanophotonics design to structural design for aircraft and aerospace components.

17.9   AM Software for Modeling and Simulation

It is becoming increasingly common to perform analyses on AM-designed components as well as to predict physics which occur during the AM process itself. Most of these analyses make use of the finite element method (FEM) and are known as finite element analyses (FEA).

FEA techniques are increasingly popular tools to predict how the outcome of various manufacturing processes changes with changing process parameters, geometry, and/or material. Commercial software packages provided by SYSWELD, COMSOL, Autodesk, Ansys, Siemens, Dassault, DEFORM, MSC, etc. are used to predict the outcomes of many manufacturing processes, including welding, forming, molding, casting, and AM processes. In some cases, existing manufacturing simulation software can give insight into AM processes. More recently, however, commercial focus has been on tools which are developed specifically for AM. Most of these commercial tools are focused on metal AM, particularly Powder Bed Fusion (PBF), Directed Energy Deposition (DED), and BJT. In 2020, at the time of this book, by far the most mature process simulation tools for AM are for metal laser PBF processes.

AM technologies are particularly difficult to simulate using predictive finite element tools. For instance, the multi-scale nature of metal PBF approaches such as metal laser sintering and electron beam melting is incredibly time-consuming to accurately simulate using physics-based FEA. AM processes are inherently multi-scale in nature, and fine-scale finite element meshes that are 10 microns or smaller in size are required to accurately capture the solidification physics around the melt pool, while the overall part size can be 10,000 times larger than the element size. In three dimensions, this means that if we apply a uniform 10 micron mesh size, we would need 10^8 elements in the first layer and more than 10^{12} elements in total to capture the physics for a single part that fills much of a powder bed. Recently, Laser-Based Powder Bed Fusion (LB-PBF) manufacturers have increased the size of the build chambers and parts, for instance, GE has Project A.T.L.A.S. with $1200 \times 1200 \times 1200$ mm. This makes the problem even more difficult to solve. Since rapid movement of a point heat source is used to create parts, capturing the physics requires a time step of 10 microseconds or less during laser/electron beam melting, which for a complete build would require more than 10^{10} total time steps.

To solve a problem with this number of elements for this many time steps on a relatively high-speed supercomputer would take billions of years. Thus, to date all AM simulation tools which seek to do multi-physics, fully couple thermo−/mechanical simulations of AM processes are limited to predictions of only a small fraction of a part or very small geometries. In order to simulate full-sized components, simplified assumptions, average behavior, or decoupled analyses which make use of periodic behavior are necessary.

Several FEA tools have been developed which make use of average assumptions, whereby they cut and paste solutions from simplified simulations or experiments to form a solution for large, complex geometries. The most common approach is through the use of "inherent strain" where an average strain (e.g., shrinkage effect) is either measured or simulated and then applied layer-by-layer to each cross-section of material being deposited. These simulation layers are typically significantly larger than the physical build layer. By accumulating this strain layer-by-layer, the distortion and residual stress for a real part can be approximated quickly. The smaller the layer thickness, the more accurate the approximation, but the slower the simulation speed. This approach has the benefit of faster solution time; however, for large, complex geometries, particularly geometries where there are multiple thick and thin regions or where multiple scan strategies are applied in the same part, these types of predictions fail to accurately capture the effects of changing scan patterns, complex accumulation of residual stresses, and localized thermal characteristics. In addition, minor changes to input conditions can make the simplified solutions invalid, requiring a new experiment or simulation to find the new inherent strain.

Another use of FEA simulation is preciting the effects of process parameters on melt pool dimensions, percent porosity, and microstructure. In LB-PBF processes, microstructures are always different from traditionally manufactured microstructures and vary based upon process parameters. To do a full thermal and microstructural simulation for a full-scale part, the problem size explodes, as described above. For this reason, it is common to decouple the problem; solve for the thermal gradients, solidification rates, and melt pool dimensions separately from the grain growth simulation; and then use periodical thermal history in the scanning to approximate the microstructure in a reasonable amount of time.

Capturing every nuance of process variations using simulation is incredibly difficult. For instance, the interface between the build plate and the first few layers of the samples has an increased cooling rate compared to higher layers, increasing the chance of distortion and cracking due to the high cooling rate. This phenomenon affects geometric features differently, depending upon sample orientation and contact area between samples and substrate. (This is one reason it is more reproducible to create parts on supports, so that the cooling rate is more consistent throughout the part.) When changing the cross-section, due to different cooling rates and stress states, the chance of cracks, keyholes, and pores increases. At the interface between supports and the part, different heat transfer scenarios lead to increases in the propensity for cracks, as can be seen in Fig. 17.19. These variations in thermal history, stress, and distortion make a general-purpose simulation infrastructure for any arbitrary geometry, input condition, and scan pattern a difficult endeavor.

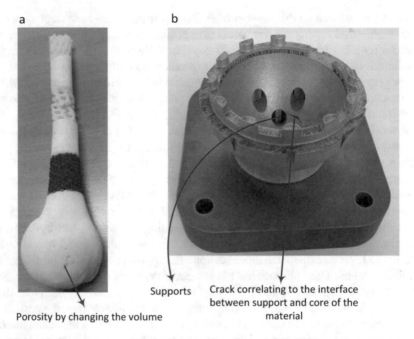

a

b

Supports   Crack correlating to the interface
            between support and core of the
Porosity by changing the volume   material

Fig. 17.19 Crack and porosity for (**a**) MEX simulated bone, (**b**) LB-PBF acetabular hip joint

Over the past decade, researchers have developed dynamic, multi-scale moving meshes to accelerate FEA analysis for AM [39–42]. These types of multi-scale simulations are many orders of magnitude faster than standard FEA simulations. However, multi-scale simulations alone are still too slow to enable a complete part simulation, even on the world's fastest supercomputers. Thus, new computational approaches for AM include not only multi-scale moving meshes but also novel numerical and periodic approaches. These approaches, commercialized as part of the Ansys Additive Suite, have reduced the solution time for large-scale AM problems by many orders of magnitude. As a result, AM simulation tools for predicting the effects of process parameter changes, predicting distortion for purposes of distortion compensation, and predicting common failure mechanisms like cracking, blade crash, and built-up-edge are proving to be invaluable tools for increasing machine productivity, improving part accuracy, tailoring part properties, and avoiding build failures.

17.10   Manufacturing Execution System Software for AM

AM manufacturing execution system (MES) software handles management of the chain of process steps needed to create parts in a commercial environment. This is particularly important for companies that are running many AM machines with different configurations. A typical MES tool [43] is one which has the following features:

   I. Import of single or multiple files in AMF, STL OBJ, G-code, and/or native CAD format, with cloud and network storage options.

  II. Slicing of 3D models into layers and conversion into operational code for 3D printing.

 III. Repairing, analyzing, and rotating files for specified printers.

 IV. STL move, rotate, and scale operations, including automatic positioning of the models on a build platform to fit the specified printer.

  V. Creation of the shell of an object to remove interior material in cases where material properties are unimportant and only the shape of the model is needed, significantly reducing the material and build time.

 VI. Distribution of files or build packages to different 3D printers in the production line with queuing or immediate print privileges.

VII. Observation of the generated operational codes and layer-by-layer breakdown of build speed. Also observation of the complete slicing settings and parameters to keep track of builds and ensure every print is correct. This option can also be used by administrators to visualize if files are printable before running jobs.

VIII. Security coding on STL files.

 IX. Creation of models with lower polygon counts and optimizing of polygon structure.

  X. Managing the workflow for the different machines that are part of the manufacturing process by using a virtual factory.

MES software also enables tracking of files through the entire workflow, including data on materials, operator, and other factors that may be used to determine build quality and potential sources of errors. In addition to the features described above, AM software can include one or more of the following functions, which are also a part of any comprehensive MES system:

Build time estimation: AM is a highly automated process, and the latest machines are very reliable and can operate unattended for long periods of time. For effective process planning, it is important to know when a build is going to be completed. Knowing this will help determine when operators will be required to change over jobs. Good estimates will also help to balance builds; adding or subtracting a part from the job batch may ensure that the machine cycle will complete within a day shift, for example, making it possible to keep machines running unattended at night. Also, if you are running multiple machines, it may be helpful to stagger the builds throughout the shift to optimize the manual work required. Early build time estimation software was extremely unreliable, performing rolling calculations of the average build time per layer. Since the layer time is dependent on the part geometry, such estimates could be imprecise and vary wildly, especially at the beginning of a build. Later software versions saw the benefit in having more precise build time estimations. A simplified build time model is discussed in Chap. 18.

Monitoring: This is a relatively new feature for most AM systems. Even though nearly every AM machine will be connected either directly or indirectly to the Internet, this has traditionally been for uploading of model files for building. Export of information from the machine to the Internet or within an Intranet has not been common except in the larger, more expensive machines. The simplest monitoring systems provide basic information concerning the status of the build, access to build log information, and estimates for how much longer before the build is complete. More complex systems may tell you about how much material is remaining; the current status parameters like temperatures, laser powers, etc.; and whether there is any need for manual intervention through an alerting system. Some monitoring systems may also provide video feedback and photographs of each layer of the build.

Planning: Having a predictive animation of the AM process running on a separate computer may be helpful to those working in process planning. Process planners may be able to determine what a build could look like, thus allowing the possibility of planning for new jobs, variability analysis, or quoting.

Although many AM system vendors offer one or more of these MES features with their machines, third-party software is more comprehensive and can be modified to suit the exact requirements of the user. A large number of companies now offer MES software for AM. It is likely that many of these will be acquired, merge, or leave the market over the next few years. As a result, the reader should do an Internet search on MES for AM to determine current commercial offerings and which software tool(s) are most appropriate for their needs.

17.11   The Additive Manufacturing File (AMF) Format

As AM technologies move forward to include multiple materials, lattice structures, textured surfaces, etc., it is likely that an alternative format to STL will be required. The ASTM Committee F42 on Additive Manufacturing Technologies released the ASTM 2915–12 AMF Standard Specification for Additive Manufacturing File Format 1.1 in May 2011 [44], which has gone through several subsequent iterations. This file format is still under development but has already been implemented in some commercial machines and software tools. Considerably more complex than the STL format, AMF aims to embrace a whole host of new part descriptions that have hindered the development of current AM technologies. The 3MF consortium, a collection of several software and hardware companies involved in AM, has also published a standard similar to AMF. ASTM and the 3MF consortium started working together in 2016 to coordinate development of these neutral file formats. The features of AMF and 3MF include:

Curved triangles: In STL, the surface normal lies on the same plane as the triangle vertices that it is connected to. However, in AMF, the start location of the normal vector does not have to lie on the same plane. If so, the corresponding triangle must be curved. The definition of curvature is such that all triangle edges meeting at that vertex are curved so that they are perpendicular to that normal and in the plane defined by the normal and the original straight edge (i.e., if the original triangle had straight edges rather than curved ones). By specifying the triangles in this way, many fewer triangles need be used for a typical CAD model. This addresses problems associated with large STL files resulting from complex geometry models for high-resolution systems. The curved triangle approach is still an approximation since the degree of curvature cannot be too high. Overall accuracy is significantly improved in terms of cusp height deviation.

Color: Color can be assigned in a nested way so that the main body of the part can be colored according to a function within the original design. Red, green, and blue coloration can be applied along with a transparency value to vertices, triangles, volumes, objects, or materials. Note that many AM processes, like Vat Photopolymerization (VPP), can make clear parts, so the transparency value can be an effective parameter. Color values may work along with other materials-based parameters to provide a versatile way of controlling an AM process.

Texture: The above color assignment cannot deal directly with image data assigned to objects. This can however be achieved using the texture operator. Texture is assigned first geometrically, by scaling it to the feature so that individual pixels apply to the object in a uniform way. These pixels will have intensity, which are then assigned color. It should be noted that this is an image texturing process similar to computer graphics rather than a physical texture, like ridges or dimples.

Material: Different volumes can be assigned to be made using different materials. Several MJT, BJT, DED, and MEX systems have the capacity to build multiple material parts. When using STL, designing parts requires a tedious redefinition process within the machine's operating system. By having a material definition within AMF, it is possible to carry this all the way through from the design stage.

Material variants: AMF operators can be used to modify the basic structure of the part to be fabricated. For example, many medical and aerospace applications may require a lattice or porous structure. An operator can be applied so that a specified volume can be constructed using an internal lattice structure or porous material. Furthermore, some AM technologies would be capable of making parts from materials that gradually blend with others. A periodic operator can be applied to a surface that will turn it into a physical texture, rather than the color mapping mentioned earlier. It is even possible to apply a random operator to provide unusual effects to the AM part.

It should be noted that a part designed and coded using AMF will almost certainly look differently when built using different machines. This will be particularly so for parts that are coded according to different materials, colors, and textures. Each machine will have the capacity to accept and interpret the AMF design according to functionality. For example, if a part is defined with a fine texture, a lower-resolution process will not be able to apply it so well. Opaque materials will not be able to make much use of the transparency function. Some machines will not be able to create parts with multiple materials and so on. It should also be noted that machines should all be able to accept the geometry definition and make something of the AMF defined part. AMF is therefore backward compatible so that it can recognize a simple STL file, but with the capacity to specify any conceivable design in the future.

17.12   Questions

1. How would you adjust Fig. 17.6 to include multiple contours?
2. Under what circumstances might you want to merge more than one STL file together?
3. Write out an ASCII STL file for a perfect cube, aligned with the Cartesian coordinate frame, starting at (0, 0, 0) and all dimensions positive. Model the same cube in a CAD system. Does it make the same STL file? What happens when you make slight changes to the CAD design?
4. Why might it be possible that a part could inadvertently be built 25 times too small or too large in any one direction?
5. Is it okay to ignore the vertex of a triangle that lies directly on an intersecting cutting plane?
6. Prove to yourself with some simple examples that the number of faces divided by the number of edges is 2/3.
7. Using the Wikipedia description of AMF [45], consider how you would code an airplane wing model that has honeycomb internal lattice for lightweight and some colored decorations on the outer skin. How might the part look using a color MJT system compared to using a VPP machine? What design considerations would you still have to make to ensure the part is properly made on these machines?
8. Search the Internet for the history of the STL file. What definitions did you find for the abbreviation "STL?" Why do you think people wanted to redefine what STL means?
9. AMF and 3MF formats have begun to show up in CAD software programs. Using an Internet search, identify five leading CAD software tools and find out which file types they support. What does this tell you about the transition from STL to new file formats?
10. List and describe at least three problems that are encountered with STL files. Name some software tools which can be used to obviate these kinds of problems.
11. Look up the most recent source you can find for the AMF and 3MF formats. What are key advantages of each compared to the other?

References

1. CADspan 3D printing solutions. (2019). http://www.cadspan.com/software/
2. 3dmtp. (2020). https://www.3dprintingbusiness.directory/company/3dmtp/
3. 3DSlicer. (2020). https://www.slicer.org/

4. 3D Tool. (2020). https://www.3d-tool.com/
5. Autodesk, Netfabb. (2020). https://www.autodesk.com/products/netfabb/overview
6. Printrun. (2020). https://www.pronterface.com/
7. Deskartes. (2020). http://www.deskartes.com/
8. MakePrintable. (2020). https://makeprintable.com/
9. Materialise, Magics. (2020). https://www.materialise.com/en/software/magics
10. Sourceforge, MeshFix. (2020). https://sourceforge.net/projects/meshfix/
11. Ultimaker, cura software. (2020). https://ultimaker.com/en/products/ultimaker-cura-software
12. ViewSTL. (2020). https://www.viewstl.com/
13. The StL Format.(1989). *Standard Data Format for Fabbers. (StereoLithography Interface Specification, 3D Systems, Inc.)*
14. Autodesk CAD. (2020). software. https://www.autodesk.com/
15. Rhinoceros CAD. (2020). *software.* http://www.rhino3d.com/
16. KISSlicer. (2020). http://www.kisslicer.com/overview.html
17. CraftWare. (2020). https://craftbot.com/craftware/
18. Repetier. (2020). https://www.repetier.com/
19. ReplicatorG. (2020). http://replicat.org/
20. Slic3r. (2020). https://slic3r.org/
21. Vatani, M., et al. (2009). An enhanced slicing algorithm using nearest distance analysis for layer manufacturing. In *Proceedings of World Academy of Science, Engineering and Technology.*
22. Simplify3D. (2020). https://www.simplify3d.com/
23. Materialise, AM software systems. (2020). http://www.materialise.com/materialise/view/en/92074-Magics.html
24. Jamieson, R., & Hacker, H. (1995). Direct slicing of CAD models for rapid prototyping. *Rapid Prototyping Journal, 1*(2), 4–12.
25. Ming, L. W., & Gibson, I. (2002). Specification of VRML in color rapid prototyping. *International Journal of CAD/CAM, 1*(1), 1–9.
26. Siu, Y., & Tan, S. (2002). Representation and CAD modeling of heterogeneous objects. *Rapid Prototyping Journal, 8*(2), 70–75.
27. Delft Spline. (2020). https://www.spline.nl/
28. Roland, desktop milling and subtractive RP. (2020). http://www.rolanddga.com/ASD/
29. Choi, S., & Samavedam, S. (2001). Visualisation of rapid prototyping. *Rapid Prototyping Journal, 7*(2), 99–114.
30. Abaqus. (2020). http://dsk.ippt.pan.pl/docs/abaqus/v6.13/books/usb/default.htm?startat=pt04ch13s01abo16.html
31. Ameba. (2020). https://www.food4rhino.com/app/ameba.
32. Altair Engineering. (2020). *solidThinking Inspire.* https://solidthinking.com/product/inspire/resources/
33. Autodesk, Within Medical. (2020). http://www.withinlab.com/
34. Dassault Systemes, Tosca Structure. (2020). https://www.3ds.com/products-services/simulia/products/tosca/structure/
35. Desktop Metals, Live Parts. (2020). https://www.desktopmetal.com/products/software/live-parts/
36. DTU, TopOpt. (2020). http://www.topopt.mek.dtu.dk/
37. Siemens, NX.(2020). https://www.plm.automation.siemens.com/global/en/products/mechanical-design/
38. Topostruct. (2020). http://grasshopperdocs.com/components/millipede/topostruct3D-Solver.html
39. Zhao, Y., et al. (2019). Molten pool behavior and effect of fluid flow on solidification conditions in selective electron beam melting (SEBM) of a biomedical Co-Cr-Mo alloy. *Additive Manufacturing, 26,* 202–214.

40. Zhao, G., et al. (2019). Numerical analysis of arc driving forces and temperature distribution in pulsed TIG welding. *Journal of the Brazilian Society of Mechanical Sciences and Engineering, 41*(1), 60.
41. Wu, D., et al. (2019). Elucidation of the weld pool convection and keyhole formation mechanism in the keyhole plasma arc welding. *International Journal of Heat and Mass Transfer, 131*, 920–931.
42. Sama, S. R., et al. (2019). Novel sprue designs in metal casting via 3D sand-printing. *Additive Manufacturing, 25*, 563–578.
43. Wohlers, T. (2019). *Wohlers report. 3D Printing and Additive Manufacturing State of the Industry*. Wohlers Associates.
44. ASTM International. (2012). *F2915–12. Standard Specification for Additive Manufacturing File Format (AMF)*.
45. dia AMF description. (2020). http://en.wikipedia.org/wiki/Additive_Manufacturing_File_Format

Chapter 18
Direct Digital Manufacturing

Abstract From the mid-1980s to the present, the percentage of parts made from Additive Manufacturing that are used as final, production parts has consistently increased. For the first decade of AM, commercial focus was on prototyping. During the second decade of AM, a significant amount of work went into illustrating the benefits of AM for producing tooling (the subject of a subsequent chapter). Starting in the mid-2000s and continuing to the present, the focus has shifted to utilizing AM for production of end-use components, which we call "Direct Digital Manufacturing" (DDM). DDM has continued to gain momentum, and today the majority of parts made using AM are for DDM and rapid tooling, rather than for use as prototypes. An overview of the types of applications that DDM is enabling as well as benefits and drawbacks of DDM compared to traditional part manufacturing operations is overviewed in this chapter.

18.1 Introduction

Generally, the industrial revolution has been described in four major stages called Industry 1.0, 2.0, 3.0, and 4.0. Industry 1.0 was the first industrial revolution based around the external combustion engine or "steam engine" that was developed mainly in Europe by inventors like the Scottish engineer James Watt in 1781 [1] that produced controllable rotary power. This enabled the rapid development of a wide range of manufacturing and machinery systems. Such inventions created the opportunity for changing from manual to machine production. From 1870 to 1914 through the harnessing of electricity, the second industrial revolution "Industry 2.0" appeared. The main advantages of Industry 2.0 were opportunities provided by distributed electrical motors toward mass production and assembly, thus creating faster production lines. Improvements in railroad and shipping systems, telegraphic communication systems, etc. fueled this development further. However, the main dilemma in Industry 2.0 was the production of high precision products [2, 3]. Therefore Industry 3.0 appeared around the 1970s where computational devices combined with sensor technology provided the opportunity to control systems thus addressing the previous

© Springer Nature Switzerland AG 2021
I. Gibson et al., *Additive Manufacturing Technologies*,
https://doi.org/10.1007/978-3-030-56127-7_18

problems on accuracy. Further digitalization improved quality, performance, throughput, and lead times by reduction of errors and defects. Another improvement in Industry 3.0 was associated with robots and similar automated systems. This provided greater flexibility in movement which enabled the manufacturer to further speed up production as well as improving quality and reliability [4, 5].

Great efforts have been made to improve processes, material usage, and supply chains in production systems. Cyber-physical systems (physical systems integrated with computational intelligence for decision-making) are now being used to generate smarter machines, storage, and materials handling that autonomously transfer information, triggering actions to control the whole system without significant human intervention. This interconnection of processes is largely referred to as Industry 4.0. To increase customer satisfaction, mass customization is a main driver of industry 4.0. Capitalizing on the industrial Internet of Things (IoT), robotics, artificial intelligence (AI), autonomous vehicles, etc., Industry 4.0 has embraced AM in the form of direct digital manufacturing (DDM) [6–8].

DDM is a term that describes the usage of Additive Manufacturing Technologies as part of the production of end-use components and products. Although it may seem that DDM is a natural extension of rapid prototyping, in practice this is not usually the case. Many additional considerations and requirements come into play for production manufacturing that are not important for prototyping. In this chapter, we explore these considerations through an examination of several DDM examples, distinctions between prototyping and production, and advantages of Additive Manufacturing for custom and low-volume production.

Many times, DDM applications have taken advantage of the geometric complexity capabilities of AM technologies to produce parts with customized geometries. In these instances, DDM is not a replacement for mass production applications, as customized geometry cannot be mass-produced using traditional manufacturing technologies. In addition, since the economics of AM technologies do not enable economically competitive high-volume production for many geometries and applications, DDM is often most economical for low-volume production applications.

Several early examples of DDM will be discussed that provide important insights into technology, operations, and business considerations necessary for production manufacturing using AM. Additional examples of current production applications illustrate the opportunities arising from AM adoption. A look at some example digital factories highlights the considerations relevant to scaling production rates using AM. This will be followed by a discussion of the unique characteristics of AM technologies that lead to DDM. Sections that describe aspects of build time and cost modeling round out the chapter.

18.2   Early DDM Examples

Three examples are presented of early adoption of AM technologies for the production manufacture of parts and products that date from the late 1990s to early 2000s. Examples are from the dental, hearing aid, and aerospace industries.

Fig. 18.1 Aligner from
Align Technology. (Photo
courtesy of Align
Technology)

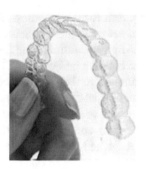

18.2.1   Align Technology

Align Technology, in Santa Clara, California, is in the business of providing orthodontic treatment devices (www.aligntech.com). Their Invisalign treatments are essentially clear braces, called aligners, that are worn on the teeth (*see* Fig. 18.1). Every 1–2 weeks, the orthodontic patient receives a new set of aligners that are intended to continue moving their teeth. That is, every 1–2 weeks, new aligners that have slightly different shapes are fabricated and shipped for the patient's use. Over the total treatment time (several months to a year typically), the aligners cause the patient's teeth to move from their initial position to the position desired by the orthodontist. If both the upper and lower teeth must be adjusted for 6 months, then 26 different aligners are needed for one patient, assuming that aligners are shipped every 2 weeks.

The need for many different geometries in a short period of time requires a mass customization approach to aligner production. Align's manufacturing process has been extensively engineered. First, the orthodontist takes an impression of the patient's mouth with a typical dental clay. The impression is shipped to Align Technology where it is scanned using a laser digitizer. The resulting point cloud is converted into a tessellation (set of triangles) that describes the geometry of the mouth. This tessellation is separated into gums and teeth, and then each tooth is separated into its own set of triangles. Since the data for each tooth can be manipulated separately, an Align Technology technician can perform treatment operations as prescribed by the patient's orthodontist. Each tooth can be positioned into its desired final position. Then, the motion of each tooth can be divided into a series of treatments (represented by different aligners). For example, if 13 different upper aligners are needed over 6 months, the total motion of a tooth can be divided into 13 increments. After manipulating the geometric information into specific treatments, aligner molds are built in one of Align's stereolithography (SL) Vat Photopolymerization (VPP) machines. At one point in the mid-2010s, Align Technology employed over 30 SLA-7000 machines. The aligners themselves are fabricated by thermal forming of a sheet of clear plastic over SL molds in the shape of the patient's teeth.

The aligner development process is geographically distributed, as well as highly engineered. Although Align headquarters and manufacturing operations have moved around over the years, an example from the midpoint of the company's evolution follows. The patient and orthodontist are separated from Align Technology headquarters, originally in California. Data processing for the aligners was performed in

Costa Rica, translating customer-specific, doctor-prescribed tooth movements into a set of aligner models. Each completed dataset was transferred electronically to Align's manufacturing facility in Juarez Mexico, where the dataset was added into a build on one of their SL machines. After building the mold from the dataset, the molds are thermal formed. After thermal forming, they are shipped back to Align and, from there, shipped to the orthodontist or the patient.

Between its founding in 1997 and 2020, over 8.3 million patients were been treated, and more than 100 million aligners were fabricated (www.aligntech.com). Align's SL machines are able to operate 24 h per day, producing approximately 100 aligner molds in one build tray, with a total production capacity of 45,000–50,000 unique aligners per day (~17 million per year) in 2014 [9]. And that number has continued to increase since 2014. As each aligner is unique, they are truly "customized." And by any measure, 45,000 components per day are mass production and not prototyping. Thus, Align Technology represents an excellent example of "mass customization" using DDM, albeit in a tooling application.

To achieve mass customization, Align needed to overcome the time-consuming pre- and post-processing steps in SL usage. A customized version of 3D Systems pre-processing software was developed to automate most of the build preparation. Aligner mold models were laid out, supports generated, process variables set, and the models sliced automatically. Typical post-processing steps, including rinsing and post-curing, can take hours. Instead, Align developed several of its own post-processing technologies. They developed a rinsing station that utilizes only warm water, instead of hazardous solvents. After rinsing, conveyors transported the platforms to the special UV post-cure station that Align developed. UV lamps provided intense energy that can post-cure an entire platform in 2 min, instead of the 30–60 min that were typical in a Post-Cure Apparatus unit. Support structures were removed manually for many years, until automated support removal technology was developed. The Align Technology example illustrates some of the growing pains experienced when trying to apply technologies developed for prototyping to production applications.

18.2.2   Siemens and Phonak

Siemens Hearing Instruments, Inc. (www.siemens-hearing.com) and Phonak Hearing Systems were competitors in the hearing aid business. In the early 2000s, they teamed up to investigate the feasibility of using polymer Powder Bed Fusion (PBF) technology in the production of shells for hearing aids [10]. A typical hearing aid is shown in Fig. 18.2. The production of hearing aid shells (housings that fit into the ear) required many manual steps. Each hearing aid must be shaped to fit into an individual's ear. Fitting problems caused up to one out of every four hearing aids to be returned to the manufacturer, a rate that would be devastating in most other industries.

Traditionally, an impression was taken of a patient's ear, which was used as a pattern to make a mold for the hearing aid shell. An acrylic material is then injected

Fig. 18.2 Siemens
LASR® hearing aid
and shell

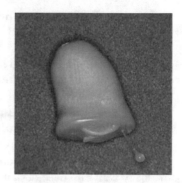

Fig. 18.3 Hearing aid
within scanned point
cloud. (Photo courtesy of
Siemens Hearing
Instruments)

into the mold to form the shell. Electronics, controls, and a cover plate are added to complete the hearing aid. To ensure proper operation and comfort, hearing aids must fit snugly, but not too tightly, into the ear and must remain in place when the patient talks and chews (which change the geometry of the ear).

To significantly reduce return rates and improve customer satisfaction, Siemens and Phonak sought to redesign their hearing aid production processes. Since AM technologies require a solid CAD model of the design to be produced, the companies had to introduce solid modeling CAD systems into the production process. Originally, impressions were taken from patients' ears and were scanned by a laser scanner, rather than used directly as a pattern. The point cloud is converted into a 3D CAD model, which is manipulated to fine-tune the shell design so that a good fit is achieved. Today, small hand-held scanners are used to scan customers' ears directly, eliminating the impression step. The CAD shell model is then exported as an STL file for processing by an AM machine. A scanned point cloud is shown superimposed on a hearing aid model in Fig. 18.3.

In the mid-2000s, Siemens developed a process to produce shells using VPP technology to complement their PBF fabrication capability. VPP has two main advantages over PBF. First, VPP has better feature detail, which makes it possible to fabricate small features on shells that aid assembly to other hearing aid components. Second, acrylate VPP materials are similar to the materials originally used in the hearing aid industry (heat setting acrylates), which are biocompatible.

In the late 2000s, Siemens Hearing Instruments produced about 250,000 hearing aids annually. In 2007, they claimed that about half of the in-the-ear hearing aids that they produced in the USA were fabricated using AM technologies. Since the introduction of AM-fabricated hearing aid shells, hearing aid manufacturers in the Western world, and most worldwide, have adopted AM in order to remain competitive. By 2012, surveys estimate that about 90 percent of all custom in-the-ear hearing aid shells are fabricated using AM, which totaled approximately two million AM-fabricated shells [9]. In 2018, the estimated number of hearing aids sold in the USA was about 4,000,000 virtually all of them with AM-fabricated shells. Furthermore, since the adoption of AM, the hearing aid return rate has fallen dramatically with improved design and manufacturing processes.

A number of technology advances have targeted this market. At present, most hearing aid shells are fabricated using Material Jetting (MJT) or laser scan or mask projection VPP. 3D Systems, EnvisionTEC, and Stratasys are the major companies that target this market, but other vendors do as well. Some of these companies market resins specifically for this application.

Hearing aid shell production is a great example of how companies can take advantage of the shape complexity capability of AM technologies to economically achieve mass customization. With improvements in scanning technology, patients' ears can be scanned directly, eliminating the need for impressions [11]. With the development of desktop AM systems and accurate hand-held scanners, the fabrication of custom hearing aids in the audiologist's office is technically feasible, rather than having to ship impressions or data sets to a central location.

18.2.3   Polymer Aerospace Parts

In the late 1990s and into the 2000s, aerospace companies were investigating the production manufacturing of non-structural polymer parts for use in non-civilian aerospace applications. During this time, hundreds of opportunities were identified to replace conventionally manufactured parts and modules with parts fabricated using polymer PBF. These explorations were performed for military aircraft, satellites, and even the space shuttle and space station.

One of the well-publicized applications was for air ducts on F-18 fighter jets. Upgrade packages became available in the 2000s and 2010s for these fighters; typically 30–50 were upgraded each year, so production volumes were small. Cooled air needed to be delivered to avionics modules mounted, for example, in the nose of the aircraft. Air ducts were typically injection molded or rotomolded and required many parts and fasteners. Tooling costs and assembly costs were high. Instead, engineers at Boeing and Northrop Grumman redesigned the components for DDM using polymer PBF which enabled significant part count reductions, more integral parts, better performance, and lower overall costs. The conceptual example shown in Fig. 18.4 illustrates the principle: a complex 16-part assembly can be replaced with one consolidated part that performs better and costs less. This idea of part consolidation will be discussed further in Chap. 19.

At about the same time, SAAB Avitronics used polymer PBF to manufacture antenna RF boxes for an unmanned aircraft. Paramount Industries, a division of 3D Systems, produced PBF parts for a helicopter, including ventilation parts and electrical enclosures, and structures for unmanned aerial vehicles. The parts were manufactured on their EOSINT P 700 machine from EOS using the PA 2210 FR (flame retardant) material. Advantages of this approach over conventional manufacturing processes include a more compact design, integral features, better performance, 30–50% reduction in mass, and similar reductions in lead times and costs. By the end of the 2000s, thousands of parts were flying and orbiting the earth. Currently, many tens of thousands of polymer parts are flying on a wide variety of aircraft, including newer models of civilian aircraft.

18.3 Applications of DDM

Recent advances in AM technologies and materials have enabled many production manufacturing applications to be realized. In general, these applications have become technically and economically feasible due to a convergence of AM's unique capabilities, market demand (e.g., production volumes), and business model innovations. Applications in the aerospace, automotive, medical, and consumer industries will be surveyed in this section.

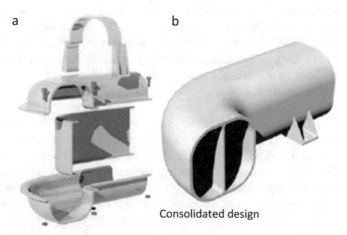

a

b

Consolidated design

Original design with 16 parts

Fig. 18.1 Aircraft duct example. (Photo courtesy of 3D Systems)

18.3.1 Aerospace and Power Generation Industries

The aerospace industry employs technologies like AM because of the possibility of manufacturing lighter structures to reduce weight with no reduction in performance. A major goal is to make the lightest practical aircraft while maintaining reliability and safety. For instance, both Airbus and Boeing have invested in lightweighting of their aircraft and are looking for AM to contribute greatly to this objective. Recently, GE and Rolls-Royce have made large investments in AM to optimize their engine components. Lightweighting is very important in the aerospace industry to increase range and to decrease corresponding fuel costs. For example, the range of the Boeing 777 200ER is 17,395 km, while the range of the 777 300ER is about 3000 km less, since it is about 3900 kg greater in mass (their fuel capacities are identical). In fact, Airbus claims that every kilogram saved prevents 25 tons of CO_2 emissions during the lifespan of an aircraft [12, 13].

The major airframe manufacturers, Airbus and Boeing, have invested heavily in AM in recent years, including fabrication technologies, materials, operations, and design tools. Airbus announced that the first titanium AM part (a bracket) was installed on a serial production aircraft in 2017, with European Union Aviation Safety Agency (EASA) approval. In 2015, an A350 model had over 1000 AM parts, with about 1000 of them printed using the Material Extrusion (MEX) process in Stratasys ULTEM 9085 material [14]. The development of flame retardant materials for PBF and MEX processes contributed greatly to the proliferation of polymer parts on commercial aircraft. For a specific example of advantages afforded by AM, Airbus uses metal PBF to fabricate latch shafts for A350 aircraft doors. Each aircraft requires 16 of these shaft parts, which are 45% lighter and 25% less expensive than conventionally manufactured shafts, for a total savings of over 4 kilograms per aircraft [15]. In 2018, Boeing created an enterprise-level organization called Boeing Additive Manufacturing to align efforts among their commercial, defense, space, and services activities. At around the same time, they received Federal Aviation Administration (FAA) certification for their first structural titanium part. From these examples, it is clear that the airframe manufacturers have adopted AM for production and are likely to increasingly utilize AM in the future.

Manufacturers of smaller aircraft, helicopters, satellites, and rocket engines are adopting AM as well. Bombardier Cessna, Gulfstream, Piper, BAE Systems, Bell Helicopter, and SpaceX, among others, have major initiatives underway to utilize AM. Many of these companies have formed partnerships with AM machine vendors, material suppliers, and parts suppliers on technology development projects, a trend that is likely to continue. It is noteworthy that both PBF and Directed Energy Deposition (DED) processes are being used for structural metal parts in airframes. Machine vendors are partnering with airlines to develop capabilities to provide production, spare, and replacement parts for airplane cabins; for example, Stratasys has announced a partnership with SIA Engineering (affiliated with Singapore Airlines).

Similarly, the aircraft engine manufacturers have demonstrated impressive results using metal AM. GE may be most well-known for its efforts, but Pratt &

Whitney, Rolls-Royce, Honeywell, and other suppliers have major initiatives in AM. GE took about 4 years to develop their fuel nozzle for the LEAP engine, learning about the technology and how to design for it. Their nozzle design consolidates 18–20 parts into 1, weighs 25% less, and exhibits about a 5X durability improvement, in large part due to the reduction in brazed and welded joints. Figure 18.5 shows the fuel nozzle which has both as-built and machined surfaces. 19 fuel nozzles are needed per engine; approximately 2000 LEAP engines were produced in 2019, with increases in deliveries projected into the future. Among other developments, GE Aviation has a plant in Alabama dedicated to metal AM with over 40 AM machines [16].

As part of their technology development efforts, Rolls-Royce flight tested an engine in 2015 with what it claims is the largest AM part on an aircraft engine, the front bearing housing for a Trent XWB-97 engine. The 1.5 meter diameter housing was fabricated in titanium [17]. Recently, they indicated that their new engine development will take significant advantage of AM to reduce weight and emissions and improve fuel efficiency [18].

Satellite manufacturers have increased their investigation of metal AM for the fabrication of lightweight components over the past decade. Thales Alenia Space, for example, announced in 2019 that they are entering series DDM production of telecom satellite components, primarily brackets and supports. Their reaction wheel brackets are large (466×367×403mm) aluminum components that are less expensive and 30% lighter than conventionally manufactured versions [20]. Previously, they described several antenna supports that were designed using topology optimization and fabricated in aluminum.

Gas turbines for power generation are quite similar in many aspects of their design to turbine engines for aircraft. As such, many of the DDM innovations used for aircraft engines are also used for power generation. Both GE and Siemens are making extensive use of metal AM for production manufacturing. For example, Siemens investigated a combustion burner head as one of the first components for DDM, as shown in Fig. 18.6 [21]. Compared to heads cast in a nickel-based alloy, the PBF parts were more accurate and had better mechanical properties. A different burner design was redesigned to consolidate 13 parts and eliminate 18 welds. To

As-built surface   Machined area

Fig. 18.5   GE LEAP engine fuel nozzle [19]

Fig. 18.6 Combustion
burner head from Siemens

support its metal AM initiatives, Siemens reconfigured a plant in Sweden to manufacture the burners and components, acquired UK-based Materials Solutions, which
has a metal AM factory, and created several innovation centers around the world.

18.3.2   Automotive Industry

In the automotive industry, examples of DDM are emerging. Formula 1 teams have
been using AM technologies extensively on their race cars for several years.
Applications include electrical housings, camera mounts, and other aerodynamic
parts. In the late 2000s, the Renault Formula 1 team used over 900 parts on race cars
each racing season. Indy and NASCAR teams also make extensive use of AM parts
on their cars.

Local Motors conducted a crowd-sourced car design project, with the requirement
that the majority of the car be fabricated by AM. Specifically, they intended to use a
hybrid additive/subtractive developed at the Oak Ridge National Laboratory. The
machine has a large diameter MEX head and subtractive machining capabilities. In
May 2014, voting on the various car modules (body, internal structure, etc.) was
completed. Fabrication of body and structural components using the ORNL machine
took place on the exhibition floor of the International Machine Technology Show in
September 2014. After assembling an electric powertrain, wheels, and interior components, the car was driven in the parking lot of the Show to demonstrate the successful 3D printing of a drivable car in a short period of time (the Strati, Fig. 6.8).

All major automakers have AM initiatives focused on production applications.
Most automotive manufacturers use AM parts on concept cars and for other purposes. Hyundai used PBF to fabricate flooring components for their QarmaQ concept car in 2007, with assistance from Freedom of Creation. Bentley uses PBF to
produce some specialty parts that are subsequently covered in leather or wood.
Others use AM to fabricate replacement parts for antique cars, including Jay Leno's
famous garage (www.jaylenosgarage.com). BMW uses MEX extensively in production for fixtures and tooling in automotive assembly. BMW has accelerated the
integration of Additive Manufacturing into its production cycle by launching a new

AM plant in Munich, Germany. They presented a 3D printed chassis for a motorcycle that is lightweight using a lattice structure. Japanese manufacturers like Toyota, Mitsubishi, Nissan, etc. have turned to AM for their metal and plastic productions [22–24]. In 2018, Bugatti introduced an eight-piston monobloc brake caliper made from titanium PBF, which they claim as the largest production brake caliper in the automotive industry, giving their cars unprecedented high-performance braking capabilities using a lightweight, optimized design.

In 2016, Daimler Trucks North America started fabricating replacement and spare parts using polymer PBF. They also created a digital warehouse from which customers can order parts. This enables Daimler to avoid physical inventory of parts. Further, for those parts with low or intermittent demand, additional benefits accrue: no need for tooling inventory and faster part delivery.

In 2017, BMW introduced a service to provide several customized parts for Mini Cooper models. Customers can order side scuttles or cockpit fascias with their names or short messages; the components are fabricated using polymer PBF.

Many consider the first production parts on a series car model to be a bracket for the retractable roof on the 2018 i8 Roadster. The bracket was designed using topology optimization to achieve 44% weight savings [25]. The evolution of the design is illustrated in Fig. 18.7.

18.3.3 Medical Industry

The medical examples presented so far all relate to body-fitting, external customized parts. However, many other opportunities exist. Many companies worldwide are investigating the use of PBF technologies for the creation of orthopedic implants. For instance, Adler Ortho Group of Italy is using Arcam's EBM system to produce stock sizes of acetabular cups for hip implants made from Ti–6Al–4V. The use of

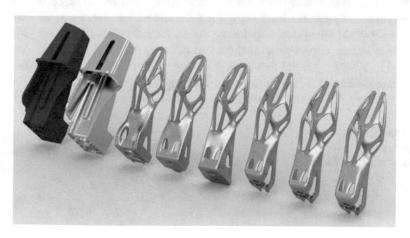

Fig. 18.7 BMW i8 Roadster roof bracket

AM techniques enables a more compact design and a better transition between the solid bearing surface and the porous bone in-growth portion of the implant. Although a porous coating of titanium beads or hydroxyapatite on an implant's surface works well, they do not provide the optimum conditions for osseointegration. The hierarchical structure capabilities of AM enable the creation of a more optimal bone in-growth structure for osseointegration. As of early 2014, more than 40,000 cups had been implanted, and more than 90,000 implants had been produced by companies such as Adler Ortho and Lima Corporate SpA. More than 20 different AM medical implant products have received FDA clearance for implantation in patients [9]. In 2020, the number of implanted orthopedic devices made using DDM is in the hundreds of thousands and is quickly on its way into the millions.

In the late 2010s, Stryker created its AMagine Institute, a research organization focused on AM applications for medical implants and products in the spine, craniomaxillofacial, and joint replacement areas. Their Tritanium® porous technology is used in, for example, spinal fixation cages to promote bone in-growth and biological fixation. They have reportedly produced over one million AM medical devices as of the beginning of 2020.

DePuy Synthes and Materialise have partnered to develop products and services for the craniomaxillofacial area. Their software can be used to design a variety of implants, surgical guides, and surgery planning models. Additionally, DePuy Synthes has developed a porous lattice structure technology, broadly similar to Stryker's, for spinal implants. They claim their porous structures have a similar elastic modulus compared to cancellous bone.

18.3.4   Consumer Industries

The footwear industry has been a pioneering adopter of AM. Starting in the late 2000s, a British company called Prior 2 Lever (P2L) claimed to be manufacturing the world's first custom soccer shoes for professional athletes. Polymer PBF was used to fabricate the outsoles, including cleats, for individual customers [26]. The one-piece leather uppers were also custom tailored. A model called the Assassin retailed in 2008 for several thousand British pounds per pair; a photo is shown in Fig. 18.8. Research on PBF outsoles for custom shoes started in the early 2000s at Loughborough University; Freedom of Creation and others contributed to the

Fig. 18.8 Assassin model soccer shoe. (Photo courtesy of Prior 2 Lever)

development of this work. Early testing demonstrated a significant reduction in peak pressures during walking and running with personalized outsoles [27]. Custom sprinting shoes and tri-athlete shoes are also available.

In 2013, Nike developed a line of football cleats called the Vapor Laser Talon. The cleat plate was specially designed to improve player performance, particularly for speed positions. It has an intricate, lattice design and was produced using PBF in a proprietary polymer material.

More recently, Adidas, New Balance, Nike, and Under Armour have adapted AM for the mass production of soles for sport shoes. New Balance started a digital sports division and fabricates mid-soles using polymer PBF. Nike announced a partnership with HP to use their polymer PBF machines. Adidas has a partnership with Carbon to use VPP for producing heels and mid-soles. One of the drivers for adopting AM is its shape complexity capability, since the shoe companies are experimenting with complex lattice and other cellular structures for parts of their soles. The speed of HP and Carbon machines is promising for economical production.

Low-volume production is often economical via AM since hard tooling does not need to be developed. This has led to the rapid adoption of DDM for low-volume components across many industries. However, the most exciting aspects of DDM are the opportunities to completely rethink how components can be shaped in order to best fulfill their functions, as discussed in Chap. 19. Integrated designs can be produced that combine several parts, eliminate assembly operations, improve performance by designing parts to utilize material efficiently, eliminate shape compromises driven by manufacturing limitations, and completely enable new styles of products to be produced. This can be true for housewares, everyday items, and even customized luxury items, as illustrated in Chap. 19 with respect to houseware and fashion products. Each of these areas will be explored briefly in this section. Many of the examples were taken, or cited, in Wohlers Report [9, 28].

AM machine manufacturers use their own machines to help build components for new machines. For instance, Stratasys developed a new class of MEX machines in 2007, the Fortus X00mc series. They introduced the Fortus 900mc in December and reported that 32 parts on the machine were fabricated on their MEX machines. Since introducing the 900mc, Stratasys has marketed several new models, using in-house fabricated parts on these models as well. This is also true for other major AM manufacturers, including EOS and 3D Systems.

In consumer-oriented industries, many specialty applications are beginning to emerge. For example, Franco Bicycles teamed with AREVO to produce using DDM a carbon fiber bike frame. The hybrid MEX and carbon fiber placement process are used to make the Emery One eBike that can be custom manufactured to the customer's requirements. In the metal area, several bicycle companies are investigating steel and titanium bike frame components (lugs and dropouts), while others are combining topology optimization and metal AM for entire frames. One example is the partnership between Renishaw and Empire Cycles which produced a titanium bike frame that is 33% lighter than the original.

Many service bureaus do production manufacturing runs for customized or other specialty components. An interesting class of applications is to bridge the virtual and physical worlds. World of Warcraft is a popular online video game. Players can design their own characters for use in the virtual world, often adding elaborate clothing, accessories, and weapons. A company called FigurePrints (www.figure-prints.com) produces models of such characters. They use Binder Jetting (BJT) machines from 3D Systems with color printing capabilities. Many start-ups have marketed similar services to print custom-designed characters, with many of these start-ups failing within a year or two. Jujups is one such example, which offered custom Christmas ornaments, printed with a person's photograph using color print-ers from 3D Systems. In many of these applications, AM can utilize the input data only after it has been converted to a usable form, as the original data was created to serve a visual purpose and not necessarily as a representation of a true 3D object.

The gaming industry alone accounts for hundreds of millions of unique 3D vir-tual creations that consumers may want to have made into physical objects. Just as the development of computer graphics has often been driven by the gaming indus-try, it is likely the further development of color DDM technologies may also be driven by market opportunities which are enabled by the gaming industry. Quite a few online games allow players to create or customize their own characters. Tools such as Maya are used to create/modify geometry, colors, textures, etc. which are then uploaded to the game site. These same files can be packaged and sent to AM service providers, such as Shapeways and Sculpteo, for printing color models of the custom characters. Make Magazine ran a very informative article on this topic in 2013 [29].

18.4   DDM Drivers

It is useful to generalize from these examples and explore how the unique capabili-ties of AM technologies may lead to new DDM applications. The factors that enable DDM applications include:

- Unique shapes: parts with customized shapes
- Complex shapes: improved performance
- Parts consolidation: merging multiple components into a single part or assembly
- Lot size of one: economical to fabricate customized parts
- Fast turnaround: save time and costs; increase customer satisfaction
- Digital manufacturing: precisely duplicate CAD model
- Digital record: have reusable, traceable dataset
- Electronic "spare parts": fabricate spare parts on demand, rather than holding inventory
- No hard tooling: no need to design, fabricate, and inventory tools; economical low-volume production

As indicated in the Align Technology and hearing aid examples, the capability to create customized, person-specific *unique geometries* is an important factor for DDM. Many AM processes are effective at fabricating platforms full of parts, essentially performing mass customization of parts. For example, 100 aligner molds fit on one SLA-7000 platform. Each has a unique geometry. Approximately 150 hearing aid shells can fit in an EnvisionTEC Perfactory machine. Upward of 4000 hearing aid shells can be built in one PBF powder bed in one build. The medical device industry is a leading – and growing – industry where DDM and rapid tooling applications are needed due to the capability of fabricating patient-specific geometries.

The capability of building parts with *complex geometries* is another benefit of DDM. Features can be built into hearing aid shells that could not have been molded in, due to constraints in removing the shells from their molds. In many cases, it is possible to combine several parts into one DDM part due to AM's complexity capabilities, as described above for the F-18 air duct example. *Parts consolidation* can lead to tremendous cost savings in assembly tooling and assembly operations that would be required if multiple parts were fabricated using conventional manufacturing processes. Complexity capabilities also enable new design paradigms, as discussed with respect to medical implants and as seen in Chap. 21. These new design concepts will be increasingly realized in the near future.

Related to the unique geometry capability of AM, economical *lot sizes of one* are another important DDM capability. Since no tooling is required in DDM, there is no need to amortize investments over many production parts. DDM also avoids the extensive process planning that can be required for machining, so time and costs are often significantly reduced. These factors and others help make small lot sizes economical for DDM.

Fast turnaround is another important benefit of DDM. Again, little time must be spent in process planning, tooling can be avoided, and AM machines build many parts at once. All these properties lead to time savings when DDM is used. It is common for hearing aid manufacturers to deliver new hearing aids in less than 1 week from the time a patient visits an audiologist. Align Technology must deliver new aligners to patients every 1–2 weeks. Rapid response to customer needs is a hallmark of AM technologies, and DDM takes advantage of this capability.

The capability of *digital manufacturing*, or precisely fabricating a mathematical model, has important applications in several areas. The medical device industry takes advantage of this; hearing aid shells must fit the patient's ear canal well, the shape of which is described mathematically. This is also important in artwork and high-end housewares, where small shape changes dictated by manufacturing limitations (e.g., draft angles for injection molding) may be unwelcome. More generally, the concept of digital manufacturing enables digital archiving of the design and manufacturing information associated with the part. This information can be transferred electronically anywhere in the world for part production, which can have important implications for global enterprises.

A *digital record* is similar in many ways to the digital manufacturing capability just discussed. The emphasis here is on the capability to archive the design information associated with a part. Consider a medical device that is unique to a patient (e.g., hearing aid, foot orthotic). The part design can be a part of the patient's digital medical records, which streamlines record keeping, sharing of records, and fabricating replacement parts.

Another way of explaining digital records and manufacturing, for engineered parts, is by using the phrase *electronic spare parts*. The air handling ducts installed on F-18 fighters as part of an avionics upgrade program may be flying for another 20 years. During that time, if replacement ducts are needed, Boeing must manufacture the spare parts. If the duct components were molded or stamped, the molds or stamps must be retrieved from a warehouse to fabricate some spares. By having digital records and no tooling, it is much easier to fabricate the spare parts using AM processes; plus the fabrication can occur wherever it is most convenient. This flexibility in selecting fabrication facilities and locations is impossible if hard tooling must be used.

As mentioned several times, the advantages are numerous and significant to not requiring *tooling* for part fabrication. Note that in cases such as Align Technology, tooling is required, but the tooling itself is fabricated when and where needed, not requiring tooling inventories. The elimination of tooling makes DDM economically competitive across many applications for small lot size production.

18.5   Manufacturing Versus Prototyping

Production manufacturing environments and practices are much more rigorous than prototyping environments and practices. Certification of equipment, materials, and personnel, quality control, and logistics are all critical in a production environment. Even small considerations like part packaging can be much different than in a prototyping environment. Table 18.1 compares and contrasts prototyping and DDM practices for several primary considerations [30].

Certification is critical in a production environment. Customers must have a dependable source of manufactured parts with guaranteed properties. The DDM company must carefully maintain their equipment, periodically calibrate the equipment, and ensure it is always running within specifications. Processes must be engineered and not left to the informal care of a small number of technicians with varying skills. Experimentation on production parts is not acceptable. Meticulous records must be kept for quality assurance and traceability concerns. Personnel must be fully trained, cross-trained to ensure some redundancy, and certified to deliver quality parts.

Most, if not all, DDM companies are compliant or certified in one of the ISO 9xxx processes. ISO 9xxx (where the "xxx" is replaced with a specific ISO #) is an international series of standards for quality systems and practices. Most customers will require such ISO 9xxx practices so that they can depend on their suppliers. Many books have been written on ISO standards so, rather than go into extensive detail here, readers should utilize these books to learn more about this topic [31].

As mentioned, personnel should be trained, certified, and periodically retrained and/or recertified. Cross-training personnel on various processes and equipment help mitigate risks of personnel being unavailable at critical times. If multiple shifts are run, these issues become more important, since the quality must be consistent across all shifts.

Table 18.1 Contrast between rapid prototyping and direct digital manufacturing

Key characteristic	RP company	DDM company
Certification		
Equipment	Mainly AM equipment	High-end AM production machines, extensive post-processing technologies, and calibration equipment
Personnel	No formal testing or certification. Informal training, on-the-job training	Ongoing need for certification, training, and competency testing. Redundant personnel needed for risk mitigation
Practices	Trial-and-error, no formal documentation of practices	Formal testing for each critical step, periodic recertification
Quality	Basic procedures; some inspection	ISO 9xxx compliance. Extensive, thorough quality system needed
Documentation system	Basic system; controls and documentation not essential	Developed system; controls and documentation required
Planning	Basic. Requires only modest part assessment	Formal planning to ensure customer requirements are met. Developed process chains, no experimentation
Scheduling and delivery	Informally managed; critical jobs can be expedited; usually only one delivery date	Sophisticated scheduling, just-in-time delivery
Vertical integration	May not have any services besides prototype manufacturer	From customer's perspective, should be a one-stop shop. Qualified suppliers must be lined up ahead of time to enable integration

[a]Much of this section was adapted from Brian Hasting's presentation at the 2007 SME RAPID Conference [30]

Vertical integration is important, since many customers will want their suppliers to be "one-stop shops" for their needs. DDM companies may rely on their own suppliers, so the supplier network may be tiered. It is up to the DDM company, however, to identify their suppliers for specialty operations, such as bonding, coating, assembly, etc., and ensure that their suppliers are certified.

As mentioned previously, several service bureaus offer production manufacturing services. In addition, many large companies have opened production facilities that incorporate AM at its core. For example, the UK-based Siemens factory was mentioned in the aerospace section of this chapter. They will supply parts to their power generation products. The capacity of their facility in 2020 is 32 metal printers with all associated post-processing equipment. Above the printer floor, the mezzanine has the powder material handling systems to deliver powder to printers below. Operations are intended to be highly automated, supported by Siemens' software suite, including PLM, NX, and their MindSphere cloud-based IoT system, as well as Ansys Additive Suite simulation software.

GE Additive opened a factory in Pittsburgh to supply GE Aviation and other customers with metal AM parts. In 2019, they received AS9100D certification for their operations. AS9100 is administered by the American System Registrar organization and is a superset of ISO9001 quality specifications. The standard provides guidelines for product safety and continuous improvement to satisfy US Department of Defense, NASA, FAA, and other aerospace applicable statutory and regulatory requirements. AS9100D refers to revision D, from 2016, that includes additional product safety, counterfeit parts prevention, risk, and human factors clauses.

GE Aviation has a dedicated factory in Auburn, AL, to manufacture jet engine components, including the LEAP engine fuel nozzle. They have over 40 metal printers and about 300 employees as of this writing.

In the polymer PBF area, Jabil manufactures HP printers in a facility in Singapore. On one floor of the factory, they have an array of HP printers that print parts for the HP printers that are being assembled on another floor of the factory. This facility is part of the new Jabil Additive Manufacturing Network that connects AM operations across about 12 factories worldwide. This network enables Jabil to consolidate and optimize supply chains, balancing demand in a distributed manufacturing context. They claim to have over 100 3D printers, including various polymer and metal printers, and certified manufacturing processes.

18.6   Cost Estimation

From a cost perspective, AM can appear to be much more expensive for part manufacture than conventional mass production processes. A single part out of a large VPP or PBF machine can cost upward of $5000, if the part fills much of the material chamber. However, if parts are smaller, the time and cost of a build can be divided among all the parts built at one time. For small parts, such as the hearing aid shell, costs can be only several dollars or less. In this section, we will develop a simple cost model that applies to DDM. A major component of costs is the time required to fabricate a set of parts; as such, a detailed build time model will be presented.

18.6.1   Cost Model

Broadly speaking, costs fall into four main categories: machine purchase, machine operation, material, and labor costs. In equation form, this high-level cost model can be expressed, on a per build basis, as:

$$\text{Cost} = P + O + M + L \tag{18.1}$$

or, on a per part basis, as:

$$\text{cost} = p + o + m + l = 1 / N \times \left(P + O + M + L \right) \tag{18.2}$$

where P = machine purchase cost allocated to the build, O = machine operation cost, M = material cost, L = labor cost, N = number of parts in the build, and the lowercase letters are the per-part costs corresponding to the per-build costs expressed using capital letters. An important assumption made in this analysis is that all parts in one build are the same kind of part, with roughly the same shape and size. This simplifies the allocation of times and costs to the parts in a build.

Machine purchase and operations costs are based on the build time of the part. We can assume a useful life of the machine, denoted Y years, and apportion the purchase price equally to all years. Note that this is a much different approach than would be taken in a cash flow model, where the actual payments on the machine would be used (assuming it was financed or leased). A typical uptime percentage needs to be assumed also. For our purposes, we will assume a 95% uptime (the machine builds parts 95% of the time during a year). Then, purchase price for one build can be calculated as:

$$P = \frac{\text{PurchasePrice} \cdot T_{b}}{0.95 \cdot 24 \cdot 365 \cdot Y} \tag{18.3}$$

where T_b is the time for the build in hours and $24 \cdot 365$ represents the number of hours in a year. Operation cost is simply the build time multiplied by the cost rate of the machine, which can be a complicated function of machine maintenance, utility costs, cost of factory floor space, and company overhead, where the operation cost rate is denoted by C_o:

$$O = T_{b} \cdot C_{o} \tag{18.4}$$

Material cost is conceptually simple to determine. It is the volume, v, of the part multiplied by the cost of the material per unit mass, C_m, and the mass density, ρ, as given in (18.5). For AM technologies that use powders, however, material cost can be considerably more difficult to determine. The recyclability of material that is used, the volume fraction of the build that is made up of parts versus loose powder (in the case of powder bed techniques), and/or the powder capture efficiency of the process (in the case of DED techniques) will result in the need to multiply the volume, v, of the part by a factor ranging from a low of 1.0 to a number as high as 7.0 to accurately capture the true cost of material consumed. Thus, for powder processes where the build material is not 100% recyclable, material cost has a complex dependency on the various factors mentioned here. The term k_r will be introduced for the purpose of modeling the additional material consumption that considers these factors. In addition, for processes that require support materials (such as MEX and VPP), the volume and cost of the supports needed to create each part must also be taken into account. The factor k_s takes this into account for such processes; typical values would range from 1.1 to 1.5 to include the extra material volume needed for supports. As a result, the model described in (18.5) will be used for material cost:

$$M = k_{s} \cdot k_{r} \cdot N \cdot v \cdot C_{m} \cdot \rho \tag{18.5}$$

Labor cost is the labor rate, C_l in \$/hour, multiplied by the time, T_l, required for workers to set up the build, remove fabricated parts, clean the parts, clean the machine, and get the machine ready for the next build:

$$L = T_l \cdot C_l \qquad\qquad (18.6)$$

18.6.2 Build Time Model

The major variable in this cost model is the build time of the parts. Build time (T_b) is a function of part size, part shape, number of parts in the build, and the machine's build speed. Viewed slightly differently, build time is the sum of scan or deposition time (T_s), transition time between layers (T_t), and delay time (T_e):

$$T_b = T_s + T_t + T_e \qquad\qquad (18.7)$$

For this analysis, we will assume that we are given the part size in terms of its volume, v, and its bounding box, aligned with the coordinate axes: bbx, bby, and bbz. Layer transition time is the easiest to deal with. The processes that build in material beds or vats have to recoat or deposit more material between layers; other processes do not need to recoat and have a T_r of 0. Recoat times for building support structures can be different than times for recoating when building parts, as indicated by (18.8):

$$T_t = L_s \cdot T_{ts} + L_p \cdot T_{tp} \qquad\qquad (18.8)$$

where L_s is the number of layers of support structure, T_{ts} is the time to recoat a layer of support structures, L_p is the number of layers for building parts ($L_p = bb_z/LT$), T_{tp} is the recoat time for a part layer, and LT is the layer thickness.

Scan/deposition time is a function of the total cross-sectional area for each layer, the scan or fill strategy utilized, and the number of layers. Cross-sectional area depends upon the part volume and the number of parts. Scan/deposition time also depends upon whether the machine has to scan vectors to build the part in a point-wise fashion, as in standard VPP, PBF, and MEX machines, or the part deposits material in a wide, line-wise swath, as in MJT processes, or as a complete layer, as in layer-based (e.g., mask projection) VPP processes. The equations are similar; we will present the build time model for scanning and leave the wide swath deposition and layer-based scanning processes for the exercises.

Now, we need to consider the part layout in the build chamber. Assuming a build platform, we have a 2D layout of parts on the platform. Parts are assumed to be of similar sizes and are laid out in a rectangular grid according to their bounding box sizes. Additionally, X and Y gaps are specified so that the parts do not touch. In the event that the parts can nest inside one another, gaps with negative values can be given. A 2D platform layout is shown in Fig. 18.9 showing the bounding boxes of

18 long, flat parts with gaps of 10 mm in the X direction and 20 mm in the Y direction. The number of parts on the platform can be computed as:

$$N = \left(\frac{PL_x + g_x - 20}{bb_x + g_x} \right) \left(\frac{PL_y + g_y - 20}{bb_y + g_y} \right) \tag{18.9}$$

where PL_x and PL_y are the platform sizes in X and Y, g_x and g_y are the X and Y gaps, and the -20 mm terms prevent parts from being built at the edges of the platform (10 mm buffer area along each platform edge). This analysis can be extended to 3D build chambers for processes which enable stacking in the z direction.

The time to scan one part depends on the part cross-sectional area, the laser or deposition head diameter d, the distance between scans h, and the average scan speed ss_{avg}. Cross-sectional area, A_{avg}, is approximated by using an area correction factor γ [32], which corrects the area based on the ratio of the actual part volume to the bounding box volume, v_{bb}, $\gamma = v/v_{bb}$. The following correction has been shown to give reasonable results in many cases:

$$A_{fn} = \gamma \cdot e^{\alpha(1-\gamma)} \tag{18.10}$$

$$A_{avg} = bb_x \cdot bb_y \cdot A_{fn} \tag{18.11}$$

where α is typically taken as 1.5.

For scanning processes, it is necessary to determine the total scan length per layer. This can be accomplished by simply dividing the cross-sectional area by the diameter of the laser beam or deposited filament. Alternatively, the scan length can be determined by dividing the cross-sectional area by the hatch spacing (distance

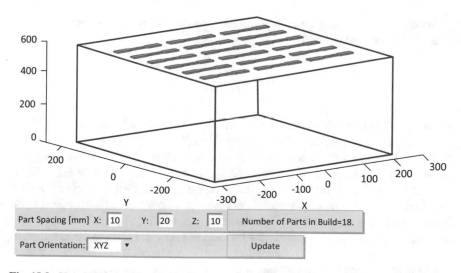

Fig. 18.9 SLA-7000 vat with 18 parts laid out on the platform

between scans). We will use the latter approach, where the hatch spacing, hr, is given as a percentage of the laser beam diameter. For support structures, we will assume that the amount of support is a constant percentage, supfac, of the cross-sectional area (assumed as about 30%). If a process does not require supports, then the constant percentage can be taken as 0. The final consideration is the number of times a layer is scanned to fabricate a layer, denoted n_{st}. For example, in stereo-lithography, both X and Y scans are performed for each layer, while in MEX, only one scan is performed to deposit material. Scan length for one part and its support structure is determined using (18.12):

$$sl = A_{avg} \left(\frac{n_{st} L_p}{hr \cdot d} + \text{supfac} \frac{L_s}{d} \right) \tag{18.12}$$

The final step in determining scan/deposition time is to determine scan speed. This is a function of how fast the laser or deposition head moves when depositing material, ss_s, as well as when moving (jumping) between scans, ss_j. In some cases, jump speed is much higher than typical scan speeds. To complicate this matter, many machines have a wide range of scan speeds that depend on several part building details. For example, new VPP machines have scan speeds that range from 100 to 25,000 mm/s. For our purposes, we can assume a typical scan speed that is half of the maximum speed, unless we have more specific information from the machine vendor. The average scan/deposition speed will be corrected using the area correction factor determined earlier [32] as:

$$ss_{avg} = ss_{max} / 2 \cdot \gamma + ss_j \cdot (1 - \gamma) \tag{18.3a}$$

$$ss_{avg} = ss_s \cdot \gamma + ss_j \cdot (1 - \gamma) \tag{18.3b}$$

With the intermediate terms determined, we can compute the scan/deposition time for all parts in the build as:

$$T_s = \frac{N \cdot sl}{3,600 \; ss_{avg}} \tag{18.14}$$

where the 3600 in the denominator converts from seconds to hours.

The final term in the build time expression (18.7) is the delay time, T_e. Many processes have delays built into their operations, such as platform move time, pre-recoat delay ($T_{predelay}$), post-recoat delay ($T_{postdelay}$), nozzle cleaning, sensor recalibration, temperature setpoint delays (waiting for the layer to heat or cool to within a specified range), and more. These delays are often user specified and depend upon build details for a particular process. For example, in VPP, if parts have many fine features, longer pre-recoat delays may be used to allow the resin to cure further, to strengthen the part, before subjecting fragile features to recoating stresses. Additionally, some processes require a start-up time, for example, to heat the build chamber or warm up a laser. This start-up time will be denoted T_{start}. For our purposes, delays will be given by (18.15), but it is important to realize that each process and machine may have additional or different delay terms:

$$T_e = L_p \left(T_{predelay} + T_{postdelay} \right) + T_{Start} \tag{18.15}$$

With the cost and build time models presented, we now turn to the application of these models to VPP.

18.6.3 Laser Scanning Vat Photopolymerization Example

The build time and cost models presented in Sect. 18.6.2 will be applied to the case of hearing aid shell manufacturing using an SLA ProX 800 stereolithography machine from 3D Systems with the smallest vat. The machine parameters are given in Table 18.2. Part information will be assumed to be as follows: bounding box = 15 × 12 × 20 mm, v = 1000 mm^3. An average cross-sectional area of 75 mm^2 will be assumed, instead of using Eqs. 18.10 and 18.11. Layer thickness for the part is 0.05 mm. Support structures are assumed to be 10 mm tall, built with 0.1 mm layer thickness. Since the shell's walls are small, most of the scans will be border vectors; thus, an average laser beam diameter of 0.21 mm is assumed. Gaps of 4 mm will be used between shells.

With these values assumed and given, the build time will be computed first, followed by the cost per shell. We start with the total number of parts on one platform:

$$N = \left(\frac{650 + 4 - 20}{15 + 4} \right) \left(\frac{750 + 4 - 20}{12 + 4} \right) = 1716$$

The numbers of layers of part and support structure are L_p = 400 and L_s = 100. The scan length and scan speed average can be computed as: s_1 = 582,143 mm, ss_s = 5635 mm/s (linearly interpolated based on d = 0.21 mm), and ss_{avg} = 19,619 mm/s. With these quantities, the scan time is:

$$T_s = \frac{1716 \cdot 582,143}{3600 \cdot 19,619} = 14.14 \, hr \tag{18.14}$$

Recoat (layer transition) time is:

$$T_t = 6 / 3,600 \times \left(400 + 100 \right) = 0.83333 \, h.$$

Delay times total:

$$T_e - 400 / 3,600 \times \left(13 + 10 \right) + 0.5 = 3.2 / 8h$$

Table 18.2 SLA ProX 800 parameters

		Short vat	
	PL_x (mm)	650	---
	PL_y (mm)	750	---
	PL_z (mm)	275	
	Purchase Price ($*1000)	700	
	C_o ($/hr)	30	
	C_l ($/hr)	20	
	Y (yrs)	7	
		Border vectors	Hatch vectors
	d (mm)	0.13	0.76
	ss_s (mm/s)	5000	15,000
	ss_j (mm/s)	25,000	$2 \times V_{scan}$
	hr (hatch) (mm)	0.5	
	LT (mm)	0.05–0.15	
	n_{st}	2	
	z_{supp} (mm)	0.10	
	Supfac	0.3	
	$T_{predelay}$ (s)	15	
	$T_{postdelay}$ (s)	10	
	T_{start} (hr)	0.5	
	C_m ($/kg)	200	
	ρ (g/cm³)	1.1	

Adding up the scan, recoat, and delay times gives a total build time of:

$$T_b = 18.25\,hr.$$

Part costs can be investigated now. Machine purchase price allocated to the build is $149. Operating cost for 18.25 hours is $548. Material and labor costs for the build are $339 and $10, respectively. The total cost for the build is computed to be $1045. With 1716 shells in the build, each shell costs about $0.61, which is pretty low considering that the hearing aid will retail for $500 to $2500. However these costs do not include support removal and finishing costs, electronics inserted into the hearing aids, professional services, nor the life-cycle costs discussed below.

18.7   Life-Cycle Costing

In addition to part costs, it is important to consider the costs incurred over the lifetime of the part, from both the customer's and the supplier's perspectives. For any manufactured part (not necessarily using AM processes), life-cycle costs associated with the part can be broken down into six main categories: equipment cost, material cost, operation cost, tooling cost, service cost, and retirement cost. As in Sect. 18.6, equipment cost includes the costs to purchase the machine(s) used to manufacture the part. Material and operation costs are related to the actual manufacturing process and are one-time costs associated only with one particular part. For most conventional manufacturing processes, tooling is required for part fabrication. This may include an injection mold, stamping dies, or machining fixtures. The final two costs, service and retirement, are costs that accrue over the lifetime of the part.

This section will focus on tooling, service, and retirement costs, since they have not been addressed yet. Service costs typically include costs associated with repairing or replacing a part, which can include costs related to taking the product out of service, disassembling the product to gain access to the part, repairing or replacing the part, re-assembling the product, and possibly testing the product. Design-for-service guidelines indicate that parts needing frequent service should be easy to access and easy to repair or replace. Service-related costs are also associated with warranty costs, which can be significant for consumer products.

Let's consider the interactions between service and tooling costs. Typically, tooling is considered for part manufacture. However, tooling is also needed to fabricate replacement parts. If a certain injection molded part starts to fail in aircraft after being in service for 25 years, it is likely that no replacement parts are available "off the shelf." As a result, new parts must be molded. This requires tooling to be located or fabricated anew, refurbished to ensure it is production-worthy, installed, and tested. Assuming the tooling is available, the company would have had to store it in a warehouse for all of those years, which necessitates the construction and maintenance of a warehouse of old tools that may never be used.

In contrast, if the parts were originally manufactured using AM, no physical tooling need be stored, located, refurbished, etc. It will be necessary to maintain an electronic model of the part, which can be a challenge since forms of media become outdated; however, maintenance of a computer file is much easier and less expensive than a large, heavy tool. This aspect of life-cycle costs heavily favors AM processes.

Retirement costs are associated with taking a product out of service, dismantling it, and disposing of it. Large product dismantling facilities exist around the world that take products apart, separate parts into different material streams, and separate materials for distribution to recyclers, incinerators, and landfills. The first challenge for such facilities is collecting the discarded products. A good example of product collection is a community run electronic waste collection event, where people can discard old electronic products at a central location, such as a school or mall parking lot. Product take-back legislation in Europe offers a different approach for the same objective. For automobiles, an infrastructure already exists to facilitate disposal and recycling of old cars. For most other industries, little organized product take-back infrastructure exists in the USA, with the exceptions of glass, metal, paper, and plastic food containers. In contrast to consumer products, recycling and disposal infrastructure exists for industrial equipment and wastes, particularly for metals, glass, and some plastics.

How recyclable are materials used in AM? Metals are very recyclable regardless of the method used to process it into a part. Thus, stainless steel, titanium alloys, and other metal parts fabricated in PBF and DED processes can be recycled. For plastics, the situation is more complicated. The nylon blends used in PBF can be recycled, in principle. However, nylon is not as easily recycled as other common thermoplastics, such as the ABS or polycarbonate materials from MEX systems. Thermoset polymers, such as photopolymers in VPP and MJT processes, cannot be recycled. These materials can only be used as fillers, landfilled, or incinerated.

In general, the issue of life-cycle costing has simple aspects to it but is also very complicated. It is clear that the elimination of hard tooling for part manufacture is a significant benefit of AM technologies, both at the time of part manufacture and over the part's lifetime since spare parts can be manufactured when needed. On the other hand, issues of material recycling and disposal become more complicated, reflecting the various industry and consumer practices across society.

18.8   Future of Direct Digital Manufacturing

There is no question that we will see increasing utilization of AM technologies in production manufacturing. In the near-term, it is likely that new applications will continue to take advantage of the shape complexity capabilities for economical low production volume manufacturing. Longer time frames will see emergence of applications that take advantage of functional complexity capabilities (e.g., mechanisms, embedded components) and material complexities.

To date, tens of thousands of parts have been manufactured for the aerospace industry. Many of these parts are flying on military aircraft, space vehicles, the International Space Station, and many satellites. AM service companies have been created to serve one or more markets. Some large companies have created production factories utilizing AM machines. The machine vendors have reconceptualized some of their machine designs to better serve manufacturing markets. Examples of this are the development of the Additive Industries PBF machine, 3D Systems Sinterstation Pro, HP polymer PBF machines, high-end MEX systems from Stratasys, and newer versions of metal PBF machines from EOS, Renishaw, and others.

Other markets will emerge:

- One need only consider the array of devices and products that are customized for our bodies to see more opportunities that are similar to aligners and hearing aids. From eye glasses and other lenses to dentures and other dental restorations, to joint replacements, the need for complex, customized geometries, hierarchical structures, and complex material compositions is widespread in medical and health-related areas.
- New design interfaces for non-experts have enabled individuals to design and purchase their own personal communication/computing device (e.g., cell phones) housings or covers in a manner similar to their current ability to have a physical

representation of their virtual gaming characters produced. File sharing sites such as thingiverse.com and storefronts such as shapeways.com are popular, and many expansions and generalizations are to be expected.

- Structural components will have embedded sensors that detect fatigue and material degradation, warning of possible failures before they occur.
- The opportunities are bounded only by the imagination of those using AM technologies.

In summary, the capability to process material in an additive manner will drastically change some industries and produce new devices that could not be manufactured using conventional technologies. This will have a lasting and profound impact upon the way products are manufactured and distributed and thus on society as a whole. A further discussion of how DDM will likely affect business models, distributed manufacturing, and entrepreneurship is contained in the final chapter of this book.

18.9   Questions

1. Estimate the build time and cost for a platform of 100 aligner mold parts in an iPro 8000 SLA Center (see Chap. 4). Assume that the bounding box for each part is $11 \times 12 \times 8$ cm and the mold volume is 75,000 mm3. Assume a scanning speed of 5000 mm/s and a jump speed of 20,000 mm/s. All remaining quantities are given in Sect.18.6. What is the estimated cost per mold (two parts)?
2. From an Internet search, identify two DW inks for conductive traces, one ink for resistors and one for dielectric traces, that are commonly used in nozzle-based systems. Make a table which lists their room-temperature properties and their primary benefits and drawbacks.
3. A vat of hearing aid shells is to be built in an SLS Pro 140 machine (build platform size: $550 \times 550 \times 460$ mm). How many hearing aid shells can fit in this build platform? Determine the estimated build time and cost for this build platform full of shells. Assume laser scan and jump speed of 5000 mm/s and 20,000 mm/s, respectively. Assume the laser spot size is 0.2 mm, layer thicknesses are 0.1 mm, and only 1 scanning pass per layer is needed (nst = 1). Assume 4 mm gaps in X, Y, and Z directions. Recall that no support structures are needed. Assume that the SLS machine needs 2 h to warm up and 2 h to cool down after the build. Assume that $T_{predelay}$ is 15 s and $T_{postdelay}$ is 2 s.
4. Develop a build time model for a jetting machine, such as the Eden models from Stratasys or the ProJets from 3D Systems. Note that this is a line-type process, in contrast to the point-wise vector scanning process used in VPP or PBF. Consider that the jetting head can print material during each traversal of the build area and n_{st} may be 2 or 3 (e.g., 2 or 3 passes of the head are required to fully cover the total build area). Assume that $T_{predelay}$ and $T_{postdelay}$ are 2 s.
5. Estimate the build time and cost for a platform of hearing aid shells in an Eden 500 V machine (see Chap. 7). What is the estimated cost per shell? You will

need to visit the Stratasys website and possibly contact Stratasys personnel in order to acquire all necessary information for computing times and costs.

6. Develop a build time model for an MEX machine from Stratasys, such as the Fortus 900mc. Note that this is a point-wise vector process without overlapping scans. Scan speeds can be up to 1000 mm/s. Assume that a warm-up time of 0.5 h is needed to heat the build chamber. Assume that $T_{predelay}$ and $T_{postdelay}$ are 1 s.

7. Estimate the build time and cost for a platform of hearing aid shells in a Fortus 900mc MEX machine. What is the estimated cost per shell? You will need to visit the Stratasys website and possibly contact Stratasys personnel in order to acquire all necessary information for computing times and costs.

8. Modify the model for purchase cost to incorporate net present value considerations. Rework the hearing aid shell example in Sect. 18.6.2 to use net present value. What is the estimated cost of a shell?

9. Bentley Motors has a production volume of 10,000 cars per year, over its 4 main models. Production volume per model per year ranges from about 200 to 4500. Since each car may sell for $120,000 to over $500,000, each car is highly customized. Write a one-page essay on the DDM implications of such a business. The engines for these cars are shared with another car manufacturer; as such, do not focus your essay on the engines. Rather, focus on the chassis, interiors, and other parts of the car that customers will see and interact with.

10. How does process selection software help enable Direct Digital Manufacturing?

11. Give an example not from the book of a company that uses DDM in high volume with mass customization. Explain why DDM is used in this manner (make sure to include specific numbers).

12. You are asked to investigate the time feasibility of an SL process for the production of a new part. The company is allowing for the use of four machines. Your company requires a production rate of 90 parts per day. Assume the machine can run 24 hours day. Assume the machine delays are negligible. Based on the data in the Table Q 18.1, is this going to work?

13. Please explain how DDM relates to the transition from Industry 1.0 up to 4.0.

Table Q 18.1 Question 11

Build box	$102 \times 78 \times 45$ mm
Build volume	95,000 mm^3
Machine build area	$650 \times 750 \times 550$ mm
Avg. cross section	1900 mm^2
Part layer thickness	0.10 mm
Support structure height	5 mm
Support layer thickness	0.5 mm
Laser bream diameter	0.20 mm
Gap between shells	5 mm
Scan length for part	14,747,800 mm
Average scan speed	12,660 mm/s
Support recoat time	80 s
Build layer recoat time	250 s

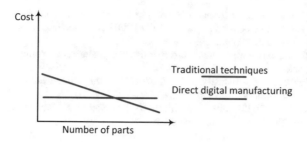

Fig. Q18.1 Question 14

14. Using the figure above, discuss how this type of chart can guide companies as they consider adopting DDM. What are the areas a company can invest R&D funds into to affect the slopes and magnitudes of these curves?

References

1. Muirhead, J. P. (1858). The life of James Watt, with selections from his correspondence. With portraits and woodcuts.
2. Engelman, R. (2015). The second industrial revolution, 1870–1914. US History Scene. 10.
3. Kanji, G. K. (1990). Total quality management: The second industrial revolution. *Total Quality Management, 1*(1), 3–12.
4. Blinder, A. S. (2006). Offshoring: The next industrial revolution? *Foreign Affairs*, 113–128.
5. Rifkin, J. (2011). The third industrial revolution: How lateral power is transforming energy, the economy, and the world. Macmillan.
6. Brettel, M., et al. (2014). How virtualization, decentralization and network building change the manufacturing landscape: An Industry 4.0 Perspective. *International Journal of Mechanical, Industrial Science and Engineering, 8*(1), 37–44.
7. Lasi, H., et al. (2014). Industry 4.0. *Business & Information Systems Engineering, 6*(4), 239–242.
8. Lee, J., Bagheri, B., & Kao, H.-A. (2015). A cyber-physical systems architecture for industry 4.0-based manufacturing systems. *Manufacturing Letters, 3*, 18–23.
9. Wohlers, T. (2014). *Wohlers report. 3D Printing and Additive Manufacturing State of the Industry, Annual Worldwide Progress Report*. Wohlers Associates.
10. Masters, M. (2002). Direct manufacturing of custom-made hearing instruments. In *SME Rapid Prototyping Conference and Exhibition*.
11. Masters, M., Velde, T., & McBagonluri, F. Rapid manufacturing in the hearing industry. In Rapid manufacturing-an industrial revolution for the digital age. 2006, Wiley Verlag. p. 195–208.
12. Airbus. (2020). https://www.airbus.com/public-affairs/brussels/our-topics/innovation/3d-printing.html.
13. Joshi, S. C., & Sheikh, A. A. (2015). 3D printing in aerospace and its long-term sustainability. *Virtual and Physical Prototyping, 10*(4), 175–185.
14. Airbus. (2020). https://3dprintingindustry.com/news/airbus-a350-xwb-takes-off-with-over-1000-3d-printed-parts-48412/

15. *Airbus Helicopters to start large-scale printing of A350 XWB components*. (2020). https://www.airbus.com/newsroom/press-releases/en/2018/09/airbus-helicopters-to-start-large-scale-printing-of-a350-compone.html
16. GE. (2020). https://www.ge.com/additive/stories/new-manufacturing-milestone-30000-additive-fuel-nozzles
17. 3D Printing Industry. (2020). https://3dprintingindustry.com/news/rolls-royce-to-3d-print-aerospace-parts-with-slm-500-157331/
18. Additive News. (2020). https://additivenews.com/boeing-rolls-royce-made-37-6-million-investment-reaction-engines-aerospace-company/
19. 3D Printing Industry. (2020). https://3dprintingindustry.com/news/3d-printed-jet-engine-certified-use-ge-concept-laser-deal-update-101792/
20. Thales. (2020). https://www.thalesgroup.com/en/worldwide/space/press-release/thales-alenia-space-takes-3d-printing-series-production
21. Siemens. (2020). https://www.industrial-lasers.com/home/article/14033779/status-of-additive-manufacturing-for-gas-turbine-components
22. Nissan Motor. (2020). https://www.3dnatives.com/en/nissan-rogue210520184/
23. Toyota. (2020). https://www.toyota-motorsport.com/en/services-en/manufacturing-en/additive-manufacturing-en
24. Mitsubishi. (2020). https://all3dp.com/4/mitsubishi-enters-metal-3d-printing-market-system/
25. BMW. (2020). https://3dprint.com/222268/bmw-3d-printed-roof-bracket/
26. Wohlers, T. (2008). *Wohlers report. State of the Industry, Annual Worldwide Progress Report*. Wohlers Associates.
27. Zarringhalam, H., et al. (2006). Effects of processing on microstructure and properties of SLS Nylon 12. *Materials Science and Engineering: A, 435*, 172–180.
28. Wohlers, T. (2013). Wohler report. 3D Printing and Additive Manufacturing State of the Industry, Annual Worldwide Progress Report. Wohlers Associates.
29. Make Magazine. (2020). http://makezine.com/2013/06/27/how-to-3d-print-a-video-game-figurine/
30. Hastings, B. (2007). The transition from rapid prototyping to direct manufacturing. In *SME RAPID Conference*. Detroit.
31. Johnson, P. (1994). *ISO 9000: Meeting the new international standards* (p. 246). McGraw-Hill.
32. Pham, D., & Wang, X. (2000). Prediction and reduction of build times for the selective laser sintering process. *Proceedings of the Institution of Mechanical Engineers, Part B: Journal of Engineering Manufacture, 214*(6), 425–430.

Chapter 19
Design for Additive Manufacturing

Abstract The benefits and drawbacks of Additive Manufacturing Technologies enable designers to think beyond traditional design for manufacture and assembly constraints. AM has unique geometric, material, and customization benefits not provided by other production techniques. Likewise, AM has need for supports, typically produces anisotropic properties, and may require considerable post-processing. These and other benefits and drawbacks of AM have led to an increased emphasis on training designers to Design for Additive Manufacturing. In this chapter, we will revisit some of the concepts from prior chapters and introduce new concepts and ways of thinking to help designers take advantage of AM without falling into design pitfalls.

19.1 Introduction

Design for manufacture and assembly (DFM[1]) has typically meant that designers should tailor their designs to eliminate manufacturing difficulties and minimize manufacturing, assembly, and logistics costs. However, the capabilities of Additive Manufacturing Technologies provide an opportunity to rethink DFM to take advantage of the unique capabilities of these technologies. As covered in Chap. 18, several companies are now using AM technologies for production manufacturing. For example, Siemens, Phonak, Widex, and the other hearing aid manufacturers use Powder Bed Fusion (PBF), Vat Photopolymerization (VPP), and Material Jetting (MJT) machines to produce hearing aid shells, Align Technology uses stereolithography to fabricate molds for producing clear dental braces ("aligners"), GE uses metal PBF for aircraft engine and power generation components, and Boeing and its suppliers use polymer PBF to produce ducts and similar parts for F-18 fighter jets. For hearing aids and dental aligners, AM machines enable manufacturing of millions of parts,

---

[1] Design for manufacturing is typically abbreviated DFM, whereas design for manufacture and assembly is typically abbreviated as DFMA. To avoid confusion with the abbreviation for design for additive manufacturing (DFAM) we have utilized the shorter abbreviation DFM to encompass both design for manufacture and design for assembly.

where each part is uniquely customized based upon person-specific geometric data. In the case of aircraft components, AM technology enables low-volume manufacturing, easy integration of design changes, optimized geometries, and, at least as importantly, piece part reductions to greatly simplify product assembly.

The unique capabilities of AM technologies enable new opportunities for customization, very significant improvements in product performance, multifunctionality, and lower overall manufacturing costs. These unique capabilities include *shape complexity*, in that it is possible to build virtually any shape; *hierarchical complexity*, in that hierarchical multi-scale structures can be designed and fabricated from the microstructure through geometric mesostructure (sizes in the millimeter range) to the part-scale macrostructure; *material complexity*, in that material can be processed one point, or one layer, at a time; and *functional complexity*, in that fully functional assemblies and mechanisms can be fabricated directly using AM processes. These capabilities will be expanded upon in Sect. 19.3.

In this chapter, we begin with a brief look at DFM to draw contrasts with Design for Additive Manufacturing (DFAM). A considerable part of the chapter is devoted to the unique capabilities of AM technologies and a variety of illustrations of these capabilities. We cover the emerging area of engineered cellular materials and relate it to AM's unique capabilities. Perhaps the most exciting aspect of AM is the design freedom that is enabled; we illustrate this with several examples from the area of industrial design (housewares, consumer products) that exhibit unique approaches to product design, resulting in geometries that can be fabricated only using AM processes. The limitations of current Computer-Aided Design (CAD) tools are discussed, and thoughts on capabilities and technologies needed for DFAM are presented. The chapter concludes with a discussion of design synthesis approaches to optimize designs.

19.2 Design for Manufacturing and Assembly

Design for manufacturing and assembly can be defined as the practice of designing products to reduce, and hopefully minimize, manufacturing and assembly difficulties and costs. This makes perfectly good sense, as why would one want to increase difficulties and costs? However, DFM requires extensive knowledge of manufacturing and assembly processes, supplier capabilities, material behavior, etc. DFM, although simple conceptually, can be difficult and time-consuming to apply. To achieve the objectives of DFM, researchers and companies have developed a large number of methods, tools, and practices. Our purpose in this chapter is not to cover the wide spectrum of DFM advances; rather, it is to convey a sense of the variety of DFM approaches so that we can compare and contrast DFAM with DFM [1].

Broadly speaking, DFM efforts can be classified into three categories:

- Industry practices, including reorganization of product development using integrated product teams, concurrent engineering, and the like

- Collections of DFM rules and practices
- University research in DFM methods, tools, and environments

During the 1980s and 1990s, much of the product development industry underwent significant changes in structuring product development organizations [2]. Companies such as Boeing, Pratt & Whitney, Ford, etc. reorganized product development into teams of designers, engineers, manufacturing personnel, and possibly other groups; these teams could have hundreds or even thousands of people. The idea was to ensure good communication among the team so that design decisions could be made with adequate information about manufacturing processes, factory floor capabilities, and customer requirements. Concurrently, manufacturing engineers could understand decision rationale and start process planning and tooling development to prepare for the in-progress designs. A significant driver of this restructuring was to identify conflicts early in the product development process and reduce the need for redesign or, even worse, retooling of manufacturing processes after production starts.

The second category of DFM work, that of DFM rules and practices, is best exemplified by the Handbook for Product Design for Manufacture [3]. The 1986 edition of this handbook was over 950 pages long, with detailed descriptions of engineering materials, manufacturing processes, and rules-of-thumb. Extensive examples of good and bad practices are offered for product design for many of these manufacturing processes, such as molding, stamping, casting, forging, machining, and assembly.

University research during the 1980s and 1990s started with the development of tools and metrics for part manufacture and assembly. The Boothroyd and Dewhurst toolkit is probably the most well-known example [4]. The main concept was to develop simple tools for designers to evaluate the manufacturability of their designs. For example, injection molding DFM tools were developed that asked designers to identify how many undercuts were in a part, how much geometric detail is in a part, how many tight tolerances were needed, and similar information. From this information, the tool provided assessments of manufacturability difficulties, cost estimates, and some suggestions about part redesign. Similar tools and metrics were developed for many manufacturing and assembly processes, based in part on the handbook mentioned above, and similar collections of information. Some of these tools and methods were manual, while others were automated; some were integrated into CAD systems and performed automated recognition of difficulties. For instance, Boothroyd Dewhurst, Inc. markets a set of software tools that help designers conceive and modify their design to achieve lower-cost parts, taking into account the specific manufacturing process being utilized. In addition, they sell software tools which help designers improve the design of assembled components through identification of the key functional requirements of an assembled component, leading the designer through a process of design modifications with the aim of minimizing the number of parts and assembly operations used to create that assembled component. The work in this area is extensive; see, for example, the ASME Journal of Mechanical Design and the ASME International Design Engineering Technical Conferences proceedings since the mid-1980s (see, e.g., [5]).

The extensive efforts on DFM over many years are an indication of the difficulty and pervasiveness of the issues surrounding DFM. In effect, DFM is about the designer understanding the constraints imposed by manufacturing processes and then designing products to minimize constraint violation. Some of these difficulties are lessened when parts are manufactured by AM technologies, but some are not. Integrated product development teams that practice concurrent engineering make sense, regardless of intended manufacturing processes. Rules, methods, and tools that assist designers in making good decisions about product manufacturability have a significant role to play. However, the nature of the rules, methods, and tools should change to assist designers in understanding the design freedom allowed by AM and, potentially, aiding the designer in exploring the resulting open design spaces while ensuring that manufacturing constraints (yes, AM technologies do have constraints) are not violated.

To illustrate the differences between DFM and DFAM, we will introduce two examples. The first involves typical injection molding considerations, that of undercuts and feature detail. Consider the camera spool part shown in Fig. 19.1 [6]. The various ribs and pockets are features that contribute to the time and cost of machining the mold in which the spools will be molded. Such feature detail is not relevant to AM processes since ribs can be added easily during processing in an AM machine. A similar result relates to undercuts. This spool design has at least one undercut, since it cannot be oriented in a mold consisting of only two mold pieces (core and cavity) while enabling the mold halves to be separated and the part removed. Most probably, the spool will be oriented so that the mold closure direction is parallel to the walls of the ribs. In this manner, the core and cavity mold halves form most of the spool features, including the ribs (or pockets), the flanges near the ends, and the groove seen at the right end. In this orientation, the hole in the right end cannot be formed using the core and/or cavity. A third moving mold section, called a side action, is needed to form the hole. In AM processes, it is not necessary to be so concerned about the relative position and orientation of features, since, again, AM machines can fabricate features regardless of their position in the part.

In design for assembly, two main considerations are often offered to reduce assembly time, cost, and difficulties: minimize the number of parts and eliminate fasteners. Both considerations translate directly to fewer assembly operations, the

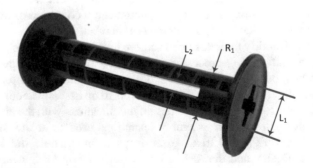

Fig. 19.1 Camera spool example

primary driver for assembly costs [4]. To minimize parts and fasteners, integrated part designs typically become much more complex and costly to manufacture. Design for manufacture and design for assembly will often be repeated, iteratively, until an optimal solution is found, one where the increasing manufacturing costs for more complex components are no longer compensated by the assembly cost savings.

The designs in Fig. 18.4 show two very different approaches to designing ducts for aircraft [7, 8]. This example represents a design concept for conveying cooling air to electronic units in military aircraft but could apply to many different applications. The first design is a typical approach using parts fabricated by conventional manufacturing processes (stamping, sheet metal forming, assembly using screws, etc.). In contrast, the approach on the right illustrates the benefits of taking design for assembly guidelines to their extreme: the best way to reduce assembly difficulties and costs is to eliminate assembly operations altogether. The resulting design replaces 16 parts and fasteners with 1 part that exhibits integrated flow vanes and other performance-enhancing features. However, this integrated design cannot be fabricated using conventional manufacturing techniques and is only manufacturable using AM.

19.3   Core DFAM Concepts and Objectives

Given unique capabilities of AM, we can articulate some core DFAM concepts and objectives. In contrast to DFM, we believe the objective of DFAM should be to:

> Maximize product performance through the synthesis of shapes, sizes, hierarchical structures, and material compositions, subject to the capabilities of AM technologies.

To realize this objective, we will identify the unique capabilities of AM and relate them to design opportunities and their consequences. First, we will make a distinction between opportunistic and restrictive designs for manufacturing.

19.3.1   Opportunistic vs. Restrictive DFAM

DFAM is a design practice to encourage designers to explore new design concepts and develop new designs by exploring the unique capabilities of AM. Of course, these designs must be manufacturable by the selected AM process. These two broad considerations frame the scope of the DFAM effort, from a "clean sheet" conceptual design activity, on one hand, to the specification of manufacturable shape details on the other. The former consideration can be described as *opportunistic design*, where the designer takes advantages of the opportunities afforded by AM processes, while the latter can be called *restrictive design*, focusing on the restrictions imposed by the AM process (Table 19.1).

Table 19.1 Opportunistic vs. restrictive design

Opportunistic design	Restrictive design
Explore the design space enabled by the manufacturing process	Understand process constraints
Identify AM characteristics to leverage (customization, part consolidation)	Determine how to satisfy constraints
Generative design and topology optimization	Design rules (avoid need for support structures)
Function sharing or integration	Feature design suggestions (min. feature size)
Multiple or graded materials	Design to minimize costs

19.3.2 AM Unique Capabilities

The layer-based additive nature of AM leads to unique design opportunities in comparison with most other manufacturing processes. After explaining these uniquenesses, several examples and classes of applications will be presented in the next section. The unique capabilities mentioned at the beginning of the chapter were:

- Shape complexity: it is possible to build virtually any shape.
- Hierarchical complexity: features can be designed with shape complexity across multiple size scales.
- Functional complexity: functional devices (not just individual piece-parts) can be produced in one build.
- Material complexity: material can be processed one point, or one layer, at a time as a single material or as a combination of materials.

To date, primarily shape complexity has been used to enable production of end-use parts, but applications taking advantage of the other capabilities, particularly material and hierarchical complexity, are becoming more popular as well. Figure 19.2 illustrates these unique capabilities and relates them to design opportunities and their consequences. AM design-related capabilities will be described in this section, while the opportunities and consequences will be presented in the next section.

19.3.3 Shape Complexity

In AM, the capability to fabricate a layer is unrelated to the layer's shape. For example, the lasers in VPP and PBF can reach any point in a part's cross-section and process material there. As such, part complexity is virtually unlimited. This is in stark contrast to the limitations imposed by machining or injection molding, two common processes. In machining, tool accessibility is a key limitation that governs part complexity. In injection molding, the need to separate mold pieces and eject parts greatly limits part complexity.

Unique Capabilities	Potential Opportunities	Example Consequences
Shape complexity	Complex geometry	Part consolidation: reduce mfg operations, assembly, inventory, simplified supply chain Efficient, lightweight products Integrated products, Compliant mechanisms
Hierarchical complexity	Custom geometry	Personalization Patient-matched products Lot size of 1
Material complexity	Multi-Materials	Multi-functionality Integrated, efficient products High performance products Embedded sensors, electronics, actuators, power
Functional complexity	No tooling	Lot size of 1 Short lead-times Agile mfg; short time-to-market; Distributed mfg

Fig. 19.2 Relationships among unique capabilities, design opportunities, and consequences

A related capability is to enable custom-designed geometries. In production using AM, it does not matter if one part has a different shape than the previously produced part. Furthermore, no hard tooling or fixtures are necessary, which implies that lot sizes of one can be economically feasible. This is tremendously powerful for medical applications, for example, since everyone's body shape is different. Also, consider the design of a high-speed robot arm. High stiffness and low weight are desired. With AM, the capability is enabled to put material where it can be utilized best. The link from a commercial Adept robot (Cobra 600) shown in Fig. 19.3 has been stiffened with a custom-designed lattice structure that conforms to the link's shape. Preliminary calculations show that weight reductions of 25% are achieved readily with this lattice structure and that much greater improvements are possible. More generally, AM processes free designers from being limited to shapes that can be fabricated using conventional manufacturing processes.

Another factor enabling lot sizes of one, and shape complexity, is the capability for automated process planning. Straightforward geometric operations can be performed on AMF or STL files (or CAD models) to decompose the part model into operations that an AM machine can perform. Although CNC has improved greatly, many more manual steps are typically utilized in process planning and generating machine code for CNC than for AM.

19.3.4 Hierarchical Complexity

Similar to shape complexity, AM enables the design of hierarchical complexity across several orders of magnitude in length scale. This includes nano-/microstructures, mesostructures, and part-scale macrostructures. We will start with material microstructures.

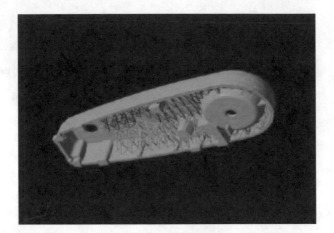

Fig. 19.3 Robot link stiffened with lattice structure

Fig. 19.4 60% CP-Ti, 40% TiC composite made using LENS. The ratio of unmelted carbides (UMCs) to resolidified carbides (RSCs) within the Ti matrix is controlled by varying LENS process parameters

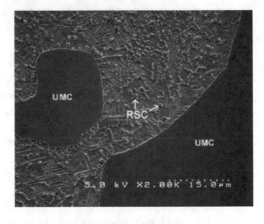

One set of processes, which has been studied extensively with respect to hierarchical complexity, are the Directed Energy Deposition (DED) processes. In LENS, for instance, the nano-/microstructure can be tailored in a particular location by controlling the size and cooling rate of the melt pool. As a result, the size and distribution of precipitates (nano-scale) and secondary particles (microscale), for example, can be changed by locally modifying the laser power and scan rate. Figure 19.4 illustrates the types of microstructural features which can be formed when using LENS to process mixtures of TiC in Ti to form a composite structure. There are several features of the microstructure which can be controlled. In cases of lower laser energy densities, there is a greater proportion of unmelted carbide (UMC) particles within the microstructure. At higher-energy densities, more of the TiC particles melt and precipitate as resolidified carbides (RSCs). In addition, as the RSCs have a different stoichiometry (TiC transforms to $TiC_{0.65}$), for a given initial mixture of TiC and Ti, the more RSC that is present in the final microstructure, the less Ti matrix material is present. The resulting microstructures can thus have very different material and mechanical properties. If sufficient RSCs are precipitated to consume the Ti matrix material, then the structure becomes very brittle. In contrast, when most of the TiC is present as UMCs, the structure is more ductile but is less resistant to abrasive wear.

In addition to the nano-/microstructure illustration above, DED technologies have been shown to be capable of producing equiaxed, columnar, directionally solidified, and single-crystal grain structures. These various types of nano-/micro-structures can be achieved by careful control of the process parameters for a particular material and can vary from point to point within a structure. In many cases, for laser or electron beam PBF processes for metals, these variations are also achievable. Similarly, by varying either the materials present (when using a multi-material AM system) or the processing of the materials, this type of nano-/microstructure control is also possible in every other AM technology as well. These related possibilities are further explored below with respect to material complexity.

The ability to change the mesostructure of a part is typically associated with the application of cellular structures, such as honeycombs, foams, or lattices, to fill certain regions of a geometry. This is often done to increase a part's strength to weight or stiffness to weight ratio. These structures are discussed in more detail in Sect. 19.5.2.

When considered together, the ability to simultaneously control a part's nano-/microstructure, mesostructure, and macrostructure simply by changing process parameters and CAD data is a capability of AM which is unparalleled using conventional manufacturing.

19.3.5   Functional Complexity

When building parts in an additive manner, one always has access to the inside of the part. Two capabilities are enabled by this. First, by carefully controlling the fabrication of each layer, it is possible to fabricate operational mechanisms in some AM processes. By ensuring that clearances between links are adequate, revolute or translational joints can be created. Second, by pausing the process, components can be inserted into parts being built, enabling in situ assembly.

A wide variety of kinematic joints has been fabricated directly in VPP, Material Extrusion (MEX) and PBF technologies, including vertical and horizontal prismatic, revolute, cylindrical, spherical, and Hooke joints. Figure 19.5 shows one example of a pulley-driven, snakelike robot with many revolute joints that was built as assembled in the SLA-3500 machine at Georgia Tech.

Similar studies have been performed using MEX and PBF processes. A research group at Rutgers University [9] demonstrated that the same joint geometries could be fabricated by both MEX and PBF machines and similar clearances were needed in both machine types. In PBF, loose powder must be removed from the joint locations to enable relative joint motion. In MEX, the usage of soluble support material ensures that joints can be movable after post-processing in a suitable solvent

In the construction of functional prototypes, it is often advantageous to embed components into parts while building them in AM machines. This avoids

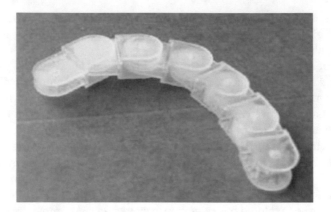

Fig. 19.5 Pulley-driven snakelike robot

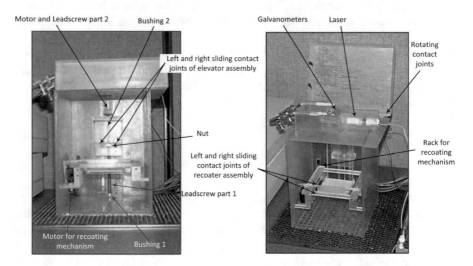

Fig. 19.6 SLA-250 model built in a SLA-250 machine with 11 embedded components

post-fabrication assembly and can greatly reduce the number of separate parts that have to be fabricated and assembled.

For example, it is possible to fabricate VPP devices with a wide range of embedded components, including small metal parts (bolts, nuts, bushing), electric motors, gears, silicon wafers, printed circuit boards, and strip sensors. Furthermore, VPP resins tend to adhere well to embedded components, reducing the need for fasteners. Shown in Fig. 19.6 is a model of a SLA-250 machine that was built in the SLA-250 at Georgia Tech [10]. This 150 × 150 × 260 mm model was built at 1:1/4 scale, with seven inserted components, four sliding contact joints, and one rotating contact joint. The recoating blade slides back-and-forth across the vat region, driven by an electric motor and gear train. Similarly, the elevator and platform translate vertically, driven by a second electric motor and leadscrew. The laser

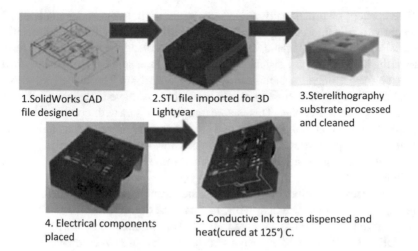

1.SolidWorks CAD
file designed

2.STL file imported for 3D
Lightyear

3.Sterelithography
substrate processed
and cleaned

4. Electrical components
placed

5. Conductive Ink traces dispensed and
heat(cured at 125°) C.

Fig. 19.7 Fabrication of a magnetic flux sensor using VPP and DW. (Photo courtesy of W.M. Keck Center for 3D Innovation at the University of Texas at El Paso)

pointer and galvanometers worked to draw patterns on the platform, but these three components were assembled after the build, rather than subjecting them to being dipped into the resin vat. Build time was approximately 75 h, including time to pause the build and insert components.

Other researchers have also demonstrated the capability of building functional devices, including at Stanford University [11]. Device complexity is greatly facilitated when the capability to fabricate kinematic joints is coupled with embedded inserts since functional mechanisms can be fabricated entirely within the VPP vat, greatly simplifying the prototyping process.

Functional complexity can also be achieved by unique combination of AM technologies to produce, for instance, 3D integrated electronics. Researchers in the W.M. Keck Center for 3D Innovation at the University of Texas at El Paso have demonstrated the ability to produce a number of working devices by novel combinations of VPP, MEX, and Direct Write (DW). Figure 19.7 illustrates the process plan for fabrication of a magnetic flux sensor using VPP and a nozzle-based DW process. Researchers have demonstrated similar capabilities with MEX, PBF, and other technologies as well.

19.3.6 Material Complexity

Since material is processed point to point in many of the AM technologies, the opportunity is available to process the material differently at different points, as illustrated above, causing different material properties in different regions of the part. In addition, many AM technologies enable changing material composition gradually or abruptly during the build process. New applications will emerge to take advantage of these characteristics.

The concept of functionally graded materials, or heterogeneous materials, has received considerable attention [12], but manufacturing useful parts from these materials has been problematic. Consider a turbine blade for a jet engine. The outside of the blade must be resistant to high temperatures and be very stiff to prevent the blade from elongating significantly during operation. The blade root must be ductile and have high fatigue life. Blade interiors must have high heat conductivity so that blades can be cooled. This is an example of a part with complex shape that requires different material properties in different regions. No single material is ideal for this range of properties. Hence, if it was possible to fabricate complex parts with varying material composition and properties, turbine blades and similar parts could benefit tremendously.

DED process machines have demonstrated capability for fabricating graded material compositions. Ongoing work in this direction is promising. Graded and multi-material compositions are used in the repair of damaged or worn components using DED machines, and the design and fabrication of new components is being explored around the world. One application that has received considerable attention is the fabrication of higher-performance orthopedic implants. In this case, certain regions of the implant require excellent bone adhesion, whereas in other regions the bearing surfaces must be optimized to minimize the implant's wear properties. Thus, by changing the composition of the material from the bone in-growth region to the bearing surface, the overall performance of the implant can be improved.

As described in Chap. 7, Objet Geometries Ltd. (now Stratasys) introduced in 2007 the first commercial AM machine, their Connex500™ system, capable of inkjet deposition of several polymer materials in one build. Their technology, called PolyJet Matrix™, is an evolution of their printing technology. Recall that Objet uses large arrays of printing nozzles (up to 3000) to quickly print parts using photopolymer materials. More recently, both Stratasys and 3D Systems have introduced full-color printing technology using MJT of photopolymers that exhibit a much wider range of mechanical properties than Objet's original materials.

For many years, MEX machines have been shipped with multiple nozzles for multi-material deposition. Although one or more nozzles are typically utilized for support materials and the other for build materials, many researchers and industrial practitioners have utilized different feedstock materials in two banks of nozzles to create multi-material constructs. As can easily be imagined, it would be quite easy, conceptually, to add more nozzles and thus easily increase the number of materials which can be deposited in a single build. In fact, this concept has been utilized by a number of researchers in their own custom-built extrusion-based machines, primarily by those investigating extrusion-based processes for biomedical materials research.

A significant issue hindering the adoption of AM's material complexity is the lack of design and CAD tools that enable representation and reasoning with multiple materials. This will be explored more completely in Sect. 19.6.

19.4   Design Opportunities

AM's unique capabilities enable design opportunities which will be explored in this section, along with their consequences. Designers should keep in mind several guidelines when designing products:

- AM enables the usage of complex geometry in achieving design goals without incurring time or cost penalties compared with simple geometry.
- As a corollary to the first guideline, it is often possible to consolidate parts, integrating features into more complex parts and avoiding assembly issues.
- AM enables the usage of customized geometry and parts by direct production from 3D data.
- With the emergence of commercial multi-material AM machines, designers should explore multifunctional part designs that combine geometric and material complexity capabilities.
- AM allows designers to ignore all of the constraints imposed by conventional manufacturing processes, although AM-specific constraints still exist (restrictive DFAM).

19.4.1   Part Consolidation Overview

As mentioned previously, part consolidation is the practice of combining parts into a smaller number of more complex parts in an effort to reduce part count and eliminate fasteners. This has several significant advantages over designs with multiple parts. First, dedicated tooling for multiple parts is not required. Potential assembly difficulties are avoided. Assembly tooling, such as fixtures, is not needed. Fasteners can often be eliminated. Finally, it is often possible to design the consolidated parts to perform better than the assemblies.

A well-known example that illustrates these advantages was shown in Fig. 18.4, that of a prototypical duct for military aircraft [7, 8]. The design shows a typical traditional design with many formed and rotomolded plastic parts, some formed sheet metal parts, and fasteners [13]. The example was from the pioneering work of the Boeing Phantom Works Advanced Direct Digital Manufacturing group in retrofitting F-18 fighter jets with dozens of parts produced using PBF. Many of these parts replaced standard ducting components to deliver cooling air to electronics modules. Significant part reductions, elimination of fasteners, and optimization of shapes are illustrative of the advances made by Boeing. Through these methods, many part manufacturing tools and assembly operations were eliminated.

A more careful investigation of part consolidation reveals the importance of considering product functionality and architecture, that is, what the product is supposed to do and how it does it in terms of module and part arrangements. In design methodology, function structures are often used to describe the high-level sequences of behaviors and flows (of energy, material, and signals) in solution-neutral terms.

From the function structures, solution principles are identified, and initial embodiments are generated as potential product concepts. At this point, relationships between the behaviors/functions in the function structure and the modules and parts in the product can be identified. If several parts are required to fulfill a single function, those parts are candidates for consolidation into a single complex part (or reduced number of parts). If one part is used to satisfy more than one function, that part is considered a multifunctional part. A third possibility is more radical: to combine several functions, which eliminates intermediate behaviors in a flow. The consequence of this function combination is to greatly change the product architecture. That is, it is likely that the overall arrangement of modules and parts in the product will be altered and the total number of parts significantly reduced.

From this perspective, three types of part consolidation can be identified:

- Part integration: combination of several parts into a smaller number, all of which implement a single function
- Function sharing: the implementation of several functions in a single part
- Function integration: combination of several functions into fewer functions, with a corresponding simplification of product architecture and part count reduction

The aircraft duct example illustrates part integration: all parts are used to convey cooling air from one place to another. Part integration is the form of part consolidation that is most common.

Function sharing [14] and integration typically result in more substantial changes to the product. Consider a motorcycle. The primary functions (from a function structure perspective) can be considered as support driver, enable steering, control speed, store chemical energy, convert chemical to mechanical energy, transfer mechanical energy, absorb significant disturbances (from road), dampen effect of significant disturbances, and dampen minor disturbances (high frequency). Such a function description is solution neutral in that it applies to either a gasoline engine or electric motor as powertrain.

With this description of functions, function sharing part consolidation can be illustrated. In the most radical approach, the functions of support driver and the absorption and dampening of all disturbances can be accomplished by the motorcycle frame, if it is designed correctly. The 3D printed Nera motorcycle shown in Fig. 19.8 is an example of part integration, function sharing, and function integration [15]. Designed by NOWLAB and printed on BigRep MEX machines, the Nera consists of 15 printed parts, the largest of which was 120 cm long, using several materials, from flexible to rigid thermoplastics. The motor, battery, electronics module, electrical connectors, and headlight were assembled into the 3D printed structure.

Regarding part consolidation characteristics, the frame and seat were integrated. The typical handlebar–stem–fork construction was reconceptualized as an integrated component with an externally visible array of eight pivot joints. The suspension functions were shared across several components, including tires, wheels, the honeycomb structure bumper under the seat, and the entire frame (which is less stiff than typical metal frames). Finally, function integration was achieved by using a low-profile electric motor that was directly mounted on the rear wheel hub, which eliminates the "transfer mechanical energy" function and the corresponding transmission (e.g., chain drive). The relationships among functions and structures are illustrated in Fig. 19.9, which demonstrates part integration, function sharing, and function integration. The arrows are intended to denote "implements."

Further, several designs for MEX practices are evident. The designers chose a "polygon/stealth design language" with which to design the eBike. Among other characteristics, this approach allowed the parts to be printed without horizontal upfacing surfaces during fabrication, which are not desirable for MEX with large deposited filaments. Most parts were printed with 12 percent fill, which was determined to satisfy structural requirements while minimizing weight and build time.

19.4.2 Design for Function

A related approach to DFAM is known as design for function. In this case, the part or product is to be designed to achieve functions optimally, using optimization or direct design methods. Optimization methods, including topology optimization and generative design, are discussed later in this chapter. An example of direct design for function is illustrated here.

This example, from Loughborough University, illustrates the advantages of reconceptualizing the design of a component based on the ability to avoid limitations of conventional manufacturing processes. Figure 19.10a shows a front plate

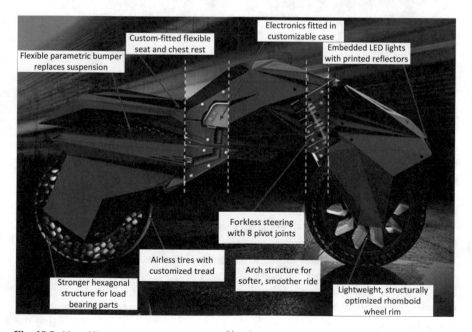

Fig. 19.8 Nora 3D printed eBike with many of its design features highlighted

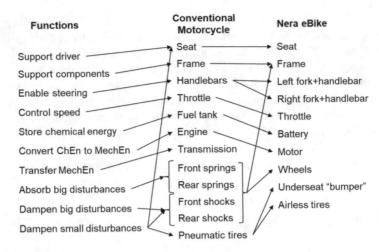

Fig. 19.9 Relations among functions and main structures for conventional motorcycles and the Nera eBike

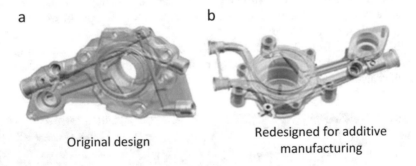

a	b
Original design	Redesigned for additive manufacturing

Fig. 19.10 Diesel front plate example

design for a diesel engine [8]. The channels through which fuel or oil flow are gun-drilled. As a result, they are straight; furthermore, plugs need to be added to plug up the holes through the housing that enabled the channels to be drilled. The redesign shown in Fig. 19.10b was developed by designing the flow channels to ensure efficient flows and then adding a minimal amount of additional material to provide structural integrity. As a result, the part is smaller and lighter and has better performance than the original design.

The layer-by-layer AM fabrication approach means that the shapes of part cross-sections can be arbitrarily complex, up to the resolution of the process. For example, VPP and PBF processes can fabricate features almost as thin as their laser spot sizes. In MJT processes, features in the layer can be the size of several printed droplets. In the Z direction (build direction), the discussion of feature complexity becomes more complicated. In principle, features can be as thin as a layer thickness; however, in practice, features typically are several layers thick. Stresses during the build, such as produced by recoating in VPP, can limit Z resolution. Also, overcure or "bonus Z" effects occur in laser-based processes and tend to create regions that

a   b

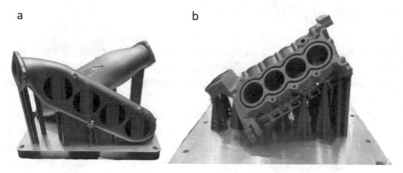

Fig. 19.11 (**a**) Manifold with support structure in internal surface from SLM Solutions. (**b**) Engine block with optimized design to avoid support structure from Lumex

are thicker than a single layer. The need to remove the support structures necessary for some AM processes may also limit geometric complexity and/or feature size. Each AM process has its individual characteristics and will take some time to learn. When designing parts, the need for support structures – and their removal – must be considered. For instance, avoiding support structures for internal surfaces can be accomplished often by changing their build orientation, which is shown in Fig. 19.11. But in general, the geometric complexity of AM processes far exceeds that of conventional manufacturing processes.

19.4.3 Part Consolidation Consequences

The capability for complex geometry enables other practices. As was demonstrated at the end of Sect. 19.2, several parts can be replaced with a single, more complex part in many cases. Even when two or more components must be able to move with respect to one another, such as in a ball-and-socket joint, AM can build these components fully assembled. These capabilities enable the integration of features from multiple parts, possibly yielding better performance. Additionally, a reduction in the number of assembly operations can have a tremendous impact on production costs and difficulties for products.

As is evident from conventional DFM practices, design changes to facilitate or eliminate assembly operations can lead to much larger reductions in production costs than changes to facilitate part manufacture [4]. This is true, at least in part, due to the elimination of any assembly tooling that may have been required. Although conventional DFM guidelines for part manufacturing are not relevant to AM, the design-for-assembly guidelines remain relevant and perhaps even more important. Other advantages exist for the consolidation of parts. For example, a reduction in part count reduces product complexity from management and production perspectives. Fewer parts need to be tracked, sourced, inspected, etc. The need for spare or replacement parts decreases. Furthermore, the need to warehouse tooling to fabricate the parts can be eliminated. In summary, part consolidation can lead to significant savings across the entire enterprise.

19.4.4   Customized Geometry

Consistent with the capability of complex geometry, AM processes can fabricate custom geometries. This has been demonstrated by a series of examples throughout this book related to direct digital manufacturing and biomedical applications. A good example is that of hearing aid shells (Sect. 18.2.2). Each shell must be customized for an individual's particular ear canal geometry. In VPP or PBF machines, hundreds or thousands of shells, each of a different geometry, can be built at the same time in a single machine. Mass customization, instead of mass production, can be realized quite readily. The lack of generic software tools for mass customization, rather than limitations of the hardware, is the key limitation when considering AM for mass customization.

19.4.5   Hierarchical Structures

The basic idea of hierarchical structures is that features at one size scale can have smaller features added to them, and each of those smaller features can have smaller features added, etc. Tailored nano-/microstructures are one example. Textures added to surfaces of parts are another example. In addition, cellular materials (materials with voids), including foams, honeycombs, and lattice structures, are a third example of hierarchical feature. To illustrate the benefits of designing with hierarchical flexibility, we will focus on cellular materials in this section.

The concept of designed cellular materials is motivated by the desire to put material only where it is needed for a specific application. From a mechanical engineering viewpoint, a key advantage offered by cellular materials is high strength accompanied by a relatively low mass. These materials can provide good energy absorption characteristics and good thermal and acoustic insulation properties as well [16]. When the characteristic lengths of the cells are in the range of 0.1–10 mm, we refer to these materials as mesostructured materials. Mesostructured materials that are not produced using stochastic processes (e.g., foaming) are called designed cellular materials.

In the past 15 years, the area of lattice materials has received considerable attention due to their inherent advantages over foams in providing light, stiff, and strong materials [17]. Lattice structures tend to have geometry variations in three dimensions, as is illustrated in Fig. 19.12. As pointed out in [18], the strength of foams scales as $\rho^{1.5}$, whereas lattice structure strength scales as ρ, where ρ is the volumetric density of the material. As a result, lattices with a $\rho = 0.1$ are about three times stronger per unit weight than a typical foam. The strength differences lie in the nature of material deformation: the foam is governed by cell wall bending, while lattice elements stretch and compress. The examples shown in Fig. 19.12 utilize the octet-truss (shown on the *left*), but many other lattice structures have been developed and studied [19, 20].

The parts shown in Fig. 19.12b and c illustrate one method of developing stiff, lightweight structures, that of using a thin part wall, or skin, and stiffening it with cellular structure. Another method could involve filling a volume with cellular structures. Using either approach results in part designs with thousands of shape elements (beams, struts, walls, etc.). Most commercial CAD systems cannot perform geometric modeling operations on designs with more than 1000–2000 elements. As a result, the design in Fig. 19.12c, which has almost 18,000 shape elements, cannot be modeled using conventional CAD software. Instead, new CAD technologies must be developed that are capable of modeling such complex geometries [21]; this is the subject of Sect. 19.6.

Several groups designed unmanned aerial vehicle (UAV) components by applying various cellular structure design approaches. Figure 19.13 shows a handheld UAV, the Streetflyer from AVID LLC, that was redesigned to utilize lattice structure reinforcement. The original design of the UAV utilized carbon fiber skins for the fuselage and wings but required many assembly operations to add stiffeners, fastening features, and mounting features to the components. In contrast, by designing for AM, the lattice structure-based design had such features and stiffeners designed in. Experts at Paramount Industries, a 3D Systems company, fabricated the fuselage and wings in Duraform using PBF. Test flights demonstrated that the PBF-fabricated UAV performed well, and, even though the UAV was not optimized, its performance approached that of the carbon fiber production version.

Another type of hierarchical structure is surface texture. AM techniques can be used to print porous structures with global morphological properties that are highly controlled by robust Computer-Aided Design. They provide a platform for design of textured parts for various applications like medical, aerospace, etc. For instance, the surface topography of different implants plays a significant role in providing cell

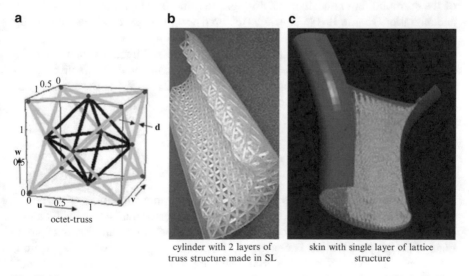

a

octet-truss

b

cylinder with 2 layers of
truss structure made in SL

c

skin with single layer of lattice
structure

Fig. 19.12 (a) Octet-truss unit cell and (b, c) example parts with octet-truss mesostructures

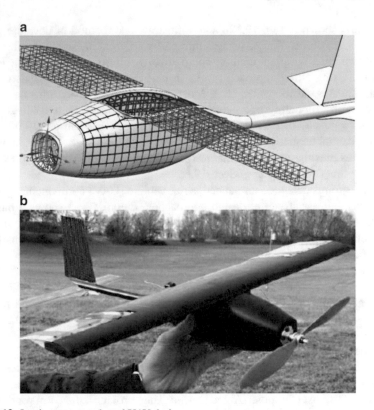

Fig. 19.13 Lattice structure-based UAV design

attachment, growth, and proliferation. Cells are able to sense the surface topography of the surrounding extracellular matrix and respond by changing their alignment and migration. Figure 19.14a illustrates textured design on an acetabular cup for hip implantation.

3D texturing can be used for different purposes and architectures such as tactile, aesthetic, mapping, etc. These features are easy to obtain and cost-effective by utilizing AM. Figure 19.14b–f shows different architectures of microwells designed for AM and printed by stereolithography.

19.4.6   Multifunctional Designs

Multifunctionality is simply the achievement of multiple functions, or purposes, with a single part. This is commonly achieved when performing part consolidation, but the capability of material complexity enables much more ambitious explorations of design possibilities. For example, if a part needs to be stiff in one location, but flexible in another, a multi-material AM process could be used to fabricate such a design simply by varying material composition. Another example is a heat

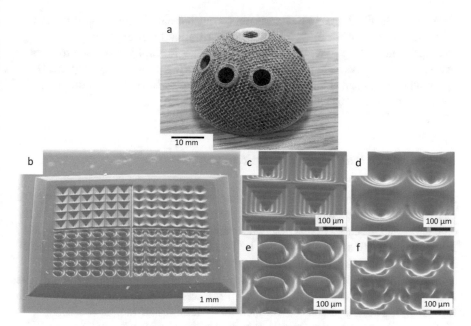

Fig. 19.14 Example textured surfaces. (**a**) Acetabular cup made by PBF [22]. (**b–f**) Microwells printed by stereolithography (Wiley license number 4638500538165) [23]

exchanger that also serves a structural purpose, which could be fabricated by grading steel and copper alloys. By combining geometric and material complexity, very high-performance devices can be fabricated. In many cases, designers will need to develop new design concepts and then explore them, since many domains will lack examples of previously successful designs.

19.4.7 Elimination of Conventional DFM Constraints

Since the 1980s, engineering design has changed considerably due to the impact of DFM, concurrent engineering, and integrated product–process teaming practices. A significant amount of time and funds were dedicated to learning about the capabilities and constraints imposed by other parts of the organization. As should be clear from this chapter, AM processes have the potential to reduce the burden on organizations to have integrated product development teams that spend large amounts of time resolving constraints and conflicts. With AM, designers have to learn far fewer manufacturing constraints. The embrace of DFM has resulted in a design culture where the design space is limited from the earliest conceptual design stage to those designs that are manufacturable using conventional techniques. With AM, these design constraints are no longer valid, and the designer can have much greater design freedom.

As such, the challenge in DFAM is not so much the understanding of the effects of manufacturing constraints. Rather, it is the difficulty in exploring new design spaces, in innovating new product structures, and in thinking about products in unconventional ways. These do not have to be difficulties, since they are really opportunities. However, the engineering community must be open to the possibilities and learn to exercise their collective creativity.

19.4.8 Industrial Design Applications

Some very intriguing approaches to product design have been demonstrated that take advantage of the shape complexity capabilities of AM, as well as some material characteristics. A leader in this field was a small company in the Netherlands called Freedom of Creation (FOC), founded by Janne Kyttanen, which was purchased by 3D Systems in the early 2010s. See http://www.freedomofcreation.com.

FOC began operations in the late 1990s. Their first commercial products were lamp shades fabricated in VPP and PBF [24], an example of which is shown in Fig. 19.15a. They have since developed many families of lampshade designs. In 2003, they partnered with Materialise to market lampshades, which retail for 300–6000 euros (as of 2009).

Many other classes of products have been developed, including chairs and stools, handbags, bowls, trays, and other specialty items. See Fig. 19.15b, c for examples of other products. Also, they have partnered with large and small organizations to develop special "giveaways" for major occasions, many of which were designed to be manufactured via AM.

In the early 2000s, they developed the concept of manufacturing textiles. Their early designs were of chain mail construction, manufactured in PBF. Since then, they have developed several lines of products using similar concepts, including handbags, other types of bags, and even shower scrubs.

More recently, quite a few other companies have demonstrated very innovative designs of housewares, clothing, fashion accessories, and even shoes. Several fashion designers have focused on AM-fabricated clothes, parts of clothing, and accessories. In 2014, Anouk Wipprecht demonstrated electrified 3D printed dresses. Other examples include hats/headpieces by Gabriela Ligenza, Ray Civello, and Stephen Ma and dresses and accessories from Iris van Herpen. For the 2019 Met Gala, Zac Posen designed the Rose Petal dress (Fig. 19.16) that had 21 uniquely shaped petals, each printed using VPP and having a mass of 450 g [25]. The petals were bolted to an electron beam PBF printed titanium cage. The entire dress had a mass of almost 14 kg.

AM parts are also being used for costumes in movies. For example, several polymer PBF parts were designed including a crown and shoulder mantle for the Black Panther movie; the costumes won an Oscar award in 2019 [26].

Another source of inspiration comes from browsing the virtual storefronts on shapeways.com and ponoko.com, where individual entrepreneurs and small companies can offer custom designs. Everything from jewelry to candle holders to bird-

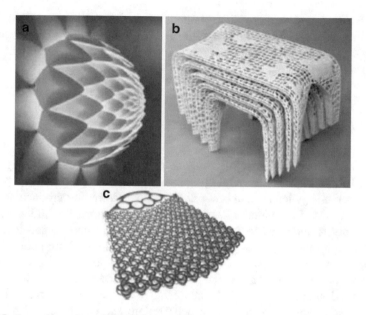

Fig. 19.15 Example products from Freedom of Creation: (**a**) a wall-mounted lampshade, Dahlia light, designed by Janne Kyttanen for Freedom of Creation, (**b**) stacking footstools, Monarch Stools, designed by Janne Kyttanen for Freedom of Creation, and (**c**) a handbag, Punch Bag, designed by Jiri Evenhuis and Janne Kyttanen for Freedom of Creation

Fig. 19.16 Rose Petal dress from Zac Posen for the 2019 Met Gala

houses can be found on these sites. Methods of manufacturing the designs offered on each storefront need to be provided, and many times the only methods are through AM. Some sites provide design guidelines, suggestions, or even specially developed CAD tools.

19.4.9 Role of Design Standards

The ASTM International and ISO partnership has developed several design guides for AM. In 2016, the first general design guide for AM was published [27]. It is meant to serve as guidance and best practices when designing products to be produced using AM. The guidelines assist designers in determining which AM capabilities apply to their design projects and how AM can best be leveraged. As a general design guide, general guidelines are presented and design considerations described, but specific design solutions and process-specific or material-specific data are not provided. As stated in the guide, the intended audience includes designers who are designing products to be fabricated in an AM system, students who are learning mechanical and Computer-Aided Design, and developers of AM design guidelines and design guidance systems.

Using the general design guide as a framework, several process-specific design guides have been developed. Guides for both metal and polymer PBF processes were published in 2019 [28, 29]. An additional design guide for electron beam PBF, for metals, should be published in 2021. A design guide for DED was published in 2020 by ASTM and will be published by ASTM and ISO jointly in the near future. Each of these guides describes specific process implementations and characteristics, characteristics of parts and features typically fabricated by the process, insights into the process-based causes of these characteristics, and an understanding of process capabilities and limitations. From this information, design rules and limitations are provided that are generally applicable. Examples are presented to illustrate typical uses of the process, how the guidelines and rules can be applied, and some of the variety that is supported by the process.

Additional design guides are under development for other processes, AM selection methods, design for post-processing, design for inspection, and related topics.

19.5 Design for Four-Dimensional (4D) Printing

As mentioned in earlier chapters, AM is often referred to as 3D Printing, particularly in mainstream media. More recently, the term 4D Printing refers to part shape change in response to an external stimulus such as heat, water, or pH [30, 31].

19.5.1   Definition of 4D Printing

The term 4D is used to demonstrate how static printed components can transform over time, thus taking time as the fourth dimension. It is mainly considered that the printed objects are dynamic in some way. In other words, 4D printed parts are produced by AM in an initial shape, and through some process, they change shape to the final shape over time in response to an external stimulus. Depending upon the 4D design, there could be a number of final shapes that are possible, and the part can move between those shapes due to varying stimuli.

Polymers are quite diverse in terms of material design-ability and active shape-changing behavior. Therefore, most 4D printing has been carried out using polymers. Figure 19.17 shows a schematic of how printing can be realized using different dimensions. In fact, active and shape memory polymers have been extensively used and studied quite some time before the term 4D printing was coined. Additionally, some investigations in AM of shape memory alloys (metals) have been conducted, mostly using nickel–titanium (Ni–Ti) alloys.

Table 19.2 shows some materials used in 4D printing along with their shape change mechanism. Multi-material structures provide more versatile shape chang-

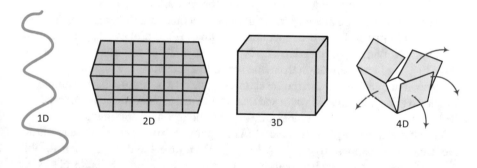

Fig. 19.17   Schematic of printing with different dimensions (1D–4D)

Table 19.2   Popular material for 4D printing [30, 32]

No	Material	Shape change mechanism
1	Shape memory polymers (SMPs)	Temperature
2	Shape memory hydrogels	Temperature, pH, or ionic strength
3	Dielectric elastomers (Des)	Electrical stimulation
4	Electroactive polymers (EAPs)	Electrical stimulation
5	Liquid crystal elastomers	Heat, light, electricity
6	Shape memory alloy (SMA)	Temperature
7	Shape memory ceramic (SMC)	Temperature
8	Shape memory composite	Temperature, light, electricity, pH, or ionic strength

ing capacity since it is possible to print active and inactive materials in different regions that can define how the part will deform over time.

19.5.2   Shape-Shifting Mechanisms and Stimuli

Depending on one or more stimuli, different geometries and shapes can be produced by 4D printing. Figure 19.18 shows the interaction and stimuli mechanisms for 4D printing. In "Unconstrained-hydro-mechanics," an expandable hydrophilic, rigid material is printed alongside an active material to provide different swelling ratios that generate shape-shifting forces. Water is the external stimulus that initiates shape-shifting. If rigid and expandable materials are designed in a proper spatial configuration, complex shapes and movements can be achieved. For "Constrained-thermo-mechanics," two levels of temperature (below (T_1) and above (T_2) a critical temperature) and a single external load are required. First, the designed structure is stretched at T_2. Then, under external stress, the structure is cooled to T_1, while the strain is maintained. Then, by removing external stress, the temporary shape is achieved. To obtain the original shape (recover), samples should be heated to T_2 [33, 34]. This is commonly known as a shape memory material.

An alternative shape memory material uses "Unconstrained-thermo-mechanics." This is again driven by two temperatures, below (T_1) and above (T_2) the critical temperature of an active material. The cycle starts at T_2, so the first step is cooling down to T_1 and transformation from one shape to another.

"Unconstrained-hydro-thermo-mechanics" has two steps. The sample is immersed and swells in cold water and then immersed and shrinks in hot water. This cycle needs both water and temperature and can be repeated continuously.

In "Unconstrained-pH-mechanics," pH-responsive hydrogels swell and shrink over time when exposed to certain pH levels. Therefore, the driving factor for this method is an aqueous environment with a different pH level. Different pH levels will affect the rate of swelling or shrinkage.

Light, heat, or a combination of these can be used to stimulate printed components in "Unconstrained-thermo-photo-mechanics." In this mechanism, photoresponsive and thermo-responsive materials are used to shape different morphologies. Photo- and thermo-responsive materials have different solubility and shrinkage behaviors when exposed to light and heat. Therefore, the material can be converted into a hydrophobic form. To recover the main shape (hydrophilic), exposure to a dark environment is needed.

In "Osmosis-mechanics," water is a stimulus and the droplet with the higher osmolarity (also known as osmotic concentration, related to the number of solute particles in a volume of fluid) swells, while the droplet with the lower osmolarity shrinks. This procedure continues until both droplets reach the same osmolarity. Water flows through the structure of the droplets, leading to self-bending. In this mechanism, the final structure is determined by the spatial arrangement and original geometry of the droplets as well as the ratio between their osmolarities.

Fig. 19.18 Shape-shifting and stimuli mechanisms for 4D printing [33]

"Dissolution-mechanics" is mainly used in bioprinting. By immersing the printed component in a solvent, a shape change happens due to the loss of non-cross-linked polymers. Then, soft phase shrinkage occurs after drying. To recover the original shape, the printed component is reimmersed in the solvent. The original shape is smaller due to the polymer loss.

19.5.3 Shape-Shifting Types and Dimensions

Shape memory materials are stimulus-responsive and the permanent shape is called "memorized." A dual shape memory effect includes a permanent shape and a temporary shape, while a triple shape memory effect has one permanent shape and two temporary shapes. In 4D printing, the type and dimensions of shape-shifting play an important role and determine how the process should be categorized [33]. Table 19.3 shows how 1D, 2D, and 3D elements can interact with each other. Different shape-shifting mechanisms are shown in Figs. 19.19 and 19.20.

Figure 19.19 illustrates how 1D strands of material can be programmed to take on a more complex shape over time for hydrogels filled with rigid materials. The stress mismatches between the rigid and flexible materials lead to the transforma

Table 19.3 Dimensions and mechanisms for 4D printing

Dimensions	Mechanism	Dimensions	Mechanism
1D-to-1D	Expansion/contraction	2D-to-3D	Twisting
1D-to-2D	Folding	2D-to-3D	Surface curling
1D-to-2D	Bending	2D-to-3D	Surface topographical change
1D-to-3D	Folding	2D-to-3D	Bending and twisting
2D-to-2D	Bending	3D-to-3D	Bending
2D-to-3D	Bending	3D-to-3D	Linear deformation
2D-to-3D	Folding	3D-to-3D	Nonlinear deformation
2D-to-2D	Force	3D-to-3D	Force

Fig. 19.19 Illustration of (**a**) 1D-to-2D shape-shifting by self-folding, (**b**) 1D-to-2D sinusoidal shape-shifting by self-bending, (**c**) 1D-to-3D strand by self-folding (Elsevier license number 4616800330874) [31, 33]

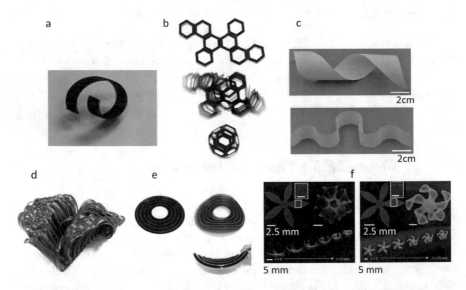

Fig. 19.20 (**a**) 2D-to-3D bending (Willey license number 4616840671090) [35, 36], (**b**) 2D-to-3D self-folding to make a truncated octahedron [31], (**c**) 2D-to-3D self-twisting helical structures [37], (**d**) 2D-to-3D surface curling hairlike shape-shifting [31], (**e**) 2D-to-3D surface topographical changes [31], (**f**) 2D-to-3D bending and twisting simple flowers. (Springer nature license number 4616811151281) [38]

tions, which are reversible. Similar transformations can also be achieved using shape memory polymers, ceramics, and metals.

By configuring the materials in even more complex geometries, 2D-to-3D and 3D-to-3D transformations can occur. Flat, 2D materials, resembling a sheet of material, can be caused to curl or fold into 3D forms. 3D models can also transform into other 3D shapes through shrinking and swelling or by inducing internal stresses to facilitate the shape change. Figure 19.20 shows different 2D-to-3D shape-shifting.

19.6   Computer-Aided Design Tools for AM

With tremendous design potential waiting for designers to explore, they need good tools to support their exploration. In this section, we present challenges and technologies associated with mechanical Computer-Aided Design systems.

19.6.1   Challenges for CAD

Current solid modeling-based CAD systems have several limitations that make them less than ideal for taking advantage of the unique capabilities of AM machines. For some applications, CAD is a bottleneck in creating novel shapes and structures, in describing desired part properties, and in specifying material compositions. These representational problems imply difficulties in driving process planning and other analysis activities. Potentially, this issue will slow the adoption of AM technologies for use in production manufacture. More specifically, the challenges for CAD can be stated as:

1. Geometric complexity – need to support models with tens and hundreds of thousands of features.
2. Physically based material representations – material compositions and distributions must be represented and must be physically meaningful.
3. Physically based property representations – desired distributions of physical and mechanical properties must be represented and tested for their physical basis.

One example of the geometric complexity issue is illustrated by the prototype textile application, from Loughborough University and Freedom of Creation, shown in Fig. 19.21 [39]. On the left is a "chain mail"-like configuration of many small rings. On the right is an example garment fabricated on a PBF machine in a Duraform material. The researchers desired to fold up the CAD model of the garment so that it occupied a very small region in the machine's build platform, which would maximize the throughput of the PBF machine for production purposes. The Loughborough researchers had great difficulty modeling the collection of thousands of rings that comprise the garment in a commercial solid modeling CAD system. Instead, they developed their own CAD system for textile and similar structured surface applica-

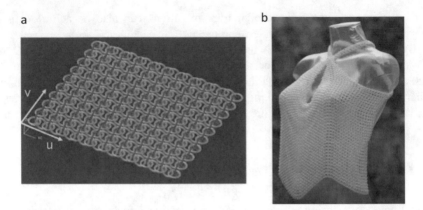

Fig. 19.21 Example of textiles produced using PBF from Loughborough University

tions over several years. However, having to develop custom CAD systems for specific applications will be a significant barrier to widespread adoption of AM.

Two CAD challenges can be illustrated by some simple examples. As presented in Chap. 7, some commercial MJT machine can deposit several different materials while building one part. To drive the machine, the vendors needed to develop new software tools that allow users to specify materials in different regions of STL files. It would be far better to be able to specify material composition in the original CAD system, so that vendor- or machine-specific tools are not needed. Binder Jetting (BJT) machines represent a second example. Depositing nanoparticle-filled droplets of binder materials can be performed readily. This enables the fabrication of functionally graded materials, modification of alloy compositions, and more generally modification of chemistries. One early example of this was the creation of gradient strength steel parts by depositing a carbon-laden binder according to a desired distribution of hardness [40]. After heat treatment, carbon diffused into the steel to vary hardness locally. More generally, one can think of varying chemistry locally through the targeted deposition of reactants. HP has presented their vision of voxel-based fabrication that utilizes such an idea. They are partnering with CAD vendors to develop the design and manufacturing tools to enable implementation of their voxel-based approach.

19.6.2   Solid Modeling CAD Technologies

Parametric, solid modeling CAD systems are used throughout much of the world for mechanical product development and are used in university education and research. Such systems, including Creo, NX, Fusion360, CATIA, and SolidWorks, are very good for representing shapes of most engineered parts. Their feature-based modeling approaches enable fast design of parts with many types of typical shape elements. Assembly modeling capabilities provide means for automatically positioning parts within assemblies and for enforcing assembly relationships when part sizes are changed.

Commercial CAD systems typically have a hybrid CSG-BRep (constructive solid geometry-boundary representation) internal representation of part geometry and topology. With the CSG part of the representation, part construction history is maintained as a sequence of feature creation, operation, and modification processes. With the BRep part of the representation, part surfaces are represented directly and exactly. Adjacencies among all points, curves, surfaces, and solids are maintained. A tremendous amount of information is represented, all of which has its purposes for providing design interactions, fast graphics, mass properties, and interfaces to other CAD/CAM/CAE tools.

For parts with dozens or hundreds of surfaces, commercial CAD systems run with interactive speeds, for most types of design operations, on typical personal computers. When more than 1000 surfaces or parts are modeled, the CAD systems tend to run very slowly and use hundreds of MB or several GB of memory. For the textile part, Fig. 19.21, thousands of rings comprise the garment. However, they have the same simple shape, that of a torus. A different type of application is that of hierarchical structures, where feature sizes span several orders of magnitude. An example is that of a multi-material mold with conformal cooling channels, where the cooling channels have small fins or other protrusions to enhance heat transfer. The fins or protrusions may have sizes of 0.01 mm, while the channels may be 10 mm in diameter, and the mold may be 400 mm long. The central region of the mold may use a high conductivity and high toughness material composition, whereas the surface of the mold may have a high hardness material composition, where a conformal, gradient transition occurs within a region near the surface of the mold. As a result, the mold model may have many thousands of small features and must also represent a gradient material composition that is derived from knowledge of the geometric features. In addition, the range of size scales may cause problems in managing internal tolerances in the CAD system. Current CAD systems are incapable of representing the thousands of features or the graded material composition of this mold example.

In summary, two main geometry-related capabilities are needed to support many emerging design applications, particularly when AM manufacturing processes will be utilized:

- Representation of tens or hundreds of thousands of features, surfaces, and parts
- Managing features, materials, surfaces, and parts across size ranges of 4–6 orders of magnitude

The ISO STEP standard provides a data exchange representation for solid geometry, material composition, and some other properties. However, it is intended for exchanging product information among CAD, CAM, and CAE systems, not for product development and manufacturing purposes. That is, the STEP representation was not developed for use within modeling and processing applications. A good assessment of its usefulness in representing parts with heterogeneous materials for

AM manufacturing is given in reference [12], although at present the standards community is revisiting the potential usage of STEP for AM.

As mentioned above, the first challenge for CAD systems is geometric complexity. The second challenge for CAD systems is to directly represent materials, to specify a part's material composition. CAD models typically cannot be used to represent parts with multiple materials or composite materials. Material composition representations are needed for parts with graded interfaces, functionally graded materials, and even simpler cases of particle or fiber filler materials. Furthermore, CAD models provide primarily geometric information for other applications, such as manufacturing or analysis, not complex multiple material information, which limits their usefulness. This type of limitation is clear when one considers the MJT and BJT examples mentioned so far.

Without a high-fidelity representation of materials, it will not be possible to directly fabricate parts using emerging AM processes. Furthermore, DFM practices will be difficult to support. Together, these limitations may prevent the adoption of AM processes for applications where fast response to orders is needed.

The third challenge, that of representing physically based property distributions, is perhaps the most challenging. The example of relating desired hardness to carbon content is a relatively simple case. More generally, the geometry, materials, processing, and property information for a design must be represented and integrated. Without such integrated CAD models, it will be very difficult to design parts with desired properties. Analysis and manufacturing applications will not be enabled. The capability of utilizing AM processes to their fullest extent will not be realized. In summary, two main issues are evident:

- Process–structure–property relationships for materials must be integrated into geometric representations of CAD models.
- CAD system capabilities must be developed that enable designers to synthesize a part, its material composition, and its manufacturing methods to meet specifications.

19.6.3   Commercial CAD Capabilities

All of the major CAD vendors have started developing AM modules for their CAD systems. Some provide DFAM analyses of part designs, build preparations, generative design, and even process simulation. CAE vendors are also developing AM analysis capabilities, focusing on AM process simulation.

Autodesk was a leader in developing CAD and CAM tools to support AM. They developed the first commercially available generative design capability, called Dreamcatcher in its early stages, that is now called "Autodesk Generative Design." They are expanding its capabilities beyond structural parts to more general engineering problems. With its purchase of the company Netfabb in 2015, they acquired unique software for design, including lattice structure generation, and build preparation that includes support structure generation, slicing, process simulation, and creation of executable build files.

Siemens NX promotes their "one integrated solution" among their NX CAD, Simcenter 3D, and NX CAM offerings AM. Within NX, they offer lattice structures, surface texturing, AM design rules, and multi-material design. Simcenter 3D has generative design and topology optimization, AM process validation, and performance validation through finite element analysis. In NX CAM, they allow users to lay out build trays and powder beds, specify 2D or 3D nesting of parts on the tray or powder bed, generate support structures, and simulate AM processes. They also offer a module on hybrid additive–subtractive manufacturing.

Dassault Systèmes and PTC have similar types of offerings for their CATIA and Creo CAD systems. Additional information is presented in Sect. 19.8.

CAE software vendors also have offerings that include build preparation and process simulation. For example, Ansys offers PBF, DED and BJT simulations that predict the final shape of the printed part as well as distortion-compensated STL files to account for that distortion. Ansys offers a suite of topology optimization and lattice optimization tools, including with AM-specific optimization constraints related to orientation and supports, to aid designers. For PBF Ansys Additive Suite offers layer-by-layer distortion and stress, orientation optimization, improved support structures, build-file creation, and identification of potential blade crashes. Their Additive Science package offers PBF porosity and melt pool analyses and predictions of thermal histories, sensor measurements, and microstructures. These capabilities enable users to determine optimum machine parameters, predict microstructures and material properties, test new materials virtually, and reduce the number of experiments needed to qualify parts.

Simufact Engineering, parent company MSC Software, offers many similar capabilities in their Additive package, which simulates both metal PBF and DED processes. They are developing metal BJT simulations as well. A partner of MSC Software, e-Xstream, has developed their Digimat product suite that includes structural analysis of parts fabricated with polymer PBF and MEX. Their underlying mean field homogenization technology models the mechanical property anisotropies that result from these processes.

19.6.4 Prototypical DFAM System

The challenges raised in this chapter are difficult and go against the directions of decades of CAD research and development. Some CAD technologies on the horizon, however, have promise in meeting these challenges. A DFAM system architecture is presented that integrates many of the technologies described in the chapter. It represents, perhaps, a simplified view of the large, complex software offerings from the CAD vendors, but with additional capabilities. A specific geometric modeling approach, implicit modeling, is offered as a potential means of achieving multi-scale, multi-material CAD.

19.6.4.1   DFAM System Architecture

Figure 19.22 shows one proposed DFAM system [41]. To the right in the figure, the designer can construct a DFAM synthesis problem, using an existing problem template if desired. For different problem types, different solution methods and algorithms will be available. Analysis codes, including FEA, boundary element, and specialty codes, will be integrated to determine design behavior. In the middle, the heterogeneous solid modeler is illustrated that consists of implicit and multi-scale modeling technologies. Heterogeneous solid modeling denotes that material and other property information will be modeled along with geometry. Libraries of materials and mesostructures enable rapid construction of design models. To the left, the manufacturing modules are shown. Both process planning and simulation modules are important in this system. After planning a manufacturing process, the idea is that the process will be simulated on the current design to determine the as-manufactured shapes, sizes, mesostructures, and microstructures. The as-manufactured model will then be analyzed to determine whether or not it actually meets design objectives.

The proposed geometric representation is a combination of implicit, nonmanifold, and parametric modeling, with the capability of generating BRep when needed. Implicit modeling is used to represent overall part geometry, while nonmanifold modeling is used to represent shape skeletons. Parametric modeling is necessary when decomposing the overall part geometry into cellular structures; each cell type will be represented as a parametric model.

19.6.4.2   Implicit Modeling

Implicit modeling has many advantages over conventional BRep, CSG, cellular decomposition, and hybrid approaches, including its conciseness, ability to model with any analytic surface models, and its avoidance of complex geometric and topological representations [42]. The primary disadvantage is that an explicit boundary representation is not maintained, making visualization and other evaluations more difficult than with some representation types. For heterogeneous solid modelers, additional advantages are apparent. Implicit modeling offers a unified approach for representing geometry, materials, and distributions of any physical quantity. A common solution method can be used to solve for material compositions, for analysis results (e.g., deflections, stresses, temperatures), and for spatial decompositions if they can be modeled as boundary value problems [43]. Furthermore, it provides a method for decomposing geometry and other properties to arbitrary resolutions which is useful for generating visualizations and manufacturing process plans.

In conventional CAD systems, parametric curves and surfaces are the primary geometric entities used in modeling typical engineered parts. For example, cubic curves are prevalent in geometric modeling; a typical 2D curve would be given by parametric equations such as:

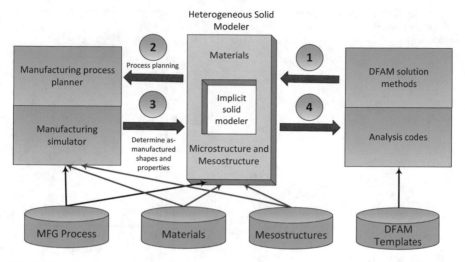

Fig. 19.22 DFAM system and overall structure

$$x(u) = au^3 + bu^2 + cu + d$$
$$y(u) = eu^3 + fu^2 + gu + h$$

$$(19.1)$$

These equations would have been simplified from their formulation as Bezier, b-spline, or NURBS (non-uniform, rational b-splines) curves [44]. In contrast, implicit functions are functions that are set equal to zero. Often, it is not possible to solve for one or more of the variables explicitly through algebraic manipulation. Rather, numerical methods must often be used to solve implicit equations. Frequently, sampling is used to visualize implicit functions or to solve them. More specifically, the general form of an implicit equation of three variables (assumed to be Cartesian coordinates) is presented along with the equation for a circle in implicit form:

$$z(x,y) = 0$$
$$z(x,y) = \frac{1}{2r}\left[(x - x_c)^2 + (y - y_c)^2 - r^2\right]$$

$$(19.2)$$

where x_c, y_c are the x and y coordinates of the circle center and r is its radius.

Shapiro and coworkers have advanced the application of the theory of R-functions to show how engineering analyses [43] and material composition [45] can be performed using implicit modeling approaches. The advantage of their approach is the unifying nature of implicit modeling to model geometry, material composition, and distributions of any physically meaningful quantity throughout a part. Furthermore, from these models of property distributions, they can perform analyses using methods akin to the boundary element method (BEM).

Fig. 19.23 Example part
to illustrate implicit
modeling

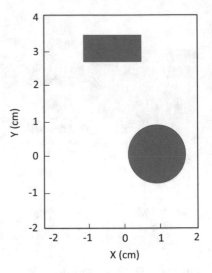

As an example, consider the 2D rectangular part shown in Fig. 19.23 with rectangular and circular holes. The implicit equations that model the boundaries of the part are presented in (19.3). Equation (19.3a, 19.3b) models the x-extents and y-extents of the part, while (19.3c, 19.3d) models the rectangular hole and (19.3e) models the circular hole ($r = 0.6$, $x_c = y_c = 0.1$). Note that the equation for each boundary feature is 0-valued at the boundary, is positive in the part interior, and is negative in the part exterior. These equations were formulated using R-functions [45]:

$$w_1(x) = \frac{4 - x^2}{4} \tag{19.3a}$$

$$w_2(y) = \frac{8 + 2y - y^2}{9} \tag{19.3b}$$

$$w_3(x) = x^2 - 0.25 \tag{19.3c}$$

$$w_4(y) = 2(y - 3)^2 - 0.125 \tag{19.3d}$$

$$w_5(x,y) = \frac{1}{2r^2}\left[(x - x_c)^2 + (y - y_c)^2 - r^2\right] \tag{19.3e}$$

An equation for the entire part can be developed by combining the boundary functions using operators $\wedge$ and $\vee$ which, in the simplest case, are functions "min" and "max," respectively; other more sophisticated expressions can be used. The part equation is:

$$\mathbf{W} = w_1 \wedge w_2 \wedge (w_3 \vee w_4) \wedge w_5 \tag{19.4}$$

Fig. 19.24 Contours of
implicit part equation

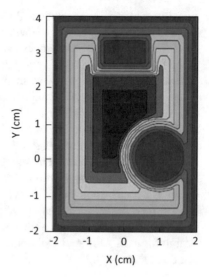

with the interpretation that the part is defined as Ω when $\mathbf{W}$ is greater than or equal
to 0: $\Omega = (\mathbf{W}(x, y) \geq 0)$.

A plot of the part function is shown in Fig. 19.24, which shows contours of constant function value (19.4). Generalizing from the example, it is always the case that
a single algebraic equation can be derived to represent a part using implicit geometry, regardless of the part complexity.

Additional, more sophisticated techniques can be applied to generate useful
parameterizations of part models for modeling multiple materials or for applications
in design, analysis, or manufacturing, but they will not be explored further here.

19.7 Design Space Exploration

New design capabilities are being integrated into CAD and CAE software to enable
designers to explore large, complex design spaces. This approach is particularly
useful when physical experiments or process simulations are time-consuming and
expensive. Instead, surrogate models of the design space are constructed and search
or optimization methods are used to explore the surrogate model. Foundational to
surrogate model construction and design space exploration (DSE) is the design of
experiments (DoE) methodology. This is discussed next, followed by DSE methods.

19.7.1 Design of Experiments

DoE methods represent a structured approach to sampling a design space or understanding process characteristics. Specific DoE methods are often related to specific
types of surrogate models. For example, orthogonal arrays are used to build response

surfaces, while more sophisticated Gaussian process regression methods have specialized sampling protocols. We will start with a consideration of typical DoE methods for characterizing AM processes.

AM is a dynamic process and many influential factors affect the quality of the printed parts. Since there are so many interdependent parameters, it is not feasible to carry out experiments to model every interaction. To ensure the best build quality and to understand the effect of changing different printing parameters, an experimental procedure has to be designed. DoE with respect to ASTM E1325 and E122 [46, 47] is used to prescribe standard terminology and show the number of tests needed to produce accurate and reliable results [48].

If the number of replications for each variable is balanced, the experimental design is called orthogonal. This is a scientific DoE, which selects the correct number of representative points, providing a reasonable number of experiments. Orthogonality results in variables analyzed independently and so interpretation is straightforward. Different DoEs are available which can be selected based on the test conditions [49].

Factorial DoE

The term factorial indicates that all combinations of factors are tested, at two or more levels each, so that main effects and interactions can be determined. If the test case is not very expensive, then full factorial DoE is the best option. However, in AM, due to the high cost of the process a fractional factorial design, such as the Plackett–Burman DoE, can be more cost-effective.

Fractional Factorial DoE

A fractional factorial design tests an adequately chosen fraction of experiments from a full factorial design. Often, only the main effects are determined. Many types of fractional factorial DoEs have been developed.

Screening DoE

A screening DoE is a specific type of a fractional factorial DoE. A screening design minimizes the number of runs required to determine the factors that significantly contribute to main effects. It is practical when it is possible to assume all factors are known, and are included, as appropriate, in the experimental design. This DoE typically covers 2 to 48 parameters with two and three levels for each.

Response Surface Method (RSM) DoE

RSM explores the relationships between responses and explanatory variables. In practice, the relationship is usually determined using regression methods to fit low-order polynomial or other convenient basis function or hypersurface as a surrogate model. Response surfaces are often used as analysis models for prediction and optimization.

Taguchi DoE

Taguchi DoE has provided major contributions for quality improvement in engineering. This method is mainly based on orthogonal array fractional factorial designs and has good flexibility for inputs. When using a Taguchi DoE, one needs to estimate which interactions are most likely to be significant and select the appropriate orthogonal array design. The advantage of Taguchi DoE is providing analysis of signal to noise, as well as mean values, to determine both the precision and accuracy of the experiments. The method identifies the significant factors and their rank ordering. Different variations of the Taguchi method are applicable to computer experiments (simulations) or physical experiments and support the construction of a surrogate model of the system being studied. As such, the method can be applied to design optimization, AM process characterization, and many other analysis tasks.

19.7.2   Design Exploration Software

Many CAD and CAE software tools have design space exploration modules. A typical capability enables the designer to establish a workflow using a CAD model with parameters that control product size and shape. An analysis model is constructed that simulates physical phenomena of interest. Typically, a DoE is set up to sample the design space that is defined by the parameters, their value ranges, and number of levels of interest. Then, the DES module automatically runs the analysis for each combination of parameter values and records the responses for the designer to explore. A surrogate model can be constructed using these responses, often in the form of a response surface. This model can then be used as the basis for optimization. Alternately, the designer can select a smaller design space region for further exploration, perhaps conducting another DoE using a narrower range of parameter values.

As one example, Ansys offers many DES capabilities. The user constructs (or imports) a geometric model of their part or system and develops a finite element model of that part or system. They can identify several size variables, such as length or width, that become parameters to be explored. A DoE is selected, ranges of each variable are identified, and the number of levels (sampling points) is chosen for each variable. For each combination of variable values, Ansys updates the part/system geometry, meshes the model, runs the analysis, and records responses of interest, such as deflections or stress values. The effects of uncertainties can also be investigated with the appropriate selection of DoE methods. This enables the assessment of response sensitivities to changes in parameters, whether size variable or mechanical property. An AM-specific example of this approach using Ansys Additive Suite and the optimization tool optiSLang would be to set up a topological optimization simulation with a constraint that the optimized geometry be self-supporting (e.g., not need secondary supports), where the parameters to be explored are the part orientation during printing and blade crash failure probability. The "optimum" answer for this designed experiment would be the orientation that results in the minimum weight topologically optimized geometry that is fully self-supporting and has a low probability of blade crash failure.

Siemens NX offers the HEEDS module with many capabilities that are similar to those from Ansys. They also offer the possibility of highly parallel simulations through connections to their cloud computing environment called MindSphere. Dassault Systèmes makes use of the iSight system, which has extensive workflow, exploration, and optimization capabilities. iSight was developed by the company Engineous Software in the 1990s which was purchased by Dassault Systèmes in 2008. The system called ModelCenter offers similar functionality and is offered by the company Phoenix Integration, Inc.

19.8   Synthesis Methods

The capabilities of AM processes have inspired many people to try to design structures so that they have minimum weight, without regard to geometric complexity. Quite a few researchers are investigating methods for synthesizing lightweight structures, with the intention of fabricating the resulting structures using AM. The work has been extended in some cases to the design of compliant mechanisms, that is, one-piece structures that move. In this section, we provide a brief survey of some research in this area. A brief exploration of optimization methods will be covered, with an emphasis on the emerging areas of generative design and topology optimization that promises to aid designers in efficiently exploring novel structures.

19.8.1   Theoretically Optimal Lightweight Structures

Several years ago, researchers rediscovered the pioneering work of AGM Michell in the early 1900s who developed the mathematical conditions under which structure weight becomes minimized [50]. He proved that structures can have minimum weight if their members are purely tension–compression members (i.e., are trusses) and derived the rules for truss layout. A typical Michell truss is shown in Fig. 19.25 for a common loaded plate structural problem. Note that the solution has a half "wagon wheel" structure.

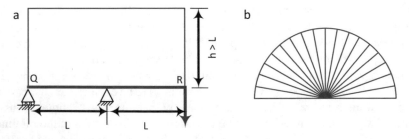

Fig. 19.25 (**a**) Simple loaded plate example for Michell truss, (**b**) Michell truss layout, (**c**) the unit cell of the octet-truss lattice structures, (**d**) octet-truss front view of the structure before loading (Elsevier license number 4638540508632)

In general, it is difficult to compute optimal Michell truss layouts for any but the simplest 2D cases. Some researchers have developed numerical procedures for computing approximate solutions. At least one research group has proposed to fabricate Michell trusses using AM processes and has investigated multiple material solution cases [51]. For proposed synthesis algorithms for large complex problems, Michell trusses provide an excellent baseline against which solutions for more complicated problems can be compared.

19.8.2 Optimization Methods

In our context, optimization methods seek to improve the design of an artifact by adjusting values of design variables in order to achieve desired objectives, typically related to structural performance or weight, as well as possible without violating constraints. A variety of optimization problem formulations has been developed that vary based on type of objectives and scope of the problem. Good textbooks [52] and many research papers have been written on the subject. The three main types of optimization problems that have been explored for design for AM include, in order of increasing complexity and scope:

- Size optimization – where values of dimensions are determined
- Shape optimization – where shapes of part surfaces are changed
- Topology optimization – where distributions of material are explored

In size optimization, the values of selected dimensions are determined that best achieve the objectives while satisfying any constraints. For typical structural optimization applications, objectives could include the minimization of maximum stress, strain energy, deflection, part volume, or weight. One or more of these quantities may also be modeled as constraints. For many mechanical parts, a small number of size dimensions will be part of the optimization problem. However, for cellular structures, such as lattices, the number of design variables could number in the tens or hundreds of thousands.

Shape optimization is a generalization of size optimization. Typically, the shape of bounding curves or surfaces is optimized to achieve similar objectives and constraints. As such, the positions of control vertices for curves or surfaces are often used as the design variables. Shape and size optimization are frequently combined in order to optimize structures that have free-form shapes, as well as standard shapes (e.g., cylinders) with dimensions.

In topology optimization, the overall shape, arrangement of shape elements, and connectivity of the design domain are determined. Again, part volume or compliance is minimized, subject to constraints on, for example, volume, compliance, stress, strain energy, and possibly additional considerations. The primary differences between topology optimization and shape or size optimization are in the starting geometric configuration and the choice of variables, which can lead to very significant improvements in structural performance. The recent interest in topology optimization as a design method for AM warrants a closer look into this technology.

19.8.3   Topology Optimization

Topology optimization (TO) is a mathematical technique that optimizes the material layout and determines the overall configuration of shape elements in a design problem. Often, TO results are used as inputs to subsequent size or shape optimization problems. As TO is a structural optimization method, finite element analyses are performed typically during each iteration of the optimization method, which means that TO can be computationally demanding. Furthermore, TO solutions should result in structures that are nearly fully stressed, or have constant strain energy, throughout the structure geometry based on the specified loading conditions. Since a wide range of shapes can result from TO, AM offers an advantageous route for part fabrication. Viewed differently, the shapes produced by TO are often too complex for conventional manufacturing. Three main approaches have been developed for TO problems: truss-based, volume-based density, and level set methods.

19.8.3.1   Truss-Based Methods

In the truss-based approach, a mesh of struts among a set of nodes is defined in a volume of interest, where sometimes the mesh represents a complete graph (e.g., ground truss) and sometimes it is based on unit cells. Topology optimization proceeds to identify which struts are most important for the problem, determine their size (e.g., diameter), and remove struts with small sizes. Result quality is often a strong function of the starting mesh of struts. Results will resemble the lattice structures presented earlier, with variations in strut diameters evident.

In the first variations of truss-based methods, a ground truss was defined over a grid of nodes, with each node connected to every other node by a truss element. Each element's diameter was used as the design variable. As optimization proceeds, those elements whose diameters become small are deleted from the design. Although the methods worked well, they tended to be computationally expensive. Recently, more sophisticated methods have been developed that utilize a different problem formulation, involving background meshes and analytical derivatives for computation of sensitivities, for truss optimization methods [53]. Good results have been achieved when both truss element size and position are used as design variables. Variations of these approaches have demonstrated the capability of achieving risk-based or reliability-related objectives [54].

Other synthesis methods utilize heuristic optimization methods in an attempt to greatly reduce the number of design variables in the optimization problem. For example, the Size Matching and Scaling (SMS) method starts with a conformal lattice structure (Sect. 19.4.5) but only requires two design variables, the minimum

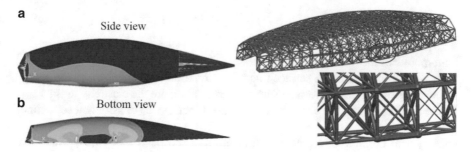

Fig. 19.26 SMS method results on UAV fuselage design problem

and maximum strut diameters, to optimize the structure [21]. The method works by performing a finite element analysis (FEA) on a solid body of the design. A conformal lattice structure is constructed that fits within the solid body. Local strain or stress values from the FEA results are used to scale struts in the lattice structure resulting in a set of relative strut size values. Size optimization is performed on the lattice structure to determine the values of the minimum and maximum strut diameters, using frame elements to model the lattice structure. Application of the SMS method to a simplified UAV fuselage design problem is illustrated in Fig. 19.26. Note that regions of high stress result in thick struts.

19.8.3.2   Volume-Based Density Methods

The second main approach is based on determining the appropriate material density in a set of voxels that comprise a spatial domain. The density-based TO method that is most common, and is used in most commercial software packages, is known as the SIMP (Solid Isotropic Material with Penalization) method. The starting geometry for the problem is typically a rectilinear block (which can have attachment holes or other simple shapes removed to or added to it) that is composed of a set of voxels. Each voxel has a density value which is used as its design variable. A density value of 1 indicates that the material is fully dense, while a value of 0 indicates that no material is present. Intermediate values indicate that the material need not be fully solid to support the local stress state in that voxel. Solutions are preferred that have voxels that are either fully dense or near 0 density, since typically partially dense materials are difficult to manufacture. Density values are used to scale voxel stiffness values in the finite element analysis models that are used during the TO process.

The typical topology optimization problem is formulated as [55]:

$$\min_{x} L(u) = \int_{\Omega} f \cdot v \, dx + \int_{\Gamma} t \cdot v \, ds \qquad (19.5)$$

$$\text{such that} \quad a(u,v) = L(v), \ \text{all } v \in U \qquad (19.6)$$

$$\text{where} \quad a(u,v) = \int_{\Omega} E_{ijkl}\varepsilon_{kl}(u)\varepsilon_{ij}(v)dx, \tag{19.7}$$

x are points in the spatial domain of interest, U are admissible displacement fields, f are body forces, and t are surface tractions. Equation 19.7 is known as the energy bilinear form. The design variables in this formulation are the elasticity tensors E_{ijkl}. In the SIMP method, the elasticity tensors are functions of density and sometimes orientation.

A typical example of topology optimization is shown in Fig. 19.27, which is a simple cantilever plate with a downward point load on its right side. Topology optimization algorithms can maintain the connectivity of material around the loading and boundary areas and also ensure that these areas are connected. They can add an arbitrary number of holes or strut regions to the design domain. However, they often produce rough or undesirable part shapes. Although the design in Fig. 19.27 could be fabricated using AM, one would probably prefer smoother shapes and transitions between major shape elements. The example was computed using the popular 99-line TO MATLAB code from Ole Sigmund [56], with inputs of 80x50 units in size, a volume fraction of 0.5, a penalization exponent of 3, and the rmin (filter size) term of 1.5.

For some multi-physics simulations, such as for optimizing structures for both heat transfer and loading conditions, truss-based and volume-density-based approaches can work. However, for some multi-physics simulations, optimizing surface characteristics of a part, such as for flow, while optimizing the structural response to load is required. Neither truss-based nor volume-density approaches are well-suited for this type of problem. Truss-based TO has a predefined surface and only the density of trusses within the interior are varied. Volume-based methods have no exact surface at any iteration, since many voxels are somewhere between a density of 1 and 0. A newer TO approach, called the level set approach (see [57]), overcomes these limitations by representing the boundary shape of the structure at every iteration. As such, fluid flow characteristics, surface pressure, and other characteristics can be calculated for each iteration, those providing opportunities for more advanced multi-physics TO than truss-based and volume-based methods. Several companies have released TO tools based upon level set approaches.

19.8.3.3   Manufacturing Considerations in TO

Although AM processes can fabricate virtually any shapes, each process does have limitations. Modifications to TO methods to account for manufacturing process constraints have been pursued since the mid-2000s. For application to AM, the most common manufacturing considerations that have been investigated include overhang control, to eliminate the need for support structures, and minimum feature size constraints. Of note, commercial software has some useful controls for manufactur-

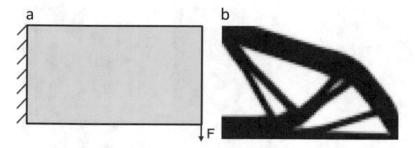

Fig. 19.27 Simple topology optimization example (**a**) schematic of cantilever plate and (**b**) cantilever plate

ability that are tailored to specific processes, for example, limiting optimized shapes to be 3-axis machinable vs. AM with no support structures vs. AM with supports (see Sect. 19.8.3.5).

To illustrate TO and overhang control, one example will be presented using the typical MBB beam design problem (simply supported beam) [58, 59]. Two general approaches have been pursued: post-TO modification of geometry to avoid unsupported overhangs and incorporation of overhang constraints into the TO algorithm. An example of the former approach uses the following steps:

- Perform TO.
- Smooth the boundaries.
- Determine local boundary angles by computing gradients.
- Add geometric features in regions where angles are too small.
- Identify optimal build orientation.

Figure 19.28a shows the comparison of samples printed with support structure, without support structure, and with the modified support-free design. Build time for no-support and support-free was four times lower than when using support structures. Also, almost half of the material was used in no-support and support-free compared to using support structures, indicating the significant build time and cost savings that can be achieved.

The latter approach to overhang control (incorporating overhang constraints into TO) typically makes use of computational filters to identify local boundary regions that violate the minimum overhang angle constraint. The objective function is modified to include the constraint using a penalty function approach to guide the TO process toward solutions without constraint violations. This has the effects of increasing strut angles and adding struts in regions with large overhangs. Research approaches differ on the specific formulations of filters and methods to integrate constraints into the TO algorithm (see [60, 61]), while others have investigated different problem formulations [62]. Other TO research of relevance to AM includes multi-material methods that seek to place materials in different regions depending upon their mechanical properties and constraints on material volumes or costs [63, 64].

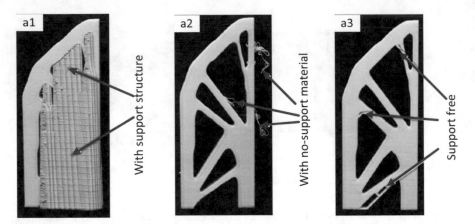

Fig. 19.28 TO printed material (**a1**) with support structure, (**a2**) with no support material, (**a3**) support-free (Elsevier license number 4639181335382) [58]

19.8.3.4   Generative Design

The term "generative design" has come to mean a design synthesis method that generates multiple, alternative geometric models that satisfy design requirements. As practiced, generative design methods are a design space exploration methodology for geometric models that utilize TO methods. A good snapshot of generative design software in 2019 is available in this engineering.com report [65].

Generative design is illustrated well in Fig. 19.29 from the Autodesk Generative Design website [66] that presents many design alternatives. The designer specifies performance and spatial requirements, along with constraints on materials, manufacturing processes, and costs. They claim to use machine learning methods so that their software learns what works as the generation process proceeds. It is likely that they are building a surrogate model of the design space.

Opportunities for generating alternatives arise due to ranges of design parameter values, which was discussed in the section on design space exploration. Additionally, in all TO algorithms, several parameters are used to control the TO method. These include the penalization exponent in the SIMP method, filter sizes, and other factors that control convergence speed. In many cases, algorithm developers set these parameters to values that tend to work well in many cases. However, one means of generating alternative designs using TO methods is to vary these parameter values. One type of filter controls the "fineness" of TO results, meaning that many small struts will be generated using a small filter value, and fewer, larger struts will be generated with a larger value.

Significant research activity is underway in this area. Researchers are exploring the combination of "deep learning" technologies with generative design. One interesting example combines generative adversarial networks (GAN), which is a type of convolutional neural network in which two subnetworks compete with each other [67]. The researchers used a combination of TO and a GAN to explore a 2D design space of wheels. TO was used to generate a set of feasible wheel designs

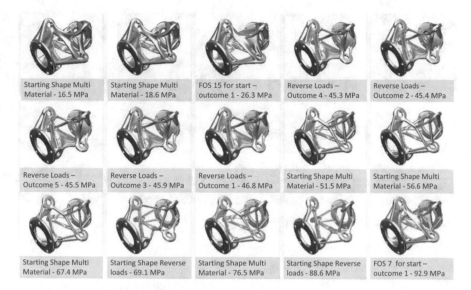

Fig. 19.29   Illustration of results from Autodesk's Generative Design software

which were used to train the GAN. Then, the GAN generated a new set of wheels, which were used as inputs for a second round of TO. Examples of generated wheel designs are shown in Fig. 19.30. It is expected that novel generative design approaches will be developed that utilize a combination of TO and deep learning methods.

19.8.3.5   TO Software

The engineering.com report on generative design and topology optimization [65] lists 15 software vendors with offerings in this area (see Table 19.4). Note that some consolidation has already occurred, for example, PTC has acquired Frustum, but the list indicates the high activity level of research and development in TO and generative design.

A straightforward 3D TO problem, the cargo sling design problem shown in Fig. 19.31, illustrates a typical application of TO. The design domain, shown in the left, is 3x3x6 m in size with a material thickness of 0.3 m (a quarter model was used to take advantage of symmetry). A pressure load of 3 kPa was applied as shown by the arrows. Symmetry boundary conditions were used. The TO solution was computed in Abaqus for a volume constraint of 15 percent of the initial volume in the design region, as shown on the right. The example demonstrates that reasonable solutions can be obtained using commercial TO systems in a reasonable amount of time (1 hour on a standard PC).

Abaqus is part of the Simulia brand of CAE software marketed by Dassault Systèmes. Abaqus is generally considered an excellent FEA package with state-of-the-art nonlinear and plasticity analysis capabilities. Multi-physics simulation is

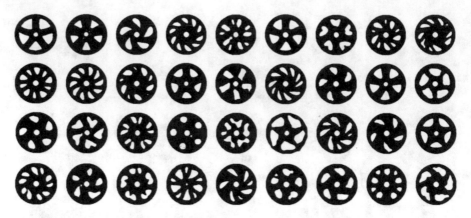

Fig. 19.30 Wheel designs generated by TO and GAN

Table 19.4 Companies and software for TO and generative design

Company	Software product
Altair	OptiStruct, inspire DesignThinking
Ansys	Ansys Discovery AIM
Autodesk	Autodesk Generative Design
Bentley	GenerativeComponents
COMSOL	Optimization Module
Dassault Systèmes	Tosca Structure, Abaqus, Solidworks xDesign
Frustum	Generate
MSC	Patran/Nastran
nTopology	Element Free/Pro
Onshape	N/A
ParaMatters	CogniCAD
PTC	Creo BMX
RUAG Space	N/A
Siemens	Solid Edge Generative Design
Vanderplaats Research and Development	GENESIS

provided with integration between structural, thermal, fluid flow, and other mechanics models. Additionally, Abaqus has an extensive library of material models that includes metals, polymers, rubbers, and even biological tissues. A wide array of physical properties is included, including standard mechanical, thermal, fluidic, acoustic, and diffusion as well as user-defined materials. The Abaqus Topology Optimization Module (ATOM) offers topology and shape optimization capabilities that utilize much of the simulation power of Abaqus. Specifically, topology and shape optimization are offered for single parts and assemblies while leveraging advanced simulation capabilities such as contact, material nonlinearity, and large deformation.

Fig. 19.31 3D cargo sling topology optimization example. (Photo courtesy of Mahmoud Alzahrani and Dr. Seung-Kyum Choi, Georgia Institute of Technology)

Ansys also has excellent topology optimization capabilities, which can interact with their design exploration modules, to enable designers to accomplish multiphysics TO. The Discovery Live simulation platform by Ansys leverages GPU computation and level set TO to enable real-time topology optimization. A user can watch the part transform in real time as they change boundary conditions and other assumptions, enabling rapid exploration of either user-driven or automated DoEs.

One of the original TO systems is OptiStruct from Altair, which is part of their HyperWorks suite of CAE software. More generally, OptiStruct is marketed as a structural analysis solver for linear and nonlinear structural problems under static or dynamic loadings. Structures can be optimized for their strength, durability, NVH (noise–vibration–harshness), thermal, and some acoustics characteristics. Altair claims that OptiStruct can solve optimization problems with thousands of design variables and can combine topology, topography (e.g., vary thickness of a sheet), size, and shape optimization capabilities. Additionally, the composites models that can be generated in HyperMesh can be optimized in OptiStruct. Further, OptiStruct and Ansys both have lattice structure generation, where regions of intermediate density (from TO) are replaced by lattice with similar density. They have also incorporate AM constraints into their TO capabilities including overhang control and minimum feature size. Also from Altair, solidThinking Inspire is a separate application that supports easy-to-use topology optimization capabilities.

Autodesk offers TO and generative design capabilities. Of the major CAD and CAE vendors, they are the market leader in generative design capabilities in both the mechanical and architectural domains. However, recent proliferation of level set methods from competitors, combined with the realization that many designers want a single optimized design that best fits their constraints rather than a chance to select from a series of designs of varying optimality, has helped competitive tools gain traction.

Siemens, Dassault Systèmes, and PTC are rolling out improvements to their TO offerings as well as generative design. The field is moving quickly with research results being commercialized on a regular basis. As indicated, TO and generative design remain very active topics of research. We can expect to see dramatic improvements in design capabilities and increasing features related to AM and other manufacturing processes.

19.9   Summary

The unique capabilities of AM technologies enable new opportunities for designers to explore new methods for customizing products, improving product performance, cutting manufacturing and assembly costs, and in general developing new ways to conceptualize products. In this chapter, we compared traditional DFM approaches to DFAM. AM enables tremendous improvements in many of the considerations that are important to DFM due to the capabilities of shape, hierarchical, functional, and material complexity. Through a series of examples, new concepts enabled by AM were presented that illustrate various methods of exploring design freedoms. No doubt, many new concepts will be developed in future years. Challenges and potential methods for new CAD tools were presented to overcome the limitations of traditional parametric, solid modeling CAD systems. A brief overview of optimization methods was given to illustrate some automated synthesis methods for designing complex structures. Several examples were given to illustrate the types of solutions that can be generated; the resulting geometries are complex enough to preclude fabrication using conventional manufacturing processes.

This chapter covered a snapshot of design concepts, examples, and research results in the broad area of DFAM. In future years, a much wider variety of concepts should emerge that lead to revolutionary ways of conceiving and developing products.

19.10   Questions

1. Describe in your own words the four AM unique design capabilities.
2. Give one example of a complex product that could be improved by the simultaneous application of all four of the DFAM design capabilities you described in problem 1. The example product cannot be one that was mentioned in this book.
3. What are three ways that current designers are trained that are at odds with the concept of DFAM?

4. Why is optimization a more challenging issue with DFAM than for DFM?
5. For the product identified in problem 2, draw in CAD the original design and your redesign based upon the application of DFAM principles.

References

1. Thompson, M. K., et al. (2016). Design for additive manufacturing: Trends, opportunities, considerations, and constraints. *CIRP Annals, 65*(2), 737–760.
2. Susman, G. I. (1992). *Integrating design and manufacturing for competitive advantage.* New York/Oxford: Oxford University Press.
3. Bralla, J. (1986). *Handbook of product design for manufacturing: A practical guide to low-cost production.* New York: McGraw-Hill.
4. Boothroyd, G., Dewhurst, P., & Knight, W. A. (2001). *Product design for manufacture and assembly, revised and expanded.* Boca Raton: CRC Press.
5. Shah, J. J., & Wright, P. K. (2000). Developing theoretical foundations of DFM. In *ASME design technical conference.*
6. Rosen, D. W., et al. (2003). The rapid tooling testbed: A distributed design-for-manufacturing system. *Rapid Prototyping Journal, 9*(3), 122–132.
7. 3D Systems, Inc. (2020). http://www.3dsystems.com
8. Hague, R. (2006). Unlocking the design potential of rapid manufacturing. In *Rapid manufacturing: An industrial revolution for the digital age.* Chichester: Wiley.
9. Mavroidis, C., et al. (2001). Fabrication of non-assembly mechanisms and robotic systems using rapid prototyping. *Journal of Mechanical Design, 123*(4), 516–524.
10. Kataria, A., & Rosen, D. W. (2001). Building around inserts: Methods for fabricating complex devices in stereolithography. *Rapid Prototyping Journal, 7*(5), 253–262.
11. Binnard, M. (2012). *Design by composition for rapid prototyping* (Vol. 525). Boston, MA: Springer Science & Business Media.
12. Patil, L., et al. (2000). Representation of heterogeneous objects in ISO 10303 (STEP). In *ASME International Mechanical Engineering Congress and Exposition*, Orlando.
13. Boeing Corp. (2020). http://www.boeing.com
14. Ulrich, K. T., & Seering, W. P. (1990). Function sharing in mechanical design. *Design Studies, 11*(4), 223–234.
15. Nera. (2020). https://bigrep.com/posts/deeper-look_into-the-fully-3d-printed-e-bike-nera/
16. Gibson, L. J., & Ashby, M. F. (1999). *Cellular solids: Structure and properties.* Cambridge: Cambridge University Press.
17. Ashby, M., et al. (2001). Metal foams: A design guide. *Applied Mechanics Reviews, 54*, B105.
18. Deshpande, V. S., Fleck, N. A., & Ashby, M. F. (2001). Effective properties of the octet-truss lattice material. *Journal of the Mechanics and Physics of Solids, 49*(8), 1747–1769.
19. Wang, A.-J., & McDowell, D. (2003). Optimization of a metal honeycomb sandwich beam-bar subjected to torsion and bending. *International Journal of Solids and Structures, 40*(9), 2085–2099.
20. Wang, J., et al. (2003). On the performance of truss panels with Kagome cores. *International Journal of Solids and Structures, 40*(25), 6981–6988.
21. Nguyen, J., Park, S.-I., & Rosen, D. (2013). Heuristic optimization method for cellular structure design of light weight components. *International Journal of Precision Engineering and Manufacturing, 14*(6), 1071–1078.
22. Lou, S., et al. (2019). Surface texture evaluation of additively manufactured metallic cellular scaffolds for acetabular implants using X-ray computed tomography. *Bio-Design and Manufacturing, 2*(2), 55–64.

23. Zhang, A. P., et al. (2012). Rapid fabrication of complex 3D extracellular microenvironments by dynamic optical projection stereolithography. *Advanced Materials, 24*(31), 4266–4270.
24. Rosen, D. W. (2007). Computer-aided design for additive manufacturing of cellular structures. *Computer-Aided Design and Applications, 4*(5), 585–594.
25. Rose Petal dress. (2020). https://www.dezeen.com/2019/05/09/zac-posen-3d-printed-rose-dress-met-gala/
26. Black panther. (2020). https://www.dezeen.com/2019/02/27/black-panther-best-costume-design-oscar-3d-printing/
27. ASTM International. (2018). *ISO/ASTM52910-18 Additive manufacturing — Design — Requirements, guidelines and recommendations*. West Conshohocken: ASTM International.
28. ASTM International. (2019). *ISO/ASTM52911-2-19 Additive manufacturing — Design — Part 2: Laser-based powder bed fusion of polymers*. West Conshohocken: ASTM International.
29. ASTM International. (2019). *ISO/ASTM52911-1-19 Additive manufacturing — Design — Part 1: Laser-based powder bed fusion of metals*. West Conshohocken: ASTM International.
30. Wu, J.J., et al. (2018). 4D printing: History and recent progress. *Chinese Journal of Polymer Science, 36*(5), 563–575.
31. Tibbits, S., et al. (2014). 4D Printing and universal transformation. In *Material agency*. New York: Springer.
32. Yang, Z., et al. (2006). Thermal and UV shape shifting of surface topography. *Journal of the American Chemical Society, 128*(4), 1074–1075.
33. Momeni, F., et al. (2017). A review of 4D printing. *Materials & Design, 122*, 42–79.
34. Monzón, M., et al. (2017). 4D printing: Processability and measurement of recovery force in shape memory polymers. *The International Journal of Advanced Manufacturing Technology, 89*(5–8), 1827–1836.
35. Jamal, M., et al. (2013). Bio-origami hydrogel scaffolds composed of photocrosslinked PEG bilayers. *Advanced Healthcare Materials, 2*(8), 1142–1150.
36. Wu, J., et al. (2016). Multi-shape active composites by 3D printing of digital shape memory polymers. *Scientific Reports, 6*, 24224.
37. Zhang, Q., Zhang, K., & Hu, G. (2016). Smart three-dimensional lightweight structure triggered from a thin composite sheet via 3D printing technique. *Scientific Reports, 6*, 22431.
38. Gladman, A. S., et al. (2016). Biomimetic 4D printing. *Nature Materials, 15*(4), 413.
39. Additive Manufacturing and 3D Printing Research Group, Nottingham University, UK. (2020). https://www.nottingham.ac.uk/research/groups/cfam/
40. Beaman, J., et al. (2004). Assessment of European research and development in additive. In *Subtractive manufacturing, final report from WTEC panel*.
41. Kytannen, J. (2006). Rapid manufacture for the retail industry. In *Rapid manufacturing: An industrial revolution for the digital age*. Chichester: Wiley.
42. Ensz, M. T., Storti, D. W., & Ganter, M. A. (1998). Implicit methods for geometry creation. *International Journal of Computational Geometry and Applications, 8*(05n06), 509–536.
43. Shapiro, V., & Tsukanov, I. (1999). Meshfree simulation of deforming domains. *Computer-Aided Design and Applications, 31*(7), 459–471.
44. Zeid, I. (2004). *Mastering CAD/CAM with engineering subscription card*. USA: McGraw-Hill.
45. Rvachev, V. L., et al. (2001). Transfinite interpolation over implicitly defined sets. *Computer Aided Geometric Design, 18*(3), 195–220.
46. ASTM International. (2016). *ASTM E1325-16, Standard terminology relating to design of experiments*. West Conshohocken: ASTM International.
47. ASTM International. (2017). *ASTM E122-17, Standard practice for calculating sample size to estimate, with specified precision, the average for a characteristic of a lot or process*. West Conshohocken: ASTM International.
48. Roy, R. K. (2010). *A primer on the Taguchi method*. USA (Michigan): Society of Manufacturing Engineers.
49. Wu, H. (2013). Application of orthogonal experimental design for the automatic software testing. In *Applied mechanics and materials*. Durnten-Zurich: Trans Tech Publications.

50. Michell, A. G. M. (1904). LVIII. The limits of economy of material in frame-structures. *The London, Edinburgh, and Dublin Philosophical Magazine and Journal of Science, 8*(47), 589–597.
51. Dewhurst, P., & Srithongchai, S. (2005). An investigation of minimum-weight dual-material symmetrically loaded wheels and torsion arms. *Journal of Applied Mechanics, 72*(2), 196–202.
52. Baldick, R. (2006). *Applied optimization: Formulation and algorithms for engineering systems.* Cambridge: Cambridge University Press.
53. Xia, Q., Wang, M. Y., & Shi, T. (2013). A method for shape and topology optimization of truss-like structure. *Structural and Multidisciplinary Optimization, 47*(5), 687–697.
54. Patel, J., & Choi, S.-K. (2012). Classification approach for reliability-based topology optimization using probabilistic neural networks. *Structural and Multidisciplinary Optimization, 45*(4), 529–543.
55. Bendsoe, M. P. (1989). Optimal shape design as a material distribution problem. *Structural Optimization, 1*(4), 193–202.
56. Sigmund, O. (2001). A 99 line topology optimization code written in Matlab. *Structural and Multidisciplinary Optimization, 21*(2), 120–127.
57. Wang, M. Y., Wang, X., & Guo, D. (2003). A level set method for structural topology optimization. *Computer Methods in Applied Mechanics and Engineering, 192*(1), 227–246.
58. Leary, M., et al. (2014). Optimal topology for additive manufacture: A method for enabling additive manufacture of support-free optimal structures. *Materials & Design, 63*, 678–690.
59. Leary, M. (2019). *Design for additive manufacturing.* Amsterdam: Elsevier.
60. Langelaar, M. (2017). An additive manufacturing filter for topology optimization of print-ready designs. *Structural and Multidisciplinary Optimization, 55*(3), 871–883.
61. Allaire, G., et al. (2017). Structural optimization under overhang constraints imposed by additive manufacturing technologies. *Journal of Computational Physics, 351*, 295–328.
62. Xian, Y., & Rosen, D. W. (2020). Morphable components topology optimization for additive manufacturing. *Structural and Multidisciplinary Optimization, 62*, 19–39.
63. Wang, M. Y., & Wang, X. (2004). "Color" level sets: A multi-phase method for structural topology optimization with multiple materials. *Computer Methods in Applied Mechanics and Engineering, 193*(6–8), 469–496.
64. Giraldo-Londoño, O., et al. (2020). Multi-material thermomechanical topology optimization with applications to additive manufacturing: Design of main composite part and its support structure. *Computer Methods in Applied Mechanics and Engineering, 363*, 112812.
65. Generative design and topology optimization: In-depth look at the two latest design technologies. (2020). https://www.engineering.com/ResourceMain.aspx?resid=826
66. Autodesk. (2020). https://www.autodesk.com/solutions/generative-design/manufacturing
67. Oh, S., et al. (2019). Deep generative design: Integration of topology optimization and generative models. *Journal of Mechanical Design, 141*(11): paper 111405.

Chapter 20
Rapid Tooling

Abstract This chapter discusses how Additive Manufacturing can be used to develop tooling solutions. Although AM is not well-suited to high-volume production for many types of geometries and materials, it does have unique benefit when producing tools for traditional volume manufacturing operations. This can be from the perspective of using AM to create patterns for parts where the materials or properties needed in those parts are not currently available using AM or for longer run tooling where AM may be able to simplify the process chain or improve the performance of the tool. Commonly referred to as rapid tooling, we discuss here how AM can contribute to tool-making for product manufacturing processes.

20.1 Introduction

The term "rapid tooling" refers to the use of AM to create production tools. The tool is therefore an impression, pattern, or mold from which a final part can be made. There are a variety of different ways in which this can be achieved, and these will be discussed in this chapter.

In recent years, as can be seen from other chapters in this book, there has been a tendency to attempt to use AM for production of parts directly from the machine. Direct Digital Manufacturing (DDM) can be a preferable approach to production for numerous reasons. However, there are still a number of reasons for creating tooling rather than DDM:

- The larger the number of parts produced, the more cost-effective it may be to make a production tool, provided it is known how many parts can be made using such a tool.
- The material requirements for the final part may be very specific and not currently available as an AM material but may however be possible through the tooling route.
- It may be that the product developer wants to understand the tooling process and thus use AM to create a prototype tool.

© Springer Nature Switzerland AG 2021
I. Gibson et al., *Additive Manufacturing Technologies*,
https://doi.org/10.1007/978-3-030-56127-7_20

– This may actually be the quickest and most effective way to create the tooling according to the required specifications. This may be particularly relevant where short lead times are important.
– AM-produced tooling can be better than traditionally produced tooling, such as tools that make use of conformal cooling.

Tooling is often broken up into two types, referred to as "short-run" and "long-run" tooling. Although discussed in numerous articles like those by Pham and Dimov [1], there are no specific definitions for either of these. Therefore we will attempt to distinguish them here.

Short-run tooling may also be referred to as prototype tooling or soft tooling. The objective is to use techniques that achieve a tool quickly, at low cost, and with few process stages. Quite often there are a number of manual steps in the process. It is understood that only a few parts are likely to result from use of the tool, possibly even just one or two parts up to around 100 or more. Every time the tool is used, it should be inspected for damage and viability. It may even be possible (or necessary) to repair the tool before it can be used again. It should be noted that if a tooling solution is required in a very short time (say in a few days), then AM-based short-run tooling may be the only way to arrive there.

Long-run tooling has greater emphasis on use of tooling for mass production purposes. Some injection molding tools can last for years and millions of parts. Although wear is always going to occur, the wear-rate is very low due to the relative hardness of the tool compared with the resulting parts that come from them. The processes required to create long-run tools from AM would still be chosen for their relative cost and lead time, but in this case they are more likely compared with conventional (subtractive) manufacturing processes. Almost every AM-based long-run tooling solution is likely to involve a metal fabrication process.

The benefits of using a rapid tooling solution may be difficult to determine, but could be immense. Very rarely is a product created from a single tool, and the more complex the product, the more difficult it is to plan. Consider the problem of bringing a new mass-produced car to the market. Some parts will already be available; some existing parts may require re-designing, while others will require design from scratch. Some of these new parts will be relatively simple, while others will have significant performance specifications that could have very long lead times. Now consider how you would create a plan to bring all these together so that the car is launched on schedule. Even the manufacture of a very simple part could delay the whole process. The use of AM-based short-run and long-run tooling can be extremely beneficial because of the short reaction times and simplified process chains. A car manufacturer may be able to plan more easily and react to disturbances in the process chain more efficiently. Even tooling that does not last very long (or, for that matter, DDM) can be used to bridge the gap to long-term tooling made using conventional methods. Delivery times can be met even though the entire mass-production facility has yet to be completed.

The majority of rapid tooling solutions are focused on the creation of injection molding (IM) tooling. This is because there are a huge number of products made from polymers using this approach. We will go on in this chapter to discuss how we can directly fabricate IM tools using AM as a replacement for subtractive machining processes. Electrical discharge machining is an alternative to the more conventional abrasive metal cutting that is worth separate consideration in this chapter. Of course, not all products are made from polymer parts. There is a huge variety of metal, ceramic, and composite-based materials and related manufacturing methods. One method that fits very well into an AM process chain is investment casting, which we will discuss here, followed by some less mainstream AM-based approaches that have found niches for some manufacturers.

20.2   Direct AM Production of Injection Molding Inserts

Injection molding is the most common modern method of manufacturing parts and is ideal to produce high volumes of the same polymer part [2]. The general principle is quite straightforward in that molten polymer is forced into a metal mold. A simple IM machine diagram can be seen in Fig. 20.1. Once the polymer has cooled and solidified, the mold splits open to reveal the part which is then ejected and the process repeats. There are many texts that cover IM in varying levels of detail. An excellent resource is from Bolur [3]. From these we can see that, similar to many processes, optimization and maximization of the output from IM can be very complex. As our demand for higher throughput, performance, quality, etc. increases, so will the need for more cost-effective solutions.

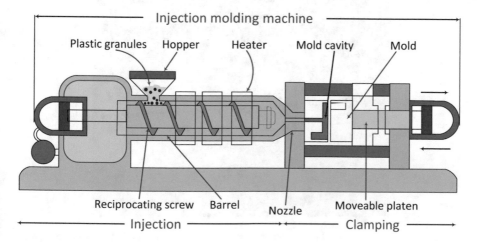

Fig. 20.1 A simple IM machine setup based upon drawing by Röckey [4]

Since the IM process requires a mold that can somehow separate for the part to be removed, there are a number of issues that require attention:

- A simple mold will have a cavity into which the polymer is injected. A core will form the other side of the mold, which is removed after the cooling process so that the part can be ejected. A mechanism (usually a set of ejector pins) is engaged to push the part out from the cavity. However, for this to be effective, the cavity walls usually have a slight slope (referred to as a "draft angle") that reduces shear forces between the polymer and the mold that would cause the part to stick.
- Not all molds can be easily split into a simple core/cavity to reveal the part. Complex geometry parts may require mold sets that separate into more than two segments. Parts may require very careful redesign so that the number of mold components is minimized. Even so, mold sets can be very complex.
- Filling the mold with molten polymer can also be problematic. The mold must be completely full before it starts to solidify, else there may be cavities. Parts that comprise many features, like thick or thin walls, ribs, bosses, etc., must be carefully analyzed to ensure the mold set is properly filled. Very complex parts may require multiple injection and venting points to ensure effective mold filling as well as fine-tuning of the temperatures, pressures, and cycle operations within the IM machine. There are numerous software tools available for mold operation analysis, like Moldflow [5].

As can be seen in Figs. 20.1 and 20.2, an IM machine has a standard plate set into which mold sets are inserted. For these inserts, it is necessary to know where to locate the injection point, the ejector pins, risers, and other features that comprise a fully functioning mold solution. It is these inserts that effectively "customize" the process and where AM can therefore contribute toward a solution.

Fig. 20.2 A core/cavity mold set showing a central injection point and channeling to regions where five different parts are formed in one cycle

Fig. 20.3 PolyJet inserts for a two-cavity mold set, showing a close-up of the ejector pins. (Photo courtesy of Stratasys)

Inserts can be made using either metal or polymer AM technology. Polymer inserts are obviously less durable, but are much quicker and cheaper to make. In a white paper published by Stratasys, the PolyJet process was demonstrated to be effective for producing inserts for a variety of applications [6]. Figure 20.3 shows PolyJet inserts for a two-cavity set, with a close-up of the ejector pin arrangement.

IM applications have been tested using the standard PolyJet materials. Best results were presented for the Digital ABS material. Parts were made in a conventional IM machines using a variety of materials, including polyamide, ABS, and polyethylene at temperatures up to 300 °C. Up to 100 cycles have been observed before the inserts broke. Similar results have been reported using Vat Photopolymerization (VPP) and polymer Powder Bed Fusion (PBF) parts. It is important to note that the IM inserts made this way should be handled carefully so that they can achieve acceptable results. Even though the IM process operates above the heat deflection temperature for the AM materials, it is still possible to get acceptable molded parts. This is possible if the IM cycle is lengthened so that the parts can cool more inside the mold before separation and ejection. Note that this only really works for relatively simple core/cavity sets. For this type of application, the costs can be around half of similar aluminum molds, with significant reductions in lead time. One can expect some hand-finishing of the resulting molded parts.

The primary concerns when making mold inserts using polymer AM are heat deflection, wear, and accuracy. Most AM processes can provide partial solutions to these problems, but generally the most accurate processes have low heat deflection temperatures, and the highest temperature materials can be found in lower accuracy processes. A number of attempts have been made to develop materials for IM inserts with polymer AM processes. One material of note is the copper–polyamide material that was developed for the polymer PBF process. Adding a copper filler to the polyamide matrix material served to improve the heat transfer away from the surface when a mold is used in the IM machine. The copper also provided additional wear resistance, which increases the life of the mold. It is interesting to note however that this is not a widely used material as the copper-polyamide is not very useful for many other applications so only appropriate where a large number of these molds are needed.

A number of chapters in this book discuss AM of metal parts. One of the initial drivers for this technology was for IM mold inserts. Significant research was performed in the 1990s and early 2000s focused on metal green part fabrication (metal powder held together by a polymer binder) in Binder Jetting (BJT) and PBF processes, followed by furnace sintering and often infiltration. These AM approaches provided a near-net shape for the metal inserts. Several materials were developed for metal green part-based AM, but the most widely used were stainless and tool steel. These processes can achieve up to an Ra surface roughness of 12–20 μm, but this would generally not be acceptable for most IM applications, and machining of the parting surfaces in particular would be necessary. If the mold surface also requires machine finishing, then very careful attention must be given to gaging so that all of the original part lies outside of the machining volume. Incorrect gaging could lead to some regions not having sufficient stock material to achieve an adequate surface. It is therefore common for designers to add material to the CAD model as a machining allowance.

Early metal powder AM machines used a green part fabrication process and were very expensive and suffered from problems with accuracy and consistent material properties, including the Rapid Steel [6] and KelTool [7] processes. While these approaches have become virtually obsolete, there were distinct advantages in that these processes could result in a fully metal part but using a conventional polymer AM machine. However, there was the need for additional furnace technology that added to the expense of the process.

More recent metal PBF technologies, pioneered by EOS and now sold by dozens of manufacturers worldwide, enable direct production of many metal alloys useful for IM tooling. As described in the PBF chapter, these alloys are fully melted during AM and can result in accuracies and surface finishes that are good enough for many tooling applications. In these machines, it is common to integrate a portion of the build plate into the mold design, which lowers the overall build time and cost needed to create tooling. Maraging steel is a widely used material in metal PBF for IM and die casting applications. Figure 20.4 shows mold inserts made using an EOS system, with a part molded from them.

Fig. 20.4 A direct metal laser-sintered tool set, with part that has been molded [8]

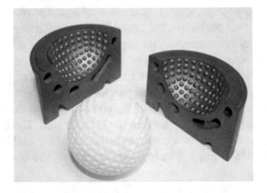

One significant benefit to the use of AM for creation of injection mold tooling is the capability of creating conformal cooling channels. It is normal to run coolant through IM molds and inserts, facilitating the cooling of the plastic part following the injection of the molten polymer. This cooling process is very dependent on the geometry of the part being molded, with larger voluminous segments cooling slower than smaller, thinner sections. Greater flow of coolant close to the larger segments can enable faster and more regular cooling, which can also improve the part quality by preventing part warpage due to thermally induced stress. More importantly, by reducing the cooling time via conformal cooling, parts can be produced at much lower overall cost by effectively increasing the number of parts that can be produced in the IM machine per day and thus increasing the overall capacity of a manufacturing operation. The geometric freedom that is a characteristic of AM can enable very complex cooling channels to be designed into molds and inserts. The use of finite element analysis, including IM process simulation, can help mold designers achieve an optimum cooling configuration. Many benefits to conformal cooling have been published [9, 10], and the use of AM components as conformally cooled inserts and molds is becoming increasingly common in industry.

Figure 20.5 shows two different cooling configurations for an injection molding tool, one made with conformal cooling and one made using gun-drilled conventional cooling channels. The use of conformal cooling channels is illustrated in Fig. 20.5a,

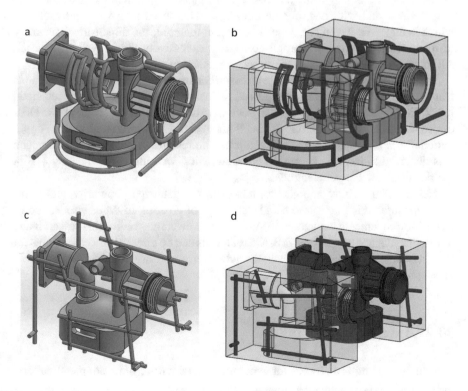

Fig. 20.5 Cooling for injection molding (**a, b**) conformal cooling method, (**c, d**) conventional cooling method

b, and an example of traditionally produced cooling channels is shown in Fig. 20.5c, d. It should be noted that plugs are required in many locations of the conventionally drilled cooling channels to create a flow path from a single input to a single output location for each channel. In the case of conformal cooling, these plugs are not needed as a single contiguous path is created with only one input and output per channel. The result of this type of conformal cooling is a part manufactured more quickly, with less distortion than an identical part made using conventional cooling channels.

20.3   EDM Electrodes

A number of attempts have been made to develop electrical discharge machining (EDM) electrodes by plating AM parts [11]. These electrodes could feasibly be used for die-sinking EDM for creating cavities for IM application. The most common method of plating polymer AM parts would be by using electroless plating of copper. There are two major drawbacks to this plating approach. The first is that electroless plating is best suited to plating a thin layer of material on a surface. For EDM, however, the electrodes are more effective with a thicker amount of conductive material deposited. It is difficult to deposit sufficiently thick material in a quick and easy manner and with controllable thickness. This leads to the second problem, which is that even if you can deposit sufficient material, the dimensional accuracy of the electrode will be compromised by this excessively thick layer of material. Although possible, it is not a very effective method of making electrodes.

While it may be possible to create an electrode using powder metallurgy methods from AM molds, possibly a more effective method would be to use direct metal fabrication. Stucker et al. [12] used a green part-based PBF approach to create electrodes using zirconium diboride (ZrB_2). This material was infiltrated with a copper material in a subsequent furnace operation. The resulting metal matrix composite was observed to have good erosion characteristics, wearing approximately 1/16th the rate of a pure copper electrode.

Neither of the above approaches, however, has achieved popularity, and traditional manufacturing appear to be a better ways of creating EDM electrodes today. Recent improvements in metal PBF systems, however, may revive this research and development since copper and ZrB_2/Cu materials can be directly processed in newer PBF machines with laser wavelengths whose absorptivity characteristics enable direct processing of these materials.

20.4   Investment Casting

Investment casting is a process for generating metal parts from a nonmetal pattern. Figure 20.6 describes the investment casting process. The patterns are in some way assembled into a structure that can be coated with ceramic to produce a shell. The

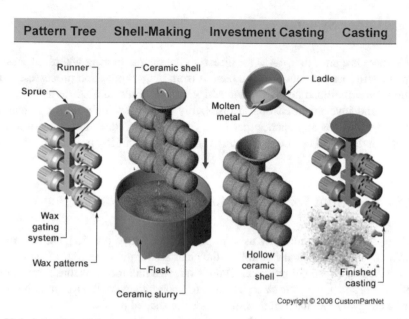

Fig. 20.6 Schematic of the investment casting process

ceramic starts as a slurry into which the structure, referred to as a "tree", is dipped to produce a closely forming skin. Once this has dried, it is strengthened by applying more coats until it is strong enough to withstand the casting process. Prior to casting, the pattern is removed by burning out the material. Care must be taken at this stage to ensure all the material has been burned out of the shell, leaving no residue. The ceramic shell can withstand the high temperature of molten metal during the pouring process, which can then be left to cool before the shell is broken from the tree. The metal replicas of the original pattern are cut from the "trunk" of the structure prior to post-treatment.

The great advantage of this is that parts can be made in a wide range of materials, specific to the application. While powder metal AM systems can produce parts directly in metal, there is a much more limited range of metals available. Furthermore, this is an approach that can result in metal parts from a nonmetal AM technology. A number of AM processes are capable of directly making parts in wax, including Material Jetting (MJT) and Material Extrusion (MEX). However, it is also possible to make investment casting patterns from other materials, including polycarbonate and ABS, which are available from a wide range of AM machines. The key is to ensure that the material does not expand rapidly during the burnout process, prior to the metal casting. One way to achieve this is to apply the honeycomb core approach, such as the stereolithography QuickCast build style, rather than using a solid fill.

20.5   Other Systems

While there is a push to use AM for direct digital manufacturing wherever possible, there are still many products which benefit from mass production processes. In this section we will illustrate ways that AM can still contribute. Although injection molding and investment casting are probably the most widely used applications, there are numerous approaches that have been considered. Below are a few other examples where AM can be used to help solve manufacturing problems.

20.5.1   Vacuum Forming Tools

Vacuum forming is commonly used in packaging, where plastic parts are formed from a flat sheet. A typical example is the clear blister packaging that is commonly used to display consumer products. Other examples include parts that form an outer shell for a product, a plastic safety helmet, for example. After the forming, it is common to cut away the material that surrounds the shaped plastic.

Heat and vacuum are applied when the sheet is placed over a tool, which has holes through which the vacuumed air is extracted. This allows the sheet to conform to the shape of the tool. If a small number of formed parts are required in a series plastic, then the tool could be fabricated using polymer AM. Locating the vacuum holes would be a straightforward process and can be included during the build. Since the heat is not directly targeted at the tool and with the pressures and other forces not being very high, it is acceptable to use polymeric materials that are commonly used in AM, like ABS or nylon. Production vacuum forming tools can be readily produced using metal laser PBF processes, but at more time and cost than polymeric materials.

20.5.2   Paper Pulp Molding Tools

It is becoming quite popular to use paper pulp molding techniques to create packaging. The pulp is made from recycled paper and therefore more environmentally friendly. The forming process is also quite sustainable since it does not require much energy to create the shapes since they are primarily created by pressing out the excess water from the pulp. Again, if the packaging is for small volume part production, AM can be used to create the forming tools. The tools can be created quite quickly using a honeycomb fill to reduce build time, weight, and material costs. Furthermore, features can be included to facilitate the excess water channeling. Figure 20.7 shows a typical mold and part made using this approach.

Fig. 20.7 A paper pulp molding tool shown with a molded packaging component. (Photo courtesy of RedEye Redeyeondemand.com)

Fig. 20.8 Polymer melt extruded AM parts used as formwork for carbon composite manufacture

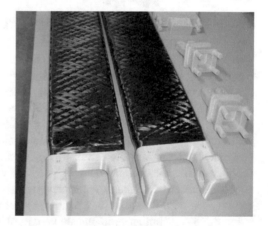

20.5.3   Formwork for Composite Manufacture

Carbon and glass fiber composites are increasingly popular materials used to manufacture high-performance items that require high strength to weight ratios. This is particularly important for vehicles, where the reduction in weight can reduce the energy requirements to move it around. Use of AM can assist in this process, particularly where complex shapes are involved. Use of honeycomb core AM build methods can assist in the creation of lightweight patterns around which fiber reinforced composites can be wound. Alternatively, AM parts can act as molds into which carbon or glass fiber reinforced polymers (CFRP or GFRP) can be placed, either pre-impregnated (prepreg) or by applying the resin later. The AM part can be kept inside the composite part in some cases or the AM pattern can be separated from the composite after the resin curing (hardening) stage. The fact that some AM materials can be dissolved away could be useful at this stage. Figure 20.8 shows

Fig. 20.9 A hard drive assembly jig. (Photo courtesy of Javelin javelin-tech.com)

some parts that have been developed for constructing high-performance UAVs (unmanned air vehicles) using CFRP. The white parts are all made using AM.

20.5.4 Assembly Tools and Metrology Registration Rigs

The majority of manufactured products are assembled in some way from multiple components. Any technique that can simplify or accelerate the assembly process can be extremely beneficial to a mass market manufacturer. We discuss the benefits of Direct Digital Manufacturing in terms of part simplification to reduce the assembly costs in Chap. 18. However, even for assembly based manufacture using conventionally made components AM can make a contribution. Some assembly processes benefit from the use of jigs that make it easier to perform the tasks by keeping some of the components in place as well as ensuring that all the components are present, like the example shown in Fig. 20.9. A variation of this approach can be seen with the metrology fixation system produced by Materialise to ensure automotive and similar moldings are kept in place during the metrology process for quality assurance purposes [13].

20.6 Questions

1. Are there any other reasons for using AM to create tooling other than the four mentioned at the beginning of this chapter?
2. What different IM flow analysis software can you find on the Internet?
3. Find three examples from the literature dated no earlier than 2020 of IM tooling made from AM. What material did they use, and why did they use AM instead of traditional manufacturing for these examples?

4. Find two examples of conformal cooling from the web. Can you identify which method is better? Why is it better?
5. Investigate the manufacture and use of EDM electrodes. What are the potential benefits and pitfalls surrounding the use of AM to directly fabricate these electrodes?
6. There are many examples of traditional manufacturing and assembly processes that are assisted by AM technology. Build a portfolio of four examples not from this book where AM is used to assist a traditional manufacturing operation. Write a two-page essay discussing these examples and how they illustrate unique benefits AM brings to a manufacturing operation.

References

1. Pham, D. T., & Dimov, S. S. (2003). Rapid prototyping and rapid tooling—The key enablers for rapid manufacturing. *Proceedings of the Institution of Mechanical Engineers, Part C: Journal of Mechanical Engineering Science, 217*(1), 1–23.
2. Injection moulding. (2020). https://en.wikipedia.org/wiki/Injection_moulding
3. Bolur, P. C. (2000). *A guide to injection moulding of plastics.* Allied Publishers Ltd, Delhi, India.
4. Rockey, B. (2009). *Injection molding machine.* University of Alberta Industrial Design.
5. Moldflow Injection Molding Simulation. (2020). https://www.autodesk.com/
6. Rapid Tooling with the SLS® Process and LaserForm™ A6 Steel Material. (2014). 3D Systems.
7. 3D Keltool, How it works. (2020). http://www.3dsystems.ru/products/productiontooling/3dkeltool/products_3dkel_howitworks.asp.htm
8. Modern Machine Shop. (2019). https://www.mmsonline.com/blog/post/this-golf-ball-mold-is-really-cool
9. Conformal cooling: Why use it now. (2020). https://www.plasticstoday.com/injection-molding/conformal-cooling-why-use-it-now/44493914113440
10. Sachs, E., et al. (2000). Production of injection molding tooling with conformal cooling channels using the three dimensional printing process. *Polymer Engineering and Science, 40*(5), 1232–1247.
11. Arthur, A., P.M. Dickens, and R.C. Cobb, Using rapid prototyping to produce electrical discharge machining electrodes. Rapid Prototyping Journal, 1996. 2(1): p. 4–12.
12. Stucker, B. E., et al. (1999). *Manufacture and use of ZrB2/Cu composite electrodes.* US Patents.
13. Materialise Metrology System. (2020) https://www.materialise.com/en

Chapter 21
Industrial Drivers for AM Adoption

Abstract Additive Manufacturing is in its fourth decade of commercial technological development. Over that time, we have experienced a number of significant
changes that have led to improvements in accuracy, better mechanical properties, a
broader range of applications, and reductions in costs of machines and the parts
made by them. In this chapter, we explore the evolution of the field and how these
developments have impacted a variety of applications over time. We note also that
different applications benefit from different aspects of AM, highlighting the versa-
tility of this technology.

21.1 Introduction

Additive Manufacturing is in its fourth decade. AM has significantly evolved over
this time, leading to improvements in accuracy, better mechanical properties, a
broader range of applications, and reductions in costs of machines and the parts
made by them. AM has found applications in design and development within almost
every consumer product sector, and the advent of low-cost AM machines has
brought these technologies into peoples' homes. As AM becomes more popular and
as technology costs inevitably decrease, this can only serve to generate more
momentum and further broaden the range of applications.

Even though AM is applicable to every major industry in one way or another,
many of the improvements to AM have been driven by the needs of specific industries. Three industries which adopted AM early, and continue to drive AM developments, are the automotive, aerospace, and medical industries. Another consistent
driver for AM has been the use of AM to produce prototypes. This chapter discusses
the use of AM for these industries and for prototyping. In aerospace and automotive
industries, AM is valued mainly because of the complex geometric capabilities and
the time that can be saved in development of products. With medicine, the benefit is
primarily in the ability to include patient-specific data from medical sources so that
customized solutions to medical problems can be found. Although many of these
techniques and applications were covered in a previous chapter, the point here is to

© Springer Nature Switzerland AG 2021
I. Gibson et al., *Additive Manufacturing Technologies*,
https://doi.org/10.1007/978-3-030-56127-7_21

put this activity in a historical context and realize how early in the AM field's development these applications were investigated. We begin with a brief survey of historical developments in Rapid Prototyping (RP), rapid tooling, and other advances, with a focus mostly on aerospace and automotive industries.

21.2   Historical Developments

In the late 1980s, 3D Systems started selling their first stereolithography (SL) machines. The first five customers of the SLA-1 beta program were AMP Incorporated, General Motors, Baxter Health Care, Eastman Kodak, and Pratt & Whitney [1]. These companies represent the four largest industrial sectors, in terms of historical AM usage, including automotive (GM and AMP, their automotive and consumer business group was the customer), health care (Baxter), consumer products (Eastman Kodak), and aerospace (Pratt & Whitney). Texas Instruments, specifically their Defense Systems and Electronics Group, was also an early adopter which applied AM to the aerospace field. Similarly, one of the first customers of DTM was BF Goodrich, which is a supplier to the aerospace and automotive industries.

Focusing on the aerospace industry, many success stories were realized by design and manufacturing engineers who used AM for Rapid Prototyping purposes. In many cases, thousands of dollars and months of product development time were saved through the use of RP, since prototype parts did not have to be fabricated using conventional manufacturing processes. Additionally, many new applications for AM parts were discovered.

21.2.1   Value of Physical Models

Early adopters discovered that AM, through the Rapid Prototyping function, provided several benefits including enhanced visualization, the ability to detect design flaws, reduced prototyping time, and significant cost reductions associated with the ability to develop correct designs quickly. Of course, there were also significant costs associated with being an early adopter. AM machines were more expensive than conventional machine tools, and people had to be hired and trained to run the AM machines. New post-processing equipment had to be installed and hazardous solvents used to clean SL parts. But, for those companies willing to take the risk, the significant investments in AM had a large return on investment when AM was integrated into their product development processes.

For example, in 1992, Texas Instruments reported several case studies demonstrating thousands of dollars and months of prototyping time saved through the use of stereolithography [1]. Furthermore, they were one of the first companies to explore the use of SL parts as patterns for investment casting.

Chrysler purchased two SLA-250 machines in early 1990 and reported that they fabricated over 1500 parts in the first 2 years of usage, with the machines running virtually 24 hours per day and 7 days per week [1]. They also reported significant time and cost savings particularly for form/fit and packaging assessments. Many other companies soon realized that they could greatly increase their chances of winning contracts to supply parts if they included RP parts with their quotes. By including physical prototypes, they can demonstrate that they understand the design requirements, and both customer and supplier can identify potential problems early on.

In the medical industry, DePuy, Inc. was another early adopter of SL. They reported on a project that began in 1990 to develop a new line of shoulder implants with dozens of models for various component sizes [1]. They used SL models, fabricated on their in-house SLA-250 machines, of the implant components during several iterations of early project reviews, saved several months of development time, and avoided costly changes before production. Furthermore, they used SL masters for urethane tooling to make wax patterns for investment casting for the first 500 pieces of each size. As they noted, this allowed them to proceed with product launch as part of their development process.

21.2.2 Functional Testing

Engineers at aerospace, automotive, and medical device companies soon discovered that AM parts could be used for a variety of functional testing applications. Specifically, flow testing was investigated by these companies, even with the early SL resins that were brittle and absorbed water easily. As one example, Chrysler tested airflow through several cylinder head designs in early 1992. They built a model of the cylinder head geometry in SL, installed steel valves and springs, and then ran the model on their flow bench. They achieved a 38% improvement in airflow.

Other companies reported similar experiences. Engineers at Pratt & Whitney pioneered several new types of flow apparatus and experiments with SL in the early years with both air and water. A report from Porsche in 1994 described water flow testing in a series of engine models to study coolant flow characteristics [2]. By using SL and an early epoxy resin, they could successfully design, fabricate, and test engine models within about 1 week per iteration.

Also in 1994, AlliedSignal reported on a study where SL models of turbine blades were used to determine their frequency spectra [2]. To study the use of SL models, they built SL models at full scale and at 3:1 scale, tested all three blades experimentally, and compared the results to finite element analysis. Theoretically, the full-scale SL models should have natural frequencies that are 35.7 percent of those of the steel blades; experimentally, they determined that the SL blades exhibited frequencies of 35 percent. Similarly, the 3:1 scale SL blades had natural frequencies that were 12 percent of the steel blade frequencies, compared to a theoretical prediction of 11.9 percent. In comparison, FEA predictions ranged from 3.6 percent lower to 19.4 percent higher than experimental results. As a consequence, AlliedSignal had much more confidence in their use of SL models than FEA, since the SL models enabled much more accurate determinations of natural frequencies.

Concurrently, aerospace companies started using AM parts to perform wind tunnel testing. Wind tunnel models are typically instrumented with arrays of pressure sensors. Standard metal models required considerable machining in order to fabricate channels for all of the wiring to the sensors. With AM, the channels and sensor mounts could be designed into the model. Automotive companies also adopted this practice. For high-speed testing, or large aerospace models, rapid tooling methods were commonly used in order to fabricate stiffer metal wind tunnel models. With proper designs, engineers could design the channels and sensor mounts into the AM patterns that were subsequently used to produce the tooling.

21.2.3   Rapid Tooling

Prior to 1992, Chrysler experimented with a variety of rapid tooling processes with stereolithography master patterns. This included vacuum forming, resin transfer molding, sand casting, squeeze molding, and silicone molding.

An area of significant effort in both the aerospace and automotive industries was the use of SL parts as investment casting patterns. Early experiments used thin-walled SL patterns or hollow parts. Because SL resins expand more than investment casting wax, when used as patterns, the SL part tended to expand and crack the ceramic shell. This led to the development of the QuickCast™ pattern style in 1992, which is a type of lattice structure that was added automatically to hollow part STL files by SL machine pre-processing software. The QuickCast style was designed to support thin walls but not to be too strong. Upon heating and thermal expansion, the QuickCast lattice struts were designed to flex, collapse inward, break, but not transfer high loads to the part skins which could crack the shell.

The QuickCast 1.0 style worked, but not as well as desired. This led to the development of QuickCast 1.1 by 3D Systems in 1995 and then QuickCast 2.0 by Phill Dickens and Richard Hague at the University of Nottingham in the late 1990s. This was quickly adopted by many manufacturers and service bureaus and, arguably, revolutionized the investment casting industry.

Another interesting development in the early 2000s was the large-frame Binder Jetting (BJT) technology by ExOne, where a sand material was developed that was suitable for use as sand casting dies. As mentioned in Chap. 8, ExOne marketed the S15 BJT machine for several years (the technology was purchased from a German company Generis GmbH in 2003). As one example, two of these sand machines were operating at the Ford Dunton Technical Center in England in the mid-2000s to support their design and development activities. Much of the Ford of Europe operations are housed here, including small car design, powertrain design and development, and some commercial vehicles. As of the end of 2005, ExOne had reportedly sold 19 S15 machines, each of which cost over $1 M.

More recently, Boeing, Northrop Grumman, and other aerospace companies have used Material Extrusion (MEX) technology to fabricate tooling. They developed tooling designs for composite part layup that were suitable for MEX fabrication. Other reported tooling applications included drill guides and various assembly tools.

21.3 The Use of AM to Support Medical Applications

AM models have been used for medical applications almost from the very start. AM could not have existed before 3D CAD since the technology is digitally driven. Computerized tomography (CT) was also a technology that developed alongside 3D representation techniques. Figure 21.1a–c shows a CT machine, a model directly generated from this machine (shown as cross-sectional slices), and a model with all segments combined into a 3D image. CT is an X-ray-based technique that moves the sensors in 3D space relative to the X-ray source so that a correlation can be made between the position and the absorption profile. By combining multiple images in this way, a 3D image can be built up. The level of absorption of the X-rays is dependent on the density of the subject matter, with bone showing up very well because it is much denser than the surrounding soft tissue. What some people don't realize is that soft tissue images can also be created using CT technology if the tissues of interest have enough contrast with surrounding tissues. Clinicians use CT technology to create 3D images for viewing the subject from any angle, so as to better understand any associated medical condition. Note that this is one of a number of medical scanning technologies working in the 3D domain, including 3D MRI, 3D ultrasound, and 3D laser scanning (for external imaging). With this increasing use of 3D medical imaging technology, the need to share and order this data across

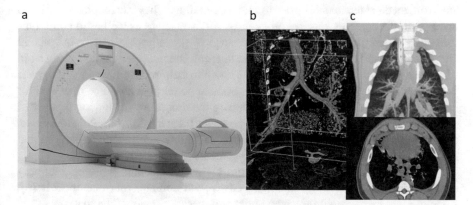

Fig. 21.1 (a) Toshiba Aquilion One CT scanner with sliced images. (Photo courtesy of Canon), (b–c) a 3D image and cross-sectional views created using this technology (Elsevier license agreement number 4650650233936) [4]

platforms has led to information exchange standards like DICOM [3], from the National Electrical Manufacturers Association in the USA, which allows users to view patient data with a variety of different software and sourced from a variety of different imaging platforms.

While originally used just for imaging and diagnostic purposes, 3D medical imaging data quickly found its way into CAD/CAM systems, with AM technology being the most effective means of realizing these models due to the complex, organic nature of the input forms. Medical data generated from patients is essentially unique to an individual. The automated form of production that AM provides makes it an obvious route for generating products from patient data.

AM-based fabrication contributes to many areas in medical diagnostics and treatment. Orthopedics was described in previous chapters. In this chapter, we will highlight the following categories of medical applications:

- Surgical and diagnostic aids
- Prosthetics and implants
- Tissue engineering
- Software tools and surgical guides

21.3.1   Surgical and Diagnostic Aids

The use of AM for diagnostic purpose was probably the first medical application of AM. Surgeons are often considered to be as much artists as they are technically proficient. Since many of their tasks involve working inside human bodies, much of their operating procedure is carried out using the sense of touch almost as much as by vision. As such, models that they can both see from any angle and feel with their hands are very useful to them.

Surgeons work in teams with support from doctors and nurses during operations and from medical technicians prior to those operations. They use models in order to understand the complex surgical procedures for themselves as well as to communicate with others in the team. Complex surgical procedures also require patient understanding and compliance, and so the surgeon can use these models to assist in this process too. AM models have been known to help reduce time in surgery for complex cases, both by allowing the surgeons to better plan ahead of time and for them to understand the situation better during the procedure (by having a physical model on hand to refer to within the operating theater). Machine vendors have, therefore, developed a range of materials that can allow sterilization of parts so that models can be brought inside the operating theater without contamination.

Most models for bony tissue result from CT data. MRI data is more commonly used for soft tissue imaging, including cases with complex vascular models [5]. Bone models are easy to produce and interpret, because many of the materials used in AM machines actually resemble bone in some way and can even respond to cutting operations in a similar manner. AM models of soft tissue are commonly used for visualizations, and multi-material AM processes as well as molded parts from

AM patterns enable practicing surgery on physical models if their compliance can be made to match the tissue of interest.

Many models may benefit from having different colors to highlight important features. Such models can display tumors, cavities, vascular tracks, etc. MEX, MJT, and BJT technologies can be used to represent this kind of part. Sometimes, these features may be buried inside bone or other tissue, and so having an opaque material encased in a transparent material can also be helpful in these situations. For this, the Stereocol resin for SLA machines [6] or Connex materials from Stratasys [7] can be used to see inside the part. The Stereocol material no longer appears to be commercially available, however. Some examples of different parts that illustrate this capability can be seen in Fig. 21.2.

Some of the most noteworthy applications of AM as medical models were from well-publicized surgeries to separate conjoined twins. Surgeons reported that having multicolored, complex models of the head or abdomen areas were invaluable in planning the surgeries, which can take 12–24 hours and involve large teams of surgeons and support staff [8].

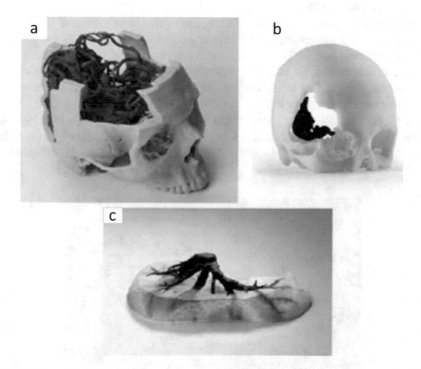

Fig. 21.2 Images of medical parts made using different colored AM systems. (**a**) 3DP used to make a skull with vascular tracks in a darker color. (**b**) A bone tumor highlighted using ABS. (c) Stratasys Connex process showing vascularity inside a human organ

21.3.2 Prosthetics and Implants

The low resolution of CT-generated 3D data combined with the low resolution of earlier AM technology led to models that may have looked anatomically correct but were perhaps not very accurate when compared with the actual patient. As technologies improved over the past few decades, models have become more precise, and it is now possible to use scanning plus AM to create close-fitting prosthetic devices. Wang [9] states that CT-based measurement can be as close as 0.2 mm from the actual value. While this is subjective, it is clear that resulting models, when built properly, can be sufficiently precise to suit many applications.

Support from CAD software can add to the process of model development by including fixtures for orientation, for tooling guidance, and for screwing into bones. For example, it is quite common for surgeons to use flexible titanium mesh as a bone replacement in cancer cases or as a method for joining pieces of broken bone together, prior to osteointegration. While described as flexible, this material still requires tools in order to bend the material. Models can be used as templates for these meshes, allowing the surgeon's technical staff to precisely bend the mesh to shape so that minimal rework is required during surgery. Figure 21.3 shows a customized biomodel implant for craniofacial reconstruction surgery that has been used for this purpose [10].

Many prosthetics are comprised of components that have a range of sizes to fit a standard population distribution. This means that precise fitting is often not possible, and so the patient may still experience some postoperative difficulties. These difficulties can further result in additional requirements for rehabilitation or even corrective surgery, thus adding to the cost of the entire treatment. Greater comfort and performance can be achieved where some of the components are customized, based on actual patient data. An example would be the socket fixation for a total hip joint replacement. While a standardized process will often return joint functionality to the patient, incorrect fixation of the socket commonly results in variable motion that

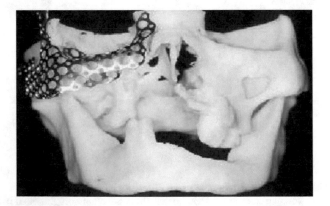

Fig. 21.3 Titanium mesh formed around a maxillofacial model

may be a discomfort and painful and require extensive physiotherapy to overcome. Customized fixtures can be made directly in titanium or cobalt–chromium (both of which are widely used for implants) using Powder Bed Fusion (PBF) technology. Such custom devices reduce the previously mentioned problems by making it possible to more precisely match the original or preferred geometry and kinematics. The use of metal systems provides considerable benefit here. While metal AM systems are not capable of producing the smooth surface finish required for effective joint articulation, the characteristic slight roughness can actually benefit osteointegration when placed inside the bone. Smooth joint articulation can be achieved through extensive polishing and use of coatings. Most metal systems may provide custom-shaped implants, but the use of highly focused energy beams will mean that the microstructure will be different and the parts may be more brittle than their equivalent cast or forged components, making brittle fracture from excessive impact loading a distinct possibility. An excellent example of a customized implant can be found in the case shown in Fig. 21.4, where Prof. J. Poukens led a multidisciplinary team to implant a complete titanium mandibular joint into an 83-year-old woman [11].

There are now examples where customized prosthetics have found their way into mainstream product manufacture. The two examples from Chap. 2 are among the most well-known in the industry: in-the-ear hearing aids and the Invisalign range of orthodontic aligners [12]. Both of these applications involve taking precise data from an individual and applying this to the basic generic design of a product. The patient data are generated by a medical specialist who is familiar with the procedure and who is able to determine whether the treatment will be beneficial. Specialized software is used that allows the patient data to be manipulated and incorporated into the medical device.

One key to success for customized prosthetics is the ability to perform the design process quickly and easily. The production process often involves AM plus numerous other conventional manufacturing tasks, and in some cases, the parts may even be more expensive to produce; but the product will perform more effectively and can sell at a premium price since it has components which suit a specific user. This

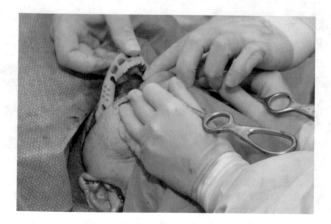

Fig. 21.4 Titanium jaw implant being located during surgery

added value can make the prosthetic less intrusive and more comfortable for the user. Additionally, the use of direct digital manufacturing makes it easier for manufacturers and practitioners.

21.3.3   Tissue Engineering and Organ Printing

The ultimate in fabrication of medical implants would be the direct fabrication of replacement body parts. This can feasibly be done using AM technology, where the materials being deposited are living cells, proteins, and other materials that assist in the generation of integrated tissue structures. There has been a great deal of active research in this area, and commercial applications are now emerging. Most of these activities use jetting and extrusion-based technology to form parts. This is because droplet-based printing technology has the ability to precisely locate very small amounts of liquid material and extrusion-based techniques are well-suited to build soft tissue scaffolding. However, ensuring that these materials are deposited under environmental conditions conducive to cell growth, differentiation, and proliferation is not a trivial task. This methodology is leading to the fabrication of complex, multicellular soft tissue structures like livers, kidneys, and even hearts. There are several 3D cell printers commercially available that can create simple layer-wise formations of cells.

An indirect approach that is appropriate to the regeneration of bony tissue is to create a scaffold from a biocompatible material that represents the shape of the final tissue construct and then add living cells at a later time. Scaffold geometry normally requires a porous structure with pores of a few hundred microns. This size permits good introduction and ingrowth of cells. Microporosity is often desirable to permit the cells to insert fibrils in order to attach firmly to the scaffold walls. Different materials and methods are used, including bioreactors to incubate the cells prior to implantation. Figure 21.5 shows a scaffold created for producing a mixture of bone

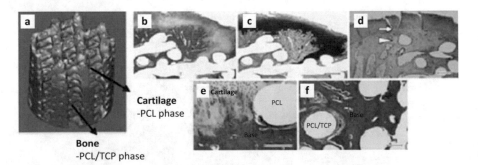

Fig. 21.5 Hybrid scaffolds composed of two phases: (**a**) Polycaprolactone (PCL) layer for cartilage tissue and bottom PCL/TCP (tricalcium phosphate) layer for bone. (**b–f**) Implantation in a rabbit for 6 months revealed formation of subchondral bone in the PCL/TCP phase and cartilage-like tissue in PCL phase. Bar is 500 μm in (**b–d**) and 200 μm in (**e**) and (**f**)

and cartilage and then implanted into a rabbit [13]. The scaffold was a mixture of polycaprolactone (PCL) which acts as a matrix material, which is also biodegradable. Mixing tricalcium phosphate (TCP) enhances the biocompatibility with bone to encourage bone regeneration and also enhances the compressive modulus of the scaffold. Even with this approach, it is still a challenge to maintain the integrity of the scaffold for sufficient lengths of time for healthy and strong bone to form. While using this approach for mass customization of soft tissue structures or load-bearing bone for everyday medical issues is some way from reality, non-load-bearing bone constructs have already been commercially proven [14], and many other commercial applications are in various stages of medical trials and adoption.

21.4   Software Tools and Surgical Guides for Medical Applications

There are a number of software tools available to assist users in preparing medical data for AM applications. Initially, such software concentrated on the translation from medical scanner systems and the creation of the standard STL files. Models made were generally replicas of the medical data. With the advent of the DICOM scanner standard, the translation tools became unnecessary, and it became necessary for such systems to add value to the data in some way. The software systems therefore evolved to include features where models could be manipulated and measured and where surgical procedures like jawbone resections could be simulated in order to determine locations for surgical implants. These have further evolved to include software tools for inclusion of CAD data in order to design prosthetic devices or support for specific surgical procedures.

Consider the application illustrated in Fig. 21.6a–c [15]. In this application, a prosthetic denture set is fixed by drilling precisely into the jawbone so that posts can be placed for anchoring the dentures. A drill guide was developed using AM,

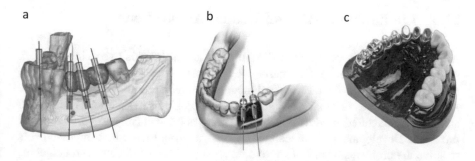

a   b   c

Fig. 21.6 (a, b) Drill guides developed using AM-related software and machines from FLEMING and DC Imaging. **(c)** Anchored printed tooth (From SLM Solutions Group, Lübeck, Germany) [16–18]

positioning the drill holes precisely so that the orthodontist could drill in the correct location and at the correct angle. The software system allows the design of the drill guide to be created, based on the patient data taken from medical scans. AM models can also be used in the development of the prosthetic itself.

Most CAD/CAM/CAE tools are used by engineers and other professionals who generally have good computer skills and an understanding of the basic principles of how such tools are constructed. Clinicians have very different backgrounds, and their basic understanding is of biological and chemical sciences with a deep knowledge of human anatomy and biological construction. Computer tools must therefore focus on being able to manipulate the anatomical data without requiring too much knowledge of CAD, graphics, or engineering construction. Software support tools for medical AM applications should therefore provide a systematic solution where different aspects of the solution can be dealt with at various stages so that the digital data are maintained and used most effectively, like the application in Fig. 21.6 where software and AM models were used at various stages to evaluate the case and to assist in the surgical procedure.

Tissue engineering software tools are likely to be very different from conventional CAD/CAM tools. This is because the data are constructed in a different form. Medical data are almost by definition freeform. If it is to be accurately reproduced, then these models require large data files. In addition, the scaffolds to be created will be highly porous, with the pores in specific locations. STL files are likely to be somewhat useless in these applications, plus if the STL files included the pore architecture, they would be inordinately large. Figure 21.5a, for example, would normally be made using an extrusion process similar to MEX, where each cross-member of the scaffold would normally correspond to an extrusion road. It would be somewhat pointless for every cross-member to be described using STL, since the slices correspond to the thickness and location of these cross-member features. Most scaffold fabrication systems, like the 3D-Bioplotter from EnvisionTEC [19], shown in Fig. 21.7, include an operating system that includes a library of scaffold fill geometries that include pore size, layer thickness, etc. rather than STL slicing systems.

21.5 Limitations of AM for Medical Applications

Although there is no doubt that medical models are useful aids to solving complex surgical problems, there are numerous deficiencies in existing AM technologies related to their use to generate medical models. Part of the reason for this is because AM equipment was originally designed to solve problems in the more widespread area of manufactured product development and not specifically to solve medical problems. Development of the technology has therefore focused on improvements to solve the problems of manufacturers rather than those of doctors and surgeons. However, recent and future improvements in AM technology may open the doors to

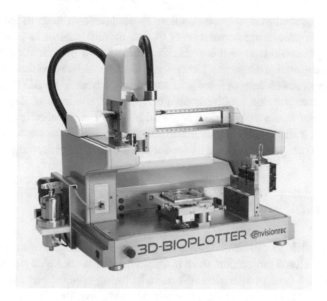

Fig. 21.7 The EnvisionTEC Bioplotter. (Photo courtesy of EnvisionTEC) [20]

a much wider range of applications in the medical industry. Key issues that may change these deficiencies in favor of using AM include:

- Speed
- Cost
- Accuracy
- Materials
- Ease of use

By analyzing these issues, we can determine which technologies may be most suitable for medical applications as well as how these technologies may develop in the future to better suit these applications.

21.5.1 Speed

AM models can often take a day or even longer to fabricate. Since medical data needs to be segmented and processed according to anatomical features, the data preparation can in fact take much longer than the AM building time. Furthermore, this process of segmentation requires considerable skill and understanding of anatomy. This means that medical models can effectively only be included in surgical procedures that involve long-term planning and cannot be used, for example, as aids for rapid diagnosis and treatment in emergency operations.

Many AM machines now have excellent throughput rate, both in terms of build speed and post-processing requirements. A few more iterations toward increasing this throughput could lead to these machines being used in critical care clinics, at least for more effective diagnosis. However, it must be understood that this use must be in conjunction with improvements in supporting software for 3D model generation that reduces the skill requirements and increases the level of data processing automation. For tissue engineering applications, the time frames are a lot longer since we must wait for cells to proliferate and combine in the bioreactors. However, the sooner we can get to the stage of seeding scaffolds with cells, the better.

21.5.2 Cost

Using AM models to solve manufacturing problems can help save millions of dollars for high-volume production, even if only a few cents are saved per unit. For the medical product (mass customization) manufacturing applications mentioned earlier, machine cost is not as important as perhaps some other factors. In comparison, the purpose of medical models for diagnosis, surgical planning, and prosthetic development is to optimize the surgeon's planning time and to improve quality, effectiveness, and efficiency. These issues are more difficult to quantify in terms of cost, but it is clear that only the more complex cases can easily justify the expense of the models. The lower the machine, materials, and operating costs, the more suitable it will be for more medical models. Some machines are very competitively priced due to the use of low-cost, high-volume technologies, like inkjet printing. Some other processes have lower-cost materials, but this relates to consumable costs, which can also be reduced with increase of volume output.

21.5.3 Accuracy

Many AM processes are being improved to create more accurate components. However, many medical applications currently do not require higher accuracy because the data from the 3D imaging systems are considerably less accurate than the AM machines they feed into. However, this does not mean that users in the medical field should be complacent. As CT and MRI technologies become more accurate and sophisticated, so the requirements for AM will become more challenging. Indeed, some CT machines appear to have very good accuracy when used properly. Also, this generally relates to medical models for communication and planning, but where devices are being manufactured, the requirements for accuracy will be more stringent. Applications which require precise fitting of implants are now becoming commonplace.

21.5.4   Materials

Only a few AM polymer materials are classified as safe for transport into the operating theater, and fewer still are capable of being placed inside the body. Those machines that provide the most suitable material properties are generally the most expensive machines. Powder-based systems are also somewhat difficult to implement due to potential contamination issues due to powder coming loose in the body or fluids entering the porous parts. This limits the range of applications for medical models.

Metal systems, on the other hand, are being used regularly to produce implants using a range of technologies, as reported by Wohlers [21]. Of these, it appears that titanium is the preferred material, but cobalt–chromium and stainless steel are both available candidates that have the necessary biocompatibility for certain applications.

21.5.5   Ease of Use

AM machines generally require a degree of technical expertise in order to achieve good quality models. This is particularly true of the larger, more complex, and more versatile machines. However, these larger machines are not particularly well-suited to medical laboratory environments. Coupled with the software skills required for data preparation, this implies a significant training investment for any medical establishment wishing to use AM. While software is a problem that all AM technologies face, it doesn't help that the machines themselves often have complex setup options, materials handling, and general maintenance requirements.

21.6   Further Development of Medical AM Applications

It is difficult to say whether a particular AM technology is more or less suited to medical applications. This is because there are numerous ways in which these machines may be applied. One can envisage that different technologies may find their way into different medical departments due the specific benefits they provide. However, the most common commercial machines certainly seem to be well-suited to being used as communication aids between surgeons, technical staff, and patients. Models can also be suitable for diagnostic aids and can assist in planning, in the development of surgical procedures, and for creating surgical tools and even the prosthetics themselves. Direct fabrication of implants and prosthetics is mostly limited to the direct metal AM technologies that can produce parts using FDA (the US Food and Drug Administration)-certified materials plus the small number of technologies that are capable of non-load-bearing polymer scaffolds.

For more of these technologies to be properly accepted in the medical arena, a number of factors must be addressed by the industry:

- Approvals
- Insurance
- Engineering training
- Location of the technology

21.6.1   Approvals

While a number of materials are now accepted by the FDA for use in medical applications, there are still questions regarding the best procedures for generating models. Little is known about the materials and processes outside of the mainstream AM industry. Approval and certification of materials and processes through ASTM and ISO will certainly help to pave the way toward FDA approval, but this can be a very long and laborious process.

Those surgeons who are aware of the processes seem to achieve excellent results and are able to present numerous successful case studies. However, the medical industry is understandably conservative about the introduction of these new technologies. Surgeons who wish to use AM generally have to resort to creative approaches based on trusting patients who sign waivers, the use of commercial AM service companies, and word of mouth promotion. Hospitals and health authorities are only just beginning to implement procedures for purchase of AM technology in the same way they might purchase a CT machine or standard medical tool.

21.6.2   Insurance

Many hospitals around the world treat patients according to their level of insurance coverage. Similar to the aforementioned issue of approvals, insurance companies need protocols for coverage using AM as a stage in the treatment process. It may be possible for some schemes to justify AM parts based on the recommendations of a surgeon, but some companies may question the purpose of the models, requiring additional paperwork that may deter some surgeons from adopting that route.

This issue will become less prominent as AM is legitimized in the medical industry. As AM has shifted from producing prototypes in the early phases of product development to mainstream manufacturing, the medical industry and consumers have taken note. Certification processes have become more common, and insurance companies are more likely to accept these technologies as part of the treatment process as more effective quality control mechanisms are now in place. Also, the increasing number of successful applications in the literature that improve patient outcomes while reducing costs makes the general use of AM more acceptable in the medical community. As such, we believe that insurance company acceptance of AM costs will continue to accelerate over the next few years.

21.6.3   Engineering Training

Creating AM models requires skills that many surgeons and technicians will not possess. While many of the newer, low-cost machines do not require significant skill to operate, preparation of the files and some post-processing requirements may require more ability. The most likely skills required for the software-based processing can be found in radiology departments since the operations for preparation of a software model are similar to manipulation and interpretation of CT and MRI models. However, technicians in this area are not used to building and manipulating physical models. These skills can however be found in prosthetics and orthotics departments. It is generally quite unusual to find radiology very closely linked with orthotics and prosthetics. The required skills are, therefore, distributed throughout a typical hospital.

21.6.4   Location of the Technology

AM machines could be located in numerous medical departments. The most likely would be to place them either in a laboratory where prosthetics are produced or in a specialist medical imaging center. If placed in the laboratories, the manual skills will be present but the accessibility will be low. If placed in imaging centers, the accessibility will be high, but the applications will probably be confined to visualization rather than fabrication of medical devices. Fortunately, most hospitals are now well equipped with high-speed intranets where patient data can be accessed quickly and easily. A separate facility that links closely to the patient data network and one that has skilled software and modeling technicians for image processing and for model post-processing (and associated downstream activities) may be a preference.

21.6.5   Service Bureaus

It can be seen that most of the hurdles for AM adoption in the medical industry are essentially procedural in nature rather than technical. A concerted effort to convince the medical industry of the value of AM models for general treatment purposes is, therefore, a key advancement that will provide a way forward.

There are a small but increasing number of companies developing excellent reputations by specializing in producing models for the medical industry. Companies like Medical Modeling LLC [8] and Anatomics [22] have demonstrated that AM companies can successfully focus on medical applications for creating models for surgeons and assisting in the development of new medical products. These service bureaus fill the skill gap between the medics and the manufacturers. AM technology is becoming better suited to a wider range of medical applications, and hospitals and clinics are beginning to have their own machines with the in-built skills to use them properly. Furthermore, the large medical product manufacturers are adopting the benefits of DDM. As AM technology becomes cheaper, easier to use, and better suited to medical application, such support companies may no longer be necessary. However, the benefits of outsourcing, as shown in many industries, still give AM services bureaus an effective role today in medical AM and likely for many years into the future.

21.7   Aerospace Applications

The aerospace industry was an early adopter of AM and continues to drive AM development today. The primary advantage for production applications in aerospace is the ability to generate complex engineered geometries with a limited number of processing steps. Aerospace companies have access to budgets significantly larger than many industries, and increased complexity in general fits well with the high-performance needs of the products being produced.

21.7.1   Characteristics Favoring AM

Significant advantages are realized when aerospace components are improved with respect to one or more of these characteristics:

Lightweight: Anything that flies requires energy to get it off the ground. The lighter the component, the less energy is required. This can be achieved by use of light-weight materials, with high strength to weight ratio. Titanium and aluminum have traditionally been materials of choice because of this. More recently, carbon fiber-reinforced composites have gained popularity. However, it is also possible to address this issue by creating lightweight structures with hollow or honeycomb internal cores. This kind of topology and lattice optimization is quite easy to achieve using AM.

High temperature: Both aircraft and spacecraft are subject to high-temperature variations, with extremes in both high and low temperatures. Engine components are subject to very high temperatures where temperature-resistant materials and innovative cooling solutions are often employed. Even components within an aircraft cabin are required to be made from flame-retardant materials. This means that AM generally requires its materials to be specially tailored to suit aerospace applications.

Complex geometry: Aerospace applications can often require components to have more than one function. For example, a structural component may also act as a conduit, or an engine turbine blade may also have an internal structure for passing coolant through it. Furthermore, geometric specifications for parts may be determined by complex mathematical formulas based on fluid flow, etc.

Economics: AM enables economical low production volumes, which are common in aerospace. Hard tooling is good for applications where the cost of the tool is amortized over a very large number of components. In the case of AM, hard tooling is not needed, and so AM is more economical than tooling-based approaches for low-volume production. In addition, designers and manufacturing engineers need not spend time designing and fabricating molds, dies, or fixtures or spend time on complex process planning (e.g., for machining) that conventional manufacturing processes require.

Digital spare parts: Many aircraft have very long useful lives (20 to 50 years or longer) which places a burden on the manufacturer to provide spare parts. Instead of warehousing spares or maintaining manufacturing tooling over the aircraft's long life, the usage of AM enables companies to maintain digital models of parts. This can be much easier and less expensive than warehousing physical parts or tools.

21.7.2 Production Manufacture

All of the major aerospace companies in the USA and Europe have pursued production applications of AM for many years. Boeing, for example, has installed tens of thousands of AM parts on their military and commercial aircraft. By 2014, over 200 different parts were flying on at least 16 models of aircraft [23]. Those numbers have dramatically increased since then. Until recently, all of these were nonstructural polymer parts for military or space applications, such as the ducts in Fig. 21.8 (c). For commercial aircraft, polymer parts need to satisfy flammability requirements, so their adoption needed to wait until flame-retardant polymer PBF materials were developed. For metals, material qualification and part certification took many years to achieve. In addition to parts manufacturing, aerospace companies are also developing new higher-performance materials in both metals and polymers, as well as processing methods.

Some of the first large-scale, metal part production manufacturing applications are emerging in the aerospace industry. GE purchased Morris Technologies in 2012 as part of a major investment in metal AM for the production of gas turbine engine components. The part that has received the most attention is a new fuel nozzle design for the CFM LEAP (Leading Edge Aviation Propulsion) turbofan engine, as shown in Fig. 21.8a [24]. The new fuel nozzle took the part consolidation concept to new levels by reportedly combining 18–20 parts into one integrated design and avoiding many brazed joints and assembly operations. This new design is projected to have a useful life five times that of the original design, a 25 percent weight

a   b   c

Fig. 21.8 (**a**) GE Aviation fuel nozzle [25], (**b**) manifold from Spartacus3D, (**c**) ducting made by SLS from Jabil

reduction, and additional cost savings realized through optimizing the design and production process. Additionally, the fuel nozzle was engineered to reduce carbon buildup, making the nozzle more efficient.

Each LEAP engine contains 19 fuel nozzles, and more than 16,300 engines had been ordered by 2018 (see [26]), so total production volume should exceed 1,000,000 total parts over the lifetime of the engine program. This single part is claimed to save 1000 lb of weight out of each engine. The nozzles are fabricated using the cobalt–chrome material fabricated in EOS metal PBF machines. Parts are stress relieved while still attached to the build plate, followed by hot isostatic pressing (HIP) to ensure that the parts are fully dense. Every part is CT scanned to ensure it is free of large porosity.

Figure 21.8b and c show a PBF metal manifold from Spartacus3D and ducting made from polymer PBF by Jabil. These both illustrate good uses of metal and polymer AM for aerospace.

Titanium alloy materials are readily available for use on almost every metal PBF platform. Ti has excellent biocompatibility and is of great interest in the medical industry. Its lightweight, high strength, and high toughness properties mean that it is a good candidate for aerospace applications as well. A recent ASTM standard addresses specifications for metal PBF parts fabricated from this alloy. Ti alloy compositions are commonly tweaked in consideration of the unique processing challenges inherent in AM. For instance, a variant of titanium called Ti–6Al–4V ELI, which denotes a titanium alloy with about 6 percent aluminum, 4 percent vanadium, and extra low interstitials (ELI), meaning the alloy has lower specified limits on iron and interstitial elements carbon and oxygen, is increasingly popular for aerospace applications. The ELI variant has better corrosion resistance, fatigue properties, and mechanical properties at cryogenic temperatures than standard

Ti–6Al–4V. Since titanium alloys are particularly susceptible to oxidation during melting, metal PBF of Ti alloys always result in as-built parts that have a higher % oxygen than the incoming powder. Using the ELI version of Ti, the as-built part still oxidizes, but since it starts out with a particularly low oxygen amount, the end result is a part that can be within specification for standard Ti64 alloys.

Airbus developed an aluminum–magnesium–scandium alloy, called Scalmalloy, for metal PBF processes. The material reportedly has mechanical properties that are twice as good as commercially available aluminum alloys, with high corrosion resistance and good fatigue properties [23].

An in-process inspection technology was developed by Sigma Labs for use in the metal PBF machines and which is utilized by several aerospace manufacturers. Called PrintRite3D, the technology is used to monitor surface characteristics during the build to help verify the build quality of metal parts while they are being fabricated. It consists of software for monitoring and data analysis to determine if parts are within specification.

Early efforts toward production manufacturing with polymer PBF systems were performed at Boeing. In 2002, a Boeing spin-off company, On Demand Manufacturing, was formed, which was subsequently purchased by RMB Products in 2005. Their first application was to manufacture environmental control system ducts to deliver cooling air to electronics instruments on F-18 military jets, as previously discussed. To ensure reliable and repeatable manufacturing, they rebuilt several SLS Sinterstation machines to meet their needs. Many aerospace companies followed their lead and made their own modifications to AM machines. In the early years of AM, machine-to-machine variations were quite large, and so each machine had to be calibrated and certified individually. As AM has moved from prototyping to DDM, machine manufacturers have had to rethink their machine designs to reduce machine-to-machine variation. While much has been accomplished in this regard, machine-to-machine variation is still too common of a problem in the AM industry today.

Airbus has investigated topology optimization applications in order to develop part designs that are significantly lighter than those from conventional manufacturing. Shown in Fig. 21.9 is an A320 nacelle hinge bracket that was originally designed as a cast steel part but was redesigned to be fabricated in a titanium alloy using PBF [27]. Reportedly, they trimmed 10 kg off the mass of the bracket, saving approximately 40 percent in weight. This study was performed as part of a larger effort to compare life-cycle environmental impacts of part design.

Aerospace remains a major driver for AM innovation. Many more production applications are expected in the near future as materials improve and production methods become standardized, repeatable, and certified. New design concepts can be expected, such as the A320 hinge bracket, for not only piece parts but entire assemblies. AM vendors are developing larger frame machines so that larger parts can be fabricated, opening up new opportunities for structural metal components and functional polymer parts.

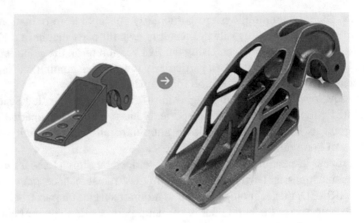

Fig. 21.9 A320 hinge bracket redesigned for AM. (Photo courtesy of EOS GmbH)

21.8   Automotive Applications

As an early adopter of AM, the automotive industry has had an outsized influence on the history of AM technology development. Automotive companies pioneered many of the AM applications in use today, and automotive-related AM expenditures remain one of the largest segments for machine and material sales in the AM industry.

Since production volumes in the automotive industry are often high (hundreds of thousands or millions of parts per year), AM has typically been evaluated as too expensive for production manufacturing, in contrast to the aerospace industry. To date, most manufacturers have not committed to AM parts on their mass-produced car models. However, AM is widely used in the automotive industry for prototype development as well as for jigs, fixtures, tooling, and many niche DDM applications.

As mentioned in the Historical Developments section, a variety of Rapid Prototyping applications were developed by automotive companies and their suppliers. In addition to RP and rapid tooling, suppliers to this industry used AM parts to debug their assembly lines. That is, they used AM parts to test assembly operations and tooling to identify potential problems before production assembly commenced. Since model line changeover involves huge investments, being able to avoid problems in production yielded very large savings.

In the metal PBF area, Concept Laser, a German company now owned by GE, developed their X line 1000R machine with a build chamber large enough to accommodate a V6 automotive engine block. This machine was developed in collaboration with Daimler AG. Although it doesn't appear that any automotive manufacturers fabricate production engine blocks in this machine, the machine was developed with production manufacture in mind. According to Concept Laser, the 1000R is capable of building at a rate of 65 cm^3 per hour. Additionally, the machine was designed with two build boxes (powder chambers) on a single turntable so that one build box could be used for part fabrication, while the other could be undergoing cool-down, part removal, preheating, or other non-part-building activities.

For specialty cars or low-volume production, AM can be economical for many more types of parts. Applications include custom parts on luxury cars or replacement parts on antique cars. Examples from Bentley Motors and Bugatti were given in Chap. 19. At Bentley, for instance, polymer PBF is used to fabricate custom interior components that are subsequently covered in leather and other materials. Typically, Bentley has production volumes of less than 10,000 cars for a given model, so this qualifies as low production volume.

Among the racing organizations, Formula 1 has been a leader in adopting AM. Originally, using AM for Rapid Prototyping, most teams started putting AM parts on their race cars in the early to mid-2000s, and today all Formula 1 teams regularly use AM for production components. Initially, these were non-structural polymer PBF parts, but they have since adopted most AM technologies as appropriate and commonly use metal PBF for structural components. Similar to the aerospace industry, Formula 1 teams utilize AM models for wind tunnel testing of scale models, as well as for full-size car models. Teams from other racing organizations, including Indy and NASCAR, have also made AM an integral aspect of their car development process.

21.9 Questions

1. How does computerized tomography actually generate 3D images? Draw a sketch to illustrate how it works, based on conventional knowledge of X-ray imaging.
2. What are the benefits of using color in production of medical models? Give several examples where color can be beneficial.
3. Why might MEX technology be particularly useful for bone tissue engineering?
4. What AM materials are already approved for medical applications and for what types of application are they suitable?
5. Consider the manufacture of metal implants using AM technology. Aside from the AM process, what other processing is likely to be needed in order to make a final part that can be implanted inside the body?
6. Why would AM be particularly useful for military applications?
7. How can the use of AM assist in the development of a new, mass-produced automobile?
8. Find some examples of parts made using AM in F1 and similar motorsports.
9. Find an example not covered in this book of both direct and indirect fabrications of biocompatible scaffolds using AM technology. Describe these examples and discuss the current commercialization status of this approach.
10. In what ways has Additive Manufacturing affected the orthodontic industry?

Table Q21.1 Question 16

AM machine utilization	2000 h/year
Depreciation time	4 years
Investment costs	500,000$
Costs for maintenance	20,000$/year
Scanning speed	1m/s
Recoat time	15 s/layer
Part volume	$1 \times 1 \times 1$ cm
Hatch space	100 μm
Layer thickness	30 μm

11. Why might it be better for a patient to use customized fixtures rather than the mass-produced standard prosthetics?
12. What are the reasons AM applications in the medical field are growing?
13. How is Additive Manufacturing involved with surgical planning? What are the advantages of this?
14. Find an article no more than 1 year old describing a recent example of tissue engineering. How close is this example to being adopted as a common medical practice, and why?
15. Describe an example of a successfully implanted prosthetic within the past 5 years. List the material, AM process, and other characteristics of the example and why this prosthetic used AM rather than a traditional manufacturing technique.
16. A company does research on AM production cost on the basis of a sample part. A simple stainless steel part has to be evaluated. In their cost model, the following information and conditions are given. The cost of machine takes up 70 percent of the total cost.

    (a) If we want to build five pieces per build, what is the cost per piece?
    (b) If we want to build ten pieces per build, what is the cost per piece?
    (c) Please draw a graph of piece versus cost per piece and give a recommendation for number of parts manufactured in one build.

17. Explain why AM is used to produce jigs and fixtures for automotive production.
18. Why is AM popular within the motorcycle industry?

References

1. Jacobs, P. F. (1992). *Rapid prototyping & manufacturing: Fundamentals of stereolithography.* Dearborn: Society of Manufacturing Engineers.
2. Jacobs, P. F. (1995). *Stereolithography and other RP&M technologies: From rapid prototyping to rapid tooling.* Dearborn: Society of Manufacturing Engineers.
3. *Digital imaging and communications in medicine.* 2020. https://www.dicomstandard.org/

4. Higgins, W. E., et al. (2008). 3D CT-video fusion for image-guided bronchoscopy. *Computerized Medical Imaging and Graphics, 32*(3), 159–173.

5. VanKoevering, K. K., Zopf, D. A., & Hollister, S. J. (2019). Tissue engineering and 3-dimensional modeling for facial reconstruction. *Facial Plastic Surgery Clinics, 27*(1), 151–161.

6. *Cordis, discussion on the use of Stereocol resin for medical applications.* 2020. https://cordis.europa.eu/projects/en

7. *Connex, multiple material AM system.* 2020. https://www.stratasys.com/

8. Medical Modeling Inc. 2020. https://www.3dsystems.com/healthcare?utm_source=medicalmodeling.com&utm_medium=301

9. Wang, J., et al. (2009). Precision of cortical bone reconstruction based on 3D CT scans. *Computerized Medical Imaging and Graphics, 33*(3), 235–241.

10. Total jaw implant. 2020. https://www.xilloc.com/patients/stories/total-mandibular-implant/

11. Gibson, I., et al. (2006). The use of rapid prototyping to assist medical applications. *Rapid Prototyping Journal, 12*(1), 53–58.

12. *Align, clear orthodontic aligners using AM technology.* 2020. https://www.invisalign.com/

13. Shao, X., et al. (2006). Repair of large articular osteochondral defects using hybrid scaffolds and bone marrow-derived mesenchymal stem cells in a rabbit model. *Advances in Tissue engineering, 12*(6), 1539–1551.

14. *Osteopore, tissue engineering technology.* 2020. http://www.osteopore.com/

15. Materialise. (2020). https://www.materialise.com/

16. DCIMAGING. (2020). https://www.dcmobileimaging.com/surgical-guides

17. *Fleming, computer-guided surgery.* 2020. https://clinicadentalfleming.es/cirugia-guiada-por-ordenador-en-que-consiste/

18. *Using 3d printing to improve dental health, AMTIL.* 2020. https://amtil.com.au/using-3d-printing-to-improve-dental-health/

19. EnvisionTEC. (2020). *3D-Bioplotter technology.* https://envisiontec.com/

20. EnvisionTEC. (2019). https://www.flickr.com/photos/envisiontec/33476682440/in/album-72157672066966280/

21. Wohlers, T. (2009). *Wohlers report. State of the industry: Annual worldwide progress report.* Wohlers Associates.

22. Anatomics. (2020). http://www.anatomics.com/

23. Wohlers, T. (2014). *Wohlers report, 3D printing and additive manufacturing state of the industry, annual worldwide progress report.* Wohlers Associates.

24. GE Additive. (2020). https://www.ge.com/additive/stories/new-manufacturing-milestone-30000-additive-fuel-nozzles

25. GE Additive. (2020). https://3dprintingindustry.com/news/ge-aviation-celebrates-30000th-3d-printed-fuel-nozzle-141165/

26. *AINonline, CFM confident leap production can catch up soon.* 2020. https://www.ainonline.com/aviation-news/aerospace/2018-07-04/cfm-confident-leap-production-can-catch-soon

27. Electro Optic Systems. (2020). http://www.eos.info/eos_airbusgroupinnovationteam_aerospace_sustainability_study

Chapter 22
Business and Societal Implications of AM

Abstract The unique attributes of Additive Manufacturing offer opportunities for new types of business enterprises. These opportunities include new types of products, organizations, and employment. In this chapter we focus our discussion on how Additive Manufacturing disrupts conventional thinking and enables a new type of entrepreneurship, called "digiproneurship." AM has already transformed the way people design, manufacture, and distribute software, hardware, products, and services; and this transformation will continue to accelerate as AM matures.

22.1 [1]Introduction

The traditional approach for manufacturing is to centralize product development, product production, and product distribution in a relatively few physical locations. These locations can concentrate employment, particularly when companies apply offshore product development, production, and/or distribution to low-cost countries/companies to take advantage of lower resource, labor, or overhead costs. For instance, Foxconn, the producer of Apple's iPhone products, has factories in China which reportedly employ hundreds of thousands of people in a single location. The history of these types of "company towns" since the advent of the industrial revolution has resulted in many negative consequences. The worker migration dynamics resulting from these types of employment concentration lead to depopulation in other regions, resulting in regions of disproportionately high underemployment and/or unemployment. As a result, nations can have regions of underpopulation with consequent national problems such as infrastructure being underutilized and long-term territorial integrity being compromised [1].

---

[1]This chapter is based on VTT Working Paper 113 *Digiproneurship: New types of physical products and sustainable employment from digital product entrepreneurship,* by Stephen Fox & Brent Stucker. The terms "Digiproneurship" and "Factory 2.0" were first introduced in this paper, which is archived at http://www.vtt.fi/inf/pdf/workingpapers/2009/W113.pdf.

© Springer Nature Switzerland AG 2021
I. Gibson et al., *Additive Manufacturing Technologies,*
https://doi.org/10.1007/978-3-030-56127-7_22

When using Additive Manufacturing, there is no fundamental reason for products to be brought to markets through centralized development, production, and distribution. Instead, products can be brought to markets through product conceptualization, product creation, and product propagation being carried out by individuals and communities in any geographical region.

In this chapter, *conceptualization* means the forming and relating of ideas, including the formation of digital versions of these ideas (e.g., CAD); *creation* means bringing an idea into physical existence (e.g., by manufacturing a component); and *propagation* means multiplying by reproduction through digital means (e.g., through digital social networks) or through physical means (e.g., by distributed AM production).

Many companies already use the Internet to collect product ideas from ordinary people from diverse locations (consider the Local Motors example discussed in Chap. 18). However, most companies feed these ideas into the centralized physical locations of their existing business operations for detailed design and creation. Distributed conceptualization, creation, and propagation can supersede concentrated development, production, and distribution by combining AM with novel human/digital interfaces which, for instance, enable non-experts to create and modify shapes. Additionally, body/place/part scanning can be used to collect data about physical features for input into digitally enabled design software and onward to AM.

Web 2.0 is considered the second generation of the Internet, where users can interact with and *transform* web content. The advent of the Internet allowed any organization, such as a newspaper publisher, to deliver information and content to anyone in the world. During the Web 2.0 transformation, social networking sites such as Facebook, or auction websites such as eBay, enabled consumers to also be content creators. These, and most new websites today, fall within the scope of Web 2.0.

AM makes it possible for digital designs to be transformed into physical products at that same location or any other location in the world (i.e., "design anywhere, build anywhere"). Moreover, the web tools associated with Web 2.0 are perfect for the propagation of product ideas and component designs that can be created through AM. The combination of Web 2.0 with AM can lead to new models of entrepreneurship.

Distributed conceptualization and propagation of digital content is known as digital entrepreneurship. However, the exploitation of AM to enable distributed creation of physical products goes beyond just digital entrepreneurship. Accordingly, the term *digiproneurship* was coined to distinguish distributed conceptualization, propagation, and creation of *physical* products from distributed conceptualization and propagation of just digital content. Thus digiproneurship is focused on transforming *digi*tal data into physical *prod*ucts using an entrepre*neurship* business model. Short definitions of the terms introduced in this section are summarized in Fig. 22.1.

Web 2.0 + AM has the potential to generate distributed, sustainable employment that is not vulnerable to off-shoring. This form of employment is not vulnerable to off-shoring because it is based on distributed networks in which resource costs are

Definition of Terms
Conceptualization
Formation and relating of ideas or concepts by individuals or communities
Creation
Bringing something into existence through digitally-enabled design and production
Propagation
Multiplying by distribution/reproduction through digitally-enabled networks
Digiproneurship
Digital to physical **prod**uct entrepre**neurship**

Fig. 22.1 Definition of terms

not a major proportion of total costs. Employment that is generated is environmentally friendly because, for example, it involves much lower energy consumption than the established concentration of product development, production, and distribution, which often involves shipping of products worldwide from centralized locations.

During the writing of this third edition, the entire world has been in the middle of the Covid-19 pandemic. The reaction of most countries to shutdown companies, transportation, and in-person commerce has been a sharp illustration of the benefits of digiproneurship. Those companies whose employees are distributed globally, who can work at home using the Internet, and whose supply chains are geographically flexible seem to be weathering the pandemic better than traditional, centralized operations. In addition, those areas of the globe with the highest concentrations of people (take, for instance, New York City) have had much higher infection rates than rural areas. Although it remains to be seen what the long-term effects of this pandemic will be, early signs indicate that digiproneurship may be a more robust business model when these types of global catastrophes occur.

As discussed throughout this book (particularly in Chaps. 19, 20, and 21), developments in AM offer possibilities for new types of products. Thus, there are many potential markets for the outputs of digiproneurship.

22.2   What Could Be New?

22.2.1   New Types of Products

Developments in AM, together with developments in advanced information and communication technologies (aICT), such as more intuitive human interfaces for design, Web 2.0, and digital scanning, are making it possible for person-specific/location-specific and/or event-specific products to be created much more quickly and at much lower cost. These products can have superior characteristics compared to products created through conventional methods. In particular, AM can enable previously intractable trade offs to be overcome. For example, design trade-offs such as manufacturing complexity versus assembly costs can be overcome (e.g., geometrically complex products can now be produced as one piece rather than having to be assembled from several pieces); material selection trade-offs such as performance requirements versus microstructures can be overcome (e.g., turbine blades can now have both high strength and high thermal performance in different locations); and economic trade-offs such as person-specific fit and/or functionality versus production time and/or cost can be overcome (e.g., customized prosthetics, such as hearing aids with person-specific fit, can be produced rapidly).

When utilizing DDM, the consumption of non-value adding resources can be radically reduced during the creation of physical goods. Further, the amount of factory equipment needed and, therefore, factory space needed is reduced. As a result, opportunities for smaller, distributed (even mobile) production facilities increase. Some examples are provided in Table 22.1. Perhaps most importantly, the potential for radically reducing the size of production facilities enables production at point-of-demand.

Although digiproneurship is probably best enabled by AM, any digitally driven technology which directly transforms digital information into a physical good can fall within the scope of digiproneurship. This can include the fabrication of structures which enclose space, such as for housing whereby each individual piece could be created using a digitally driven cutting operation and then assembled at the point of need into a usable dwelling.

It is very important to note that the limitations of manufacturing equipment and the need for expert knowledge of microstructures and material performance have previously restricted the value of direct consumer control over content. Thus, most examples of consumer-produced content are for non-physical products [2]. For example, a person who reads a newspaper (consumer of the newspaper) walks down a street and sees something newsworthy. The person takes a photograph of it. The person sends the image to the newspaper. The photograph is included in the newspaper, and hence the person becomes a partial producer of what they consume. While such forms of consumer input are established, it is only recently that developments in aICT and AM make possible consumer input into a wide range of physical goods.

From an engineering and design standpoint, AM technologies are becoming more accurate; they can directly build small products (micron-sized) and very large products (building-sized). New materials have been developed for these processes,

Table 22.1 Radical reductions in the consumption of non-value adding resources

Example	1st order effect	2nd order effect
No need for molds/dies	Less material consumption	Lower start-up costs
Fewer parts to join	Less joining equipment	Less capital tied up in infrastructure
Fewer parts to assemble	Less labor and less assembly equipment	No need to offshore production to low-labor-cost markets
No spare parts are stocked	Less storage space	Reduced factory and warehousing size

and new approaches to AM are being introduced into the marketplace. From a business-strategies standpoint, AM technologies are becoming faster, cheaper, safer, more reliable, and environmentally friendly. As each of these advancements becomes available within the marketplace, new categories of physical goods become competitive for production using AM versus conventional manufacturing. Combination of aICT with AM thus offers a wide range of opportunities for innovation in products and product services. Opportunities exist for individuals (e.g., at home), B2B (business to business), and B2C (business to consumer). Further, opportunities exist for creation of designs or creation of physical components. Thus a digiproneur could be someone who (1) creates digital tools for use by consumers or other digiproneurs; (2) creates designs which are bought by consumers or businesses; (3) creates physical products from digital data; or (4) licenses or operates enabling software or machinery in support of digiproneurship.

By replacing concentrated product development, production, and distribution with distributed product conceptualization, creation, and propagation, it is possible for individuals or communities to bring products to different types of consumers without needing to make large investment in market research, design facilities, production facilities, or distribution networks. The reasons for this are further explained in the following subsection.

22.2.2 New Types of Organizations

Web 2.0 technologies have spawned a convergence of traditional craft with technologies. One need only attend a local Maker Faire or browse etsy.com to see a gamut of entrepreneurs offering products made traditionally, by hand, with lots of electronics and/or with AM content. To support the emerging communities of craftsmen and women, online portals, blogs, and repositories have proliferated. For example, some portals have been established to focus on 3D printing (www.3ders. org) or more broadly on making (www.instructables.com). The number of blogs focused on 3D printing, AM, and making is too numerous to do justice by listing only one or two. Since the creation of 3D digital content can be challenging, several repositories of 3D content have been created, the most well-known being Thingiverse (www.thingiverse.com). Even traditional craft-based media have changed with, for example, Make Magazine adopting a synergistic combination of traditional paper distribution with online content and interaction. Each of these examples represents a business entity that was created by an entrepreneur who wanted to leverage Web 2.0 and AM.

In traditional manufacturing industries, companies such as MFG.com have become successful as industry matchmakers, finding suppliers or customers for companies around the world. They provide services for establishing supply chains and handling logistics for companies. At the time of the writing of this book, their search engine shows more than 225 companies offering "3D printing" in North America. This enables AM service bureaus to adapt traditional manufacturing business models by joining existing networks of parts suppliers. But as AM producers, there is no reason why they can't service both traditional manufacturing supply chains while becoming more consumer-focused, possibly becoming a supplier to a virtual store-front company such as Shapeways.com, or joining new supplier networks like 3dhubs.com that started as a method for joining AM providers with those nearby who needed AM services and has subsequently expanded to showcase providers offering CNC machining, sheet metal fabrication, and injection molding services worldwide.

From a different perspective, companies can utilize Web 2.0 technologies to engage with their customers to a much greater extent. Customer co-design and crowdsourcing are new terms that relate to this customer focus. Some consumer companies, such as Nike, Dell, and Home Depot, have been pioneers in providing web-based tools that enable customers to configure their own products. We can expect this trend to continue to grow. New opportunities will emerge for unprecedented levels of customer engagement. We are seeing new companies created to provide customer-designed products, for example, sunglasses, that are fabricated locally using AM. One could image kiosks at local shopping malls that are equipped with 3D printers for near real-time fabrication.

The Local Motors crowdsourcing example has been mentioned in earlier chapters. They have been an early adopter of crowdsourcing for automotive vehicles. Many other organizations and companies are experimenting with crowdsourcing technologies and practices for the development of products. It will be interesting to see how a highly technical and integrated product (such as a car) can be developed by hundreds of geographically dispersed individuals who are contributing informally on irregular schedules. From marketing and decision-making perspectives, however, having all of these individuals critique and vote on design alternatives and become invested in the outcome of the group activity can have tremendous benefits in terms of sales. Products may become successful simply because they "went viral" due to high levels of involvement from vocal online communities. The benefits and difficulties of this type of business model can be seen in kickstarter.com successes and failures. For every well-publicized failure, there are many entrepreneurial successes.

Going beyond Web 2.0 technologies, the area of cloud computing has enabled the emergence of cloud-based design and manufacturing (CBDM) concepts. Traditional CAD companies, such as Dassault Systemes and Autodesk, offer cloud-based CAD and engineering systems. Several individuals responsible for building SolidWorks into a successful CAD company created Onshape, a cloud-based CAD environment, purchased by PTC in late 2019.

A large percentage of manufacturing companies offer some type of cloud-based part ordering and quoting service. One challenge for cloud-based manufacturing is the need for hard tooling for part manufacture and product assembly. It is difficult to provide flexible, scalable, fast-turnaround "produce anywhere" services if one has to first fabricate a lot of tooling. On the other hand, AM offers a flexible, scalable, "produce anywhere" solution for CBDM. Traditional cloud-based CAD and supply chain service providers (e.g., MFG.com) have had to respond quickly to new

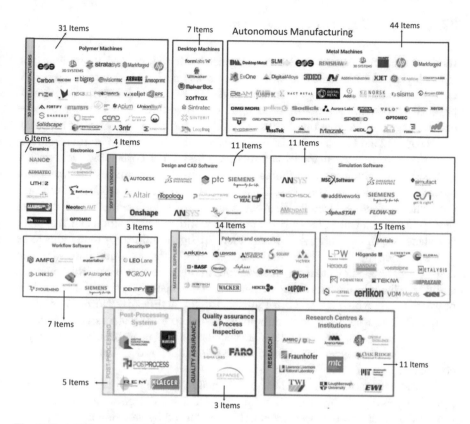

Fig. 22.2 Additive manufacturing landscape: 171 companies driving the AM industry forward (April 2019 by AMFG)

competitors (e.g., 3DHubs.com) so that their services don't become obsolete. The next evolution of this business model might be one where engineering designers can consider manufacturing, vendor, and supply chain consequences of design decisions quickly in a seamless online CAD/CAE/CAM/ordering environment.

As of 2019 the AM industry is estimated to be worth over $9 billion dollars and is growing rapidly. The impact of AM on product development and manufacturing costs is certainly orders of magnitude larger than that number. As new organizations/companies continue to enter the AM market, the market landscape and industry leaders will continue to evolve. Autonomous Manufacturing (AMFG), a provider of workflow automation software for AM, published an interesting additive manufacturing landscape, showcasing the companies they believe are the "key players" shaping the future of the AM industry (Fig. 22.2). Many of these companies were founded on business models that would have been technologically impossible just a decade ago. And all of them are taking advantage of new opportunities AM provides to streamline, strengthen, and grow their business.

22.2.3   New Types of Employment

Innovative combinations of AM and advanced information and communication technologies help eliminate non-value adding consumption of resources and reduce energy consumption arising from transportation of finished goods. Creation of physical products at point-of-demand can overturn current comparative disadvantages in the creation of physical products for global markets. As an example, today Finland has the comparative *disadvantages* of limited natural resources, far distance from mass markets, and relatively high labor costs. However, aICT + AM has the potential to make Finland's comparative disadvantages become *unimportant* in global value networks. This is because centralized models of physical production can be replaced by distributed models of value creation. In distributed models, design can take place anywhere in the world, and production can take place anywhere else in the world. As a result, there are opportunities for many jobs to be created in Finland by meeting "derived demand" for the software, hardware, and consultancy needed by creation organizations in other parts of the world. This is in addition to the jobs that can be created in Finland by meeting "primary demand" for physical goods which are used in Finland, Russia, and nearby Nordic regions or unique designs which can be electronically delivered to consumers worldwide for their creation. Thus, the resources that become important in digiproneurship are creativity, technological savvy, and access to aICT, rather than proximity to customers.

Innovative combinations of AM and aICT make it possible for creation of diverse product types by people without prior knowledge of design and/or production. Regions of persistent unemployment could be reduced by enabling a dynamic network of aICT + AM micro-businesses and small- and medium-sized enterprises (SMEs). These could be distributed among local individuals working from their homes, from their garages, from their small workshops, or from light industrial premises. They could be distributed among families and communities that have a generational investment and an abiding commitment to the regions in which they live. Accordingly, the jobs generated by digiproneurship are resistant to concentration and outsourcing.

The labor cost component of aICT + AM products is relatively low. Thus, these combinations of high technology and low labor input mean that there is little incentive to outsource to low-labor-cost economies. A summary is provided in Table 22.2

Table 22.2 Overturning regional disadvantages in the creation of physical products

Typical location-based disadvantages	aICT + AM potential
Lack of natural resources	Products make use of relatively small quantities of high-quality engineered materials procurable worldwide
High labor costs	Labor content is smaller, but networking and technology integration content is higher
Distance from markets	aICT + AM products can be designed anywhere, propagated digitally, and produced at the point of need. Shipping costs are minimized

of the factors that can enable overturning of regional disadvantages which might occur in the creation of physical products for global markets.

A diverse range of people and businesses could offer products via digiproneurship. Some examples of these people and business are:

- Artistic individuals who want to create unique physical goods
- Hobby enthusiasts who understand niche market needs
- IT savvy people who are interested in developing novel aICT software tools
- Landowners living in remote areas wanting to diversify beyond offering B&B to the occasional tourist
- Underemployed persons looking to provide supplemental income for their families
- Unemployed people who are reluctant to uproot to major cities to look for work
- Machine shops wanting to diversify and/or better utilize their skilled workforce
- SMEs that want to introduce more customer-specific versions of their product offerings
- Multinational corporations seeking to streamline the design and supply of goods which will be integrated into their products

22.3 Digiproneurship

Entrepreneurship involves individuals starting new enterprises or breathing new energy into mature enterprises through the introduction of new ideas. Entrepreneurship is associated with uncertainty because it involves introducing a new idea [3]. Well-known examples of digital entrepreneurship include Facebook, Google, and YouTube. By taking digital entrepreneurship one step further, into the production of physical goods, digiproneurship represents the next logical step.

Distributed conceptualization and propagation can reduce the risks traditionally associated with entrepreneurship. In particular, digitally enabled conceptualization and propagation of new concepts and designs for physical products can eliminate the need for costly conventional market research. Further, digitally enabled propagation of product designs to point-of-demand AM facilities can eliminate the need for physical distribution facilities such as large warehouses, costly tooling such as injection molds, and difficult to manage distribution networks. Together, digitally enabled conceptualization, propagation, and creation can eliminate many of the uncertainties and up-front expenses that have traditionally caused many entrepreneurial ventures to fail.

Digiproneurship transcends traditional design paradigms by facilitating the emergence of enterprise through the self-expression of personal feelings and opinions. New digital interfaces which enable non-experts to capture their design intent as physically producible designs could radically transform the way products are conceived and produced.

One of the earliest enterprises that could be considered a digiproneurship enterprise was Freedom of Creation (www.freedomofcreation.com). Subsequently to Freedom of Creation, numerous other digiproneurship activities have been started, including FigurePrints (www.figureprints.com) for creation of World of Warcraft figures and virtual store fronts such as Shapeways (www.shapeways.com) and the i.materialise shop (i.materialise.com/en/shop), which are online communities where digiproneurs can sell designs, services, and products.

Digiproneurship opportunities are now being considered early in the conceptualization stage for new products. The Spore game and Spore Creature Creator (www.spore.com) were designed such that Spore creatures, created by the game players, are represented by 3D digital data that can be transferred to an AM machine for direct printing using a color AM process. This is unlike the original World of Warcraft figures, which appear 3D on-screen but are not 3D solid models and thus require data manipulation in order to prepare the figure for AM.

Inexpensive, intuitive solid modeling tools, such as SketchUp (sketchup.com), are becoming widely used by consumers and some companies to design their own products. A key feature of SketchUp is its 3D Warehouse, where users may upload and download models for free. Since it's a cloud-based tool, files "downloaded" into the program from the 3D Warehouse appear quickly since nothing is transferred to the user's computer. For many products, safety or intellectual property concerns will likely lead to software which will enable consumers to modify products within expert-defined constraints so that consumers can directly make meaningful changes to products while maintaining safety or other features that are necessary in the end product.

The success of digiproneurship enterprises is due to their recognition of market needs which can be fulfilled by imaginative product offerings enabled through innovative combinations of aICT and AM. Although pioneers have demonstrated that successful enterprises can be established, the potential for digiproneurship extends significantly beyond the scope of today's technological capabilities and business networks. In particular, as aICT and AM progress, and new business networks are established, the opportunities for successful digiproneurship will expand. One example of the success of aICT + AM to rapidly fill a market need was demonstrated during the Covid-19 pandemic. It was reported that in many locations worldwide, the AM community helped overcome local personal protective equipment (PPE) shortages. AM users (companies and individuals) shared PPE designs online for specific machine/material combinations. Local AM users downloaded the designs and printed PPE for local use. As a result, lives were undoubtedly saved.

Several research and development priorities for aICT and AM are crucial for further expansion of digiproneurship:

- Further development of geometric manipulation tools with intuitively understandable interfaces which can be used readily by non-experts.
- Application of expert-defined constraints (such as through shape grammars and computational semantics) to enable experts to create versatile parameters for digiproneurship products. These parameters conform to criteria, e.g., safety and

brand, but facilitate the creation of person-specific, location-specific, and/or event-specific versions by non-experts.

- Web-based digiproneurship tools which can enable non-experts to set up and operate their own digitally driven enterprise. These web-based tools encompass market opportunities and business issues as well as technology characteristics and material properties.
- Continuing the current trend to lower-cost equipment and materials.
- Automating and minimizing post-processing, so that parts can go directly from a machine to the end customer with little or no human interaction.
- Continuing the current trend to increasing diversification of machine sizes, speeds, accuracies, and materials.
- Interfaces to automatically convert multi-material and multi-color user-specified requirements directly into digital manufacturing instructions without human intervention.

There are an increasing number of creation facilities that enable digiproneurs to reach specific types of customers. It remains to be seen which types of co-location facilities will generate enough business to be viable long term. For instance, there can be AM machines located within department stores (e.g., for customer-specific exclusive goods such as jewelry); large hospitals (e.g., for patient-specific prosthetics); home improvement stores (e.g., for family-specific furnishings); and/or industrial wholesalers (e.g., for plant-specific upgrade fittings). Competition and cooperation among creation facilities that provide services to digiproneurs will be enabled by aICT. Those who establish these creation facilities will themselves be digiproneurs and aid other digiproneurs in creating physical products. Development of digiproneurship infrastructure is leading to an increasing ability by digiproneurs to conceptualize, create, and propagate competitive new products, resulting in a sustainable model for distributed employment wherever digiproneurship is embraced. This, then, will be "Factory 2.0." As Web 2.0 has seen the move from static web pages to user-driven content, Factory 2.0 will see the move from static factories to user-driven product creation. To make this possible, Factory 2.0 will draw upon Web 2.0 and the distributed conceptualization and propagation which it and AM enables. Thus, digiproneurship represents the intersection of conceptualization, creation, and propagation to enable Factory 2.0, as illustrated in Fig. 22.3.

Since the advent of the industrial revolution, the creation of physical goods has become an ever more specialized domain requiring extensive knowledge and investment. This type of highly concentrated and meticulously planned factory production will continue. However, Factory 2.0 will likely flourish alongside it. This will enable production by consumers, as envisioned 40 years ago [4]. Thus, the innate potential of people to create physical goods will be realized by fulfilling the latent potential of Web 2.0 combined with AM in ever more imaginative ways. Additionally, for the first time since the industrial revolution began, the trends toward increasing urbanization to support increasingly centralized production may begin to reverse when the opportunities afforded by Factory 2.0 are fully realized.

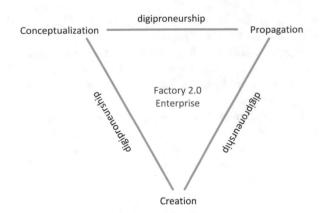

Fig. 22.3 Digiproneurship involves the creation of a business enterprise by connecting conceptualization, propagation, and/or creation, thus enabling Factor 2.0 enterprises

22.4   Summary

There is no longer any fundamental reason for products to be brought to markets through centralized product development, production, and distribution. Instead, products can be brought to markets through product conceptualization, creation, and propagation in any geographical region. This form of digiproneurship is built around combinations of advanced information and communication technologies and advanced manufacturing technologies.

Digiproneurship offers many opportunities for a reduction in the consumption of non-value adding resources during the creation of physical goods. Further, the amount of factory equipment needed and, therefore, factory space is reduced. As a result, opportunities for smaller, distributed, and mobile production facilities will increase. Digiproneurship can eliminate the need for costly conventional market research, large warehouses, distribution centers, and large capital investments in infrastructure and tooling for many types of goods.

Creation of physical products at point-of-demand can make regional disadvantages unimportant. A wide range of people and businesses could offer digiproneurship products, including artists; hobby enthusiasts; IT savvy programmers; underemployed and unemployed people who are reluctant to uproot to major cities to look for work; and others.

Novel combinations of aICT and AM have already made it possible for enterprises to be established based on digitally driven conceptualization, creation, and/or propagation. The success of these existing enterprises is due to their recognition of market needs which can be fulfilled by imaginative, digitally enabled product offerings. As aICT and AM progress, and new creation networks are established, cloud-based design and manufacturing will continue to grow, the opportunities for successful digiproneurship will expand, and Factory 2.0 will become a growing reality.

As digiproneurship expands, AM is having a substantial impact on industry, and may dramatically impact the way society is structured and interacts. In much the same way that the proliferation of digital content since the advent of the Internet has affected the way that people work, recreate, and communicate around the world, AM could one day affect the distribution of employment, resources, and opportunities worldwide.

22.5 Questions

1. Do you think AM has the potential to change the world significantly? If so, how? If not, why not?
2. In what ways could AM's future development mirror the development of the Internet?
3. Find and describe three examples of digiproneurship enterprises which are not mentioned in this book
4. How would you define Factory 2.0?
5. Based upon your interests, hobbies, or background, describe one type of digiproneurship opportunity that is not discussed in this chapter.
6. Consider milestones for 2D printing history (see Wikipedia). Take four key historical milestones from 2D printing that you think give us insight into how 3D printing will evolve and impact society. Describe these four milestones and the analogy between 2D printing and 3D printing that you see.
7. Additive Manufacturing technologies have been heavily used by artists. Find an example of an artist not mentioned in this book that uses AM to make artwork. What is this artist's reason for choosing AM? Is this artist a digiproneur? Why or why not?
8. Imagine that you plan to create a digiproneurship enterprise using your answer to problem 5.

   (a) Give a detail description of your idea (at least 1 page).
   (b) Give a list of tools/knowledge you would need to gain (inside and outside of the topics of this book) to start your enterprise.
   (c) List the primary difficulties you think you would encounter when seeking to make a profitable business out of this idea.

References

1. Beale, C. L. (2000). Nonmetro population growth recedes in a time of unprecedented national prosperity. *Rural Conditions and Trends, 11*(2), 27–31.
2. Fox, S. (2003). Recognising materials power. *Manufacturing Engineer, 82*(2), 36–39.
3. Drucker, P. (2007). *Innovation and entrepreneurship: Practice and principles.* Elsevier, Oxford (UK).
4. Toffler, A. (1990). *Future shock.* Bantam. New York (USA).

Index

Printed in the United States
by Baker & Taylor Publisher Services